M000033691

Applied Calculus for the Managerial, Life, and Social Sciences

EIGHTH EDITION

Soo T. Tan
Stonehill College

Prepared by

Soo T. Tan
Stonehill College

BROOKS/COLE
CENGAGE Learning

Australia • Brazil • Japan • Korea • Mexico • Singapore • Spain • United Kingdom • United States

BROOKS/COLE
CENGAGE Learning

© 2011 Brooks/Cole, Cengage Learning

ALL RIGHTS RESERVED. No part of this work covered by the copyright herein may be reproduced, transmitted, stored, or used in any form or by any means graphic, electronic, or mechanical, including but not limited to photocopying, recording, scanning, digitizing, taping, Web distribution, information networks, or information storage and retrieval systems, except as permitted under Section 107 or 108 of the 1976 United States Copyright Act, without the prior written permission of the publisher.

For product information and technology assistance, contact us at
Cengage Learning Customer & Sales Support, 1-800-354-9706

For permission to use material from this text or product, submit all requests online at **www.cengage.com/permissions**
Further permissions questions can be emailed to **permissionrequest@cengage.com**

ISBN-13: 978-0-538-73526-1
ISBN-10: 0-538-73526-0

Brooks/Cole
20 Davis Drive
Belmont, CA 94002-3098
USA

Cengage Learning is a leading provider of customized learning solutions with office locations around the globe, including Singapore, the United Kingdom, Australia, Mexico, Brazil, and Japan. Locate your local office at: **www.cengage.com/global**

Cengage Learning products are represented in Canada by Nelson Education, Ltd.

To learn more about Brooks/Cole, visit **www.cengage.com/brookscole**

Purchase any of our products at your local college store or at our preferred online store **www.cengagebrain.com**

Printed in the United States of America
1 2 3 4 5 6 7 14 13 12 11 10

CONTENTS

© 2011 Cengage Learning. All Rights Reserved. May not be scanned, copied or duplicated, or posted to a publicly accessible website, in whole or in part.

© 2011 Cengage Learning. All Rights Reserved. May not be scanned, copied or duplicated, or posted to a publicly accessible website, in whole or in part.

© 2011 Cengage Learning. All Rights Reserved. May not be scanned, copied or duplicated, or posted to a publicly accessible website, in whole or in part.

© 2011 Cengage Learning. All Rights Reserved. May not be scanned, copied or duplicated, or posted to a publicly accessible website, in whole or in part.

1 PRELIMINARIES

1. The interval $(3, 6)$ is shown on the number line below. Note that this is an open interval indicated by "(" and ")".

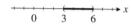

3. The interval $[-1, 4)$ is shown on the number line below. Note that this is a half-open interval indicated by "[" (closed) and ")" (open).

5. The infinite interval $(0, \infty)$ is shown on the number line below.

7. $27^{2/3} = \left(3^3\right)^{2/3} = 3^2 = 9.$

9. $\left(\frac{1}{\sqrt{3}}\right)^0 = 1.$ Recall that any number raised to the zeroth power is 1.

11. $\left[\left(\frac{1}{8}\right)^{1/3}\right]^{-2} = \left(\frac{1}{2}\right)^{-2} = (2^2) = 4.$

13. $\left(\frac{7^{-5} \cdot 7^2}{7^{-2}}\right)^{-1} = \left(7^{-5+2+2}\right)^{-1} = \left(7^{-1}\right)^{-1} = 7^1 = 7.$

15. $\left(125^{2/3}\right)^{-1/2} = 125^{(2/3)(-1/2)} = 125^{-1/3} = \frac{1}{125^{1/3}}$
$= \frac{1}{5}.$

17. $\frac{\sqrt{32}}{\sqrt{8}} = \sqrt{\frac{32}{8}} = \sqrt{4} = 2.$

19. $\frac{16^{5/8} 16^{1/2}}{16^{7/8}} = 16^{(5/8)+(1/2)-(7/8)} = 16^{1/4} = 2.$

21. $16^{1/4} \cdot 8^{-1/3} = 2 \cdot \left(\frac{1}{8}\right)^{1/3} = 2 \cdot \frac{1}{2} = 1.$

23. True.

25. False. $x^3 \times 2x^2 = 2x^{3+2} = 2x^5 \neq 2x^6.$

27. False. $\frac{2^{4x}}{1^{3x}} = \frac{2^{4x}}{1} = 2^{4x}.$

29. False. $\frac{1}{4^{-3}} = 4^3 = 64.$

31. False. $\left(1.2^{1/2}\right)^{-1/2} = (1.2)^{-1/4} \neq 1.$

33. $(xy)^{-2} = \frac{1}{(xy)^2}.$

35. $\frac{x^{-1/3}}{x^{1/2}} = x^{(-1/3)-(1/2)} = x^{-5/6} = \frac{1}{x^{5/6}}.$

37. $12^0 (s + t)^{-3} = 1 \cdot \frac{1}{(s+t)^3} = \frac{1}{(s+t)^3}.$

39. $\frac{x^{7/3}}{x^{-2}} = x^{(7/3)+2} = x^{(7/3)+(6/3)} = x^{13/3}.$

41. $\left(x^2 y^{-3}\right)\left(x^{-5} y^3\right) = x^{2-5} y^{-3+3} = x^{-3} y^0 = x^{-3} = \frac{1}{x^3}.$

43. $\frac{x^{3/4}}{x^{-1/4}} = x^{(3/4)-(-1/4)} = x^{4/4} = x.$

1

© 2011 Cengage Learning. All Rights Reserved. May not be scanned, copied or duplicated, or posted to a publicly accessible website, in whole or in part.

45. $\left(\dfrac{x^3}{-27y^{-6}}\right)^{-2/3} = x^{3(-2/3)} \left(-\dfrac{1}{27}\right)^{-2/3} y^{6(-2/3)}$

$\qquad = x^{-2} \left(-\dfrac{1}{3}\right)^{-2} y^{-4} = \dfrac{9}{x^2 y^4}.$

47. $\left(\dfrac{x^{-3}}{y^{-2}}\right)^2 \left(\dfrac{y}{x}\right)^4 = \dfrac{x^{-3\cdot 2}y^4}{y^{-2\cdot 2}x^4} = \dfrac{y^{4+4}}{x^{4+6}} = \dfrac{y^8}{x^{10}}.$

49. $\sqrt[3]{x^{-2}} \cdot \sqrt{4x^5} = x^{-2/3} \cdot 4^{1/2} \cdot x^{5/2} = x^{(-2/3)+(5/2)} \cdot 2$

$\qquad = 2x^{11/6}.$

51. $-\sqrt[4]{16x^4 y^8} = -\left(16^{1/4} \cdot x^{4/4} \cdot y^{8/4}\right) = -2xy^2.$

53. $\sqrt[6]{64x^8 y^3} = 64^{1/6} \cdot x^{8/6} y^{3/6} = 2x^{4/3} y^{1/2}.$

55. $2^{3/2} = 2\left(2^{1/2}\right) \approx 2\,(1.414) = 2.828.$

57. $9^{3/4} = \left(3^2\right)^{3/4} = 3^{6/4} = 3^{3/2} = 3 \cdot 3^{1/2}$

$\qquad \approx 3\,(1.732) \approx 5.196.$

59. $10^{3/2} = 10^{1/2} \cdot 10 \approx (3.162)\,(10) = 31.62.$

61. $10^{2.5} = 10^2 \cdot 10^{1/2} \approx 100\,(3.162) = 316.2.$

63. $\dfrac{3}{2\sqrt{x}} \cdot \dfrac{\sqrt{x}}{\sqrt{x}} = \dfrac{3\sqrt{x}}{2x}.$

65. $\dfrac{2y}{\sqrt{3y}} \cdot \dfrac{\sqrt{3y}}{\sqrt{3y}} = \dfrac{2y\sqrt{3y}}{3y} = \dfrac{2\sqrt{3y}}{3}.$

67. $\dfrac{1}{\sqrt[3]{x}} \cdot \dfrac{\sqrt[3]{x^2}}{\sqrt[3]{x^2}} = \dfrac{\sqrt[3]{x^2}}{\sqrt[3]{x^3}} = \dfrac{\sqrt[3]{x^2}}{x}.$

69. $\dfrac{2\sqrt{x}}{3} \cdot \dfrac{\sqrt{x}}{\sqrt{x}} = \dfrac{2x}{3\sqrt{x}}.$

71. $\sqrt{\dfrac{2y}{x}} = \dfrac{\sqrt{2y}}{\sqrt{x}} \cdot \dfrac{\sqrt{2y}}{\sqrt{2y}} = \dfrac{2y}{\sqrt{2xy}}.$

73. $\dfrac{\sqrt[3]{x^2 z}}{y} \cdot \dfrac{\sqrt[3]{xz^2}}{\sqrt[3]{xz^2}} = \dfrac{\sqrt[3]{x^3 z^3}}{y\sqrt[3]{xz^2}} = \dfrac{xz}{y\sqrt[3]{xz^2}}.$

75. $\left(7x^2 - 2x + 5\right) + \left(2x^2 + 5x - 4\right) = 7x^2 - 2x + 5 + 2x^2 + 5x - 4 = 9x^2 + 3x + 1.$

77. $\left(5y^2 - 2y + 1\right) - \left(y^2 - 3y - 7\right) = 5y^2 - 2y + 1 - y^2 + 3y + 7 = 4y^2 + y + 8.$

79. $x - \{2x - [-x - (1 - x)]\} = x - \{2x - [-x - 1 + x]\} = x - (2x + 1) = x - 2x - 1 = -x - 1.$

81. $\left(\dfrac{1}{3} - 1 + e\right) - \left(-\dfrac{1}{3} - 1 + e^{-1}\right) = \dfrac{1}{3} - 1 + e + \dfrac{1}{3} + 1 - \dfrac{1}{e} = \dfrac{2}{3} + e - \dfrac{1}{e} = \dfrac{3e^2 + 2e - 3}{3e}$

83. $3\sqrt{8} + 8 - 2\sqrt{y} + \tfrac{1}{2}\sqrt{x} - \tfrac{3}{4}\sqrt{y} = 3\sqrt{8} + 8 + \tfrac{1}{2}\sqrt{x} - \tfrac{11}{4}\sqrt{y} = 6\sqrt{2} + 8 + \tfrac{1}{2}\sqrt{x} - \tfrac{11}{4}\sqrt{y}.$

85. $(x + 8)(x - 2) = x\,(x - 2) + 8\,(x - 2) = x^2 - 2x + 8x - 16 = x^2 + 6x - 16.$

87. $(a + 5)^2 = (a + 5)(a + 5) = a\,(a + 5) + 5\,(a + 5) = a^2 + 5a + 5a + 25 = a^2 + 10a + 25.$

89. $(x + 2y)^2 = (x + 2y)(x + 2y) = x\,(x + 2y) + 2y\,(x + 2y) = x^2 + 2xy + 2yx + 4y^2 = x^2 + 4xy + 4y^2.$

91. $(2x + y)(2x - y) = 2x\,(2x - y) + y\,(2x - y) = 4x^2 - 2xy + 2xy - y^2 = 4x^2 - y^2.$

93. $\left(x^2 - 1\right)(2x) - x^2\,(2x) = 2x^3 - 2x - 2x^3 = -2x.$

95. $2\left(t + \sqrt{t}\right)^2 - 2t^2 = 2\left(t + \sqrt{t}\right)\left(t + \sqrt{t}\right) - 2t^2 = 2\left(t^2 + 2t\sqrt{t} + t\right) - 2t^2 = 2t^2 + 4t\sqrt{t} + 2t - 2t^2$

$\qquad = 4t\sqrt{t} + 2t = 2t\left(2\sqrt{t} + 1\right).$

© 2011 Cengage Learning. All Rights Reserved. May not be scanned, copied or duplicated, or posted to a publicly accessible website, in whole or in part.

97. $4x^5 - 12x^4 - 6x^3 = 2x^3 \left(2x^2 - 6x - 3\right)$.

99. $7a^4 - 42a^2b^2 + 49a^3b = 7a^2 \left(a^2 + 7ab - 6b^2\right)$.

101. $e^{-x} - xe^{-x} = e^{-x} \left(1 - x\right)$.

103. $2x^{-5/2} - \frac{3}{2}x^{-3/2} = \frac{1}{2}x^{-5/2} \left(4 - 3x\right)$.

105. $6ac + 3bc - 4ad - 2bd = 3c \left(2a + b\right) - 2d \left(2a + b\right) = \left(2a + b\right) \left(3c - 2d\right)$.

107. $4a^2 - b^2 = \left(2a + b\right) \left(2a - b\right)$, a difference of two squares.

109. $10 - 14x - 12x^2 = -2 \left(6x^2 + 7x - 5\right) = -2 \left(3x + 5\right) \left(2x - 1\right)$.

111. $3x^2 - 6x - 24 = 3 \left(x^2 - 2x - 8\right) = 3 \left(x - 4\right) \left(x + 2\right)$.

113. $12x^2 - 2x - 30 = 2 \left(6x^2 - x - 15\right) = 2 \left(3x - 5\right) \left(2x + 3\right)$.

115. $9x^2 - 16y^2 = \left(3x\right)^2 - \left(4y\right)^2 = \left(3x - 4y\right) \left(3x + 4y\right)$.

117. $x^6 + 125 = \left(x^2\right)^3 + \left(5\right)^3 = \left(x^2 + 5\right) \left(x^4 - 5x^2 + 25\right)$.

119. $\left(x^2 + y^2\right) x - xy \left(2y\right) = x^3 + xy^2 - 2xy^2 = x^3 - xy^2$.

121. $2 \left(x - 1\right) \left(2x + 2\right)^3 \left[4 \left(x - 1\right) + \left(2x + 2\right)\right] = 2 \left(x - 1\right) \left(2x + 2\right)^3 \left(4x - 4 + 2x + 2\right)$

$$= 2 \left(x - 1\right) \left(2x + 2\right)^3 \left(6x - 2\right) = 4 \left(x - 1\right) \left(3x - 1\right) \left(2x + 2\right)^3$$

$$= 32 \left(x - 1\right) \left(3x - 1\right) \left(x + 1\right)^3.$$

123. $4 \left(x - 1\right)^2 \left(2x + 2\right)^3 \left(2\right) + \left(2x + 2\right)^4 \left(2\right) \left(x - 1\right) = 2 \left(x - 1\right) \left(2x + 2\right)^3 \left[4 \left(x - 1\right) + \left(2x + 2\right)\right]$

$$= 2 \left(x - 1\right) \left(2x + 2\right)^3 \left(6x - 2\right) = 4 \left(x - 1\right) \left(3x - 1\right) \left(2x + 2\right)^3$$

$$= 32 \left(x - 1\right) \left(3x - 1\right) \left(x + 1\right)^3.$$

125. $\left(x^2 + 2\right)^2 \left[5 \left(x^2 + 2\right)^2 - 3\right] \left(2x\right) = \left(x^2 + 2\right)^2 \left[5 \left(x^4 + 4x^2 + 4\right) - 3\right] \left(2x\right) = \left(2x\right) \left(x^2 + 2\right)^2 \left(5x^4 + 20x^2 + 17\right)$.

127. We factor the left-hand side of $x^2 + x - 12 = 0$ to obtain $\left(x + 4\right) \left(x - 3\right) = 0$, so $x = -4$ or $x = 3$. We conclude that the roots are $x = -4$ and $x = 3$.

129. $4t^2 + 2t - 2 = \left(2t - 1\right) \left(2t + 2\right) = 0$. Thus, the roots are $t = \frac{1}{2}$ and $t = -1$.

131. $\frac{1}{4}x^2 - x + 1 = \left(\frac{1}{2}x - 1\right) \left(\frac{1}{2}x - 1\right) = 0$. Thus $\frac{1}{2}x = 1$, and so $x = 2$ is a double root of the equation.

133. We use the quadratic formula to solve the equation $4x^2 + 5x - 6 = 0$. In this case, $a = 4$, $b = 5$, and $c = -6$.

Therefore, $x = \dfrac{-b \pm \sqrt{b^2 - 4ac}}{2a} = \dfrac{-5 \pm \sqrt{5^2 - 4 \left(4\right) \left(-6\right)}}{2 \left(4\right)} = \dfrac{-5 \pm \sqrt{121}}{8} = \dfrac{-5 \pm 11}{8}$. Thus, $x = -\frac{16}{8} = -2$

and $x = \frac{6}{8} = \frac{3}{4}$ are the roots of the equation.

© 2011 Cengage Learning. All Rights Reserved. May not be scanned, copied or duplicated, or posted to a publicly accessible website, in whole or in part.

135. We use the quadratic formula to solve the equation $8x^2 - 8x - 3 = 0$. Here $a = 8$, $b = -8$, and $c = -3$, so

$$x = \frac{-b \pm \sqrt{b^2 - 4ac}}{2a} = \frac{-(-8) \pm \sqrt{(-8)^2 - 4(8)(-3)}}{2(8)} = \frac{8 \pm \sqrt{160}}{16} = \frac{8 \pm 4\sqrt{10}}{16} = \frac{2 \pm \sqrt{10}}{4}. \text{ Thus,}$$

$x = \frac{1}{2} + \frac{1}{4}\sqrt{10}$ and $x = \frac{1}{2} - \frac{1}{4}\sqrt{10}$ are the roots of the equation.

137. We use the quadratic formula to solve $2x^2 + 4x - 3 = 0$. Here $a = 2$, $b = 4$, and $c = -3$, so

$$x = \frac{-b \pm \sqrt{b^2 - 4ac}}{2a} = \frac{-4 \pm \sqrt{4^2 - 4(2)(-3)}}{2(2)} = \frac{-4 \pm \sqrt{40}}{4} = \frac{-4 \pm 2\sqrt{10}}{4} = \frac{-2 \pm \sqrt{10}}{2}. \text{ Thus,}$$

$x = -1 + \frac{1}{2}\sqrt{10}$ and $x = -1 - \frac{1}{2}\sqrt{10}$ are the roots of the equation.

139. a. $f(30{,}000) = (2.8 \times 10^{11})(30{,}000)^{-1.5} \approx 53{,}886$, or 53,886 families.

    b. $f(60{,}000) = (2.8 \times 10^{11})(60{,}000)^{-1.5} \approx 19{,}052$ or 19,052 families.

    c. $f(150{,}000) = (2.8 \times 10^{11})(150{,}000)^{-1.5} \approx 4820$, or 4820 families.

141. True. If $b^2 - 4ac < 0$, then $\sqrt{b^2 - 4ac}$ is not a real number.

1.2   Precalculus Review II   page 23

1. $\dfrac{x^2 + x - 2}{x^2 - 4} = \dfrac{(x+2)(x-1)}{(x+2)(x-2)} = \dfrac{x-1}{x-2}.$

3. $\dfrac{12t^2 + 12t + 3}{4t^2 - 1} = \dfrac{3(4t^2 + 4t + 1)}{4t^2 - 1} = \dfrac{3(2t+1)(2t+1)}{(2t+1)(2t-1)} = \dfrac{3(2t+1)}{2t-1}.$

5. $\dfrac{(4x-1)(3) - (3x+1)(4)}{(4x-1)^2} = \dfrac{12x - 3 - 12x - 4}{(4x-1)^2} = -\dfrac{7}{(4x-1)^2}.$

7. $\dfrac{2a^2 - 2b^2}{b-a} \cdot \dfrac{4a + 4b}{a^2 + 2ab + b^2} = \dfrac{2(a+b)(a-b)\,4(a+b)}{-(a-b)(a+b)(a+b)} = -8.$

9. $\dfrac{3x^2 + 2x - 1}{2x + 6} \div \dfrac{x^2 - 1}{x^2 + 2x - 3} = \dfrac{(3x-1)(x+1)}{2(x+3)} \cdot \dfrac{(x+3)(x-1)}{(x+1)(x-1)} = \dfrac{3x-1}{2}.$

11. $\dfrac{58}{3(3t+2)} + \dfrac{1}{3} = \dfrac{58 + 3t + 2}{3(3t+2)} = \dfrac{3t + 60}{3(3t+2)} = \dfrac{t + 20}{3t + 2}.$

13. $\dfrac{2x}{2x-1} - \dfrac{3x}{2x+5} = \dfrac{2x(2x+5) - 3x(2x-1)}{(2x-1)(2x+5)} = \dfrac{4x^2 + 10x - 6x^2 + 3x}{(2x-1)(2x+5)} = \dfrac{-2x^2 + 13x}{(2x-1)(2x+5)}$

    $= -\dfrac{x(2x - 13)}{(2x-1)(2x+5)}.$

15. $\dfrac{4}{x^2 - 9} - \dfrac{5}{x^2 - 6x + 9} = \dfrac{4}{(x+3)(x-3)} - \dfrac{5}{(x-3)^2} = \dfrac{4(x-3) - 5(x+3)}{(x-3)^2(x+3)} = -\dfrac{x + 27}{(x-3)^2(x+3)}.$

17. $\dfrac{1 + \dfrac{1}{x}}{1 - \dfrac{1}{x}} = \dfrac{\dfrac{x+1}{x}}{\dfrac{x-1}{x}} = \dfrac{x+1}{x} \cdot \dfrac{x}{x-1} = \dfrac{x+1}{x-1}.$

© 2011 Cengage Learning. All Rights Reserved. May not be scanned, copied or duplicated, or posted to a publicly accessible website, in whole or in part.

19. $\dfrac{4x^2}{2\sqrt{2x^2+7}} + \sqrt{2x^2+7} = \dfrac{4x^2 + 2\sqrt{2x^2+7}\sqrt{2x^2+7}}{2\sqrt{2x^2+7}} = \dfrac{4x^2 + 4x^2 + 14}{2\sqrt{2x^2+7}} = \dfrac{4x^2+7}{\sqrt{2x^2+7}}.$

21. $\dfrac{2x\,(x+1)^{-1/2} - (x+1)^{1/2}}{x^2} = \dfrac{(x+1)^{-1/2}\,(2x - x - 1)}{x^2} = \dfrac{(x+1)^{-1/2}\,(x-1)}{x^2} = \dfrac{x-1}{x^2\sqrt{x+1}}.$

23. $\dfrac{(2x+1)^{1/2} - (x+2)\,(2x+1)^{-1/2}}{2x+1} = \dfrac{(2x+1)^{-1/2}\,(2x+1 - x - 2)}{2x+1} = \dfrac{(2x+1)^{-1/2}\,(x-1)}{2x+1} = \dfrac{x-1}{(2x+1)^{3/2}}.$

25. $\dfrac{1}{\sqrt{3}-1} \cdot \dfrac{\sqrt{3}+1}{\sqrt{3}+1} = \dfrac{\sqrt{3}+1}{3-1} = \dfrac{\sqrt{3}+1}{2}.$ **27.** $\dfrac{1}{\sqrt{x}-\sqrt{y}} \cdot \dfrac{\sqrt{x}+\sqrt{y}}{\sqrt{x}+\sqrt{y}} = \dfrac{\sqrt{x}+\sqrt{y}}{x-y}.$

29. $\dfrac{\sqrt{a}+\sqrt{b}}{\sqrt{a}-\sqrt{b}} \cdot \dfrac{\sqrt{a}+\sqrt{b}}{\sqrt{a}+\sqrt{b}} = \dfrac{\left(\sqrt{a}+\sqrt{b}\right)^2}{a-b}.$ **31.** $\dfrac{\sqrt{x}}{3} \cdot \dfrac{\sqrt{x}}{\sqrt{x}} = \dfrac{x}{3\sqrt{x}}.$

33. $\dfrac{1-\sqrt{3}}{3} \cdot \dfrac{1+\sqrt{3}}{1+\sqrt{3}} = \dfrac{1^2 - \left(\sqrt{3}\right)^2}{3\left(1+\sqrt{3}\right)} = -\dfrac{2}{3\left(1+\sqrt{3}\right)}.$

35. $\dfrac{1+\sqrt{x+2}}{\sqrt{x+2}} \cdot \dfrac{1-\sqrt{x+2}}{1-\sqrt{x+2}} = \dfrac{1-(x+2)}{\sqrt{x+2}\left(1-\sqrt{x+2}\right)} = -\dfrac{x+1}{\sqrt{x+2}\left(1-\sqrt{x+2}\right)}.$

37. The statement is false because -3 is greater than -20. See the number line below.

39. The statement is false because $\frac{2}{3} = \frac{4}{6}$ is less than $\frac{5}{6}$.

41. We are given $2x + 4 < 8$. Add -4 to each side of the inequality to obtain $2x < 4$, then multiply each side of the inequality by $\frac{1}{2}$ to obtain $x < 2$. We write this in interval notation as $(-\infty, 2)$.

43. We are given the inequality $-4x \geq 20$. Multiply both sides of the inequality by $-\frac{1}{4}$ and reverse the sign of the inequality to obtain $x \leq -5$. We write this in interval notation as $(-\infty, -5]$.

45. We are given the inequality $-6 < x - 2 < 4$. First add 2 to each member of the inequality to obtain $-6 + 2 < x < 4 + 2$ and $-4 < x < 6$, so the solution set is the open interval $(-4, 6)$.

47. We want to find the values of x that satisfy at least one of the inequalities $x + 1 > 4$ and $x + 2 < -1$. Adding -1 to both sides of the first inequality, we obtain $x + 1 - 1 > 4 - 1$, so $x > 3$. Similarly, adding -2 to both sides of the second inequality, we obtain $x + 2 - 2 < -1 - 2$, so $x < -3$. Therefore, the solution set is $(-\infty, -3) \cup (3, \infty)$.

49. We want to find the values of x that satisfy the inequalities $x + 3 > 1$ and $x - 2 < 1$. Adding -3 to both sides of the first inequality, we obtain $x + 3 - 3 > 1 - 3$, or $x > -2$. Similarly, adding 2 to each side of the second inequality, we obtain $x - 2 + 2 < 1 + 2$, so $x < 3$. Because both inequalities must be satisfied, the solution set is $(-2, 3)$.

© 2011 Cengage Learning. All Rights Reserved. May not be scanned, copied or duplicated, or posted to a publicly accessible website, in whole or in part.

51. We want to find the values of x that satisfy the inequality
$(x + 3)(x - 5) \leq 0$. From the sign diagram, we see that the given
inequality is satisfied when $-3 \leq x \leq 5$, that is, when the signs of
the two factors are different or when one of the factors is equal to
zero.

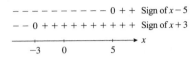

53. We want to find the values of x that satisfy the inequality
$(2x - 3)(x - 1) \geq 0$. From the sign diagram, we see that the given
inequality is satisfied when $x \leq 1$ or $x \geq \frac{3}{2}$; that is, when the signs
of both factors are the same, or one of the factors is equal to zero.

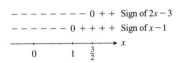

55. We want to find the values of x that satisfy the
inequalities $\dfrac{x + 3}{x - 2} \geq 0$. From the sign diagram, we see that the
given inequality is satisfied when $x \leq -3$ or $x > 2$, that is, when
the signs of the two factors are the same. Notice that $x = 2$ is not
included because the inequality is not defined at that value of x.

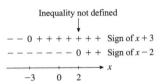

57. We want to find the values of x that satisfy the inequality
$\dfrac{x - 2}{x - 1} \leq 2$. Subtracting 2 from each side of the given inequality
and simplifying gives $\dfrac{x - 2}{x - 1} - 2 \leq 0$,

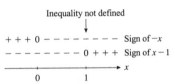

$\dfrac{x - 2 - 2(x - 1)}{x - 1} \leq 0$, and $-\dfrac{x}{x - 1} \leq 0$. From the sign diagram, we see that the given inequality is satisfied when
$x \leq 0$ or $x > 1$; that is, when the signs of the two factors differ. Notice that $x = 1$ is not included because the
inequality is undefined at that value of x.

59. $|-6 + 2| = 4$.

61. $\dfrac{|-12 + 4|}{|16 - 12|} = \dfrac{|-8|}{|4|} = 2$.

63. $\sqrt{3}\,|-2| + 3\left|-\sqrt{3}\right| = \sqrt{3}\,(2) + 3\sqrt{3} = 5\sqrt{3}$.

65. $|\pi - 1| + 2 = \pi - 1 + 2 = \pi + 1$.

67. $\left|\sqrt{2} - 1\right| + \left|3 - \sqrt{2}\right| = \sqrt{2} - 1 + 3 - \sqrt{2} = 2$.

69. False. If $a > b$, then $-a < -b$, $-a + b < -b + b$, and $b - a < 0$.

71. False. Let $a = -2$ and $b = -3$. Then $a^2 = 4$ and $b^2 = 9$, and $4 < 9$. Note that we need only to provide a
counterexample to show that the statement is not always true.

73. True. There are three possible cases.
Case 1: If $a > 0$ and $b > 0$, then $a^3 > b^3$, since $a^3 - b^3 = (a - b)\left(a^2 + ab + b^2\right) > 0$.
Case 2: If $a > 0$ and $b < 0$, then $a^3 > 0$ and $b^3 < 0$, and it follows that $a^3 > b^3$.
Case 3: If $a < 0$ and $b < 0$, then $a^3 - b^3 = (a - b)\left(a^2 + ab + b^2\right) > 0$, and we see that $a^3 > b^3$. (Note that
$a - b > 0$ and $ab > 0$.)

© 2011 Cengage Learning. All Rights Reserved. May not be scanned, copied or duplicated, or posted to a publicly accessible website, in whole or in part.

75. False. If we take $a = -2$, then $|-a| = |-(-2)| = |2| = 2 \neq a$.

77. True. If $a - 4 < 0$, then $|a - 4| = 4 - a = |4 - a|$. If $a - 4 > 0$, then $|4 - a| = a - 4 = |a - 4|$.

79. False. If we take $a = 3$ and $b = -1$, then $|a + b| = |3 - 1| = 2 \neq |a| + |b| = 3 + 1 = 4$.

81. If the car is driven in the city, then it can be expected to cover $(18.1)(20) = 362 \frac{\text{miles}}{\text{gal}} \cdot$ gal, or 362 miles, on a full tank. If the car is driven on the highway, then it can be expected to cover $(18.1)(27) = 488.7 \frac{\text{miles}}{\text{gal}} \cdot$ gal, or 488.7 miles, on a full tank. Thus, the driving range of the car may be described by the interval $[362, 488.7]$.

83. $6(P - 2500) \leq 4(P + 2400)$ can be rewritten as $6P - 15{,}000 \leq 4P + 9600$, $2P \leq 24{,}600$, or $P \leq 12{,}300$. Therefore, the maximum profit is $12{,}300.

85. Let x represent the salesman's monthly sales in dollars. Then $0.15(x - 12{,}000) \geq 3000$, $15(x - 12{,}000) \geq 300{,}000$, $15x - 180{,}000 \geq 300{,}000$, $15x \geq 480{,}000$, and $x \geq 32{,}000$. We conclude that the salesman must have sales of at least $32{,}000 to reach his goal.

87. The rod is acceptable if $0.49 \leq x \leq 0.51$ or $-0.01 \leq x - 0.5 \leq 0.01$. This gives the required inequality, $|x - 0.5| \leq 0.01$.

89. We want to solve the inequality $-6x^2 + 30x - 10 \geq 14$. (Remember that x is expressed in thousands.) Adding -14 to both sides of this inequality, we have $-6x^2 + 30x - 10 - 14 \geq 14 - 14$, or $-6x^2 + 30x - 24 \geq 0$. Dividing both sides of the inequality by -6 (which reverses the sign of the inequality), we have $x^2 - 5x + 4 \leq 0$. Factoring this last expression, we have $(x - 4)(x - 1) \leq 0$.

From the sign diagram, we see that x must lie between 1 and 4. (The inequality is satisfied only when the two factors have opposite signs.) Because x is expressed in thousands of units, we see that the manufacturer must produce between 1000 and 4000 units of the commodity.

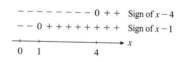

91. We solve the inequalities $25 \leq \frac{0.5x}{100 - x} \leq 30$, obtaining $2500 - 25x \leq 0.5x \leq 3000 - 30x$, which is equivalent to $2500 - 25x \leq 0.5x$ and $0.5x \leq 3000 - 30x$. Simplifying further, $25.5x \geq 2500$ and $30.5x \leq 3000$, so $x \geq \frac{2500}{25.5} \approx 98.04$ and $x \leq \frac{3000}{30.5} \approx 98.36$. Thus, the city could expect to remove between 98.04% and 98.36% of the toxic pollutant.

93. We solve the inequality $\frac{136}{1 + 0.25(t - 4.5)^2} + 28 \geq 128$ or $\frac{136}{1 + 0.25(t - 4.5)^2} \geq 100$. Next, $136 \geq 100[1 + 0.25(t - 4.5)^2]$, so $136 \geq 100 + 25(t - 4.5)^2$, $36 \geq 25(t - 4.5)^2$, $(t - 4.5)^2 \leq \frac{36}{25}$, and $t - 4.5 \leq \pm\frac{6}{5}$. Solving this last inequality, we have $t \leq 5.7$ and $t \geq 3.3$. Thus, the amount of nitrogen dioxide is greater than or equal to 128 PSI between 10:18 a.m. and 12:42 p.m.

95. False. Take $a = 1$, $b = 2$, and $c = 3$. Then $a < b$, but $a - c = 1 - 3 = -2 \not> 2 - 3 = -1 = b - c$.

97. True. $|a - b| = |a + (-b)| \leq |a| + |-b| = |a| + |b|$.

© 2011 Cengage Learning. All Rights Reserved. May not be scanned, copied or duplicated, or posted to a publicly accessible website, in whole or in part.

1.3   Problem-Solving Tips

Suppose you are asked to determine whether a given statement is true or false, and you are also asked to explain your answer. How would you answer the question?

If you think the statement is true, then prove it. On the other hand, if you think the statement is false, then give an example that disproves the statement. For example, the statement "If a and b are real numbers, then $a - b = b - a$" is false, and an example that disproves it may be constructed by taking $a = 3$ and $b = 5$. For these values of a and b, we find $a - b = 3 - 5 = -2$, but $b - a = 5 - 3 = 2$, and this shows that $a - b \neq b - a$. Such an example is called a **counterexample**.

1.3   Concept Questions   page 29

1. a. $a < 0$ and $b > 0$ **b.** $a < 0$ and $b < 0$ **c.** $a > 0$ and $b < 0$

1.3   The Cartesian Coordinate System   page 30

1. The coordinates of A are $(3, 3)$ and it is located in Quadrant I.

3. The coordinates of C are $(2, -2)$ and it is located in Quadrant IV.

5. The coordinates of E are $(-4, -6)$ and it is located in Quadrant III.

7. A **9.** $E, F,$ and G **11.** F

For Exercises 13–19, refer to the following figure.

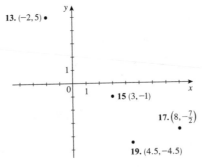

21. Using the distance formula, we find that $\sqrt{(4 - 1)^2 + (7 - 3)^2} = \sqrt{3^2 + 4^2} = \sqrt{25} = 5$.

23. Using the distance formula, we find that $\sqrt{[4 - (-1)]^2 + (9 - 3)^2} = \sqrt{5^2 + 6^2} = \sqrt{25 + 36} = \sqrt{61}$.

25. The coordinates of the points have the form $(x, -6)$. Because the points are 10 units away from the origin, we have $(x - 0)^2 + (-6 - 0)^2 = 10^2$, $x^2 = 64$, or $x = \pm 8$. Therefore, the required points are $(-8, -6)$ and $(8, -6)$.

© 2011 Cengage Learning. All Rights Reserved. May not be scanned, copied or duplicated, or posted to a publicly accessible website, in whole or in part.

27. The points are shown in the diagram. To show that the four sides are equal, we compute

$$d(A, B) = \sqrt{(-3-3)^2 + (7-4)^2} = \sqrt{(-6)^2 + 3^2} = \sqrt{45},$$

$$d(B, C) = \sqrt{[-6-(-3)]^2 + (1-7)^2} = \sqrt{(-3)^2 + (-6)^2} = \sqrt{45},$$

$$d(C, D) = \sqrt{[0-(-6)]^2 + [(-2)-1]^2} = \sqrt{(6)^2 + (-3)^2} = \sqrt{45},$$

and $d(A, D) = \sqrt{(0-3)^2 + (-2-4)^2} = \sqrt{(3)^2 + (-6)^2} = \sqrt{45}.$

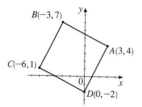

Next, to show that $\triangle ABC$ is a right triangle, we show that it satisfies the Pythagorean Theorem. Thus, $d(A, C) = \sqrt{(-6-3)^2 + (1-4)^2} = \sqrt{(-9)^2 + (-3)^2} = \sqrt{90} = 3\sqrt{10}$ and $[d(A, B)]^2 + [d(B, C)]^2 = 90 = [d(A, C)]^2$. Similarly, $d(B, D) = \sqrt{90} = 3\sqrt{10}$, so $\triangle BAD$ is a right triangle as well. It follows that $\angle B$ and $\angle D$ are right angles, and we conclude that $ADCB$ is a square.

29. The equation of the circle with radius 5 and center $(2, -3)$ is given by $(x-2)^2 + [y - (-3)]^2 = 5^2$, or $(x-2)^2 + (y+3)^2 = 25.$

31. The equation of the circle with radius 5 and center $(0, 0)$ is given by $(x-0)^2 + (y-0)^2 = 5^2$, or $x^2 + y^2 = 25.$

33. The distance between the points $(5, 2)$ and $(2, -3)$ is given by $d = \sqrt{(5-2)^2 + [2-(-3)]^2} = \sqrt{3^2 + 5^2} = \sqrt{34}$. Therefore $r = \sqrt{34}$ and the equation of the circle passing through $(5, 2)$ and $(2, -3)$ is $(x-2)^2 + [y-(-3)]^2 = 34$, or $(x-2)^2 + (y+3)^2 = 34.$

35. Referring to the diagram on page 30 of the text, we see that the distance from A to B is given by $d(A, B) = \sqrt{400^2 + 300^2} = \sqrt{250,000} = 500$. The distance from B to C is given by $d(B, C) = \sqrt{(-800-400)^2 + (800-300)^2} = \sqrt{(-1200)^2 + (500)^2} = \sqrt{1,690,000} = 1300$. The distance from C to D is given by $d(C, D) = \sqrt{[-800-(-800)]^2 + (800-0)^2} = \sqrt{0 + 800^2} = 800$. The distance from D to A is given by $d(D, A) = \sqrt{[(-800)-0]^2 + (0-0)} = \sqrt{640,000} = 800$. Therefore, the total distance covered on the tour is $d(A, B) + d(B, C) + d(C, D) + d(D, A) = 500 + 1300 + 800 + 800 = 3400$, or 3400 miles.

© 2011 Cengage Learning. All Rights Reserved. May not be scanned, copied or duplicated, or posted to a publicly accessible website, in whole or in part.

37.

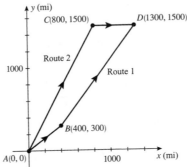

Referring to the diagram, we see that the distance the salesman would cover if he took Route 1 is given by

$$d\,(A,\,B) + d\,(B,\,D) = \sqrt{400^2 + 300^2} + \sqrt{(1300 - 400)^2 + (1500 - 300)^2}$$

$$= \sqrt{250{,}000} + \sqrt{2{,}250{,}000} = 500 + 1500 = 2000$$

or 2000 miles. On the other hand, the distance he would cover if he took Route 2 is given by

$$d\,(A,\,C) + d\,(C,\,D) = \sqrt{800^2 + 1500^2} + \sqrt{(1300 - 800)^2} = \sqrt{2{,}890{,}000} + \sqrt{250{,}000}$$

$$= 1700 + 500 = 2200$$

or 2200 miles. Comparing these results, we see that he should take Route 1.

39. To determine the VHF requirements, we calculate $d = \sqrt{25^2 + 35^2} = \sqrt{625 + 1225} = \sqrt{1850} \approx 43.01$. Models B, C, and D satisfy this requirement.

To determine the UHF requirements, we calculate $d = \sqrt{20^2 + 32^2} = \sqrt{400 + 1024} = \sqrt{1424} \approx 37.74$. Models C and D satisfy this requirement.

Therefore, Model C allows him to receive both channels at the least cost.

41. a. Let the positions of ships A and B after t hours be $A\,(0,\,y)$ and $B\,(x,\,0)$, respectively. Then $x = 30t$ and $y = 20t$. Therefore, the distance in miles between the two ships is $D = \sqrt{(30t)^2 + (20t)^2} = \sqrt{900t^2 + 400t^2} = 10\sqrt{13}\,t$.

b. The required distance is obtained by letting $t = 2$, giving $D = 10\sqrt{13}\,(2)$, or approximately 72.11 miles.

43. a. The distance in feet is given by $\sqrt{(4000)^2 + x^2} = \sqrt{16{,}000{,}000 + x^2}$.

b. Substituting the value $x = 20{,}000$ into the above expression gives $\sqrt{16{,}000{,}000 + (20{,}000)^2} \approx 20{,}396$, or 20,396 ft.

45. True. Plot the points.

47. True. $kx^2 + ky^2 = a^2$ gives $x^2 + y^2 = \dfrac{a^2}{k} < a^2$ if $k > 1$. So the radius of the circle with equation $kx^2 + ky^2 = a^2$ is a circle of radius smaller than a centered at the origin if $k > 1$. Therefore, it lies inside the circle of radius a with equation $x^2 + y^2 = a^2$.

© 2011 Cengage Learning. All Rights Reserved. May not be scanned, copied or duplicated, or posted to a publicly accessible website, in whole or in part.

49. a. Suppose that $P = (x_1, y_1)$ and $Q = (x_2, y_2)$ are endpoints of the line segment and that the point $M = \left(\dfrac{x_1 + x_2}{2}, \dfrac{y_1 + y_2}{2}\right)$ is the midpoint of the line segment PQ. The distance between P and Q is $\sqrt{(x_2 - x_1)^2 + (y_2 - y_1)^2}$. The distance between P and M is

$$\sqrt{\left(\frac{x_1 + x_2}{2} - x_1\right)^2 + \left(\frac{y_1 + y_2}{2} - y_1\right)^2} = \sqrt{\left(\frac{x_2 - x_1}{2}\right)^2 + \left(\frac{y_2 - y_1}{2}\right)^2} = \tfrac{1}{2}\sqrt{(x_2 - x_1)^2 + (y_2 - y_1)^2},$$

which is one-half the distance from P to Q. Similarly, we obtain the same expression for the distance from M to P.

b. The midpoint is given by $\left(\dfrac{4 - 3}{2}, \dfrac{-5 + 2}{2}\right)$, or $\left(\dfrac{1}{2}, -\dfrac{3}{2}\right)$.

1.4 Problem-Solving Tips

When you solve a problem in the exercises that follow each section, first read the problem. Before you start computing or writing out a solution, try to formulate a strategy for solving the problem. Then proceed by using your strategy to solve the problem.

Here we summarize some general problem-solving techniques that are covered in this section.

1. **To show that two lines are parallel**, you need to show that the slopes of the two lines are equal or that their slopes are both undefined.

2. **To show that two lines L_1 and L_2 are perpendicular**, you need to show that the slope m_1 of L_1 is the negative reciprocal of the slope m_2 of L_2; that is, $m_1 = -1/m_2$.

3. **To find the equation of a line**, you need the slope of the line and a point lying on the line. You can then find the equation of the line using the point-slope form of the equation of a line: $(y - y_1) = m(x - x_1)$.

1.4 Concept Questions   page 41

1. The slope is $m = \dfrac{y_2 - y_1}{x_2 - x_1}$, where $P(x_1, y_1)$ and $P(x_2, y_2)$ are any two distinct points on the nonvertical line. The slope of a vertical line is undefined.

3. a. $m_1 = m_2$ 　　　　　　　　　　　　　**b.** $m_2 = -\dfrac{1}{m_1}$

1.4 Straight Lines   page 41

1. (e) 　　　　　　　　　**3.** (a) 　　　　　　　　　**5.** (f)

7. Referring to the figure shown in the text, we see that $m = \dfrac{2 - 0}{0 - (-4)} = \dfrac{1}{2}$.

© 2011 Cengage Learning. All Rights Reserved. May not be scanned, copied or duplicated, or posted to a publicly accessible website, in whole or in part.

9. This is a vertical line, and hence its slope is undefined.

11. $m = \dfrac{y_2 - y_1}{x_2 - x_1} = \dfrac{8 - 3}{5 - 4} = 5.$

13. $m = \dfrac{y_2 - y_1}{x_2 - x_1} = \dfrac{8 - 3}{4 - (-2)} = \dfrac{5}{6}.$

15. $m = \dfrac{y_2 - y_1}{x_2 - x_1} = \dfrac{d - b}{c - a}$, provided $a \neq c$.

17. Because the equation is already in slope-intercept form, we read off the slope $m = 4$.

    a. If x increases by 1 unit, then y increases by 4 units.

    b. If x decreases by 2 units, then y decreases by $4(-2) = -8$ units.

19. The slope of the line through A and B is $\dfrac{-10 - (-2)}{-3 - 1} = \dfrac{-8}{-4} = 2.$ The slope of the line through C and D is

$\dfrac{1 - 5}{-1 - 1} = \dfrac{-4}{-2} = 2.$ Because the slopes of these two lines are equal, the lines are parallel.

21. The slope of the line through A and B is $\dfrac{2 - 5}{4 - (-2)} = -\dfrac{3}{6} = -\dfrac{1}{2}.$ The slope of the line through C and D is

$\dfrac{6 - (-2)}{3 - (-1)} = \dfrac{8}{4} = 2.$ Because the slopes of these two lines are the negative reciprocals of each other, the lines are

perpendicular.

23. The slope of the line through the point $(1, a)$ and $(4, -2)$ is $m_1 = \dfrac{-2 - a}{4 - 1}$ and the slope of the line through

$(2, 8)$ and $(-7, a + 4)$ is $m_2 = \dfrac{a + 4 - 8}{-7 - 2}.$ Because these two lines are parallel, m_1 is equal to m_2. Therefore,

$\dfrac{-2 - a}{3} = \dfrac{a - 4}{-9}$, $-9(-2 - a) = 3(a - 4)$, $18 + 9a = 3a - 12$, and $6a = -30$, so $a = -5$.

25. An equation of a horizontal line is of the form $y = b$. In this case $b = -3$, so $y = -3$ is an equation of the line.

27. We use the point-slope form of an equation of a line with the point $(3, -4)$ and slope $m = 2$. Thus

    $y - y_1 = m(x - x_1)$ becomes $y - (-4) = 2(x - 3)$. Simplifying, we have $y + 4 = 2x - 6$, or $y = 2x - 10$.

29. Because the slope $m = 0$, we know that the line is a horizontal line of the form $y = b$. Because the line passes

through $(-3, 2)$, we see that $b = 2$, and an equation of the line is $y = 2$.

31. We first compute the slope of the line joining the points $(2, 4)$ and $(3, 7)$ to be $m = \dfrac{7 - 4}{3 - 2} = 3$. Using the point-slope

form of an equation of a line with the point $(2, 4)$ and slope $m = 3$, we find $y - 4 = 3(x - 2)$, or $y = 3x - 2$.

33. We first compute the slope of the line joining the points $(1, 2)$ and $(-3, -2)$ to be $m = \dfrac{-2 - 2}{-3 - 1} = \dfrac{-4}{-4} = 1$. Using

the point-slope form of an equation of a line with the point $(1, 2)$ and slope $m = 1$, we find $y - 2 = x - 1$, or

$y = x + 1$.

35. We use the slope-intercept form of an equation of a line: $y = mx + b$. Because $m = 3$ and $b = 4$, the equation is

$y = 3x + 4$.

© 2011 Cengage Learning. All Rights Reserved. May not be scanned, copied or duplicated, or posted to a publicly accessible website, in whole or in part.

37. We use the slope-intercept form of an equation of a line: $y = mx + b$. Because $m = 0$ and $b = 5$, the equation is $y = 5$.

39. We first write the given equation in the slope-intercept form: $x - 2y = 0$, so $-2y = -x$, or $y = \frac{1}{2}x$. From this equation, we see that $m = \frac{1}{2}$ and $b = 0$.

41. We write the equation in slope-intercept form: $2x - 3y - 9 = 0$, $-3y = -2x + 9$, and $y = \frac{2}{3}x - 3$. From this equation, we see that $m = \frac{2}{3}$ and $b = -3$.

43. We write the equation in slope-intercept form: $2x + 4y = 14$, $4y = -2x + 14$, and $y = -\frac{2}{4}x + \frac{14}{4} = -\frac{1}{2}x + \frac{7}{2}$. From this equation, we see that $m = -\frac{1}{2}$ and $b = \frac{7}{2}$.

45. We first write the equation $2x - 4y - 8 = 0$ in slope-intercept form: $2x - 4y - 8 = 0$, $4y = 2x - 8$, $y = \frac{1}{2}x - 2$. Now the required line is parallel to this line, and hence has the same slope. Using the point-slope form of an equation of a line with $m = \frac{1}{2}$ and the point $(-2, 2)$, we have $y - 2 = \frac{1}{2}[x - (-2)]$ or $y = \frac{1}{2}x + 3$.

47. A line parallel to the x-axis has slope 0 and is of the form $y = b$. Because the line is 6 units below the axis, it passes through $(0, -6)$ and its equation is $y = -6$.

49. We use the point-slope form of an equation of a line to obtain $y - b = 0\,(x - a)$, or $y = b$.

51. Because the required line is parallel to the line joining $(-3, 2)$ and $(6, 8)$, it has slope $m = \dfrac{8 - 2}{6 - (-3)} = \dfrac{6}{9} = \dfrac{2}{3}$. We also know that the required line passes through $(-5, -4)$. Using the point-slope form of an equation of a line, we find $y - (-4) = \frac{2}{3}[x - (-5)]$, $y = \frac{2}{3}x + \frac{10}{3} - 4$, and finally $y = \frac{2}{3}x - \frac{2}{3}$.

53. Because the point $(-3, 5)$ lies on the line $kx + 3y + 9 = 0$, it satisfies the equation. Substituting $x = -3$ and $y = 5$ into the equation gives $-3k + 15 + 9 = 0$, or $k = 8$.

55. $3x - 2y + 6 = 0$. Setting $y = 0$, we have $3x + 6 = 0$ or $x = -2$, so the x-intercept is -2. Setting $x = 0$, we have $-2y + 6 = 0$ or $y = 3$, so the y-intercept is 3.

57. $x + 2y - 4 = 0$. Setting $y = 0$, we have $x - 4 = 0$ or $x = 4$, so the x-intercept is 4. Setting $x = 0$, we have $2y - 4 = 0$ or $y = 2$, so the y-intercept is 2.

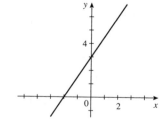

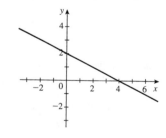

© 2011 Cengage Learning. All Rights Reserved. May not be scanned, copied or duplicated, or posted to a publicly accessible website, in whole or in part.

59. $y + 5 = 0$. Setting $y = 0$, we have $0 + 5 = 0$, which has no solution, so there is no x-intercept. Setting $x = 0$, we have $y + 5 = 0$ or $y = -5$, so the y-intercept is -5.

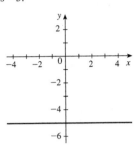

61. Because the line passes through the points $(a, 0)$ and $(0, b)$, its slope is $m = \dfrac{b - 0}{0 - a} = -\dfrac{b}{a}$. Then, using the point-slope form of an equation of a line with the point $(a, 0)$, we have $y - 0 = -\dfrac{b}{a}(x - a)$ or

$y = -\dfrac{b}{a}x + b$, which may be written in the form

$\dfrac{b}{a}x + y = b$. Multiplying this last equation by $\dfrac{1}{b}$, we

have $\dfrac{x}{a} + \dfrac{y}{b} = 1$.

63. Using the equation $\dfrac{x}{a} + \dfrac{y}{b} = 1$ with $a = -2$ and $b = -4$, we have $-\dfrac{x}{2} - \dfrac{y}{4} = 1$. Then $-4x - 2y = 8$, $2y = -8 - 4x$, and finally $y = -2x - 4$.

65. Using the equation $\dfrac{x}{a} + \dfrac{y}{b} = 1$ with $a = 4$ and $b = -\frac{1}{2}$, we have $\dfrac{x}{4} + \dfrac{y}{-1/2} = 1$, $\frac{1}{4}x + 2y = -1$, $2y = \frac{1}{4}x - 1$, and so $y = \frac{1}{8}x - \frac{1}{2}$.

67. The slope of the line passing through A and B is $m = \dfrac{7 - 1}{1 - (-2)} = \dfrac{6}{3} = 2$, and the slope of the line passing through B and C is $m = \dfrac{13 - 7}{4 - 1} = \dfrac{6}{3} = 2$. Because the slopes are equal, the points lie on the same line.

69. a.

y (% of total capacity)

```
100 |
 80 |                    _____
 60 |         _____/
 40 |
 20 |
  0 |____|____|____|_____
      5    10   15   t (years)
```

b. The slope is 1.9467 and the y-intercept is 70.082.

c. The output is increasing at the rate of 1.9467% per year. The output at the beginning of 1990 was 70.082%.

d. We solve the equation $1.9467t + 70.082 = 100$, obtaining $t \approx 15.37$. We conclude that the plants will be generating at maximum capacity shortly after the beginning of 2005.

71. a. $y = 0.55x$

b. Solving the equation $1100 = 0.55x$ for x, we have $x = \dfrac{1100}{0.55} = 2000$.

73. Using the points $(0, 0.68)$ and $(10, 0.80)$, we see that the slope of the required line is

$m = \dfrac{0.80 - 0.68}{10 - 0} = \dfrac{0.12}{10} = 0.012$. Next, using the point-slope form of the equation of a line, we have

$y - 0.68 = 0.012(t - 0)$ or $y = 0.012t + 0.68$. Therefore, when $t = 18$, we have $y = 0.012(18) + 0.68 = 0.896$, or 89.6%. That is, in 2008 women's wages were expected to be 89.6% of men's wages.

© 2011 Cengage Learning. All Rights Reserved. May not be scanned, copied or duplicated, or posted to a publicly accessible website, in whole or in part.

75. a, b.

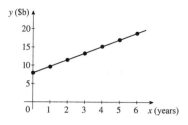

c. $m = \dfrac{18.8 - 7.9}{6 - 0} \approx 1.82$, so $y - 7.9 = 1.82\,(x - 0)$, or

$y = 1.82x + 7.9$.

d. $y = 1.82\,(5) + 7.9 \approx 17$ or \$17 billion. This agrees with the actual data for that year.

77. True. The slope of the line is given by $-\frac{2}{4} = -\frac{1}{2}$.

79. False. Let the slope of L_1 be $m_1 > 0$. Then the slope of L_2 is $m_2 = -\dfrac{1}{m_1} < 0$.

81. True. Set $y = 0$ and we have $Ax + C = 0$ or $x = -C/A$, and this is where the line intersects the x-axis.

83. Writing each equation in the slope-intercept form, we have $y = -\dfrac{a_1}{b_1}x - \dfrac{c_1}{b_1}$ $(b_1 \neq 0)$ and $y = -\dfrac{a_2}{b_2}x - \dfrac{c_2}{b_2}$

$(b_2 \neq 0)$. Because two lines are parallel if and only if their slopes are equal, we see that the lines are parallel if and only if $-\dfrac{a_1}{b_1} = -\dfrac{a_2}{b_2}$, or $a_1b_2 - b_1a_2 = 0$.

CHAPTER 1   Concept Review   page 46

1. ordered, abscissa or x-coordinate, ordinate or y-coordinate

3. $\sqrt{(c - a)^2 + (d - b)^2}$

5. a. $\dfrac{y_2 - y_1}{x_2 - x_1}$ **b.** undefined   **c.** 0   **d.** positive

7. a. $y - y_1 = m\,(x - x_1)$, point-slope form   **b.** $y = mx + b$, slope-intercept

CHAPTER 1   Review   page 47

1. Adding x to both sides yields $3 \leq 3x + 9$, $3x \geq -6$, or $x \geq -2$. We conclude that the solution set is $[-2, \infty)$.

3. The inequalities imply $x > 5$ or $x < -4$, so the solution set is $(-\infty, -4) \cup (5, \infty)$.

5. $|-5 + 7| + |-2| = |2| + |-2| = 2 + 2 = 4$.   **7.** $|2\pi - 6| - \pi = 2\pi - 6 - \pi = \pi - 6$.

9. $\left(\dfrac{9}{4}\right)^{3/2} = \dfrac{9^{3/2}}{4^{3/2}} = \dfrac{27}{8}$.   **11.** $(3 \cdot 4)^{-2} = 12^{-2} = \dfrac{1}{12^2} = \dfrac{1}{144}$.

13. $\dfrac{(3 \cdot 2^{-3})(4 \cdot 3^5)}{2 \cdot 9^3} = \dfrac{3 \cdot 2^{-3} \cdot 2^2 \cdot 3^5}{2 \cdot (3^2)^3} = \dfrac{2^{-1} \cdot 3^6}{2 \cdot 3^6} = \dfrac{1}{4}$.   **15.** $\dfrac{4(x^2 + y)^3}{x^2 + y} = 4(x^2 + y)^2$.

© 2011 Cengage Learning. All Rights Reserved. May not be scanned, copied or duplicated, or posted to a publicly accessible website, in whole or in part.

17. $\dfrac{\sqrt[4]{16x^5yz}}{\sqrt[4]{81xyz^5}} = \dfrac{\left(2^4x^5yz\right)^{1/4}}{\left(3^4xyz^5\right)^{1/4}} = \dfrac{2x^{5/4}y^{1/4}z^{1/4}}{3x^{1/4}y^{1/4}z^{5/4}} = \dfrac{2x}{3z}.$

19. $\left(\dfrac{3xy^2}{4x^3y}\right)^{-2}\left(\dfrac{3xy^3}{2x^2}\right)^3 = \left(\dfrac{3y}{4x^2}\right)^{-2}\left(\dfrac{3y^3}{2x}\right)^3 = \left(\dfrac{4x^2}{3y}\right)^2\left(\dfrac{3y^3}{2x}\right)^3 = \dfrac{\left(16x^4\right)\left(27y^9\right)}{\left(9y^2\right)\left(8x^3\right)} = 6xy^7.$

21. $-2\pi^2r^3 + 100\pi r^2 = -2\pi r^2\left(\pi r - 50\right).$

23. $16 - x^2 = 4^2 - x^2 = \left(4 - x\right)\left(4 + x\right).$

25. $8x^2 + 2x - 3 = \left(4x + 3\right)\left(2x - 1\right) = 0$, so $x = -\frac{3}{4}$ and $x = \frac{1}{2}$ are the roots of the equation.

27. $-x^3 - 2x^2 + 3x = -x\left(x^2 + 2x - 3\right) = -x\left(x + 3\right)\left(x - 1\right) = 0$, and so the roots of the equation are $x = 0$, $x = -3$, and $x = 1$.

29. Factoring the given expression, we have $\left(2x - 1\right)\left(x + 2\right) \le 0$. From the sign diagram, we conclude that the given inequality is satisfied when $-2 \le x \le \frac{1}{2}$.

```
- - - - - - - - 0 + +   Sign of 2x − 1
- - 0 + + + + + + + +   Sign of x + 2
————————————————————→ x
        −2      0 ½
```

31. The given inequality is equivalent to $|2x - 3| < 5$ or $-5 < 2x - 3 < 5$. Thus, $-2 < 2x < 8$, or $-1 < x < 4$.

33. Here we use the quadratic formula to solve the equation $x^2 - 2x - 5 = 0$. Here $a = 1$, $b = -2$, and $c = -5$, so

$$x = \dfrac{-b \pm \sqrt{b^2 - 4ac}}{2a} = \dfrac{-(-2) \pm \sqrt{(-2)^2 - 4(1)(-5)}}{2(1)} = \dfrac{2 \pm \sqrt{24}}{2} = 1 \pm \sqrt{6}.$$

35. $\dfrac{\left(t + 6\right)\left(60\right) - \left(60t + 180\right)}{\left(t + 6\right)^2} = \dfrac{60t + 360 - 60t - 180}{\left(t + 6\right)^2} = \dfrac{180}{\left(t + 6\right)^2}.$

37. $\dfrac{2}{3}\left(\dfrac{4x}{2x^2 - 1}\right) + 3\left(\dfrac{3}{3x - 1}\right) = \dfrac{8x}{3\left(2x^2 - 1\right)} + \dfrac{9}{3x - 1} = \dfrac{8x\left(3x - 1\right) + 27\left(2x^2 - 1\right)}{3\left(2x^2 - 1\right)\left(3x - 1\right)} = \dfrac{78x^2 - 8x - 27}{3\left(2x^2 - 1\right)\left(3x - 1\right)}.$

39. $\dfrac{\sqrt{x} - 1}{x - 1} = \dfrac{\sqrt{x} - 1}{x - 1} \cdot \dfrac{\sqrt{x} + 1}{\sqrt{x} + 1} = \dfrac{\left(\sqrt{x}\right)^2 - 1}{\left(x - 1\right)\left(\sqrt{x} + 1\right)} = \dfrac{x - 1}{\left(x - 1\right)\left(\sqrt{x} + 1\right)} = \dfrac{1}{\sqrt{x} + 1}.$

41. The distance is $d = \sqrt{\left[1 - (-2)\right]^2 + \left[-7 - (-3)\right]^2} = \sqrt{3^2 + (-4)^2} = \sqrt{9 + 16} = \sqrt{25} = 5.$

43. An equation is $x = -2$.

45. The slope of L is $m = \dfrac{\frac{7}{2} - 4}{3 - (-2)} = -\dfrac{1}{10}$, and an equation of L is $y - 4 = -\frac{1}{10}\left[x - (-2)\right] = -\frac{1}{10}x - \frac{1}{5}$, or $y = -\frac{1}{10}x + \frac{19}{5}$. The general form of this equation is $x + 10y - 38 = 0$.

47. Writing the given equation in the form $y = \frac{5}{2}x - 3$, we see that the slope of the given line is $\frac{5}{2}$. Thus, an equation is $y - 4 = \frac{5}{2}\left(x + 2\right)$, or $y = \frac{5}{2}x + 9$. The general form of this equation is $5x - 2y + 18 = 0$.

© 2011 Cengage Learning. All Rights Reserved. May not be scanned, copied or duplicated, or posted to a publicly accessible website, in whole or in part.

49. Rewriting the given equation in slope-intercept form, we have $4y = -3x + 8$ or $y = -\frac{3}{4}x + 2$. We conclude that the slope of the required line is $-\frac{3}{4}$. Using the point-slope form of the equation of a line with the point $(2, 3)$ and slope $-\frac{3}{4}$, we obtain $y - 3 = -\frac{3}{4}(x - 2)$, and so $y = -\frac{3}{4}x + \frac{6}{4} + 3 = -\frac{3}{4}x + \frac{9}{2}$. The general form of this equation is $3x + 4y - 18 = 0$.

51. The slope of the line passing through $(-2, -4)$ and $(1, 5)$ is $m = \dfrac{5 - (-4)}{1 - (-2)} = \dfrac{9}{3} = 3$, so the required line is $y - (-2) = 3[x - (-3)]$. Simplifying, this is equivalent to $y + 2 = 3x + 9$, or $y = 3x + 7$.

53. Setting $x = 0$ gives $y = -6$ as the y-intercept. Setting $y = 0$ gives $x = 8$ as the x-intercept. The graph of $3x - 4y = 24$ is shown.

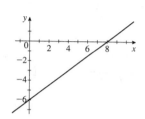

55. Simplifying $2(1.5C + 80) \leq 2(2.5C - 20)$, we obtain $1.5C + 80 \leq 2.5C - 20$, so $C \geq 100$ and the minimum cost is $100.

57. We solve the inequality $-16t^2 + 64t + 80 \geq 128$, obtaining $-16t^2 + 64t - 48 \geq 0$, $t^2 - 4t + 3 \leq 0$, and $(t - 3)(t - 1) \leq 0$. From the sign diagram, we see that the required solution is $[1, 3]$. Thus, the stone is 128 ft or higher off the ground between 1 and 3 seconds after it was thrown.

$$- - - - - - 0 + + \quad \text{Sign of } t - 3$$
$$- - 0 + + + + + + \quad \text{Sign of } t - 1$$

| CHAPTER 1 | Before Moving On... | page 48 |

1. a. $\left|\pi - 2\sqrt{3}\right| - \left|\sqrt{3} - \sqrt{2}\right| = -\left(\pi - 2\sqrt{3}\right) - \left(\sqrt{3} - \sqrt{2}\right) = \sqrt{3} + \sqrt{2} - \pi$.

b. $\left[\left(-\frac{1}{3}\right)^{-3}\right]^{1/3} = \left(-\frac{1}{3}\right)^{(-3)\left(\frac{1}{3}\right)} = \left(-\frac{1}{3}\right)^{-1} = -3$.

2. a. $\sqrt[3]{64x^6} \cdot \sqrt{9y^2x^6} = (4x^2)(3yx^3) = 12x^5y$.

b. $\left(\dfrac{a^{-3}}{b^{-4}}\right)^2 \left(\dfrac{b}{a}\right)^{-3} = \dfrac{a^{-6}}{b^{-8}} \cdot \dfrac{b^{-3}}{a^{-3}} = \dfrac{b^8}{a^6} \cdot \dfrac{a^3}{b^3} = \dfrac{b^5}{a^3}$.

3. a. $\dfrac{2x}{3\sqrt{y}} \cdot \dfrac{\sqrt{y}}{\sqrt{y}} = \dfrac{2x\sqrt{y}}{3y}$.

b. $\dfrac{x}{\sqrt{x} - 4} \cdot \dfrac{\sqrt{x} + 4}{\sqrt{x} + 4} = \dfrac{x(\sqrt{x} + 4)}{x - 16}$.

4. a. $\dfrac{(x^2 + 1)\left(\frac{1}{2}x^{-1/2}\right) - x^{1/2}(2x)}{(x^2 + 1)^2} = \dfrac{\frac{1}{2}x^{-1/2}\left[(x^2 + 1) - 4x^2\right]}{(x^2 + 1)^2} = \dfrac{1 - 3x^2}{2x^{1/2}(x^2 + 1)^2}$.

b. $-\dfrac{3x}{\sqrt{x + 2}} + 3\sqrt{x + 2} = \dfrac{-3x + 3(x + 2)}{\sqrt{x + 2}} = \dfrac{6}{\sqrt{x + 2}} = \dfrac{6\sqrt{x + 2}}{x + 2}$.

© 2011 Cengage Learning. All Rights Reserved. May not be scanned, copied or duplicated, or posted to a publicly accessible website, in whole or in part.

5. $\dfrac{\sqrt{x}+\sqrt{y}}{\sqrt{x}-\sqrt{y}} = \dfrac{\sqrt{x}+\sqrt{y}}{\sqrt{x}-\sqrt{y}} \cdot \dfrac{\sqrt{x}-\sqrt{y}}{\sqrt{x}-\sqrt{y}} = \dfrac{x-y}{\left(\sqrt{x}-\sqrt{y}\right)^2}$.

6. a. $12x^3 - 10x^2 - 12x = 2x\left(6x^2 - 5x - 6\right) = 2x\left(2x - 3\right)\left(3x + 2\right)$.

   b. $2bx - 2by + 3cx - 3cy = 2b\left(x - y\right) + 3c\left(x - y\right) = \left(2b + 3c\right)\left(x - y\right)$

7. a. $12x^2 - 9x - 3 = 0$, so $3\left(4x^2 - 3x - 1\right) = 0$ and $3\left(4x + 1\right)\left(x - 1\right) = 0$. Thus, $x = -\frac{1}{4}$ or $x = 1$.

   b. $3x^2 - 5x + 1 = 0$. Using the quadratic formula with $a = 3$, $b = -5$, and $c = 1$, we have
   $$x = \frac{-(-5) \pm \sqrt{25 - 12}}{2(3)} = \frac{5 \pm \sqrt{13}}{6}.$$

8. $d = \sqrt{[6 - (-2)]^2 + (8 - 4)^2} = \sqrt{64 + 16} = \sqrt{80} = 4\sqrt{5}$.

9. $m = \dfrac{5 - (-2)}{4 - (-1)} = \dfrac{7}{5}$, so $y - (-2) = \frac{7}{5}[x - (-1)]$, $y + 2 = \frac{7}{5}x + \frac{7}{5}$, or $y = \frac{7}{5}x - \frac{3}{5}$.

10. $m = -\frac{1}{3}$ and $b = \frac{4}{3}$, so an equation is $y = -\frac{1}{3}x + \frac{4}{3}$.

© 2011 Cengage Learning. All Rights Reserved. May not be scanned, copied or duplicated, or posted to a publicly accessible website, in whole or in part.

FUNCTIONS, LIMITS, AND THE DERIVATIVE

2.1 Problem-Solving Tips

New mathematical terms in each section appear in **boldface** type along with their definition, or they are defined in green boxes. Each time you encounter a new term, read through the definition and then try to express the definition in your own words without looking at the book. Once you understand these definitions, it will be easier for you to work the exercise sets that follow each section.

Here are some tips for solving the problems in the exercises that follow:

1. **To find the domain of a function** $f(x)$, find all values of x for which $f(x)$ is a real number.

   a. **If the function involves a quotient**, check to see if there are any values of x at which the denominator is equal to zero. (Remember, division by zero is not allowed.) Then exclude those points from the domain.

   b. **If the function involves the root of a real number**, check to see if the root is an even or an odd root. If n is even, the nth root of a negative number is not defined, so values of x yielding the nth root of a negative number must be excluded from the domain of f. For example, $\sqrt{x-1}$ is defined only for $x \geq 1$, so the domain of $f(x) = \sqrt{x-1}$ is $[1, \infty)$.

2. **To evaluate a piecewise-defined function** $f(x)$ at a specific value of x, check to see which subdomain x lies in. Then evaluate the function using the rule for that subdomain.

3. **To determine whether a curve is the graph of a function**, use the vertical line test. If you can draw a vertical line through the curve that intersects the curve at more than one point, then the curve is not the graph of a function.

2.1 Concept Questions   page 57

1. a. A function is a rule that associates with each element in a set A exactly one element in a set B.
   b. The domain of a function f is the set of all elements x in the set such that $f(x)$ is an element in B. The range of f is the set of all elements $f(x)$ whenever x is an element in its domain.
   c. An independent variable is a variable in the domain of a function f. The dependent variable is $y = f(x)$.

3. a. Yes, every vertical line intersects the curve in at most one point.
   b. No, a vertical line intersects the curve at more than one point.
   c. No, a vertical line intersects the curve at more than one point.
   d. Yes, every vertical line intersects the curve in at most one point.

© 2011 Cengage Learning. All Rights Reserved. May not be scanned, copied or duplicated, or posted to a publicly accessible website, in whole or in part.

2.1   Functions and Their Graphs   page 57

1. $f(x) = 5x + 6$. Therefore $f(3) = 5(3) + 6 = 21$, $f(-3) = 5(-3) + 6 = -9$, $f(a) = 5(a) + 6 = 5a + 6$, $f(-a) = 5(-a) + 6 = -5a + 6$, and $f(a+3) = 5(a+3) + 6 = 5a + 15 + 6 = 5a + 21$.

3. $g(x) = 3x^2 - 6x - 3$, so $g(0) = 3(0) - 6(0) - 3 = -3$, $g(-1) = 3(-1)^2 - 6(-1) - 3 = 3 + 6 - 3 = 6$, $g(a) = 3(a)^2 - 6(a) - 3 = 3a^2 - 6a - 3$, $g(-a) = 3(-a)^2 - 6(-a) - 3 = 3a^2 + 6a - 3$, and $g(x+1) = 3(x+1)^2 - 6(x+1) - 3 = 3(x^2 + 2x + 1) - 6x - 6 - 3 = 3x^2 + 6x + 3 - 6x - 9 = 3x^2 - 6$.

5. $f(x) = 2x + 5$, so $f(a+h) = 2(a+h) + 5 = 2a + 2h + 5$, $f(-a) = 2(-a) + 5 = -2a + 5$, $f(a^2) = 2(a^2) + 5 = 2a^2 + 5$, $f(a - 2h) = 2(a - 2h) + 5 = 2a - 4h + 5$, and $f(2a - h) = 2(2a - h) + 5 = 4a - 2h + 5$

7. $s(t) = \dfrac{2t}{t^2 - 1}$. Therefore, $s(4) = \dfrac{2(4)}{(4)^2 - 1} = \dfrac{8}{15}$, $s(0) = \dfrac{2(0)}{0^2 - 1} = 0$, $s(a) = \dfrac{2(a)}{a^2 - 1} = \dfrac{2a}{a^2 - 1}$; $s(2+a) = \dfrac{2(2+a)}{(2+a)^2 - 1} = \dfrac{2(2+a)}{a^2 + 4a + 4 - 1} = \dfrac{2(2+a)}{a^2 + 4a + 3}$, and $s(t+1) = \dfrac{2(t+1)}{(t+1)^2 - 1} = \dfrac{2(t+1)}{t^2 + 2t + 1 - 1} = \dfrac{2(t+1)}{t(t+2)}$.

9. $f(t) = \dfrac{2t^2}{\sqrt{t - 1}}$. Therefore, $f(2) = \dfrac{2(2^2)}{\sqrt{2} - 1} = 8$, $f(a) = \dfrac{2a^2}{\sqrt{a} - 1}$, $f(x+1) = \dfrac{2(x+1)^2}{\sqrt{(x+1) - 1}} = \dfrac{2(x+1)^2}{\sqrt{x}}$, and $f(x - 1) = \dfrac{2(x - 1)^2}{\sqrt{(x-1) - 1}} = \dfrac{2(x - 1)^2}{\sqrt{x - 2}}$.

11. Because $x = -2 \le 0$, we calculate $f(-2) = (-2)^2 + 1 = 4 + 1 = 5$. Because $x = 0 \le 0$, we calculate $f(0) = (0)^2 + 1 = 1$. Because $x = 1 > 0$, we calculate $f(1) = \sqrt{1} = 1$.

13. Because $x = -1 < 1$, $f(-1) = -\frac{1}{2}(-1)^2 + 3 = \frac{5}{2}$. Because $x = 0 < 1$, $f(0) = -\frac{1}{2}(0)^2 + 3 = 3$. Because $x = 1 \ge 1$, $f(1) = 2(1^2) + 1 = 3$. Because $x = 2 \ge 1$, $f(2) = 2(2^2) + 1 = 9$.

15. **a.** $f(0) = -2$.

 b. (i) $f(x) = 3$ when $x \approx 2$.   **(ii)** $f(x) = 0$ when $x = 1$.

 c. $[0, 6]$

 d. $[-2, 6]$

17. $g(2) = \sqrt{2^2 - 1} = \sqrt{3}$, so the point $\left(2, \sqrt{3}\right)$ lies on the graph of g.

19. $f(-2) = \dfrac{|-2 - 1|}{-2 + 1} = \dfrac{|-3|}{-1} = -3$, so the point $(-2, -3)$ does lie on the graph of f.

21. Because the point $(1, 5)$ lies on the graph of f it satisfies the equation defining f. Thus, $f(1) = 2(1)^2 - 4(1) + c = 5$, or $c = 7$.

23. Because $f(x)$ is a real number for any value of x, the domain of f is $(-\infty, \infty)$.

25. $f(x)$ is not defined at $x = 0$ and so the domain of f is $(-\infty, 0) \cup (0, \infty)$.

© 2011 Cengage Learning. All Rights Reserved. May not be scanned, copied or duplicated, or posted to a publicly accessible website, in whole or in part.

27. $f(x)$ is a real number for all values of x. Note that $x^2 + 1 \geq 1$ for all x. Therefore, the domain of f is $(-\infty, \infty)$.

29. Because the square root of a number is defined for all real numbers greater than or equal to zero, we have $5 - x \geq 0$, or $-x \geq -5$ and so $x \leq 5$. (Recall that multiplying by -1 reverses the sign of an inequality.) Therefore, the domain of g is $(-\infty, 5]$.

31. The denominator of f is zero when $x^2 - 1 = 0$, or $x = \pm 1$. Therefore, the domain of f is $(-\infty, -1) \cup (-1, 1) \cup (1, \infty)$.

33. f is defined when $x + 3 \geq 0$, that is, when $x \geq -3$. Therefore, the domain of f is $[-3, \infty)$.

35. The numerator is defined when $1 - x \geq 0$, $-x \geq -1$ or $x \leq 1$. Furthermore, the denominator is zero when $x = \pm 2$. Therefore, the domain is the set of all real numbers in $(-\infty, -2) \cup (-2, 1]$.

37. a. The domain of f is the set of all real numbers.

b. $f(x) = x^2 - x - 6$, so

$$f(-3) = (-3)^2 - (-3) - 6 = 9 + 3 - 6 = 6,$$
$$f(-2) = (-2)^2 - (-2) - 6 = 4 + 2 - 6 = 0,$$
$$f(-1) = (-1)^2 - (-1) - 6 = 1 + 1 - 6 = -4,$$
$$f(0) = (0)^2 - (0) - 6 = -6,$$

c.

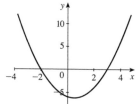

$f\left(\frac{1}{2}\right) = \left(\frac{1}{2}\right)^2 - \left(\frac{1}{2}\right) - 6 = \frac{1}{4} - \frac{2}{4} - \frac{24}{4} = -\frac{25}{4}$, $f(1) = (1)^2 - 1 - 6 = -6$, $f(2) = (2)^2 - 2 - 6 = 4 - 2 - 6 = -4$, and $f(3) = (3)^2 - 3 - 6 = 9 - 3 - 6 = 0$.
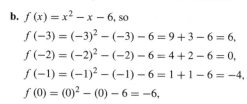

39. $f(x) = 2x^2 + 1$ has domain $(-\infty, \infty)$ and range $[1, \infty)$.

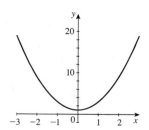

41. $f(x) = 2 + \sqrt{x}$ has domain $[0, \infty)$ and range $[2, \infty)$.

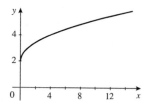

43. $f(x) = \sqrt{1 - x}$ has domain $(-\infty, 1]$ and range $[0, \infty)$

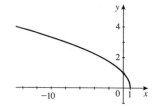

45. $f(x) = |x| - 1$ has domain $(-\infty, \infty)$ and range $[-1, \infty)$.

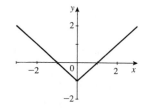

© 2011 Cengage Learning. All Rights Reserved. May not be scanned, copied or duplicated, or posted to a publicly accessible website, in whole or in part.

47. $f(x) = \begin{cases} x & \text{if } x < 0 \\ 2x + 1 & \text{if } x \geq 0 \end{cases}$ has domain

$(-\infty, \infty)$ and range $(-\infty, 0) \cup [1, \infty)$.

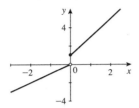

49. If $x \leq 1$, the graph of f is the half-line $y = -x + 1$. For $x > 1$, we calculate a few points: $f(2) = 3$, $f(3) = 8$, and $f(4) = 15$. f has domain $(-\infty, \infty)$ and range $[0, \infty)$.

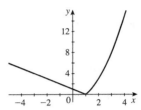

51. Each vertical line cuts the given graph at exactly one point, and so the graph represents y as a function of x.

53. Because there is a vertical line that intersects the graph at three points, the graph does not represent y as a function of x.

55. Each vertical line intersects the graph of f at exactly one point, and so the graph represents y as a function of x.

57. Each vertical line intersects the graph of f at exactly one point, and so the graph represents y as a function of x.

59. The circumference of a circle with a 5-inch radius is given by $C(5) = 2\pi(5) = 10\pi$, or 10π inches.

61. $\frac{4}{3}(\pi)(2r)^3 = \frac{4}{3}\pi 8r^3 = 8\left(\frac{4}{3}\pi r^3\right)$. Therefore, the volume of the tumor is increased by a factor of 8.

63. a. From $t = 0$ to $t = 5$, the graph for cassettes lies above that for CDs, so from 1985 to 1990, the value of prerecorded cassettes sold was greater than that of CDs.

b. Sales of prerecorded CDs were greater than those of prerecorded cassettes from 1990 onward.

c. The graphs intersect at the point with coordinates $x = 5$ and $y \approx 3.5$, and this tells us that the sales of the two formats were the same in 1990 at the sales level of approximately \$3.5 billion.

65. a. The slope of the straight line passing through the points $(0, 0.58)$ and $(20, 0.95)$ is $m_1 = \dfrac{0.95 - 0.58}{20 - 0} = 0.0185$, so an equation of the straight line passing through these two points is $y - 0.58 = 0.0185(t - 0)$ or $y = 0.0185t + 0.58$. Next, the slope of the straight line passing through the points $(20, 0.95)$ and $(30, 1.1)$ is $m_2 = \dfrac{1.1 - 0.95}{30 - 20} = 0.015$, so an equation of the straight line passing through the two points is $y - 0.95 = 0.015(t - 20)$ or $y = 0.015t + 0.65$. Therefore, a rule for f is

$$f(t) = \begin{cases} 0.0185t + 0.58 & \text{if } 0 \leq t \leq 20 \\ 0.015t + 0.65 & \text{if } 20 < t \leq 30 \end{cases}$$

b. The ratios were changing at the rates of 0.0185/yr from 1960 through 1980 and 0.015/yr from 1980 through 1990.

c. The ratio was 1 when $t \approx 20.3$. This shows that the number of bachelor's degrees earned by women equaled the number earned by men for the first time around 1983.

© 2011 Cengage Learning. All Rights Reserved. May not be scanned, copied or duplicated, or posted to a publicly accessible website, in whole or in part.

67. a. $T(x) = 0.06x$

    b. $T(200) = 0.06(200) = 12$, or $12.00 and $T(5.65) = 0.06(5.65) = 0.34$, or $0.34.

69. The child should receive $D(4) = \frac{2}{25}(500)(4) = 160$, or 160 mg.

71. a. Take $m = 7.5$ and $b = 20$. Then $f(t) = 7.5t + 20$ for $0 \le t \le 6$.

    b. $f(6) = 7.5(6) + 20 = 65$, or 65 million households.

73. a. The graph of the function is a straight line passing through
    $(0, 120000)$ and $(10, 0)$. Its slope is

$$m = \frac{0 - 120{,}000}{10 - 0} = -12{,}000.$$ The required equation is
$V = -12{,}000n + 120{,}000.$

    c. $V = -12{,}000(6) + 120{,}000 = 48{,}000$, or $48,000.

    d. This is given by the slope, that is, $12,000 per year.

b.

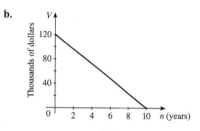

75. The domain of the function f is the set of all real positive numbers
where $V \ne 0$; that is, $(0, \infty)$.

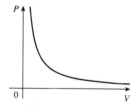

77. a. $N(0) = 3.6$, or 3.6 million people; $N(25) = 0.0031(25)^2 + 0.16(25) + 3.6 = 9.5375$, or approximately
    9.5 million people.

    b. $N(30) = 0.0031(30)^2 + 0.16(30) + 3.6 = 11.19$, or approximately 11.2 million people.

79. $N(t) = -t^3 + 6t^2 + 15t$. Between 8 a.m. and 9 a.m., the average worker can be expected to assemble
$N(1) - N(0) = (-1 + 6 + 15) - 0 = 20$, or 20 walkie-talkies. Between 9 a.m. and 10 a.m., we expect that
$N(2) - N(1) = \left[-2^3 + 6(2^2) + 15(2)\right] - (-1 + 6 + 15) = 46 - 20 = 26$, or 26 walkie-talkies can be assembled
by the average worker.

81. The amount spent in 2004 was $S(0) = 5.6$, or $5.6 billion. The amount spent in 2008 was
$S(4) = -0.03(4)^3 + 0.2(4)^2 + 0.23(4) + 5.6 = 7.8$, or $7.8 billion.

83. a. The assets at the beginning of 2002 were $0.6 trillion. At the beginning of 2003, they were $f(1) = 0.6$, or
    $0.6 trillion.

    b. The assets at the beginning of 2005 were $f(3) = 0.96$, or $0.96 trillion. At the beginning of 2007, they were
    $f(5) \approx 1.20$, or $1.2 trillion.

© 2011 Cengage Learning. All Rights Reserved. May not be scanned, copied or duplicated, or posted to a publicly accessible website, in whole or in part.

85. a. The amount of solids discharged in 1989 ($t = 0$) was 130 tons/day; **b.**
in 1992 ($t = 3$), it was 100 tons/day; and in 1996 ($t = 7$), it was
$f(7) = 1.25\,(7)^2 - 26.25\,(7) + 162.5 = 40$, or 40 tons/day.

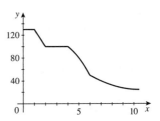

87. a. The passenger ship travels a distance given by $14t$ miles east and the cargo ship travels a distance of
$10\,(t - 2)$ miles north. After two hours have passed, the distance between the two ships is given by

$$\sqrt{[10\,(t-2)]^2 + (14t)^2} = \sqrt{296t^2 - 400t + 400} \text{ miles, so } D(t) = \begin{cases} 14t & \text{if } 0 \le t \le 2 \\ 2\sqrt{74t^2 - 100t + 100} & \text{if } t > 2 \end{cases}$$

b. Three hours after the cargo ship leaves port the value of t is 5. Therefore,
$$D = 2\sqrt{74\,(5)^2 - 100\,(5) + 100} \approx 76.16, \text{ or } 76.16 \text{ miles.}$$

89. False. Take $f(x) = x^2$, $a = 1$, and $b = -1$. Then $f(1) = 1 = f(-1)$, but $a \ne b$.

91. False. It intersects the graph of a function in at most one point.

93. False. Take $f(x) = x^2$ and $k = 2$. Then $f(x) = (2x)^2 = 4x^2 \ne 2x^2 = 2f(x)$.

2.1   Using Technology   page 66

1. a.

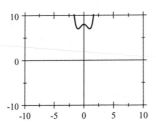

b.

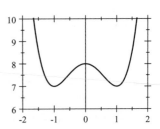

3. a.

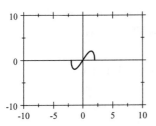

b.

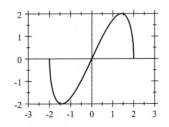

© 2011 Cengage Learning. All Rights Reserved. May not be scanned, copied or duplicated, or posted to a publicly accessible website, in whole or in part.

5.

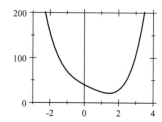

7.

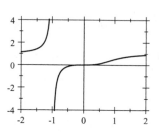

9. $f(2.145) \approx 18.5505$.

11. $f(2.41) \approx 4.1616$.

13. a.

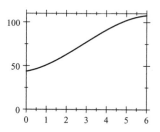

15. a.

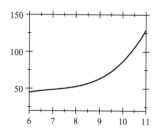

b. The amount spent in the year 2000 was $f(2) \approx 62.96$, or approximately \$62.96 million. In 2004, it was $f(6) \approx 107.66$, or approximately \$107.7 million.

b. $f(6) = 44.7$, $f(8) = 52.7$, and $f(11) = 129.2$.

| **2.2** | **Problem-Solving Tips** |

When you come across new notation, make sure that you understand that notation. If you can't express the notation verbally, you haven't yet grasped its meaning. For example, in this section we introduced the notation $g \circ f$, read "g circle f." We use this notation to describe the composition of the functions g and f. Note that $g \circ f$ is not the same as $f \circ g$.

Here are some tips for solving the problems in the exercises that follow:

1. If f and g are functions with domains A and B, respectively, then the domain of $f + g$, $f - g$, and fg is $A \cap B$. The domain of the quotient f/g is $A \cap B$ excluding all numbers x such that $g(x) = 0$.

2. **To find the rule for the composite function** $g \circ f$, evaluate the function g at $f(x)$. Similarly, to find $f \circ g$, evaluate the function f at $g(x)$.

© 2011 Cengage Learning. All Rights Reserved. May not be scanned, copied or duplicated, or posted to a publicly accessible website, in whole or in part.

2.2   Concept Questions   page 71

1. a. $(f + g)(x) = f(x) + g(x)$, $(f - g)(x) = f(x) - g(x)$, and $(fg)(x) = f(x)g(x)$; all have domain $A \cap B$.

$(f/g)(x) = \dfrac{f(x)}{g(x)}$ has domain $A \cap B$ excluding $x \in A \cap B$ such that $g(x) = 0$.

b. $(f + g)(2) = f(2) + g(2) = 3 + (-2) = 1$, $(f - g)(2) = f(2) - g(2) = 3 - (-2) = 5$,

$(fg)(2) = f(2)g(2) = 3(-2) = -6$, and $(f/g)(2) = \dfrac{f(2)}{g(2)} = \dfrac{3}{-2} = -\dfrac{3}{2}$.

3. a. The domain of $(f \circ g)(x) = f(g(x))$ is the set of all x in the domain of g such that $g(x)$ is in the domain of f.
   The domain of $(g \circ f)(x) = g(f(x))$ is the set of all x in the domain of f such that $f(x)$ is in the domain of g.

b. $(g \circ f)(2) = g(f(2)) = g(3) = 8$. We cannot calculate $(f \circ g)(3)$ because $(f \circ g)(3) = f(g(3)) = f(8)$,
   and we don't know the value of $f(8)$.

2.2   The Algebra of Functions   page 72

1. $(f + g)(x) = f(x) + g(x) = (x^3 + 5) + (x^2 - 2) = x^3 + x^2 + 3$.

3. $fg(x) = f(x)g(x) = (x^3 + 5)(x^2 - 2) = x^5 - 2x^3 + 5x^2 - 10$.

5. $\dfrac{f}{g}(x) = \dfrac{f(x)}{g(x)} = \dfrac{x^3 + 5}{x^2 - 2}$.

7. $\dfrac{fg}{h}(x) = \dfrac{f(x)g(x)}{h(x)} = \dfrac{(x^3 + 5)(x^2 - 2)}{2x + 4} = \dfrac{x^5 - 2x^3 + 5x^2 - 10}{2x + 4}$.

9. $(f + g)(x) = f(x) + g(x) = x - 1 + \sqrt{x + 1}$.

11. $(fg)(x) = f(x)g(x) = (x - 1)\sqrt{x + 1}$.

13. $\dfrac{g}{h}(x) = \dfrac{g(x)}{h(x)} = \dfrac{\sqrt{x + 1}}{2x^3 - 1}$.

15. $\dfrac{fg}{h}(x) = \dfrac{(x - 1)(\sqrt{x + 1})}{2x^3 - 1}$.

17. $\dfrac{f - h}{g}(x) = \dfrac{x - 1 - (2x^3 - 1)}{\sqrt{x + 1}} = \dfrac{x - 2x^3}{\sqrt{x + 1}}$.

19. $(f + g)(x) = x^2 + 5 + \sqrt{x} - 2 = x^2 + \sqrt{x} + 3$, $(f - g)(x) = x^2 + 5 - (\sqrt{x} - 2) = x^2 - \sqrt{x} + 7$,

$(fg)(x) = (x^2 + 5)(\sqrt{x} - 2)$, and $\left(\dfrac{f}{g}\right)(x) = \dfrac{x^2 + 5}{\sqrt{x} - 2}$.

21. $(f + g)(x) = \sqrt{x + 3} + \dfrac{1}{x - 1} = \dfrac{(x - 1)\sqrt{x + 3} + 1}{x - 1}$, $(f - g)(x) = \sqrt{x + 3} - \dfrac{1}{x - 1} = \dfrac{(x - 1)\sqrt{x + 3} - 1}{x - 1}$,

$(fg)(x) = \sqrt{x + 3}\left(\dfrac{1}{x - 1}\right) = \dfrac{\sqrt{x + 3}}{x - 1}$, and $\left(\dfrac{f}{g}\right) = \sqrt{x + 3}(x - 1)$.

© 2011 Cengage Learning. All Rights Reserved. May not be scanned, copied or duplicated, or posted to a publicly accessible website, in whole or in part.

23. $(f + g)(x) = \dfrac{x + 1}{x - 1} + \dfrac{x + 2}{x - 2} = \dfrac{(x + 1)(x - 2) + (x + 2)(x - 1)}{(x - 1)(x - 2)} = \dfrac{x^2 - x - 2 + x^2 + x - 2}{(x - 1)(x - 2)}$

$$= \dfrac{2x^2 - 4}{(x - 1)(x - 2)} = \dfrac{2(x^2 - 2)}{(x - 1)(x - 2)},$$

$(f - g)(x) = \dfrac{x + 1}{x - 1} - \dfrac{x + 2}{x - 2} = \dfrac{(x + 1)(x - 2) - (x + 2)(x - 1)}{(x - 1)(x - 2)} = \dfrac{x^2 - x - 2 - x^2 - x + 2}{(x - 1)(x - 2)}$

$$= \dfrac{-2x}{(x - 1)(x - 2)},$$

$(fg)(x) = \dfrac{(x + 1)(x + 2)}{(x - 1)(x - 2)}$, and $\left(\dfrac{f}{g}\right)(x) = \dfrac{(x + 1)(x - 2)}{(x - 1)(x + 2)}$.

25. $(f \circ g)(x) = f(g(x)) = f(x^2) = (x^2)^2 + x^2 + 1 = x^4 + x^2 + 1$ and

$(g \circ f)(x) = g(f(x)) = g(x^2 + x + 1) = (x^2 + x + 1)^2$.

27. $(f \circ g)(x) = f(g(x)) = f(x^2 - 1) = \sqrt{x^2 - 1} + 1$ and

$(g \circ f)(x) = g(f(x)) = g(\sqrt{x} + 1) = (\sqrt{x} + 1)^2 - 1 = x + 2\sqrt{x} + 1 - 1 = x + 2\sqrt{x}$.

29. $(f \circ g)(x) = f(g(x)) = f\left(\dfrac{1}{x}\right) = \dfrac{1}{x} \div \left(\dfrac{1}{x^2} + 1\right) = \dfrac{1}{x} \cdot \dfrac{x^2}{x^2 + 1} = \dfrac{x}{x^2 + 1}$ and

$(g \circ f)(x) = g(f(x)) = g\left(\dfrac{x}{x^2 + 1}\right) = \dfrac{x^2 + 1}{x}$.

31. $h(2) = g(f(2))$. But $f(2) = 2^2 + 2 + 1 = 7$, so $h(2) = g(7) = 49$.

33. $h(2) = g(f(2))$. But $f(2) = \dfrac{1}{2(2) + 1} = \dfrac{1}{5}$, so $h(2) = g\left(\dfrac{1}{5}\right) = \dfrac{1}{\sqrt{5}} = \dfrac{\sqrt{5}}{5}$.

35. $f(x) = 2x^3 + x^2 + 1$, $g(x) = x^5$.   **37.** $f(x) = x^2 - 1$, $g(x) = \sqrt{x}$.

39. $f(x) = x^2 - 1$, $g(x) = \dfrac{1}{x}$.   **41.** $f(x) = 3x^2 + 2$, $g(x) = \dfrac{1}{x^{3/2}}$.

43. $f(a + h) - f(a) = [3(a + h) + 4] - (3a + 4) = 3a + 3h + 4 - 3a - 4 = 3h$.

45. $f(a + h) - f(a) = 4 - (a + h)^2 - (4 - a^2) = 4 - a^2 - 2ah - h^2 - 4 + a^2 = -2ah - h^2 = -h(2a + h)$.

47. $\dfrac{f(a + h) - f(a)}{h} = \dfrac{[(a + h)^2 + 1] - (a^2 + 1)}{h} = \dfrac{a^2 + 2ah + h^2 + 1 - a^2 - 1}{h} = \dfrac{2ah + h^2}{h}$

$$= \dfrac{h(2a + h)}{h} = 2a + h.$$

49. $\dfrac{f(a + h) - f(a)}{h} = \dfrac{[(a + h)^3 - (a + h)] - (a^3 - a)}{h} = \dfrac{a^3 + 3a^2h + 3ah^2 + h^3 - a - h - a^3 + a}{h}$

$$= \dfrac{3a^2h + 3ah^2 + h^3 - h}{h} = 3a^2 + 3ah + h^2 - 1.$$

51. $\dfrac{f(a + h) - f(a)}{h} = \dfrac{\dfrac{1}{a + h} - \dfrac{1}{a}}{h} = \dfrac{\dfrac{a - (a + h)}{a(a + h)}}{h} = -\dfrac{1}{a(a + h)}$.

© 2011 Cengage Learning. All Rights Reserved. May not be scanned, copied or duplicated, or posted to a publicly accessible website, in whole or in part.

53. $F(t)$ represents the total revenue for the two restaurants at time t.

55. $f(t)g(t)$ represents the dollar value of Nancy's holdings at time t.

57. $g \circ f$ is the function giving the amount of carbon monoxide pollution from cars at time t.

59. $C(x) = 0.6x + 12{,}100$.

61. a. $f(t) = 267$; $g(t) = 2t^2 + 46t + 733$.

   b. $h(t) = (f + g)(t) = f(t) + g(t) = 267 + \left(2t^2 + 46t + 733\right) = 2t^2 + 46t + 1000$.

   c. $h(13) = 2(13)^2 + 46(13) + 1000 = 1936$, or 1936 tons.

63. a. $(g \circ f)(1) = g(f(1)) = g(406) = 23$. So in 2002, the percentage of reported serious crimes that end in arrests or in the identification of suspects was 23.

   b. $(g \circ f)(6) = g(f(6)) = g(326) = 18$. In 2007, 18% of reported serious crimes ended in arrests or in the identification of suspects.

   c. Between 2002 and 2007, the total number of detectives had dropped from 406 to 326 and as a result, the percentage of reported serious crimes that ended in arrests or in the identification of suspects dropped from 23 to 18.

65. a. $P(x) = R(x) - C(x) = -0.1x^2 + 500x - \left(0.000003x^3 - 0.03x^2 + 200x + 100{,}000\right)$

   $\qquad = -0.000003x^3 - 0.07x^2 + 300x - 100{,}000$.

   b. $P(1500) = -0.000003(1500)^3 - 0.07(1500)^2 + 300(1500) - 100{,}000 = 182{,}375$, or \$182,375.

67. a. The gap is $G(t) - C(t) = \left(3.5t^2 + 26.7t + 436.2\right) - (24.3t + 365) = 3.5t^2 + 2.4t + 71.2$.

   b. At the beginning of 1983, the gap was $G(0) = 3.5(0)^2 + 2.4(0) + 71.2 = 71.2$, or 71,200.

   At the beginning of 1986, the gap was $G(3) = 3.5(3)^2 + 2.4(3) + 71.2 = 109.9$, or 109,900.

69. a. The occupancy rate at the beginning of January is $r(0) = \frac{10}{81}(0)^3 - \frac{10}{3}(0)^2 + \frac{200}{9}(0) + 55 = 55$, or 55%.

   $r(5) = \frac{10}{81}(5)^3 - \frac{10}{3}(5)^2 + \frac{200}{9}(5) + 55 \approx 98.2$, or approximately 98.2%.

   b. The monthly revenue at the beginning of January is $R(55) = -\frac{3}{5000}(55)^3 + \frac{9}{50}(55)^2 \approx 444.68$, or approximately \$444,700.

   The monthly revenue at the beginning of June is $R(98.2) = -\frac{3}{5000}(98.2)^3 + \frac{9}{50}(98.2)^2 \approx 1167.6$, or approximately \$1,167,600.

71. a. $s = f + g + h = (f + g) + h = f + (g + h)$. This suggests we define the sum s by

   $s(x) = (f + g + h)(x) = f(x) + g(x) + h(x)$.

   b. Let f, g, and h define the revenue (in dollars) in week t of three branches of a store. Then its total revenue (in dollars) in week t is $s(t) = (f + g + h)(t) = f(t) + g(t) + h(t)$.

73. True. $(f + g)(x) = f(x) + g(x) = g(x) + f(x) = (g + f)(x)$.

75. False. Take $f(x) = \sqrt{x}$ and $g(x) = x + 1$. Then $(g \circ f)(x) = \sqrt{x} + 1$, but $(f \circ g)(x) = \sqrt{x + 1}$.

© 2011 Cengage Learning. All Rights Reserved. May not be scanned, copied or duplicated, or posted to a publicly accessible website, in whole or in part.

2.3   Problem-Solving Tips

When you solve a problem involving a function, it is helpful to identify the type of function you are working with. For example, there are no restrictions on the domain of a polynomial function. If you want to find the domain of a rational function, you have to check to see if there are any values for which the denominator is equal to 0.

Here are some tips for solving the problems in the exercises that follow:

1. **To find the market equilibrium of a commodity**, find the point of intersection of the supply and demand equations for the commodity. (Market equilibrium prevails when the quantity produced is equal to the quantity demanded.).

2. **To construct a mathematical model**, follow the guidelines given in the text on page 83. First try solving Examples 5 and 6 in the text without looking at the solutions. Then go on to try a few similar problems (Exercises 72–80 on pages 90–91 of the text).

2.3   Concept Questions   page 85

1. See page 75 of the text. Answers will vary.

3. **a.** A demand function $p = D(x)$ gives the relationship between the unit price of a commodity p and the quantity x demanded. A supply function $p = S(x)$ gives the relationship between the unit price of a commodity p and the quantity x the supplier will make available in the marketplace.

   b. Market equilibrium occurs when the quantity produced is equal to the quantity demanded. To find the market equilibrium, we solve the equations $p = D(x)$ and $p = S(x)$ simultaneously.

2.3   Functions and Mathematical Models   page 85

1. Yes. $2x + 3y = 6$ and so $y = -\frac{2}{3}x + 2$.

3. Yes. $2y = x + 4$ and so $y = \frac{1}{2}x + 2$.

5. Yes. $4y = 2x + 9$ and so $y = \frac{1}{2}x + \frac{9}{4}$.

7. No, because of the term x^2.

9. f is a polynomial function in x of degree 6.

11. Expanding $G(x) = 2(x^2 - 3)^3$, we have $G(x) = 2x^6 - 18x^4 + 54x^2 - 54$, and we conclude that G is a polynomial function of degree 6 in x.

13. f is neither a polynomial nor a rational function.

15. $f(0) = 2$ gives $f(0) = m(0) + b = b = 2$. Next, $f(3) = -1$ gives $f(3) = m(3) + b = -1$. Substituting $b = 2$ in this last equation, we have $3m + 2 = -1$, or $3m = -3$, and therefore, $m = -1$ and $b = 2$.

17. **a.** $C(x) = 8x + 40,000$.

   b. $R(x) = 12x$.

   c. $P(x) = R(x) - C(x) = 12x - (8x + 40,000) = 4x - 40,000$.

© 2011 Cengage Learning. All Rights Reserved. May not be scanned, copied or duplicated, or posted to a publicly accessible website, in whole or in part.

d. $P(8000) = 4(8000) - 40,000 = -8000$, or a loss of $8000. $P(12,000) = 4(12,000) - 40,000 = 8000$, or a profit of $8000.

19. The individual's disposable income is $D = (1 - 0.28) \cdot 60,000 = 43,200$, or $43,200.

21. The child should receive $D(4) = \left(\dfrac{4+1}{24}\right)(500) \approx 104.17$, or approximately 104 mg.

23. $P(28) = -\frac{1}{8}(28)^2 + 7(28) + 30 = 128$, or $128,000.

25. $S(6) = 0.73(6)^2 + 15.8(6) + 2.7 = 123.78$ million kilowatt-hr.
$S(8) = 0.73(8)^2 + 15.8(8) + 2.7 = 175.82$ million kilowatt-hr.

27. a. The revenue of the company in 2005 is given by $R(0) = 3.25$, or $3.25 billion.

b. $R(1) = -0.06(1)^2 + 0.69(1) + 3.25 = 3.88$. This says that the revenue of the company in 2006 was $3.88 billion. The revenue in 2007 was $R(2) = -0.06(2)^2 + 0.69(2) + 3.25 = 4.39$, or $4.39 billion. The revenue in 2008 was $R(3) = -0.06(3)^2 + 0.69(3) + 3.25 = 4.78$, or $4.78 billion.

c.

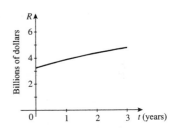

29. a. The average amount paid in 2001 is given by $C(0) = 751.5$, or $751.50/yr.

b. If the trend continued, the average amount paid in 2008 was $C(7) = 2.16(7)^3 + 40(7) + 751.5 = 1772.38$, or $1772.38/yr.

31. a. $N(0) = 0.32$ or 320,000.

b. $N(4) = -0.0675(4)^4 + 0.5083(4)^3 - 0.893(4)^2 + 0.66(4) + 0.32 = 3.9232$, or 3,923,200.

33. a. We first construct a table.

t	$N(t)$	t	$N(t)$
1	52	6	135
2	75	7	146
3	93	8	157
4	109	9	167
5	122	10	177

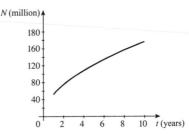

b. The number of viewers in 2012 is given by $N(10) = 52(10)^{0.531} \approx 176.61$, or approximately 176 million viewers.

35. $N(5) = 0.0018425(10)^{2.5} \approx 0.58265$, or approximately 0.583 million. $N(10) = 0.0018425(15)^{2.5} \approx 1.6056$, or approximately 1.606 million.

37. $A(0) = \dfrac{699}{1^{0.94}} = 699$ or $699. $A(5) = \dfrac{699}{6^{0.94}} \approx 129.722$, or approximately $130.

© 2011 Cengage Learning. All Rights Reserved. May not be scanned, copied or duplicated, or posted to a publicly accessible website, in whole or in part.

39. a. $f(0) = 6.85$, $g(0) = 16.58$. Because $g(0) > f(0)$, we see that more film cameras were sold in 2001 (when $t = 0$).

   b. We solve the equation $f(t) = g(t)$, that is, $3.05t + 6.85 = -1.85t + 16.58$, so $4.9t = 9.73$ and $t = 1.99 \approx 2$. So sales of digital cameras first exceed those of film cameras in approximately 2003.

41. $h(t) = f(t) - g(t) = \dfrac{110}{\frac{1}{2}t + 1} - 26\left(\frac{1}{4}t^2 - 1\right)^2 - 52.$

   $h(0) = f(0) - g(0) = \dfrac{110}{\frac{1}{2}(0) + 1} - 26\left[\frac{1}{4}(0)^2 - 1\right]^2 - 52 = 110 - 26 - 52 = 32$, or \$32.

   $h(1) = f(1) - g(1) = \dfrac{110}{\frac{1}{2}(1) + 1} - 26\left[\frac{1}{4}(1)^2 - 1\right]^2 - 52 \approx 6.71$, or approximately \$6.71.

   $h(2) = f(2) - g(2) = \dfrac{110}{\frac{1}{2}(2) + 1} - 26\left[\frac{1}{4}(2)^2 - 1\right]^2 - 52 = 3$, or \$3. We conclude that the price gap was narrowing.

43. The slope of the line is $m = \dfrac{S - C}{n}$. Therefore, an equation of the line is $y - C = \dfrac{S - C}{n}(t - 0)$. Letting $y = V(t)$, we have $V(t) = C - \dfrac{C - S}{n}t$.

45. The average U.S. credit card debt at the beginning of 1994 was $D(0) = 4.77(1 + 0)^{0.2676} = 4.77$, or \$4770. At the beginning of 1996, it was $D(2) = 4.77(1 + 2)^{0.2676} \approx 6.400$, or approximately \$6400. At the beginning of 1999, it was $D(5) = 5.6423(5^{0.1818}) \approx 7.560$, or approximately \$7560.

47. a.

   b. $f(0) = 8.37(0) + 7.44 = 7.44$, or \$7.44/kilo.
   $f(20) = 2.84(20) + 51.68 = 108.48$, or \$108.48/kilo.

49. a.

   b. $f(5) = \frac{2}{7}(5) + 12 = \frac{10}{7} + 12 \approx 13.43$, or approximately 13.43%. $f(15) = \frac{1}{3}(15) + \frac{41}{3} = 18\frac{2}{3}$, or $18\frac{2}{3}$%.

© 2011 Cengage Learning. All Rights Reserved. May not be scanned, copied or duplicated, or posted to a publicly accessible website, in whole or in part.

51. a.

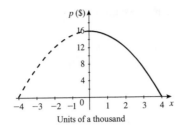

Units of a thousand

b. If $p = 7$, we have $7 = -x^2 + 16$, or $x^2 = 9$, so that $x = \pm 3$. Therefore, the quantity demanded when the unit price is $7 is 3000 units.

53. a.

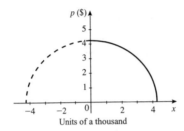

Units of a thousand

b. If $p = 3$, then $3 = \sqrt{18 - x^2}$, and $9 = 18 - x^2$, so that $x^2 = 9$ and $x = \pm 3$. Therefore, the quantity demanded when the unit price is $3 is 3000 units.

55. a.

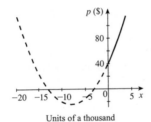

Units of a thousand

b. If $x = 2$, then $p = 2^2 + 16(2) + 40 = 76$, or $76.

57. a.

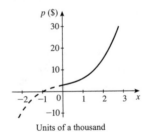

Units of a thousand

b. $p = 2^3 + 2(2) + 3 = 15$, or $15.

59. The slope of L_2 is greater than that of L_1. This means that for each drop of a dollar in the price of a clock radio, the quantity demanded of model B clock radios is greater than that of model A clock radios.

61. a.

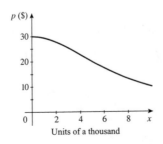

Units of a thousand

b. Substituting $x = 10$ into the demand function, we have

$$p = \frac{30}{0.02(10)^2 + 1} = \frac{30}{3} = 10, \text{ or } 10.$$

© 2011 Cengage Learning. All Rights Reserved. May not be scanned, copied or duplicated, or posted to a publicly accessible website, in whole or in part.

63. If $x = 5$, then $p = 0.1\,(5)^2 + 0.5\,(5) + 15 = 20$, or $20.

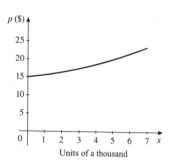

65. a. We solve the system of equations $p = cx + d$ and $p = ax + b$. Substituting the first equation into the second gives $cx + d = ax + d$, so $(c - a)\,x = b - d$ and $x = \dfrac{b - d}{c - a}$. Because $a < 0$ and $c > 0$, $c - a \neq 0$ and x is well-defined. Substituting this value of x into the second equation, we obtain

$$p = a\left(\frac{b - d}{c - a}\right) + b = \frac{ab - ad + bc - ab}{c - a} = \frac{bc - ad}{c - a}.$$ Therefore, the equilibrium quantity is $\dfrac{b - d}{c - a}$ and the equilibrium price is $\dfrac{bc - ad}{c - a}$.

b. If c is increased, the denominator in the expression for x increases and so x gets smaller. At the same time, the first term in the first equation for p decreases and so p gets larger. This analysis shows that if the unit price for producing the product is increased then the equilibrium quantity decreases while the equilibrium price increases.

c. If b is decreased, the numerator of the expression for x decreases while the denominator stays the same. Therefore, x decreases. The expression for p also shows that p decreases. This analysis shows that if the (theoretical) upper bound for the unit price of a commodity is lowered, then both the equilibrium quantity and the equilibrium price drop.

67. We solve the equation $-2x^2 + 80 = 15x + 30$, or $2x^2 + 15x - 50 = 0$ for x. Thus, $(2x - 5)\,(x + 10) = 0$, and so $x = \frac{5}{2}$ or $x = -10$. Rejecting the negative root, we have $x = \frac{5}{2}$. The corresponding value of p is $p = -2\left(\frac{5}{2}\right)^2 + 80 = 67.5$. We conclude that the equilibrium quantity is 2500 and the equilibrium price is $67.50.

69. Solving both equations for x, we have $x = -\frac{11}{3}p + 22$ and $x = 2p^2 + p - 10$. Equating the right-hand sides, we have $-\frac{11}{3}p + 22 = 2p^2 + p - 10$, or $-11p + 66 = 6p^2 + 3p - 30$, and so $6p^2 + 14p - 96 = 0$. Dividing this last equation by 2 and then factoring, we have $(3p + 16)\,(p - 3) = 0$, so $p = 3$ is the only valid solution. The corresponding value of x is $2\,(3)^2 + 3 - 10 = 11$. We conclude that the equilibrium quantity is 11,000 and the equilibrium price is $3.

71. Equating the right-hand sides of the two equations, we have $144 - x^2 = 48 + \frac{1}{2}x^2$, so $288 - 2x^2 = 96 + x^2$, $3x^2 = 192$, and $x^2 = 64$. Therefore, $x = \pm 8$. We take $x = 8$, and the corresponding value of p is $144 - 8^2 = 80$. We conclude that the equilibrium quantity is 8000 tires and the equilibrium price is $80.

73. The area of Juanita's garden is 250 ft². Therefore $xy = 250$ and $y = \dfrac{250}{x}$. The amount of fencing needed is given by $2x + 2y$. Therefore, $f = 2x + 2\left(\dfrac{250}{x}\right) = 2x + \dfrac{500}{x}$. The domain of f is $x > 0$.

© 2011 Cengage Learning. All Rights Reserved. May not be scanned, copied or duplicated, or posted to a publicly accessible website, in whole or in part.

75. Because the volume of the box is the area of the base times the height of the box, we have $V = x^2 y = 20$. Thus, we have $y = \dfrac{20}{x^2}$. Next, the amount of material used in constructing the box is given by the area of the base of the box, plus the area of the four sides, plus the area of the top of the box; that is, $A = x^2 + 4xy + x^2$. Then, the cost of constructing the box is given by $f(x) = 0.30x^2 + 0.40x \cdot \dfrac{20}{x^2} + 0.20x^2 = 0.5x^2 + \dfrac{8}{x}$, where $f(x)$ is measured in dollars and $f(x) > 0$.

77. The average yield of the apple orchard is 36 bushels/tree when the density is 22 trees/acre. Let x be the unit increase in tree density beyond 22. Then the yield of the apple orchard in bushels/acre is given by $(22 + x)(36 - 2x)$.

79. a. Let x denote the number of bottles sold beyond 10,000 bottles. Then
$$P(x) = (10{,}000 + x)(5 - 0.0002x) = -0.0002x^2 + 3x + 50{,}000.$$
b. He can expect a profit of $P(6000) = -0.0002\left(6000^2\right) + 3(6000) + 50{,}000 = 60{,}800$, or \$60,800.

81. False. $f(x) = 3x^{3/4} + x^{1/2} + 1$ is not a polynomial function. The powers of x must be nonnegative integers.

83. False. $f(x) = x^{1/2}$ is not defined for negative values of x.

2.3   Using Technology   page 94

1. $(-3.0414, 0.1503)$, $(3.0414, 7.4497)$.

3. $(-2.3371, 2.4117)$, $(6.0514, -2.5015)$.

5. $(-1.0219, -6.3461)$, $(1.2414, -1.5931)$, and $(5.7805, 7.9391)$.

7. a.

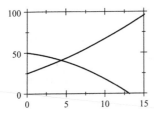

b. 438 wall clocks; \$40.92.

9. a. $f(t) = 1.85t + 16.9$.

b.

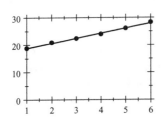

c.

t	y
1	18.8
2	20.6
3	22.5
4	24.3
5	26.2
6	28.0

These values are close to the given data.

d. $f(8) = 1.85(8) + 16.9 = 31.7$ gallons.

© 2011 Cengage Learning. All Rights Reserved. May not be scanned, copied or duplicated, or posted to a publicly accessible website, in whole or in part.

11. a. $f(t) = -0.221t^2 + 4.14t + 64.8.$

b.

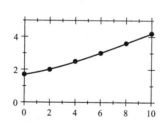

c. 77.8 million

13. a.

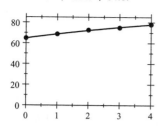

b. $f(t) = 2.94t^2 + 38.75t + 188.5$

c. The spending in 2007 was

$f(t) = 2.94\left(7^2\right) + 38.75\left(7\right) + 188.5 = 603.81,$

or approximately $604 billion.

15. a. $f(t) = -0.00081t^3 + 0.0206t^2 + 0.125t + 1.69.$

b.

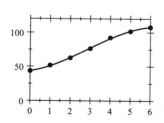

c.

t	y
1	1.8
5	2.7
10	4.2

The revenues were $1.8 trillion in 2001, $2.7 trillion in 2005, and $4.2 trillion in 2010.

17. a. $f(t) = -0.425t^3 + 3.6571t^2$
$\qquad\qquad + 4.0179t + 43.6643.$

b.

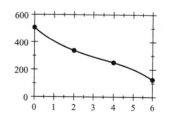

c.

t	0	3	6
$f(t)$	43.7	77.2	107.7

19. a. $f(t) = -2.4167t^3 + 24.5t^2$
$\qquad\qquad - 123.33t + 506.$

b.

c. $f(0) = 506$, or 506,000; $f(2) = 338$, or 338,000; and $f(6) = 126$, or 126,000.

© 2011 Cengage Learning. All Rights Reserved. May not be scanned, copied or duplicated, or posted to a publicly accessible website, in whole or in part.

21. a. $f(t) = 0.00125t^4 - 0.0051t^3$
$$- 0.0243t^2 + 0.129t + 1.71.$$

b.

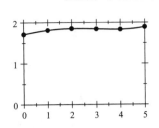

c.

t	$f(t)$
0	1.71
1	1.81
2	1.85
3	1.84
4	1.83
5	1.89

d. The average amount of nicotine in 2005 is $f(6) = 1.887$, or approximately 2.13 mg/cigarette.

2.4   Problem-Solving Tips

In this section, an important theorem was introduced (properties of limits on page 102). After you read Theorem 1, try to express the theorem in your own words. While you will not usually be required to prove the theorem in this course, you will be asked to understand the results of the theorem. For example, Theorem 1 gives us the properties of limits that allow us to evaluate sums, differences, products, quotients, powers, and constant multiples of functions at specified values, with certain restrictions. You should be able to use limit notation to write out each of these properties. You should also be able to use these properties to evaluate the limits of functions.

Here are some tips for solving the problems in the exercises that follow:

1. **To find the limit of a function** $f(x)$ **as** $x \to a$, where a is a real number, first try substituting a for x in the rule for f and simplify the result.

2. **To evaluate the limit of a quotient** that has the indeterminate form $0/0$:

   a. Replace the given function with an appropriate one that takes on the same values as the original function everywhere except at $x = a$.
   b. Evaluate the limit of this function as x approaches a.

2.4   Concept Questions   page 111

1. The values of $f(x)$ can be made as close to 3 as we please by taking x sufficiently close to $x = 2$.

3. a. $\lim\limits_{x \to 4} \sqrt{x} \, (2x^2 + 1) = \lim\limits_{x \to 4} (\sqrt{x}) \lim\limits_{x \to 4} (2x^2 + 1)$ (Rule 4)

$= \sqrt{4} \left[2(4)^2 + 1 \right]$ (Rules 1 and 3)

$= 66$

© 2011 Cengage Learning. All Rights Reserved. May not be scanned, copied or duplicated, or posted to a publicly accessible website, in whole or in part.

b. $\lim\limits_{x \to 1} \left(\dfrac{2x^2 + x + 5}{x^4 + 1} \right)^{3/2} = \left(\lim\limits_{x \to 1} \dfrac{2x^2 + x + 5}{x^4 + 1} \right)^{3/2}$ (Rule 1)

$= \left(\dfrac{2 + 1 + 5}{1 + 1} \right)^{3/2}$ (Rules 2, 3, and 5)

$= 4^{3/2} = 8$

5. $\lim\limits_{x \to \infty} f(x) = L$ means $f(x)$ can be made as close to L as we please by taking x sufficiently large.

$\lim\limits_{x \to -\infty} f(x) = M$ means $f(x)$ can be made as close to M as we please by taking negative x as large as we please in absolute value.

2.4   Limits   page 111

1. $\lim\limits_{x \to -2} f(x) = 3$.

3. $\lim\limits_{x \to 3} f(x) = 3$.

5. $\lim\limits_{x \to -2} f(x) = 3$.

7. The limit does not exist. If we consider any value of x to the right of $x = -2$, $f(x) \le 2$. If we consider values of x to the left of $x = -2$, $f(x) \ge -2$. Because $f(x)$ does not approach any one number as x approaches $x = -2$, we conclude that the limit does not exist.

9.

x	1.9	1.99	1.999	2.001	2.01	2.1
$f(x)$	4.61	4.9601	4.9960	5.004	5.0401	5.41

$\lim\limits_{x \to 2} \left(x^2 + 1 \right) = 5$.

11.

x	-0.1	-0.01	-0.001	0.001	0.01	0.1
$f(x)$	-1	-1	-1	1	1	1

The limit does not exist.

13.

x	0.9	0.99	0.999	1.001	1.01	1.1
$f(x)$	100	10,000	1,000,000	1,000,000	10,000	100

The limit does not exist.

15.

x	0.9	0.99	0.999	1.001	1.01	1.1
$f(x)$	2.9	2.99	2.999	3.001	3.01	3.1

$\lim\limits_{x \to 1} \dfrac{x^2 + x - 2}{x - 1} = 3$.

© 2011 Cengage Learning. All Rights Reserved. May not be scanned, copied or duplicated, or posted to a publicly accessible website, in whole or in part.

17.

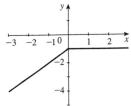

$$\lim_{x \to 0} f(x) = -1.$$

19.

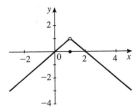

$$\lim_{x \to 1} f(x) = 1.$$

21.

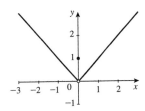

$$\lim_{x \to 0} f(x) = 0.$$

23. $\lim\limits_{x \to 2} 3 = 3.$

25. $\lim\limits_{x \to 3} x = 3.$

27. $\lim\limits_{x \to 1} (1 - 2x^2) = 1 - 2(1)^2 = -1.$

29. $\lim\limits_{x \to 1} (2x^3 - 3x^2 + x + 2) = 2(1)^3 - 3(1)^2 + 1 + 2$

$$= 2.$$

31. $\lim\limits_{s \to 0} (2s^2 - 1)(2s + 4) = (-1)(4) = -4.$

33. $\lim\limits_{x \to 2} \dfrac{2x + 1}{x + 2} = \dfrac{2(2) + 1}{2 + 2} = \dfrac{5}{4}.$

35. $\lim\limits_{x \to 2} \sqrt{x + 2} = \sqrt{2 + 2} = 2.$

37. $\lim\limits_{x \to -3} \sqrt{2x^4 + x^2} = \sqrt{2(-3)^4 + (-3)^2} = \sqrt{162 + 9}$

$$= \sqrt{171} = 3\sqrt{19}.$$

39. $\lim\limits_{x \to -1} \dfrac{\sqrt{x^2 + 8}}{2x + 4} = \dfrac{\sqrt{(-1)^2 + 8}}{2(-1) + 4} = \dfrac{\sqrt{9}}{2} = \dfrac{3}{2}.$

41. $\lim\limits_{x \to a} [f(x) - g(x)] = \lim\limits_{x \to a} f(x) - \lim\limits_{x \to a} g(x)$

$$= 3 - 4 = -1.$$

43. $\lim\limits_{x \to a} [2f(x) - 3g(x)] = \lim\limits_{x \to a} 2f(x) - \lim\limits_{x \to a} 3g(x)$

$$= 2(3) - 3(4) = -6.$$

45. $\lim\limits_{x \to a} \sqrt{g(x)} = \lim\limits_{x \to a} \sqrt{4} = 2.$

47. $\lim\limits_{x \to a} \dfrac{2f(x) - g(x)}{f(x)g(x)} = \dfrac{2(3) - (4)}{(3)(4)} = \dfrac{2}{12} = \dfrac{1}{6}.$

49. $\lim\limits_{x \to 1} \dfrac{x^2 - 1}{x - 1} = \lim\limits_{x \to 1} \dfrac{(x - 1)(x + 1)}{x - 1} = \lim\limits_{x \to 1} (x + 1)$

$$= 1 + 1 = 2.$$

© 2011 Cengage Learning. All Rights Reserved. May not be scanned, copied or duplicated, or posted to a publicly accessible website, in whole or in part.

51. $\lim\limits_{x\to 0} \dfrac{x^2 - x}{x} = \lim\limits_{x\to 0} \dfrac{x\,(x-1)}{x} = \lim\limits_{x\to 0} (x-1)$

$= 0 - 1 = -1.$

53. $\lim\limits_{x\to -5} \dfrac{x^2 - 25}{x + 5} = \lim\limits_{x\to -5} \dfrac{(x+5)\,(x-5)}{x+5}$

$= \lim\limits_{x\to -5} (x-5) = -10.$

55. $\lim\limits_{x\to 1} \dfrac{x}{x-1}$ does not exist.

57. $\lim\limits_{x\to -2} \dfrac{x^2 - x - 6}{x^2 + x - 2} = \lim\limits_{x\to -2} \dfrac{(x-3)\,(x+2)}{(x+2)\,(x-1)} = \lim\limits_{x\to -2} \dfrac{x-3}{x-1} = \dfrac{-2-3}{-2-1} = \dfrac{5}{3}.$

59. $\lim\limits_{x\to 1} \dfrac{\sqrt{x}-1}{x-1} = \lim\limits_{x\to 1} \dfrac{\sqrt{x}-1}{x-1} \cdot \dfrac{\sqrt{x}+1}{\sqrt{x}+1} = \lim\limits_{x\to 1} \dfrac{x-1}{(x-1)\,(\sqrt{x}+1)} = \lim\limits_{x\to 1} \dfrac{1}{\sqrt{x}+1} = \dfrac{1}{2}.$

61. $\lim\limits_{x\to 1} \dfrac{x-1}{x^3 + x^2 - 2x} = \lim\limits_{x\to 1} \dfrac{x-1}{x\,(x-1)\,(x+2)} = \lim\limits_{x\to 1} \dfrac{1}{x\,(x+2)} = \dfrac{1}{3}.$

63. $\lim\limits_{x\to \infty} f\,(x) = \infty$ (does not exist) and $\lim\limits_{x\to -\infty} f\,(x) = \infty$ (does not exist).

65. $\lim\limits_{x\to \infty} f\,(x) = 0$ and $\lim\limits_{x\to -\infty} f\,(x) = 0$.

67. $\lim\limits_{x\to \infty} f\,(x) = -\infty$ (does not exist) and $\lim\limits_{x\to -\infty} f\,(x) = -\infty$ (does not exist).

69. $f\,(x) = \dfrac{1}{x^2 + 1}.$

x	1	10	100	1000
$f\,(x)$	0.5	0.009901	0.0001	0.000001

x	-1	-10	-100	-1000
$f\,(x)$	0.5	0.009901	0.0001	0.000001

$\lim\limits_{x\to \infty} f\,(x) = \lim\limits_{x\to -\infty} f\,(x) = 0.$

71. $f\,(x) = 3x^3 - x^2 + 10.$

x	1	5	10	100	1000
$f\,(x)$	12	360	2910	2.99×10^6	2.999×10^9

x	-1	-5	-10	-100	-1000
$f\,(x)$	6	-390	-3090	-3.01×10^6	-3.0×10^9

$\lim\limits_{x\to \infty} f\,(x) = \infty$ (does not exist) and $\lim\limits_{x\to -\infty} f\,(x) = -\infty$ (does not exist).

73. $\lim\limits_{x\to \infty} \dfrac{3x+2}{x-5} = \lim\limits_{x\to \infty} \dfrac{3 + \dfrac{2}{x}}{1 - \dfrac{5}{x}} = \dfrac{3}{1} = 3.$

75. $\lim\limits_{x\to -\infty} \dfrac{3x^3 + x^2 + 1}{x^3 + 1} = \lim\limits_{x\to -\infty} \dfrac{3 + \dfrac{1}{x} + \dfrac{1}{x^3}}{1 + \dfrac{1}{x^3}} = 3.$

© 2011 Cengage Learning. All Rights Reserved. May not be scanned, copied or duplicated, or posted to a publicly accessible website, in whole or in part.

77. $\lim\limits_{x \to -\infty} \dfrac{x^4+1}{x^3-1} = \lim\limits_{x \to -\infty} \dfrac{x+\dfrac{1}{x^3}}{1-\dfrac{1}{x^3}} = -\infty$; that is, the limit does not exist.

79. $\lim\limits_{x \to \infty} \dfrac{x^5 - x^3 + x - 1}{x^6 + 2x^2 + 1} = \lim\limits_{x \to \infty} \dfrac{\dfrac{1}{x} - \dfrac{1}{x^3} + \dfrac{1}{x^5} - \dfrac{1}{x^6}}{1 + \dfrac{2}{x^4} + \dfrac{1}{x^6}} = 0.$

81. a. The cost of removing 50% of the pollutant is $C(50) = \dfrac{0.5(50)}{100-50} = 0.5$, or \$500,000. Similarly, we find that the cost of removing 60%, 70%, 80%, 90%, and 95% of the pollutant is \$750,000, \$1,166,667, \$2,000,000, \$4,500,000, and \$9,500,000, respectively.

b. $\lim\limits_{x \to 1} \dfrac{0.5x}{100-x} = \infty$, which means that the cost of removing the pollutant increases without bound if we wish to remove almost all of the pollutant.

83. $\lim\limits_{x \to \infty} \overline{C}(x) = \lim\limits_{x \to \infty} 2.2 + \dfrac{2500}{x} = 2.2$, or \$2.20 per DVD. In the long run, the average cost of producing x DVDs approaches \$2.20/disc.

85. a. $T(1) = \dfrac{120}{1+4} = 24$, or \$24 million. $T(2) = \dfrac{120(2)^2}{8} = 60$, or \$60 million. $T(3) = \dfrac{120(3)^2}{13} = 83.1$, or \$83.1 million.

b. In the long run, the movie will gross $\lim\limits_{x \to \infty} \dfrac{120x^2}{x^2+4} = \lim\limits_{x \to \infty} \dfrac{120}{1+\dfrac{4}{x^2}} = 120$, or \$120 million.

87. a. The average cost of driving 5000 miles per year is

$C(5) = \dfrac{2010}{5^{2.2}} + 17.80 = 76.07$, or 76.1 cents per mile.

Similarly, we see that the average cost of driving 10, 15, 20, and 25 thousand miles per year is 30.5, 23, 20.6, and 19.5 cents per mile, respectively.

c. It approaches 17.80 cents per mile.

b.

89. False. Let $f(x) = \begin{cases} -1 & \text{if } x < 0 \\ 1 & \text{if } x \ge 0 \end{cases}$. Then $\lim\limits_{x \to 0} f(x) = 1$, but $f(1)$ is not defined.

91. True. Division by zero is not permitted.

93. True. Each limit in the sum exists. Therefore, $\lim\limits_{x \to 2} \left(\dfrac{x}{x+1} + \dfrac{3}{x-1} \right) = \lim\limits_{x \to 2} \dfrac{x}{x+1} + \lim\limits_{x \to 2} \dfrac{3}{x-1} = \dfrac{2}{3} + \dfrac{3}{1} = \dfrac{11}{3}.$

95. $\lim\limits_{x \to \infty} \dfrac{ax}{x+b} = \lim\limits_{x \to \infty} \dfrac{a}{1+\dfrac{b}{x}} = a$. As the amount of substrate becomes very large, the initial speed approaches the constant a moles per liter per second.

© 2011 Cengage Learning. All Rights Reserved. May not be scanned, copied or duplicated, or posted to a publicly accessible website, in whole or in part.

97. Consider the functions $f(x) = \begin{cases} -1 & \text{if } x < 0 \\ 1 & \text{if } x \geq 0 \end{cases}$ and $g(x) = \begin{cases} 1 & \text{if } x < 0 \\ -1 & \text{if } x \geq 0 \end{cases}$ Then $\lim_{x \to 0} f(x)$ and $\lim_{x \to 0} g(x)$

do not exist, but $\lim_{x \to 0} [f(x)g(x)] = \lim_{x \to 0} (-1) = -1$. This example does not contradict Theorem 1 because the

hypothesis of Theorem 1 is that $\lim_{x \to 0} f(x)$ and $\lim_{x \to 0} g(x)$ both exist. It does not say anything about the situation

where one or both of these limits fails to exist.

2.4   Using Technology   page 117

1. 5   **3.** 3   **5.** $\frac{2}{3}$ **7.** $e^2 \approx 7.38906$

9.

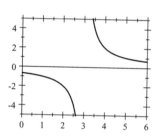

From the graph we see that $f(x)$ does not approach any finite number as x approaches 3.

11. a.

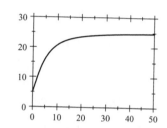

b. $\lim_{t \to \infty} \dfrac{25t^2 + 125t + 200}{t^2 + 5t + 40} = 25$, so in the long run the population will approach 25,000.

2.5   Problem-Solving Tips

The problem-solving skills that you learned in earlier sections are building blocks for the rest of the course. You can't skip a section or a concept and hope to understand the material in a new section. It just won't work. If you don't build a strong foundation, you won't be able to understand the later concepts. For example, in this section we discussed one-sided limits. You need to understand the definition of a limit before you can understand what is meant by a one-sided limit. That means you should be able to express the definition of a limit in your own words. If you can't grasp a new concept, it may well be that you still don't understand a previous concept. If so, you need to go back and review the earlier section before you go on.

As another example, the continuity of polynomial and rational functions is discussed on page 121. If you don't remember how to identify polynomial and rational functions, go back to Section 2.3 and review this material.

Here are some tips for solving the problems in the exercises that follow:

1. To evaluate the limit of a piecewise-defined function at a real number a, follow the same procedure that you used to evaluate a piecewise-defined function. First find the subdomain that a lies in, then use the rule for that subdomain to find the limit of f at a.

© 2011 Cengage Learning. All Rights Reserved. May not be scanned, copied or duplicated, or posted to a publicly accessible website, in whole or in part.

2. **To determine the values of x at which a function is continuous,** check to see if the function is a polynomial or rational function. A polynomial function $y = P(x)$ is continuous at every value of x and a rational function is continuous at every value of x where the denominator is nonzero.

2.5 Concept Questions page 125

1. $\lim\limits_{x \to 3^-} f(x) = 2$ means $f(x)$ can be made as close to 2 as we please by taking x sufficiently close to but to the left of $x = 3$. $\lim\limits_{x \to 3^+} f(x) = 4$ means $f(x)$ can be made as close to 4 as we please by taking x sufficiently close to but to the right of $x = 3$.

3. a. f is continuous at a if $\lim\limits_{x \to a} f(x) = f(a)$.

  b. f is continuous on an interval I if f is continuous at each point in I.

5. Refer to page 123 in the text. Answers will vary.

2.5 One-Sided Limits and Continuity page 126

1. $\lim\limits_{x \to 2^-} f(x) = 3$ and $\lim\limits_{x \to 2^+} f(x) = 2$, so $\lim\limits_{x \to 2} f(x)$ does not exist.

3. $\lim\limits_{x \to -1^-} f(x) = \infty$ and $\lim\limits_{x \to -1^+} f(x) = 2$, so $\lim\limits_{x \to -1} f(x)$ does not exist.

5. $\lim\limits_{x \to 1^-} f(x) = 0$ and $\lim\limits_{x \to 1^+} f(x) = 2$, so $\lim\limits_{x \to 1} f(x)$ does not exist.

7. $\lim\limits_{x \to 0^-} f(x) = -2$ and $\lim\limits_{x \to 0^+} f(x) = 2$, so $\lim\limits_{x \to 0} f(x)$ does not exist.

9. True. **11.** True. **13.** False. **15.** True. **17.** False. **19.** True.

21. $\lim\limits_{x \to 1^+} (2x + 4) = 6$.

23. $\lim\limits_{x \to 2^-} \dfrac{x-3}{x+2} = \dfrac{2-3}{2+2} = -\dfrac{1}{4}$.

25. $\lim\limits_{x \to 0^+} \dfrac{1}{x}$ does not exist because $\dfrac{1}{x} \to \infty$ as $x \to 0$ from the right.

27. $\lim\limits_{x \to 0^+} \dfrac{x-1}{x^2+1} = \dfrac{-1}{1} = -1$.

29. $\lim\limits_{x \to 0^+} \sqrt{x} = \sqrt{\lim\limits_{x \to 0^+} x} = 0$.

31. $\lim\limits_{x \to -2^+} \left(2x + \sqrt{2+x}\right) = \lim\limits_{x \to -2^+} 2x + \lim\limits_{x \to -2^+} \sqrt{2+x} = -4 + 0 = -4$.

33. $\lim\limits_{x \to 1^-} \dfrac{1+x}{1-x} = \infty$, that is, the limit does not exist.

35. $\lim\limits_{x \to 2^-} \dfrac{x^2-4}{x-2} = \lim\limits_{x \to 2^-} \dfrac{(x+2)(x-2)}{x-2} = \lim\limits_{x \to 2^-} (x+2) = 4$.

© 2011 Cengage Learning. All Rights Reserved. May not be scanned, copied or duplicated, or posted to a publicly accessible website, in whole or in part.

37. $\lim\limits_{x \to 0^-} f(x) = \lim\limits_{x \to 0^-} 2x = 0$ and $\lim\limits_{x \to 0^+} f(x) = \lim\limits_{x \to 0^+} x^2 = 0$.

39. The function is discontinuous at $x = 0$. Conditions 2 and 3 are violated.

41. The function is continuous everywhere.

43. The function is discontinuous at $x = 0$. Condition 3 is violated.

45. f is continuous for all values of x.

47. f is continuous for all values of x. Note that $x^2 + 1 \geq 1 > 0$.

49. f is discontinuous at $x = \frac{1}{2}$, where the denominator is 0.

51. Observe that $x^2 + x - 2 = (x + 2)(x - 1) = 0$ if $x = -2$ or $x = 1$, so f is discontinuous at these values of x.

53. f is continuous everywhere since all three conditions are satisfied.

55. f is continuous everywhere since all three conditions are satisfied.

57. Because the denominator $x^2 - 1 = (x - 1)(x + 1) = 0$ if $x = -1$ or 1, we see that f is discontinuous at -1 and 1.

59. Because $x^2 - 3x + 2 = (x - 2)(x - 1) = 0$ if $x = 1$ or 2, we see that the denominator is zero at these points and so f is discontinuous at these numbers.

61. The function f is discontinuous at $x = 1, 2, 3, \ldots, 12$ because the limit of f does not exist at these points.

63. Having made steady progress up to $x = x_1$, Michael's progress comes to a standstill at that point. Then at $x = x_2$ a sudden breakthrough occurs and he then continues to solve the problem.

65. Conditions 2 and 3 are not satisfied at either of these points.

67.

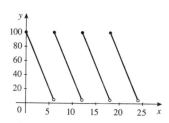

f is discontinuous at $x = 6, 12, 18,$ and 24.

69.

$$f(x) = \begin{cases} 2 & \text{if } 0 < x \leq \frac{1}{2} \\ 3 & \text{if } \frac{1}{2} < x \leq 1 \\ \vdots & \vdots \\ 10 & \text{if } 4\frac{1}{2} < x \leq 5 \end{cases}$$

f is discontinuous at $x = \frac{1}{2}, 1, 1\frac{1}{2}, \ldots, 4$.

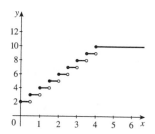

© 2011 Cengage Learning. All Rights Reserved. May not be scanned, copied or duplicated, or posted to a publicly accessible website, in whole or in part.

71. a. $\lim\limits_{t \to 0^+} S(t) = \lim\limits_{t \to 0^+} \dfrac{a}{t} + b = \infty$. As the time taken to excite the tissue is made shorter and shorter, the strength of the electric current gets stronger and stronger.

b. $\lim\limits_{t \to \infty} \dfrac{a}{t} + b = b$. As the time taken to excite the tissue is made longer and longer, the strength of the electric current gets smaller and smaller and approaches b.

73. We require that $f(1) = 1 + 2 = 3 = \lim\limits_{x \to 1^+} kx^2 = k$, so $k = 3$.

75. a. Yes, because if $f + g$ were continuous at a, then $g = (f + g) - f$ would be continuous (the difference of two continuous functions is continuous), and this would imply that g is continuous, a contradiction.

b. No. Consider the functions f and g defined by $f(x) = \begin{cases} -1 & \text{if } x < 0 \\ 1 & \text{if } x \geq 0 \end{cases}$ and $g(x) = \begin{cases} 1 & \text{if } x < 0 \\ -1 & \text{if } x \geq 0 \end{cases}$ Both f and g are discontinuous at $x = 0$, but $f + g$ is continuous everywhere.

77. a. f is a polynomial of degree 2 and is therefore continuous everywhere, including the interval $[1, 3]$.

b. $f(1) = 3$ and $f(3) = -1$ and so f must have at least one zero in $(1, 3)$.

79. a. f is a polynomial of degree 3 and is therefore continuous on $[-1, 1]$.

b. $f(-1) = (-1)^3 - 2(-1)^2 + 3(-1) + 2 = -1 - 2 - 3 + 2 = -4$ and $f(1) = 1 - 2 + 3 + 2 = 4$. Because $f(-1)$ and $f(1)$ have opposite signs, we see that f has at least one zero in $(-1, 1)$.

81. $f(0) = 6$, $f(3) = 3$, and f is continuous on $[0, 3]$. Thus, the Intermediate Value Theorem guarantees that there is at least one value of x for which $f(x) = 4$. Solving $f(x) = x^2 - 4x + 6 = 4$, we find $x^2 - 4x + 2 = 0$. Using the quadratic formula, we find that $x = 2 \pm \sqrt{2}$. Because $2 + \sqrt{2}$ does not lie in $[0, 3]$, we see that $x = 2 - \sqrt{2} \approx 0.59$.

83. $x^5 + 2x - 7 = 0$

Step	Interval in which a root lies
1	$(1, 2)$
2	$(1, 1.5)$
3	$(1.25, 1.5)$
4	$(1.25, 1.375)$
5	$(1.3125, 1.375)$
6	$(1.3125, 1.34375)$
7	$(1.328125, 1.34375)$
8	$(1.3359375, 1.34375)$
9	$(1.33984375, 1.34375)$

We see that a root is approximately 1.34.

85. a. $h(0) = 4 + 64(0) - 16(0) = 4$ and $h(2) = 4 + 64(2) - 16(4) = 68$.

b. The function h is continuous on $[0, 2]$. Furthermore, the number 32 lies between 4 and 68. Therefore, the Intermediate Value Theorem guarantees that there is at least one value of t in $(0, 2]$ such that $h(t) = 32$, that is, Joan must see the ball at least once during the time the ball is in the air.

© 2011 Cengage Learning. All Rights Reserved. May not be scanned, copied or duplicated, or posted to a publicly accessible website, in whole or in part.

c. We solve $h(t) = 4 + 64t - 16t^2 = 32$, obtaining $16t^2 - 64t + 28 = 0$, $4t^2 - 16t + 7 = 0$, and $(2t - 1)(2t - 7) = 0$. Thus, $t = \frac{1}{2}$ or $t = \frac{7}{2}$. Joan sees the ball on its way up half a second after it was thrown and again 3 seconds later when it is on its way down. Note that the ball hits the ground when $t \approx 4.06$, but Joan sees it approximately half a second before it hits the ground.

87. False. Take $f(x) = \begin{cases} -1 & \text{if } x < 2 \\ 4 & \text{if } x = 2 \\ 1 & \text{if } x > 2 \end{cases}$ Then $f(2) = 4$, but $\lim\limits_{x \to 2} f(x)$ does not exist.

89. False. Consider $f(x) = \begin{cases} 0 & \text{if } x < 2 \\ 3 & \text{if } x \geq 2 \end{cases}$ Then $\lim\limits_{x \to 2^+} f(x) = f(2) = 3$, but $\lim\limits_{x \to 2^-} f(x) = 0$.

91. False. Consider $f(x) = \begin{cases} 2 & \text{if } x < 5 \\ 3 & \text{if } x > 5 \end{cases}$ Then $f(5)$ is not defined, but $\lim\limits_{x \to 5^-} f(x) = 2$.

93. False. Let $f(x) = \begin{cases} x & \text{if } x \neq 0 \\ 1 & \text{if } x = 0 \end{cases}$ Then $\lim\limits_{x \to 0^+} f(x) = \lim\limits_{x \to 0^-} f(x)$, but $f(0) = 1$.

95. False. Let $f(x) = \begin{cases} 1/x & \text{if } x \neq 0 \\ 0 & \text{if } x = 0 \end{cases}$ Then f is continuous for all $x \neq 0$, but $\lim\limits_{x \to 0} f(x)$ does not exist.

97. The statement is false. The Intermediate Value Theorem says that there is at least one number c in $[a, b]$ such that $f(c) = M$ if M is a number between $f(a)$ and $f(b)$.

99. a. Both $g(x) = x$ and $h(x) = \sqrt{1 - x^2}$ are continuous on $[-1, 1]$ and so $f(x) = x - \sqrt{1 - x^2}$ is continuous on $[-1, 1]$.
b. $f(-1) = -1$ and $f(1) = 1$, and so f has at least one zero in $(-1, 1)$.
c. Solving $f(x) = 0$, we have $x = \sqrt{1 - x^2}$, $x^2 = 1 - x^2$, and $2x^2 = 1$, so $x = \frac{\pm\sqrt{2}}{2}$.

101. Consider the function f defined by $f(x) = \begin{cases} -1 & \text{if } -1 \leq x < 0 \\ 1 & \text{if } 0 \leq x < 1 \end{cases}$ Then $f(-1) = -1$ and $f(1) = 1$, but if we take the number $\frac{1}{2}$, which lies between $y = -1$ and $y = 1$, there is no value of x such that $f(x) = \frac{1}{2}$.

© 2011 Cengage Learning. All Rights Reserved. May not be scanned, copied or duplicated, or posted to a publicly accessible website, in whole or in part.

2.5   Using Technology   page 132

1.

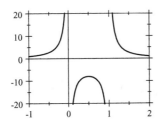

The function is discontinuous at $x = 0$ and $x = 1$.

3.

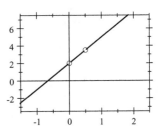

The function is discontinuous at $x = 0$ and $\frac{1}{2}$.

5.

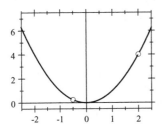

The function is discontinuous at $x = -\frac{1}{2}$ and 2.

7.

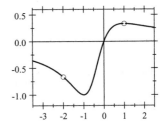

The function is discontinuous at $x = -2$ and 1.

9.

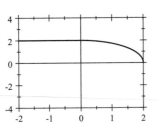

11.

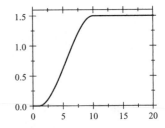

2.6   Problem-Solving Tips

When you solve an applied problem, it is important to understand the question in mathematical terms. For example, if you are given a function $f(t)$ describing the size of a country's population at any time t and asked to find the rate of change of that country's population at any time t, this means that you need to find the derivative of the given function; that is, find $f'(t)$. If you are then asked to find the population of the country at a specified time, say $t = 2$, you need to evaluate the function at $t = 2$; that is, find $f(2)$. On the other hand, if you are asked to find the rate of change of the population at time $t = 2$, then you need to evaluate the derivative of the function at the value $t = 2$; that is find $f'(2)$. Here again, the key is to be familiar with the terminology and notation introduced in the chapter.

© 2011 Cengage Learning. All Rights Reserved. May not be scanned, copied or duplicated, or posted to a publicly accessible website, in whole or in part.

Here are some tips for solving the problems in the exercises that follow:

1. **To find the slope of the tangent line to the graph of a function at an arbitrary point** on the graph of that function, find the derivative of f; that is, find $f'(x)$.

2. **To find the slope of the tangent line to the graph of a function at a given point** (x_0, y_0) on the graph of that function, find $f'(x)$ and then evaluate $f'(x_0)$.

2.6   Concept Questions   page 145

1. a. $m = \dfrac{f(2+h) - f(2)}{h}$

   b. The slope of the tangent line is $\lim\limits_{h \to 0} \dfrac{f(2+h) - f(2)}{h}$.

3. a. The expression $\dfrac{f(x+h) - f(x)}{h}$ gives (i) the slope of the secant line passing through the points $(x, f(x))$ and $(x+h, f(x+h))$, and (ii) the average rate of change of f over the interval $[x, x+h]$.

   b. The expression $\lim\limits_{h \to 0} \dfrac{f(x+h) - f(x)}{h}$ gives (i) the slope of the tangent line to the graph of f at the point $(x, f(x))$, and (ii) the instantaneous rate of change of f at x.

2.6   The Derivative   page 145

1. The rate of change of the average infant's weight when $t = 3$ is $\frac{7.5}{5}$, or 1.5 lb/month. The rate of change of the average infant's weight when $t = 18$ is $\frac{3.5}{6}$, or approximately 0.58 lb/month. The average rate of change over the infant's first year of life is $\frac{22.5 - 7.5}{12}$, or 1.25 lb/month.

3. The rate of change of the percentage of households watching television at 4 p.m. is $\frac{12.3}{4}$, or approximately 3.1 percent per hour. The rate at 11 p.m. is $\frac{-42.3}{2} = -21.15$, that is, it is dropping off at the rate of 21.15 percent per hour.

5. a. Car A is travelling faster than Car B at t_1 because the slope of the tangent line to the graph of f is greater than the slope of the tangent line to the graph of g at t_1.

   b. Their speed is the same because the slope of the tangent lines are the same at t_2.

   c. Car B is travelling faster than Car A.

   d. They have both covered the same distance and are once again side by side at t_3.

7. a. P_2 is decreasing faster at t_1 because the slope of the tangent line to the graph of g at t_1 is greater than the slope of the tangent line to the graph of f at t_1.

   b. P_1 is decreasing faster than P_2 at t_2.

   c. Bactericide B is more effective in the short run, but bactericide A is more effective in the long run.

© 2011 Cengage Learning. All Rights Reserved. May not be scanned, copied or duplicated, or posted to a publicly accessible website, in whole or in part.

9. $f(x) = 13$.

  Step 1 $f(x + h) = 13$.

  Step 2 $f(x + h) - f(x) = 13 - 13 = 0$.

  Step 3 $\dfrac{f(x + h) - f(x)}{h} = \dfrac{0}{h} = 0$.

  Step 4 $f'(x) = \lim\limits_{h \to 0} \dfrac{f(x + h) - f(x)}{h} = \lim\limits_{h \to 0} 0 = 0$.

11. $f(x) = 2x + 7$.

  Step 1 $f(x + h) = 2(x + h) + 7$.

  Step 2 $f(x + h) - f(x) = 2(x + h) + 7 - (2x + 7) = 2h$.

  Step 3 $\dfrac{f(x + h) - f(x)}{h} = \dfrac{2h}{h} = 2$.

  Step 4 $f'(x) = \lim\limits_{h \to 0} \dfrac{f(x + h) - f(x)}{h} = \lim\limits_{h \to 0} 2 = 2$.

13. $f(x) = 3x^2$.

  Step 1 $f(x + h) = 3(x + h)^2 = 3x^2 + 6xh + 3h^2$.

  Step 2 $f(x + h) - f(x) = (3x^2 + 6xh + 3h^2) - 3x^2 = 6xh + 3h^2 = h(6x + 3h)$.

  Step 3 $\dfrac{f(x + h) - f(x)}{h} = \dfrac{h(6x + 3h)}{h} = 6x + 3h$.

  Step 4 $f'(x) = \lim\limits_{h \to 0} \dfrac{f(x + h) - f(x)}{h} = \lim\limits_{h \to 0} (6x + 3h) = 6x$.

15. $f(x) = -x^2 + 3x$.

  Step 1 $f(x + h) = -(x + h)^2 + 3(x + h) = -x^2 - 2xh - h^2 + 3x + 3h$.

  Step 2 $f(x + h) - f(x) = (-x^2 - 2xh - h^2 + 3x + 3h) - (-x^2 + 3x) = -2xh - h^2 + 3h$

  $$= h(-2x - h + 3).$$

  Step 3 $\dfrac{f(x + h) - f(x)}{h} = \dfrac{h(-2x - h + 3)}{h} = -2x - h + 3$.

  Step 4 $f'(x) = \lim\limits_{h \to 0} \dfrac{f(x + h) - f(x)}{h} = \lim\limits_{h \to 0} (-2x - h + 3) = -2x + 3$.

17. $f(x) = 2x + 7$.

  Step 1 $f(x + h) = 2(x + h) + 7 = 2x + 2h + 7$.

  Step 2 $f(x + h) - f(x) = 2x + 2h + 7 - 2x - 7 = 2h$.

  Step 3 $\dfrac{f(x + h) - f(x)}{h} = \dfrac{2h}{h} = 2$.

  Step 4 $f'(x) = \lim\limits_{h \to 0} \dfrac{f(x + h) - f(x)}{h} = \lim\limits_{h \to 0} 2 = 2$.

  Therefore, $f'(x) = 2$. In particular, the slope at $x = 2$ is 2. Therefore, an equation of the tangent line is $y - 11 = 2(x - 2)$ or $y = 2x + 7$.

19. $f(x) = 3x^2$. We first compute $f'(x) = 6x$ (see Exercise 13). Because the slope of the tangent line is $f'(1) = 6$, we use the point-slope form of the equation of a line and find that an equation is $y - 3 = 6(x - 1)$, or $y = 6x - 3$.

© 2011 Cengage Learning. All Rights Reserved. May not be scanned, copied or duplicated, or posted to a publicly accessible website, in whole or in part.

21. $f(x) = -1/x$. We first compute $f'(x)$ using the four-step process:

Step 1 $f(x+h) = -\dfrac{1}{x+h}$.

Step 2 $f(x+h) - f(x) = -\dfrac{1}{x+h} + \dfrac{1}{x} = \dfrac{-x+(x+h)}{x(x+h)} = \dfrac{h}{x(x+h)}$.

Step 3 $\dfrac{f(x+h) - f(x)}{h} = \dfrac{\frac{h}{x(x+h)}}{h} = \dfrac{1}{x(x+h)}$.

Step 4 $f'(x) = \lim\limits_{h\to 0} \dfrac{f(x+h) - f(x)}{h} = \lim\limits_{h\to 0} \dfrac{1}{x(x+h)} = \dfrac{1}{x^2}$.

The slope of the tangent line is $f'(3) = \frac{1}{9}$. Therefore, an equation is $y - \left(-\frac{1}{3}\right) = \frac{1}{9}(x-3)$, or $y = \frac{1}{9}x - \frac{2}{3}$.

23. a. $f(x) = 2x^2 + 1$.

Step 1 $f(x+h) = 2(x+h)^2 + 1 = 2x^2 + 4xh + 2h^2 + 1$.

Step 2 $f(x+h) - f(x) = (2x^2 + 4xh + 2h^2 + 1) - (2x^2 + 1) = 4xh + 2h^2 = h(4x + 2h)$.

Step 3 $\dfrac{f(x+h) - f(x)}{h} = \dfrac{h(4x+2h)}{h} = 4x + 2h$.

Step 4 $f'(x) = \lim\limits_{h\to 0} \dfrac{f(x+h) - f(x)}{h} = \lim\limits_{h\to 0} (4x + 2h) = 4x$.

c.

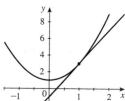

b. The slope of the tangent line is $f'(1) = 4(1) = 4$. Therefore, an equation is $y - 3 = 4(x-1)$ or $y = 4x - 1$.

25. a. $f(x) = x^2 - 2x + 1$. We use the four-step process:

Step 1 $f(x+h) = (x+h)^2 - 2(x+h) + 1 = x^2 + 2xh + h^2 - 2x - 2h + 1$.

Step 2 $f(x+h) - f(x) = (x^2 + 2xh + h^2 - 2x - 2h + 1) - (x^2 - 2x + 1) = 2xh + h^2 - 2h$

$$= h(2x + h - 2).$$

Step 3 $\dfrac{f(x+h) - f(x)}{h} = \dfrac{h(2x+h-2)}{h} = 2x + h - 2$.

Step 4 $f'(x) = \lim\limits_{h\to 0} \dfrac{f(x+h) - f(x)}{h} = \lim\limits_{h\to 0} (2x + h - 2)$

$$= 2x - 2.$$

c.

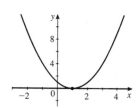

b. At a point on the graph of f where the tangent line to the curve is horizontal, $f'(x) = 0$. Then $2x - 2 = 0$, or $x = 1$. Because $f(1) = 1 - 2 + 1 = 0$, we see that the required point is $(1, 0)$.

d. It is changing at the rate of 0 units per unit change in x.

© 2011 Cengage Learning. All Rights Reserved. May not be scanned, copied or duplicated, or posted to a publicly accessible website, in whole or in part.

27. a. $f(x) = x^2 + x$, so $\dfrac{f(3) - f(2)}{3 - 2} = \dfrac{(3^2 + 3) - (2^2 + 2)}{1} = 6$,

$\dfrac{f(2.5) - f(2)}{2.5 - 2} = \dfrac{(2.5^2 + 2.5) - (2^2 + 2)}{0.5} = 5.5$, and $\dfrac{f(2.1) - f(2)}{2.1 - 2} = \dfrac{(2.1^2 + 2.1) - (2^2 + 2)}{0.1} = 5.1$.

b. We first compute $f'(x)$ using the four-step process.

Step 1 $f(x + h) = (x + h)^2 + (x + h) = x^2 + 2xh + h^2 + x + h$.

Step 2 $f(x + h) - f(x) = (x^2 + 2xh + h^2 + x + h) - (x^2 + x) = 2xh + h^2 + h = h(2x + h + 1)$.

Step 3 $\dfrac{f(x + h) - f(x)}{h} = \dfrac{h(2x + h + 1)}{h} = 2x + h + 1$.

Step 4 $f'(x) = \lim\limits_{h \to 0} \dfrac{f(x + h) - f(x)}{h} = \lim\limits_{h \to 0} (2x + h + 1) = 2x + 1$.

The instantaneous rate of change of y at $x = 2$ is $f'(2) = 2(2) + 1$, or 5 units per unit change in x.

c. The results of part (a) suggest that the average rates of change of f at $x = 2$ approach 5 as the interval $[2, 2 + h]$ gets smaller and smaller ($h = 1, 0.5$, and 0.1). This number is the instantaneous rate of change of f at $x = 2$ as computed in part (b).

29. a. $f(t) = 2t^2 + 48t$. The average velocity of the car over the time interval $[20, 21]$ is

$\dfrac{f(21) - f(20)}{21 - 20} = \dfrac{[2(21)^2 + 48(21)] - [2(20)^2 + 48(20)]}{1} = 130 \,\dfrac{\text{ft}}{\text{s}}$. Its average velocity over $[20, 20.1]$ is

$\dfrac{f(20.1) - f(20)}{20.1 - 20} = \dfrac{[2(20.1)^2 + 48(20.1)] - [2(20)^2 + 48(20)]}{0.1} = 128.2 \,\dfrac{\text{ft}}{\text{s}}$. Its average velocity over

$[20, 20.01]$ is $\dfrac{f(20.01) - f(20)}{20.01 - 20} = \dfrac{[2(20.01)^2 + 48(20.01)] - [2(20)^2 + 48(20)]}{0.01} = 128.02 \,\dfrac{\text{ft}}{\text{s}}$.

b. We first compute $f'(t)$ using the four-step process.

Step 1 $f(t + h) = 2(t + h)^2 + 48(t + h) = 2t^2 + 4th + 2h^2 + 48t + 48h$.

Step 2 $f(t + h) - f(t) = (2t^2 + 4th + 2h^2 + 48t + 48h) - (2t^2 + 48t) = 4th + 2h^2 + 48h$

$= h(4t + 2h + 48)$.

Step 3 $\dfrac{f(t + h) - f(t)}{h} = \dfrac{h(4t + 2h + 48)}{h} = 4t + 2h + 48$.

Step 4 $f'(t) = \lim\limits_{t \to 0} \dfrac{f(t + h) - f(t)}{h} = \lim\limits_{t \to 0} (4t + 2h + 48) = 4t + 48$.

The instantaneous velocity of the car at $t = 20$ is $f'(20) = 4(20) + 48$, or 128 ft/s.

c. Our results show that the average velocities do approach the instantaneous velocity as the intervals over which they are computed decreases.

31. a. We solve the equation $16t^2 = 400$ and find $t = 5$, which is the time it takes the screwdriver to reach the ground.

b. The average velocity over the time interval $[0, 5]$ is $\dfrac{f(5) - f(0)}{5 - 0} = \dfrac{16(25) - 0}{5} = 80$, or 80 ft/s. [Let

$s = f(t) = 16t^2$.]

c. The velocity of the screwdriver at time t is

$v(t) = \lim\limits_{h \to 0} \dfrac{f(t + h) - f(t)}{h} = \lim\limits_{h \to 0} \dfrac{16(t + h)^2 - 16t^2}{h} = \lim\limits_{h \to 0} \dfrac{16t^2 + 32th + 16h^2 - 16t^2}{h}$

$= \lim\limits_{h \to 0} \dfrac{(32t + 16h)h}{h} = 32t$.

In particular, the velocity of the screwdriver when it hits the ground (at $t = 5$) is $v(5) = 32(5) = 160$, or 160 ft/s.

© 2011 Cengage Learning. All Rights Reserved. May not be scanned, copied or duplicated, or posted to a publicly accessible website, in whole or in part.

33. a. We write $V = f(p) = \dfrac{1}{p}$. The average rate of change of V is $\dfrac{f(3) - f(2)}{3-2} = \dfrac{\frac{1}{3} - \frac{1}{2}}{1} = -\dfrac{1}{6}$, a decrease of $\frac{1}{6}$ liter/atmosphere.

b. $V'(t) = \lim\limits_{h \to 0} \dfrac{f(p+h) - f(p)}{h} = \lim\limits_{h \to 0} \dfrac{\frac{1}{p+h} - \frac{1}{p}}{h} = \lim\limits_{h \to 0} \dfrac{p - (p+h)}{hp(p+h)} = \lim\limits_{h \to 0} -\dfrac{1}{p(p+h)} = -\dfrac{1}{p^2}$. In particular, the rate of change of V when $p = 2$ is $V'(2) = -\dfrac{1}{2^2}$, a decrease of $\frac{1}{4}$ liter/atmosphere.

35. a. Using the four-step process, we find that

$$P'(x) = \lim_{h \to 0} \dfrac{P(x+h) - P(x)}{h} = \lim_{h \to 0} \dfrac{-\frac{1}{3}(x^2 + 2xh + h^2) + 7x + 7h + 30 - \left(-\frac{1}{3}x^2 + 7x + 30\right)}{h}$$

$$= \lim_{h \to 0} \dfrac{-\frac{2}{3}xh - \frac{1}{3}h^2 + 7h}{h} = \lim_{h \to 0} \left(-\tfrac{2}{3}x - \tfrac{1}{3}h + 7\right) = -\tfrac{2}{3}x + 7.$$

b. $P'(10) = -\frac{2}{3}(10) + 7 \approx 0.333$, or approximately $333 per $1000 spent on advertising.

$P'(30) = -\frac{2}{3}(30) + 7 = -13$, a decrease of $13,000 per $1000 spent on advertising.

37. $N(t) = t^2 + 2t + 50$. We first compute $N'(t)$ using the four-step process.

Step 1 $N(t+h) = (t+h)^2 + 2(t+h) + 50 = t^2 + 2th + h^2 + 2t + 2h + 50$.

Step 2 $N(t+h) - N(t) = (t^2 + 2th + h^2 + 2t + 2h + 50) - (t^2 + 2t + 50) = 2th + h^2 + 2h = h(2t + h + 2)$.

Step 3 $\dfrac{N(t+h) - N(t)}{h} = 2t + h + 2$.

Step 4 $N'(t) = \lim\limits_{h \to 0} (2t + h + 2) = 2t + 2$.

The rate of change of the country's GNP two years from now is $N'(2) = 2(2) + 2$, or $6 billion/yr. The rate of change four years from now is $N'(4) = 2(4) + 2$, or $10 billion/yr.

39. a. $f'(h)$ gives the instantaneous rate of change of the temperature with respect to height at a given height h, in °F per foot.

b. Because the temperature decreases as the altitude increases, the sign of $f'(h)$ is negative.

c. Because $f'(1000) = -0.05$, the change in the air temperature as the altitude changes from 1000 ft to 1001 ft is approximately $-0.05°$ F.

41. $\dfrac{f(a+h) - f(a)}{h}$ gives the average rate of change of the seal population over the time interval $[a, a+h]$.

$\lim\limits_{h \to 0} \dfrac{f(a+h) - f(a)}{h}$ gives the instantaneous rate of change of the seal population at $x = a$.

43. $\dfrac{f(a+h) - f(a)}{h}$ gives the average rate of change of the country's industrial production over the time interval $[a, a+h]$. $\lim\limits_{h \to 0} \dfrac{f(a+h) - f(a)}{h}$ gives the instantaneous rate of change of the country's industrial production at $x = a$.

45. $\dfrac{f(a+h) - f(a)}{h}$ gives the average rate of change of the atmospheric pressure over the altitudes $[a, a+h]$.

$\lim\limits_{h \to 0} \dfrac{f(a+h) - f(a)}{h}$ gives the instantaneous rate of change of the atmospheric pressure with respect to altitude at $x = a$.

© 2011 Cengage Learning. All Rights Reserved. May not be scanned, copied or duplicated, or posted to a publicly accessible website, in whole or in part.

47. a. f has a limit at $x = a$.

  b. f is not continuous at $x = a$ because $f(a)$ is not defined.

  c. f is not differentiable at $x = a$ because it is not continuous there.

49. a. f has a limit at $x = a$.

  b. f is continuous at $x = a$.

  c. f is not differentiable at $x = a$ because f has a kink at the point $x = a$.

51. a. f does not have a limit at $x = a$ because it is unbounded in the neighborhood of a.

  b. f is not continuous at $x = a$.

  c. f is not differentiable at $x = a$ because it is not continuous there.

53. Our computations yield the following results: 32.1, 30.939, 30.814, 30.8014, 30.8001, and 30.8000. The motorcycle's instantaneous velocity at $t = 2$ is approximately 30.8 ft/s.

55. False. Let $f(x) = |x|$. Then f is continuous at $x = 0$, but is not differentiable there.

57. Observe that the graph of f has a kink at $x = -1$. We have

$$\frac{f(-1+h) - f(-1)}{h} = 1 \text{ if } h > 0, \text{ and } -1 \text{ if } h < 0, \text{ so that}$$

$$\lim_{h \to 0} \frac{f(-1+h) - f(-1)}{h} \text{ does not exist.}$$

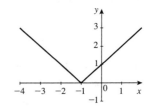

59. For continuity, we require that

$$f(1) = 1 = \lim_{x \to 1^+}(ax + b) = a + b, \text{ or } a + b = 1. \text{ In order that}$$

the derivative exist at $x = 1$, we require that $\lim_{x \to 1^-} 2x = \lim_{x \to 1} a$, or

$2 = a$. Therefore, $b = -1$ and so $f(x) = \begin{cases} x^2 & \text{if } x \le 1 \\ 2x - 1 & \text{if } x > 1 \end{cases}$

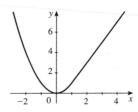

61. We have $f(x) = x$ if $x > 0$ and $f(x) = -x$ if $x < 0$. Therefore, when $x > 0$,

$$f'(x) = \lim_{h \to 0} \frac{f(x+h) - f(x)}{h} = \lim_{h \to 0} \frac{x + h - x}{h} = \lim_{h \to 0} \frac{h}{h} = 1, \text{ and when } x < 0,$$

$$f'(x) = \lim_{h \to 0} \frac{f(x+h) - f(x)}{h} = \lim_{h \to 0} \frac{-x - h - (-x)}{h} = \lim_{h \to 0} \frac{-h}{h} = -1. \text{ Because the right-hand limit does not}$$

equal the left-hand limit, we conclude that $\lim_{h \to 0} f(x)$ does not exist.

© 2011 Cengage Learning. All Rights Reserved. May not be scanned, copied or duplicated, or posted to a publicly accessible website, in whole or in part.

2.6   Using Technology   page 151

1. a. $y = 9x - 11$

b.

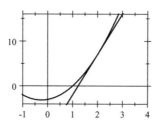

3. a. $y = \frac{1}{4}x + 1$

b.

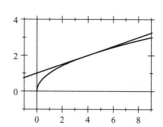

5. a. 4   **b.** $y = 4x - 1$

c.

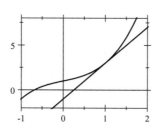

7. a. 4.02   **b.** $y = 4.02x - 3.57$

c.

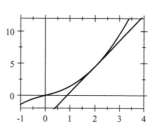

9. a.

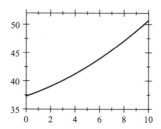

b. 41.22 cents/mile

c. 1.22 cents/mile/yr

CHAPTER 2   Concept Review   page 152

1. domain, range, B

3. $f(x) \pm g(x)$, $f(x)g(x)$, $\dfrac{f(x)}{g(x)}$, $A \cap B, 0$

5. a. $P(x) = a_n x^n + a_{n-1} x^{n-1} + \cdots + a_1 x + a_0$, where $a_n \neq 0$ and n is a positive integer

b. linear, quadratic, cubic   **c.** quotient, polynomials   **d.** x^r, where r is a real number

7. a. L^r **b.** $L \pm M$ **c.** LM **d.** $\dfrac{L}{M}$, $M \neq 0$

© 2011 Cengage Learning. All Rights Reserved. May not be scanned, copied or duplicated, or posted to a publicly accessible website, in whole or in part.

9. a. right   **b.** left   **c.** L, L

11. a. $a, a, g(a)$ **b.** everywhere   **c.** Q

13. a. $f'(a)$ **b.** $y - f(a) = m(x - a)$

---

1. a. $9 - x \geq 0$ gives $x \leq 9$, and the domain is $(-\infty, 9]$.

b. $2x^2 - x - 3 = (2x - 3)(x + 1)$, and $x = \frac{3}{2}$ or -1. Because the denominator of the given expression is zero at these points, we see that the domain of f cannot include these points and so the domain of f is $(-\infty, -1) \cup \left(-1, \frac{3}{2}\right) \cup \left(\frac{3}{2}, \infty\right)$.

3. a. $f(-2) = 3(-2)^2 + 5(-2) - 2 = 0$.

b. $f(a + 2) = 3(a + 2)^2 + 5(a + 2) - 2 = 3a^2 + 12a + 12 + 5a + 10 - 2 = 3a^2 + 17a + 20$.

c. $f(2a) = 3(2a)^2 + 5(2a) - 2 = 12a^2 + 10a - 2$.

d. $f(a + h) = 3(a + h)^2 + 5(a + h) - 2 = 3a^2 + 6ah + 3h^2 + 5a + 5h - 2$.

5. a.

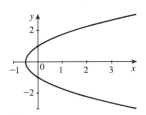

b. For each value of $x > 0$, there are two values of y. We conclude that y is not a function of x. (We could also note that the function fails the vertical line test.)

c. Yes. For each value of y, there is only one value of x.

7. a. $f(x)g(x) = \dfrac{2x + 3}{x}$.

b. $\dfrac{f(x)}{g(x)} = \dfrac{1}{x(2x + 3)}$.

c. $f(g(x)) = \dfrac{1}{2x + 3}$.

d. $g(f(x)) = 2\left(\dfrac{1}{x}\right) + 3 = \dfrac{2}{x} + 3$.

9. a. Take $f(x) = 2x^2 + x + 1$ and $g(x) = \dfrac{1}{x^3}$.

b. Take $f(x) = x^2 + x + 4$ and $g(x) = \sqrt{x}$.

11. $\lim\limits_{x \to 0} (5x - 3) = 5(0) - 3 = -3$.

13. $\lim\limits_{x \to -1} (3x^2 + 4)(2x - 1) = [3(-1)^2 + 4][2(-1) - 1] = -21$.

15. $\lim\limits_{x \to 2} \dfrac{x + 3}{x^2 - 9} = \dfrac{2 + 3}{4 - 9} = -1$.

© 2011 Cengage Learning. All Rights Reserved. May not be scanned, copied or duplicated, or posted to a publicly accessible website, in whole or in part.

17. $\lim\limits_{x \to 3} \sqrt{2x^3 - 5} = \sqrt{2(27) - 5} = 7.$

19. $\lim\limits_{x \to 1+} \dfrac{x - 1}{x(x - 1)} = \lim\limits_{x \to 1+} \dfrac{1}{x} = 1.$

21. $\lim\limits_{x \to \infty} \dfrac{x^2}{x^2 - 1} = \lim\limits_{x \to \infty} \dfrac{1}{1 - \dfrac{1}{x^2}} = 1.$

23. $\lim\limits_{x \to \infty} \dfrac{3x^2 + 2x + 4}{2x^2 - 3x + 1} = \lim\limits_{x \to \infty} \dfrac{3 + \dfrac{2}{x} + \dfrac{4}{x^2}}{2 - \dfrac{3}{x} + \dfrac{1}{x^2}} = \dfrac{3}{2}.$

25. $\lim\limits_{x \to 2+} f(x) = \lim\limits_{x \to 2+} (-x + 3) = -2 + 3 = 1$ and

$\lim\limits_{x \to 2-} f(x) = \lim\limits_{x \to 2-} (2x - 3) = 2(2) - 3 = 4 - 3 = 1.$ Therefore,

$\lim\limits_{x \to 2} f(x) = 1.$

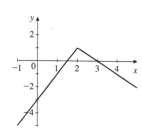

27. The function is discontinuous at $x = 2$.

29. Because $\lim\limits_{x \to -1} f(x) = \lim\limits_{x \to -1} \dfrac{1}{(x + 1)^2} = \infty$ (does not exist), we see that f is discontinuous at $x = -1$.

31. a. Let $f(x) = x^2 + 2$. Then the average rate of change of y over $[1, 2]$ is $\dfrac{f(2) - f(1)}{2 - 1} = \dfrac{(4 + 2) - (1 + 2)}{1} = 3.$

Over $[1, 1.5]$, it is $\dfrac{f(1.5) - f(1)}{1.5 - 1} = \dfrac{(2.25 + 2) - (1 + 2)}{0.5} = 2.5.$ Over $[1, 1.1]$, it is

$\dfrac{f(1.1) - f(1)}{1.1 - 1} = \dfrac{(1.21 + 2) - (1 + 2)}{0.1} = 2.1.$

b. Computing $f'(x)$ using the four-step process., we obtain

$f'(x) = \lim\limits_{h \to 0} \dfrac{f(x + h) - f(x)}{h} = \lim\limits_{h \to 0} \dfrac{h(2x + h)}{h} = \lim\limits_{h \to 0} (2x + h) = 2.$ Therefore, the instantaneous rate of

change of f at $x = 1$ is $f'(1) = 2$, or 2 units per unit change in x.

33. $f(x) = -\dfrac{1}{x}$. We use the four-step process:

Step 1 $f(x + h) = -\dfrac{1}{x + h}.$

Step 2 $f(x + h) - f(x) = -\dfrac{1}{x + h} - \left(-\dfrac{1}{x}\right) = -\dfrac{1}{x + h} + \dfrac{1}{x} = \dfrac{h}{x(x + h)}.$

Step 3 $\dfrac{f(x + h) - f(x)}{h} = -\dfrac{1}{x(x + h)}.$

Step 4 $f'(x) = \lim\limits_{h \to 0} \dfrac{f(x + h) - f(x)}{h} = \lim\limits_{h \to 0} \dfrac{1}{x(x + h)} = \dfrac{1}{x^2}.$

© 2011 Cengage Learning. All Rights Reserved. May not be scanned, copied or duplicated, or posted to a publicly accessible website, in whole or in part.

35. $f(x) = -x^2$. We use the four-step process:

Step 1 $f(x+h) = -(x+h)^2 = -x^2 - 2xh - h^2$.

Step 2 $f(x+h) - f(x) = (-x^2 - 2xh - h^2) - (-x^2) = -2xh - h^2 = h(-2x - h)$.

Step 3 $\dfrac{f(x+h) - f(x)}{h} = -2x - h$.

Step 4 $f'(x) = \displaystyle\lim_{h \to 0} \dfrac{f(x+h) - f(x)}{h} = \lim_{h \to 0} (-2x - h) = -2x$.

The slope of the tangent line is $f'(2) = -2(2) = -4$. An equation of the tangent line is $y - (-4) = -4(x - 2)$, or $y = -4x + 4$.

37. $S(4) = 6000(4) + 30{,}000 = 54{,}000$.

39. a. $C(x) = 6x + 30{,}000$.

b. $R(x) = 10x$.

c. $P(x) = R(x) - C(x) = 10x - (6x + 30{,}000) = 4x - 30{,}000$.

d. $P(6000) = 4(6000) - 30{,}000 = -6000$, or a loss of $6000. $P(8000) = 4(8000) - 30{,}000 = 2000$, or a profit of $2000. $P(12{,}000) = 4(12{,}000) - 30{,}000 = 18{,}000$, or a profit of $18,000.

41. The profit function is given by $P(x) = R(x) - C(x) = 20x - (12x + 20{,}000) = 8x - 20{,}000$.

43. The child should receive $D(35) = \dfrac{500(35)}{150} \approx 117$, or approximately 117 mg.

45. $R(30) = -\frac{1}{2}(30)^2 + 30(30) = 450$, or $45,000.

47. The population will increase by $P(9) - P(0) = \left[50{,}000 + 30(9)^{3/2} + 20(9)\right] - 50{,}000$, or 990, during the next 9 months. The population will increase by $P(16) - P(0) = \left[50{,}000 + 30(16)^{3/2} + 20(16)\right] - 50{,}000$, or 2240 during the next 16 months.

49. We need to find the point of intersection of the two straight lines representing the given linear functions. We solve the equation $2.3 + 0.4t = 1.2 + 0.6t$, obtaining $1.1 = 0.2t$ and thus $t = 5.5$. This tells us that the annual sales of the Cambridge Drug Store first surpasses that of the Crimson Drug store $5\frac{1}{2}$ years from now.

51. $f(t) = 0.1714t^2 + 0.6657t + 0.7143$.

a. $f(0) = 0.7143$ or 714,300.

b. $f(5) = 0.1714(5)^2 + 0.6657(5) + 0.7143 = 8.3278$, or 8.33 million.

53. a. $A(0) = 16.4$, or $16.4 billion; $A(1) = 16.4(1+1)^{0.1} \approx 17.58$, or $17.58 billion; $A(2) = 16.4(2+1)^{0.1} \approx 18.30$, or $18.3 billion; $A(3) = 16.4(3+1)^{0.1} \approx 18.84$, or $18.84 billion; and $A(4) = 16.4(4+1)^{0.1} \approx 19.26$, or $19.26 billion. The nutritional market grew over the years 1999 to 2003.

b.

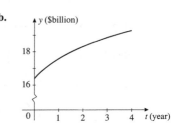

© 2011 Cengage Learning. All Rights Reserved. May not be scanned, copied or duplicated, or posted to a publicly accessible website, in whole or in part.

55. a. $f(r) = \pi r^2$.

    b. $g(t) = 2t$.

    c. $h(t) = (f \circ g)(t) = f(g(t)) = \pi [g(t)]^2 = 4\pi t^2$.

    d. $h(30) = 4\pi (30^2) = 3600\pi$, or 3600π ft^2.

57. Let h denote the height of the box. Then its volume is $V = (x)(2x)h = 30$, so that $h = \dfrac{15}{x^2}$. Thus, the cost is

$$C(x) = 30(x)(2x) + 15[2xh + 2(2x)h] + 20(x)(2x)$$

$$= 60x^2 + 15(6xh) + 40x^2 = 100x^2 + (15)(6)x\left(\frac{15}{x^2}\right)$$

$$= 100x^2 + \frac{1350}{x}.$$

59. $\displaystyle\lim_{x \to \infty} \overline{C}(x) = \lim_{x \to \infty}\left(20 + \frac{400}{x}\right) = 20$. As the level of production increases without bound, the average cost of producing the commodity steadily decreases and approaches \$20 per unit.

CHAPTER 2   Before Moving On...   page 156

1. a. $f(-1) = -2(-1) + 1 = 3$.

    b. $f(0) = 2$.

    c. $f\left(\frac{3}{2}\right) = \left(\frac{3}{2}\right)^2 + 2 = \frac{17}{4}$.

2. a. $(f + g)(x) = f(x) + g(x) = \dfrac{1}{x + 1} + x^2 + 1$.

    b. $(fg)(x) = f(x)g(x) = \dfrac{x^2 + 1}{x + 1}$.

    c. $(f \circ g)(x) = f(g(x)) = \dfrac{1}{g(x) + 1} = \dfrac{1}{x^2 + 2}$.

    d. $(g \circ f)(x) = g(f(x)) = [f(x)]^2 + 1 = \dfrac{1}{(x + 1)^2} + 1$.

3. $4x + h = 108$, so $h = 108 - 4x$. The volume is $V = x^2 h = x^2(108 - 4x) = 108x^2 - 4x^3$.

4. $\displaystyle\lim_{x \to -1} \frac{x^2 + 4x + 3}{x^2 + 3x + 2} = \lim_{x \to -1} \frac{(x + 3)(x + 1)}{(x + 2)(x + 1)} = 2$.

5. a. $\displaystyle\lim_{x \to 1^-} f(x) = \lim_{x \to 1^-} (x^2 - 1) = 0$.

    b. $\displaystyle\lim_{x \to 1^+} f(x) = \lim_{x \to 1^+} x^3 = 1$.

    Because $\displaystyle\lim_{x \to 1^-} f(x) \neq \lim_{x \to 1^+} f(x)$, f is not continuous at 1.

© 2011 Cengage Learning. All Rights Reserved. May not be scanned, copied or duplicated, or posted to a publicly accessible website, in whole or in part.

6. The slope of the tangent line at any point is

$$\lim_{h \to 0} \frac{f(x+h) - f(x)}{h} = \lim_{h \to 0} \frac{(x+h)^2 - 3(x+h) + 1 - (x^2 - 3x + 1)}{h}$$

$$= \lim_{h \to 0} \frac{x^2 + 2xh + h^2 - 3x - 3h - x^2 + 3x - 1}{h}$$

$$= \lim_{h \to 0} \frac{h(2x + h - 3)}{h} = \lim_{h \to 0} (2x + h - 3) = 2x - 3.$$

Therefore, the slope at 1 is $2(1) - 3 = -1$. An equation of the tangent line is $y - (-1) = -1(x - 1)$, or $y + 1 = -x + 1$, or $y = -x$.

© 2011 Cengage Learning. All Rights Reserved. May not be scanned, copied or duplicated, or posted to a publicly accessible website, in whole or in part.

3 DIFFERENTIATION

3.1 Problem-Solving Tips

In this section, you are given four basic rules for finding the derivative of a function. As you work through the exercises that follow, first decide which rule(s) you need to find the derivative of the given function. Then write out your solution. After doing this a few times, you should have the formulas memorized. The key here is to try not to look at the formula in the text, and to work the problem just as if you were taking a test. If you train yourself to work in this manner, writing tests will become a lot easier. Also, make sure to distinguish between the notation dy/dx and d/dx. The first notation is used for the derivative of a function y, where as the second notation tells us to find the derivative of the function that follows with respect to x.

Here are some tips for solving the problems in the exercises that follow:

1. **To find the derivative of a function involving radicals,** first rewrite the expression in exponential form. For example, if $f(x) = 2x - 5\sqrt{x}$, rewrite the function in the form $f(x) = 2x - 5x^{1/2}$.

2. **To find the point on the graph of f where the tangent line is horizontal,** set $f'(x) = 0$ and solve for x. (Here we are making use of the fact that the slope of a horizontal line is zero.) This yields the x-value of the point on the graph where the tangent line is horizontal. To find the corresponding y-value, evaluate the function f at this value of x.

3.1 Concept Questions   page 164

1. **a.** The derivative of a constant is zero.
   b. The derivative of $f(x) = x^n$ is n times x raised to the $(n-1)$th power.
   c. The derivative of a constant times a function is the constant times the derivative of the function.
   d. The derivative of the sum is the sum of the derivatives.

3. **a.** $F'(x) = \dfrac{d}{dx}\left[af(x) + bg(x)\right] = \dfrac{d}{dx}\left[af(x)\right] + \dfrac{d}{dx}\left[bg(x)\right] = af'(x) + bg'(x)$.
   b. $F'(x) = \dfrac{d}{dx}\left[\dfrac{f(x)}{a}\right] = \dfrac{1}{a}\dfrac{d}{dx}\left[f(x)\right] = \dfrac{f'(x)}{a}$.

3.1 Basic Rules of Differentiation   page 164

1. $f'(x) = \frac{d}{dx}(-3) = 0$.

3. $f'(x) = \frac{d}{dx}(x^5) = 5x^4$.

5. $f'(x) = \frac{d}{dx}(x^{2.1}) = 2.1x^{1.1}$.

7. $f'(x) = \frac{d}{dx}(3x^2) = 6x$.

59

© 2011 Cengage Learning. All Rights Reserved. May not be scanned, copied or duplicated, or posted to a publicly accessible website, in whole or in part.

9. $f'(r) = \frac{d}{dr}\left(\pi r^2\right) = 2\pi r.$

11. $f'(x) = \frac{d}{dx}\left(9x^{1/3}\right) = \frac{1}{3}(9)x^{(1/3-1)} = 3x^{-2/3}.$

13. $f'(x) = \frac{d}{dx}\left(3\sqrt{x}\right) = \frac{d}{dx}\left(3x^{1/2}\right) = \frac{1}{2}(3)x^{-1/2} = \frac{3}{2}x^{-1/2} = \frac{3}{2\sqrt{x}}.$

15. $f'(x) = \frac{d}{dx}\left(7x^{-12}\right) = (-12)(7)x^{-12-1} = -84x^{-13}.$

17. $f'(x) = \frac{d}{dx}\left(5x^2 - 3x + 7\right) = 10x - 3.$

19. $f'(x) = \frac{d}{dx}\left(-x^3 + 2x^2 - 6\right) = -3x^2 + 4x.$

21. $f'(x) = \frac{d}{dx}\left(0.03x^2 - 0.4x + 10\right) = 0.06x - 0.4.$

23. $f(x) = \dfrac{x^3 - 4x^2 + 3}{x} = x^2 - 4x + \dfrac{3}{x},$ so $f'(x) = \dfrac{d}{dx}\left(x^2 - 4x + 3x^{-1}\right) = 2x - 4 - \dfrac{3}{x^2}.$

25. $f'(x) = \frac{d}{dx}\left(4x^4 - 3x^{5/2} + 2\right) = 16x^3 - \frac{15}{2}x^{3/2}.$

27. $f'(x) = \frac{d}{dx}\left(3x^{-1} + 4x^{-2}\right) = -3x^{-2} - 8x^{-3}.$

29. $f'(t) = \frac{d}{dt}\left(4t^{-4} - 3t^{-3} + 2t^{-1}\right) = -16t^{-5} + 9t^{-4} - 2t^{-2}.$

31. $f'(x) = \frac{d}{dx}\left(2x - 5x^{1/2}\right) = 2 - \frac{5}{2}x^{-1/2} = 2 - \frac{5}{2\sqrt{x}}.$

33. $f'(x) = \dfrac{d}{dx}\left(2x^{-2} - 3x^{-1/3}\right) = -4x^{-3} + x^{-4/3} = -\dfrac{4}{x^3} + \dfrac{1}{x^{4/3}}.$

35. $f'(x) = \dfrac{d}{dx}\left(2x^3 - 4x\right) = 6x^2 - 4.$

    a. $f'(-2) = 6(-2)^2 - 4 = 20.$
    b. $f'(0) = 6(0) - 4 = -4.$
    c. $f'(2) = 6(2)^2 - 4 = 20.$

37. The given limit is $f'(1)$, where $f(x) = x^3$. Because $f'(x) = 3x^2$, we have $\displaystyle\lim_{h\to 0}\dfrac{(1+h)^3 - 1}{h} = f'(1) = 3.$

39. Let $f(x) = 3x^2 - x$. Then $\displaystyle\lim_{h\to 0}\dfrac{3(2+h)^2 - (2+h) - 10}{h} = \lim_{h\to 0}\dfrac{f(2+h) - f(2)}{h}$ because

$f(2+h) - f(2) = 3(2+h)^2 - (2+h) - [3(4) - 2] = 3(2+h)^2 - (2+h) - 10.$ But the last limit is simply

$f'(2)$. Because $f'(x) = 6x - 1$, we have $f'(2) = 11$. Therefore, $\displaystyle\lim_{h\to 0}\dfrac{3(2+h)^2 - (2+h) - 10}{h} = 11.$

41. $f(x) = 2x^2 - 3x + 4$. The slope of the tangent line at any point $(x, f(x))$ on the graph of f is $f'(x) = 4x - 3$. In particular, the slope of the tangent line at the point $(2, 6)$ is $f'(2) = 4(2) - 3 = 5$. An equation of the required tangent line is $y - 6 = 5(x - 2)$ or $y = 5x - 4$.

43. $f(x) = x^4 - 3x^3 + 2x^2 - x + 1$, so $f'(x) = 4x^3 - 9x^2 + 4x - 1$. The slope is $f'(1) = 4 - 9 + 4 - 1 = -2$. An equation of the tangent line is $y - 0 = -2(x - 1)$ or $y = -2x + 2$.

© 2011 Cengage Learning. All Rights Reserved. May not be scanned, copied or duplicated, or posted to a publicly accessible website, in whole or in part.

45. a. $f'(x) = 3x^2$. At a point where the tangent line is horizontal, $f'(x) = 0$, or $3x^2 = 0$, and so $x = 0$. Therefore, the point is $(0, 0)$.

b.

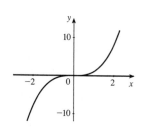

47. a. $f(x) = x^3 + 1$. The slope of the tangent line at any point $(x, f(x))$ on the graph of f is $f'(x) = 3x^2$. At the point(s) where the slope is 12, we have $3x^2 = 12$, so $x = \pm 2$. The required points are $(-2, -7)$ and $(2, 9)$.

b. The tangent line at $(-2, -7)$ has equation $y - (-7) = 12[x - (-2)]$, or $y = 12x + 17$, and the tangent line at $(2, 9)$ has equation $y - 9 = 12(x - 2)$, or $y = 12x - 15$.

c.

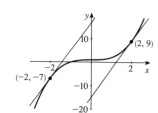

49. $f(x) = \frac{1}{4}x^4 - \frac{1}{3}x^3 - x^2$, so $f'(x) = x^3 - x^2 - 2x$.

a. $f'(x) = x^3 - x^2 - 2x = -2x$ implies $x^3 - x^2 = 0$, so $x^2(x - 1) = 0$. Thus, $x = 0$ or $x = 1$. $f(1) = \frac{1}{4}(1)^4 - \frac{1}{3}(1)^3 - (1)^2 = -\frac{13}{12}$ and $f(0) = \frac{1}{4}(0)^4 - \frac{1}{3}(0)^3 - (0)^2 = 0$. We conclude that the corresponding points on the graph are $\left(1, -\frac{13}{12}\right)$ and $(0, 0)$.

b. $f'(x) = x^3 - x^2 - 2x = 0$ implies $x(x^2 - x - 2) = 0$, $x(x - 2)(x + 1) = 0$, and so $x = 0, 2,$ or -1. $f(0) = 0$, $f(2) = \frac{1}{4}(2)^4 - \frac{1}{3}(2)^3 - (2)^2 = 4 - \frac{8}{3} - 4 = -\frac{8}{3}$, and $f(-1) = \frac{1}{4}(-1)^4 - \frac{1}{3}(-1)^3 - (-1)^2 = \frac{1}{4} + \frac{1}{3} - 1 = -\frac{5}{12}$. We conclude that the corresponding points are $(0, 0)$, $\left(2, -\frac{8}{3}\right)$, and $\left(-1, -\frac{5}{12}\right)$.

c. $f'(x) = x^3 - x^2 - 2x = 10x$ implies $x^3 - x^2 - 12x = 0$, $x(x^2 - x - 12) = 0$, $x(x - 4)(x + 3) = 0$, so $x = 0, 4,$ or -3. $f(0) = 0$, $f(4) = \frac{1}{4}(4)^4 - \frac{1}{3}(4)^3 - (4)^2 = 48 - \frac{64}{3} = \frac{80}{3}$, and $f(-3) = \frac{1}{4}(-3)^4 - \frac{1}{3}(-3)^3 - (-3)^2 = \frac{81}{4} + 9 - 9 = \frac{81}{4}$. We conclude that the corresponding points are $(0, 0)$, $\left(4, \frac{80}{3}\right)$, and $\left(-3, \frac{81}{4}\right)$.

51. $V(r) = \frac{4}{3}\pi r^3$, so $V'(r) = 4\pi r^2$.

a. $V'\left(\frac{2}{3}\right) = 4\pi\left(\frac{4}{9}\right) = \frac{16}{9}\pi$ cm³/cm.

b. $V'\left(\frac{5}{4}\right) = 4\pi\left(\frac{25}{16}\right) = \frac{25}{4}\pi$ cm³/cm.

53. a. $N(1) = 16.3\left(1^{0.8766}\right) = 16.3(1) = 16.3$, or 16.3 million cameras.

b. $N'(t) = (16.3)(0.8766)t^{-0.1234}$, so $N'(1) \approx 14.29$, or approximately 14.3 million cameras per year.

c. $N(5) = 16.3\left(5^{0.8766}\right) \approx 66.82$, or approximately 66.8 million cameras.

d. $N'(5) = (16.3)(0.8766)\left(5^{-0.1234}\right) \approx 11.71$, or approximately 11.7 million cameras per year.

© 2011 Cengage Learning. All Rights Reserved. May not be scanned, copied or duplicated, or posted to a publicly accessible website, in whole or in part.

55. a.

1970 ($t = 1$)	1980 ($t = 2$)	1990 ($t = 3$)	2000 ($t = 4$)
49.6%	41.1%	36.9%	34.1%

b. $P'(t) = (49.6)\left(-0.27t^{-1.27}\right) = -\dfrac{13.392}{t^{1.27}}$. In 1980, $P'(2) \approx -5.5$, or decreasing at 5.5%/decade. In 1990,

$P'(3) \approx -3.3$, or decreasing at 3.3%/decade.

57. a. The number of viewers will be $N(8) = 52(8)^{0.531} = 156.87$, or approximately 157 million.

b. The projected number is changing at the rate of $N'(8) = 52(0.531)t^{-0.469}\big|_{t=8} = 52(0.531)8^{-0.469} \approx 10.41$, or approximately 10.4 million viewers/year.

59. a. $f(t) = 120t - 15t^2$, so $v = f'(t) = 120 - 30t$.

b. $v(0) = 120$ ft/sec

c. Setting $v = 0$ gives $120 - 30t = 0$, or $t = 4$. Therefore, the stopping distance is $f(4) = 120(4) - 15(16)$ or 240 ft.

61. a. At the beginning of 1980, $P(0) = 5\%$. At the beginning of 1990,

$P(10) = -0.0105(10^2) + 0.735(10) + 5 = 11.3$, or 11.3%. At the beginning of 2000,

$P(20) = -0.0105(20)^2 + 0.735(20) + 5 = 15.5$, or 15.5%.

b. $P'(t) = -0.021t + 0.735$. At the beginning of 1985, $P'(5) = -0.021(5) + 0.735 = 0.63$, or 0.63%/yr. At the beginning of 1990, $P'(10) = -0.021(10) + 0.735 = 0.525$, or 0.525%/yr.

63. a. $f(t) = 5.303t^2 - 53.977t + 253.8$. The rate of change of the groundfish population at any time t is given by $f'(t) = 10.606t - 53.977$. The rate of change at the beginning of 1994 is given by $f'(5) = 10.606(5) - 53.977 = -0.947$, so the population is decreasing at the rate of 0.9 thousand metric tons/yr. At the beginning of 1996, the rate of change is $f'(7) = 10.606(7) - 53.977 = 20.265$, so the population is increasing at the rate of approximately 20.3 thousand metric tons/yr.

b. Yes.

65. $I'(t) = -0.6t^2 + 6t$.

a. In 2005, it is changing at a rate of $I'(5) = -0.6(25) + 6(5)$, or 15 points/yr. In 2007, it is $I'(7) = -0.6(49) + 6(7)$, or 12.6 points/yr. In 2010, it is $I'(10) = -0.6(100) + 6(10)$, or 0 points/yr.

b. The average rate of increase of the CPI over the period from 2005 to 2010 is

$\dfrac{I(10) - I(5)}{5} = \dfrac{[-0.2(1000) + 3(100) + 100] - [-0.2(125) + 3(25) + 100]}{5} = \dfrac{200 - 150}{5} = 10$, or 10 points/yr.

67. a. $f'(x) = \frac{d}{dx}\left[0.0001x^{5/4} + 10\right] = \frac{5}{4}\left(0.0001x^{1/4}\right) = 0.000125x^{1/4}$.

b. $f'(10{,}000) = 0.000125(10{,}000)^{1/4} = 0.00125$, or $0.00125/radio.

69. a. $f(t) = 20t - 40\sqrt{t} + 50$, so $f'(t) = 20 - 40\left(\frac{1}{2}\right)t^{-1/2} = 20\left(1 - \dfrac{1}{\sqrt{t}}\right)$.

b. $f(0) = 20(0) - 40\sqrt{0} + 50 = 50$, so $f(1) = 20(1) - 40\sqrt{1} + 50 = 30$ and $f(2) = 20(2) - 40\sqrt{2} + 50 \approx 33.43$. The average velocities at 6, 7, and 8 a.m. are 50, 30, and 33.43 mph, respectively.

© 2011 Cengage Learning. All Rights Reserved. May not be scanned, copied or duplicated, or posted to a publicly accessible website, in whole or in part.

c. $f'\left(\frac{1}{2}\right) = 20 - 20\left(\frac{1}{2}\right)^{-1/2} \approx -8.28$, $f'(1) = 20 - 20(1)^{-1/2} \approx 0$, and $f'(2) = 20 - 20(2)^{-1/2} \approx 5.86$. At 6:30 a.m. the average velocity is decreasing at the rate of 8.28 mph/hr, at 7 a.m. it is not changing, and at 8 a.m. it is increasing at the rate of 5.86 mph.

71. $N(t) = 2t^3 + 3t^2 - 4t + 1000$, so $N'(t) = 6t^2 + 6t - 4$. $N'(2) = 6(4) + 6(2) - 4 = 32$, or 32 turtles/yr; and $N'(8) = 6(64) + 6(8) - 4 = 428$, or 428 turtles/yr. The population ten years after implementation of the conservation measures will be $N(10) = 2(10^3) + 3(10^2) - 4(10) + 1000$, or 3260 turtles.

73. a. At the beginning of 1991, $P(0) = 12\%$. At the beginning of 2004,
$P(13) = 0.0004(13)^3 + 0.0036(13)^2 + 0.8(13) + 12 \approx 23.9\%$.

b. $P'(t) = 0.0012t^2 + 0.0072t + 0.8$. At the beginning of 1991, $P'(0) = 0.8$, or 0.8%/yr. At the beginning of 2004, $P'(13) = 0.0012(13)^2 + 0.0072(13) + 0.8 \approx 1.1$, or approximately 1.1%/yr.

75. a. At any time t, the function $D = g + f$ at t, $D(t) = (g + f)(t) = g(t) + f(t)$, gives the total population aged 65 and over of the developed and the underdeveloped/emerging countries.

b. $D(t) = g(t) + f(t) = (0.46t^2 + 0.16t + 287.8) + (3.567t + 175.2) = 0.46t^2 + 3.727t + 463$, so $D'(t) = 0.92t + 3.727$. Therefore, $D'(10) = 0.92(10) + 3.727 = 12.927$, which says that the combined population is growing at the rate of approximately 13 million people per year in 2010.

77. True. $\dfrac{d}{dx}[2f(x) - 5g(x)] = \dfrac{d}{dx}[2f(x)] - \dfrac{d}{dx}[5g(x)] = 2f'(x) - 5g'(x)$.

79. $\dfrac{d}{dx}(x^3) = \lim\limits_{h \to 0} \dfrac{(x+h)^3 - x^3}{h} = \lim\limits_{h \to 0} \dfrac{x^3 + 3x^2h + 3xh^2 + h^3 - x^3}{h} = \lim\limits_{h \to 0} \dfrac{h(3x^2 + 3xh + h^2)}{h}$
$= \lim\limits_{h \to 0}(3x^2 + 3xh + h^2) = 3x^2$.

3.1 Using Technology   page 170

1. 1

3. 0.4226

5. 0.1613

7. a.

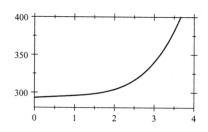

b. 3.4295 parts/million per 40 years;
105.4332 parts/million per 40 years

© 2011 Cengage Learning. All Rights Reserved. May not be scanned, copied or duplicated, or posted to a publicly accessible website, in whole or in part.

9. a. $f(t) = 0.611t^3 + 9.702t^2 + 32.544t + 473.5$ **b.**

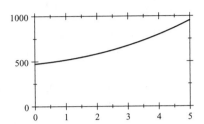

c. At the beginning of 2000, the assets of the
hedge funds were increasing at the rate of
$53.781 billion/yr, and at the beginning of
2003, they were increasing at the rate of
$139.488 billion/yr.

3.2   Problem-Solving Tips

The answers at the back of the book for the exercises in this section are given in both simplified and unsimplified terms.
Here, as with all of your homework, you should make it a practice to analyze your errors. If you do not get the right
answer for the unsimplified form, it means that you are not applying the rules for differentiating correctly. In this case you
need to review the rules, making sure that you can write out each rule. If you have the correct answer for the unsimplified
form but the incorrect answer for the simplified form, it probably means that you have made an algebraic error. You may
need to review the rules for simplifying algebraic expressions given on page 12 of the text and then work some of the
exercises given in section 1.2 to get back into practice. In any case, you will need to simplify your answers when you
work the problems on the applications of the derivative in the next chapter, so you should get in the habit of doing so
now.

Here are some tips for solving the problems in the exercises that follow:

1. **To find the derivative of a function involving radicals**, first rewrite the expression in exponential form. For
   example, if $f(x) = 2x - 5\sqrt{x}$, rewrite the function in the form $f'^{1/2}$.

2. **To find the point on the graph of f where the tangent line is horizontal**, set $f'(x) = 0$ and solve for x. (Here we
   are making use of the fact that the slope of a horizontal line is zero.) This yields the x-value of the point on the graph
   where the tangent line is horizontal. To find the corresponding y-value, evaluate the function f at this value of x.

3.2   Concept Questions   page 177

1. a. The derivative of the product of two functions is equal to the first function times the derivative of the second
   function plus the second function times the derivative of the first function.

   b. The derivative of the quotient of two functions is equal to the quotient whose numerator is given by the
   denominator times the derivative of the numerator minus the numerator times the derivative of the denominator
   and whose denominator is the square of the denominator of the quotient.

3.2   The Product and Quotient Rules   page 177

1. $f(x) = 2x(x^2 + 1)$, so $f'(x) = 2x\frac{d}{dx}(x^2 + 1) + (x^2 + 1)\frac{d}{dx}(2x) = 2x(2x) + (x^2 + 1)(2) = 6x^2 + 2.$

© 2011 Cengage Learning. All Rights Reserved. May not be scanned, copied or duplicated, or posted to a publicly accessible website, in whole or in part.

3. $f(t) = (t-1)(2t+1)$, so

$f'(t) = (t-1)\frac{d}{dt}(2t+1) + (2t+1)\frac{d}{dt}(t-1) = (t-1)(2) + (2t+1)(1) = 4t - 1.$

5. $f(x) = (3x+1)(x^2-2)$, so

$f'(x) = (3x+1)\frac{d}{dx}(x^2-2) + (x^2-2)\frac{d}{dx}(3x+1) = (3x+1)(2x) + (x^2-2)(3) = 9x^2 + 2x - 6.$

7. $f(x) = (x^3-1)(x+1)$, so

$f'(x) = (x^3-1)\frac{d}{dx}(x+1) + (x+1)\frac{d}{dx}(x^3-1) = (x^3-1)(1) + (x+1)(3x^2) = 4x^3 + 3x^2 - 1.$

9. $f(w) = (w^3 - w^2 + w - 1)(w^2 + 2)$, so

$f'(w) = (w^3 - w^2 + w - 1)\frac{d}{dw}(w^2+2) + (w^2+2)\frac{d}{dw}(w^3 - w^2 + w - 1)$

$= (w^3 - w^2 + w - 1)(2w) + (w^2+2)(3w^2 - 2w + 1)$

$= 2w^4 - 2w^3 + 2w^2 - 2w + 3w^4 - 2w^3 + w^2 + 6w^2 - 4w + 2 = 5w^4 - 4w^3 + 9w^2 - 6w + 2.$

11. $f(x) = (5x^2+1)(2\sqrt{x}-1)$, so

$f'(x) = (5x^2+1)\frac{d}{dx}(2x^{1/2}-1) + (2x^{1/2}-1)\frac{d}{dx}(5x^2+1) = (5x^2+1)(x^{-1/2}) + (2x^{1/2}-1)(10x)$

$= 5x^{3/2} + x^{-1/2} + 20x^{3/2} - 10x = \dfrac{25x^2 - 10x\sqrt{x} + 1}{\sqrt{x}}.$

13. $f(x) = (x^2 - 5x + 2)\left(x - \dfrac{2}{x}\right)$, so

$f'(x) = (x^2 - 5x + 2)\dfrac{d}{dx}\left(x - \dfrac{2}{x}\right) + \left(x - \dfrac{2}{x}\right)\dfrac{d}{dx}(x^2 - 5x + 2)$

$= \dfrac{(x^2 - 5x + 2)(x^2 + 2)}{x^2} + \dfrac{(x^2 - 2)(2x - 5)}{x} = \dfrac{(x^2 - 5x + 2)(x^2 + 2) + x(x^2 - 2)(2x - 5)}{x^2}$

$= \dfrac{x^4 + 2x^2 - 5x^3 - 10x + 2x^2 + 4 + 2x^4 - 5x^3 - 4x^2 + 10x}{x^2} = \dfrac{3x^4 - 10x^3 + 4}{x^2}.$

15. $f(x) = \dfrac{1}{x-2}$, so $f'(x) = \dfrac{(x-2)\frac{d}{dx}(1) - (1)\frac{d}{dx}(x-2)}{(x-2)^2} = \dfrac{0 - 1(1)}{(x-2)^2} = -\dfrac{1}{(x-2)^2}.$

17. $f(x) = \dfrac{x-1}{2x+1}$, so $f'(x) = \dfrac{(2x+1)\frac{d}{dx}(x-1) - (x-1)\frac{d}{dx}(2x+1)}{(2x+1)^2} = \dfrac{2x+1 - (x-1)(2)}{(2x+1)^2} = \dfrac{3}{(2x+1)^2}.$

19. $f(x) = \dfrac{1}{x^2+1}$, so $f'(x) = \dfrac{(x^2+1)\frac{d}{dx}(1) - (1)\frac{d}{dx}(x^2+1)}{(x^2+1)^2} = \dfrac{(x^2+1)(0) - 1(2x)}{(x^2+1)^2} = -\dfrac{2x}{(x^2+1)^2}.$

21. $f(s) = \dfrac{s^2 - 4}{s+1}$, so

$f'(s) = \dfrac{(s+1)\frac{d}{ds}(s^2-4) - (s^2-4)\frac{d}{ds}(s+1)}{(s+1)^2} = \dfrac{(s+1)(2s) - (s^2-4)(1)}{(s+1)^2} = \dfrac{s^2 + 2s + 4}{(s+1)^2}.$

© 2011 Cengage Learning. All Rights Reserved. May not be scanned, copied or duplicated, or posted to a publicly accessible website, in whole or in part.

23. $f(x) = \dfrac{\sqrt{x}+1}{x^2+1}$, so

$$f'(x) = \frac{(x^2+1)\frac{d}{dx}\left(x^{1/2}\right) - \left(x^{1/2}+1\right)\frac{d}{dx}\left(x^2+1\right)}{\left(x^2+1\right)^2} = \frac{(x^2+1)\left(\frac{1}{2}x^{-1/2}\right) - \left(x^{1/2}+1\right)(2x)}{\left(x^2+1\right)^2}$$

$$= \frac{\left(\frac{1}{2}x^{-1/2}\right)\left[\left(x^2+1\right)-4x^2\right]-2x}{\left(x^2+1\right)^2} = \frac{1-3x^2-4x^{3/2}}{2\sqrt{x}\left(x^2+1\right)^2}.$$

25. $f(x) = \dfrac{x^2+2}{x^2+x+1}$, so

$$f'(x) = \frac{(x^2+x+1)\frac{d}{dx}\left(x^2+2\right) - \left(x^2+2\right)\frac{d}{dx}\left(x^2+x+1\right)}{\left(x^2+x+1\right)^2}$$

$$= \frac{(x^2+x+1)(2x) - \left(x^2+2\right)(2x+1)}{\left(x^2+x+1\right)^2} = \frac{2x^3+2x^2+2x-2x^3-x^2-4x-2}{\left(x^2+x+1\right)^2} = \frac{x^2-2x-2}{\left(x^2+x+1\right)^2}.$$

27. $f(x) = \dfrac{(x+1)\left(x^2+1\right)}{x-2} = \dfrac{\left(x^3+x^2+x+1\right)}{x-2}$, so

$$f'(x) = \frac{(x-2)\frac{d}{dx}\left(x^3+x^2+x+1\right) - \left(x^3+x^2+x+1\right)\frac{d}{dx}(x-2)}{(x-2)^2}$$

$$= \frac{(x-2)\left(3x^2+2x+1\right) - \left(x^3+x^2+x+1\right)}{(x-2)^2}$$

$$= \frac{3x^3+2x^2+x-6x^2-4x-2-x^3-x^2-x-1}{(x-2)^2} = \frac{2x^3-5x^2-4x-3}{(x-2)^2}.$$

29. $f(x) = \dfrac{x}{x^2-4} - \dfrac{x-1}{x^2+4} = \dfrac{x\left(x^2+4\right)-(x-1)\left(x^2-4\right)}{\left(x^2-4\right)\left(x^2+4\right)} = \dfrac{x^2+8x-4}{\left(x^2-4\right)\left(x^2+4\right)}$, so

$$f'(x) = \frac{\left(x^2-4\right)\left(x^2+4\right)\frac{d}{dx}\left(x^2+8x-4\right) - \left(x^2+8x-4\right)\frac{d}{dx}\left(x^4-16\right)}{\left(x^2-4\right)^2\left(x^2+4\right)^2}$$

$$= \frac{\left(x^2-4\right)\left(x^2+4\right)(2x+8) - \left(x^2+8x-4\right)\left(4x^3\right)}{\left(x^2-4\right)^2\left(x^2+4\right)^2}$$

$$= \frac{2x^5+8x^4-32x-128-4x^5-32x^4+16x^3}{\left(x^2-4\right)^2\left(x^2+4\right)^2} = \frac{-2x^5-24x^4+16x^3-32x-128}{\left(x^2-4\right)^2\left(x^2+4\right)^2}.$$

31. $h'(x) = f(x)g'(x) + f'(x)g(x)$ by the Product Rule. Therefore,
$h'(1) = f(1)g'(1) + f'(1)g(1) = (2)(3) + (-1)(-2) = 8$.

33. Using the Quotient Rule followed by the Product Rule, we have

$$h'(x) = \frac{[x+g(x)]\frac{d}{dx}[xf(x)] - xf(x)\frac{d}{dx}[x+g(x)]}{[x+g(x)]^2} = \frac{[x+g(x)]\left[xf'(x)+f(x)\right] - xf(x)\left[1+g'(x)\right]}{[x+g(x)]^2}.$$

Therefore,

$$h'(1) = \frac{[1+g(1)]\left[f'(1)+f(1)\right] - f(1)\left[1+g'(1)\right]}{[1+g(1)]^2} = \frac{(1-2)(-1+2)-2(1+3)}{(1-2)^2} = \frac{-1-8}{1} = -9.$$

© 2011 Cengage Learning. All Rights Reserved. May not be scanned, copied or duplicated, or posted to a publicly accessible website, in whole or in part.

35. $f(x) = (2x - 1)(x^2 + 3)$, so

$$f'(x) = (2x - 1)\tfrac{d}{dx}(x^2 + 3) + (x^2 + 3)\tfrac{d}{dx}(2x - 1) = (2x - 1)(2x) + (x^2 + 3)(2)$$

$$= 6x^2 - 2x + 6 = 2(3x^2 - x + 3).$$

At $x = 1$, $f'(1) = 2[3(1)^2 - (1) + 3] = 2(5) = 10.$

37. $f(x) = \dfrac{x}{x^4 - 2x^2 - 1}$, so

$$f'(x) = \frac{(x^4 - 2x^2 - 1)\tfrac{d}{dx}(x) - x\tfrac{d}{dx}(x^4 - 2x^2 - 1)}{(x^4 - 2x^2 - 1)^2} = \frac{(x^4 - 2x^2 - 1)(1) - x(4x^3 - 4x)}{(x^4 - 2x^2 - 1)^2}$$

$$= \frac{-3x^4 + 2x^2 - 1}{(x^4 - 2x^2 - 1)^2}.$$

Therefore, $f'(-1) = \dfrac{-3 + 2 - 1}{(1 - 2 - 1)^2} = -\dfrac{2}{4} = -\dfrac{1}{2}.$

39. $f(x) = (x^3 + 1)(x^2 - 2)$, so

$$f'(x) = (x^3 + 1)\tfrac{d}{dx}(x^2 - 2) + (x^2 - 2)\tfrac{d}{dx}(x^3 + 1) = (x^3 + 1)(2x) + (x^2 - 2)(3x^2).$$ The slope of the tangent line at $(2, 18)$ is $f'(2) = (8 + 1)(4) + (4 - 2)(12) = 60$. An equation of the tangent line is $y - 18 = 60(x - 2)$, or $y = 60x - 102$.

41. $f(x) = \dfrac{x + 1}{x^2 + 1}$, so

$$f'(x) = \frac{(x^2 + 1)\tfrac{d}{dx}(x + 1) - (x + 1)\tfrac{d}{dx}(x^2 + 1)}{(x^2 + 1)^2} = \frac{(x^2 + 1)(1) - (x + 1)(2x)}{(x^2 + 1)^2} = \frac{-x^2 - 2x + 1}{(x^2 + 1)^2}.$$ At

$x = 1$, $f'(1) = \dfrac{-1 - 2 + 1}{4} = -\dfrac{1}{2}$. Therefore, the slope of the tangent line at $x = 1$ is $-\dfrac{1}{2}$ and an equation is $y - 1 = -\dfrac{1}{2}(x - 1)$ or $y = -\dfrac{1}{2}x + \dfrac{3}{2}$.

43. Using the Product Rule, we find

$$g'(x) = \tfrac{d}{dx}[x^2 f(x)] = x^2 \tfrac{d}{dx}[f(x)] + f(x)\tfrac{d}{dx}(x^2) = x^2 f'(x) + 2x f(x).$$ Therefore,

$$g'(2) = 2^2 f'(2) + 2(2)f(2) = (4)(-1) + 4(3) = 8.$$

45. $f(x) = (x^3 + 1)(3x^2 - 4x + 2)$, so

$$f'(x) = (x^3 + 1)\tfrac{d}{dx}(3x^2 - 4x + 2) + (3x^2 - 4x + 2)\tfrac{d}{dx}(x^3 + 1)$$

$$= (x^3 + 1)(6x - 4) + (3x^2 - 4x + 2)(3x^2)$$

$$= 6x^4 + 6x - 4x^3 - 4 + 9x^4 - 12x^3 + 6x^2 = 15x^4 - 16x^3 + 6x^2 + 6x - 4.$$

At $x = 1$, $f'(1) = 15(1)^4 - 16(1)^3 + 6(1) + 6(1) - 4 = 7$. Thus, the slope of the tangent line at the point $x = 1$ is 7 and an equation is $y - 2 = 7(x - 1)$, or $y = 7x - 5$.

47. $f(x) = (x^2 + 1)(2 - x)$, so

$$f'(x) = (x^2 + 1)\tfrac{d}{dx}(2 - x) + (2 - x)\tfrac{d}{dx}(x^2 + 1) = (x^2 + 1)(-1) + (2 - x)(2x) = -3x^2 + 4x - 1.$$ At a point where the tangent line is horizontal, we have $f'(x) = -3x^2 + 4x - 1 = 0$ or $3x^2 - 4x + 1 = (3x - 1)(x - 1) = 0$, giving $x = \dfrac{1}{3}$ or $x = 1$. Because $f\left(\dfrac{1}{3}\right) = \left(\dfrac{1}{9} + 1\right)\left(2 - \dfrac{1}{3}\right) = \dfrac{50}{27}$ and $f(1) = 2(2 - 1) = 2$, we see that the required points are $\left(\dfrac{1}{3}, \dfrac{50}{27}\right)$ and $(1, 2)$.

© 2011 Cengage Learning. All Rights Reserved. May not be scanned, copied or duplicated, or posted to a publicly accessible website, in whole or in part.

49. $f(x) = (x^2 + 6)(x - 5)$, so

$$f'(x) = (x^2 + 6)\frac{d}{dx}(x - 5) + (x - 5)\frac{d}{dx}(x^2 + 6) = (x^2 + 6)(1) + (x - 5)(2x)$$

$$= x^2 + 6 + 2x^2 - 10x = 3x^2 - 10x + 6.$$

At a point where the slope of the tangent line is -2, we have $f'(x) = 3x^2 - 10x + 6 = -2$. This gives $3x^2 - 10x + 8 = (3x - 4)(x - 2) = 0$, so $x = \frac{4}{3}$ or $x = 2$. Because $f\left(\frac{4}{3}\right) = \left(\frac{16}{9} + 6\right)\left(\frac{4}{3} - 5\right) = -\frac{770}{27}$ and $f(2) = (4 + 6)(2 - 5) = -30$, the required points are $\left(\frac{4}{3}, -\frac{770}{27}\right)$ and $(2, -30)$.

51. $y = \dfrac{1}{1 + x^2}$, so $y' = \dfrac{(1 + x^2)\frac{d}{dx}(1) - (1)\frac{d}{dx}(1 + x^2)}{(1 + x^2)^2} = \dfrac{-2x}{(1 + x^2)^2}$. Thus, the slope of the tangent line at $\left(1, \frac{1}{2}\right)$ is $y'\big|_{x=1} = \dfrac{-2x}{(1 + x^2)^2}\bigg|_{x=1} = \dfrac{-2}{4} = -\dfrac{1}{2}$ and an equation of the tangent line is $y - \frac{1}{2} = -\frac{1}{2}(x - 1)$, or $y = -\frac{1}{2}x + 1$. Next, the slope of the required normal line is 2 and its equation is $y - \frac{1}{2} = 2(x - 1)$, or $y = 2x - \frac{3}{2}$.

53. $C(x) = \dfrac{0.5x}{100 - x}$, so $C'(x) = \dfrac{(100 - x)(0.5) - 0.5x(-1)}{(100 - x)^2} = \dfrac{50}{(100 - x)^2}$. $C'(80) = \dfrac{50}{20^2} = 0.125$, $C'(90) = \dfrac{50}{10^2} = 0.5$, $C'(95) = \dfrac{50}{5^2} = 2$, and $C'(99) = \dfrac{50}{1} = 50$. The rates of change of the cost of removing 80%, 90%, 95%, and 99% of the toxic waste are 0.125, 0.5, 2, and 50 million dollars per 1% increase in waste removed. It is too costly to remove all of the pollutant.

55. $N(t) = \dfrac{10,000}{1 + t^2} + 2000$, so $N'(t) = \dfrac{d}{dt}\left[10,000(1 + t^2)^{-1} + 2000\right] = -\dfrac{10,000}{(1 + t^2)^2}(2t) = -\dfrac{20,000t}{(1 + t^2)^2}$. The rates of change after 1 minute and 2 minutes are $N'(1) = -\dfrac{20,000}{(1 + 1^2)^2} = -5000$ and $N'(2) = -\dfrac{20,000(2)}{(1 + 2^2)^2} = -1600$.

The population of bacteria after one minute is $N(1) = \dfrac{10,000}{1 + 1} + 2000 = 7000$, and the population after two minutes is $N(2) = \dfrac{10,000}{1 + 4} + 2000 = 4000$.

57. a. $N(t) = \dfrac{60t + 180}{t + 6}$, so

$$N'(t) = \dfrac{(t + 6)\frac{d}{dt}(60t + 180) - (60t + 180)\frac{d}{dt}(t + 6)}{(t + 6)^2} = \dfrac{(t + 6)(60) - (60t + 180)(1)}{(t + 6)^2} = \dfrac{180}{(t + 6)^2}.$$

b. $N'(1) = \dfrac{180}{(1 + 6)^2} \approx 3.7$, $N'(3) = \dfrac{180}{(3 + 6)^2} \approx 2.2$,

$N'(4) = \dfrac{180}{(4 + 6)^2} = 1.8$, and $N'(7) = \dfrac{180}{(7 + 6)^2} \approx 1.1$. We conclude that the rates at which the average student is increasing his or her speed one week, three weeks, four weeks, and seven weeks into the course are approximately 3.7, 2.2, 1.8, and 1.1 words per minute, respectively.

c. Yes.

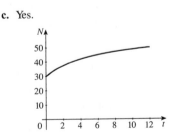

d. $N(12) = \dfrac{60(12) + 180}{12 + 6} = 50$, or 50 words/minute.

© 2011 Cengage Learning. All Rights Reserved. May not be scanned, copied or duplicated, or posted to a publicly accessible website, in whole or in part.

59. $f(t) = \dfrac{0.055t + 0.26}{t + 2}$, so $f'(t) = \dfrac{(t + 2)(0.055) - (0.055t + 0.26)(1)}{(t + 2)^2} = -\dfrac{0.15}{(t + 2)^2}$. At the beginning, the

formaldehyde level is changing at the rate of $f'(0) = -\dfrac{0.15}{4} = -0.0375$; that is, it is decreasing at the rate of

0.0375 parts per million per year. Next, $f'(3) = -\dfrac{0.15}{5^2} = -0.006$, and so the level is decreasing at the rate of

0.006 parts per million per year at the beginning of the fourth year (when $t = 3$).

61. False. Take $f(x) = x$ and $g(x) = x$. Then $f(x)g(x) = x^2$, so
$\dfrac{d}{dx}[f(x)g(x)] = \dfrac{d}{dx}(x^2) = 2x \neq f'(x)g'(x) = 1$.

63. False. Let $f(x) = x^3$. Then $\dfrac{d}{dx}\left[\dfrac{f(x)}{x^2}\right] = \dfrac{d}{dx}\left(\dfrac{x^3}{x^2}\right) = \dfrac{d}{dx}(x) = 1 \neq \dfrac{f'(x)}{2x} = \dfrac{3x^2}{2x} = \dfrac{3}{2}x$.

65. Let $f(x) = u(x)v(x)$ and $g(x) = w(x)$. Then $h(x) = f(x)g(x)$. Therefore, $h'(x) = f'(x)g(x) + f(x)g'(x)$.
But $f'(x) = u(x)v'(x) + u'(x)v(x)$, so
$$h'(x) = \left[u(x)v'(x) + u'(x)v(x)\right]g(x) + u(x)v(x)w'(x)$$
$$= u(x)v(x)w'(x) + u(x)v'(x)w(x) + u'(x)v(x)w(x).$$

3.2 Using Technology page 181

1. 0.8750   **3.** 0.0774   **5.** -0.5000 **7.** 31,312 per year

3.3 Problem-Solving Tips

1. It is often easier to find the derivative of a quotient when the numerator is a constant by using the general power rule instead of the quotient rule. For example, to find the derivative of $f(x) = -\dfrac{1}{\sqrt{2x^2 - 1}}$ in Self-Check Exercise 1 of this section, we first rewrite the function in the form $f(x) = -(2x^2 - 1)^{-1/2}$ and then use the general power rule to find the derivative.

2. To simplify a function involving the powers of an expression, factor out the lowest power of the expression. For example, to simplify $5(x + 1)^{1/2} - 3(x + 1)^{-1/2}$, factor out $(x + 1)^{-1/2}$, which is the lowest power of $x + 1$ in the expression.

3.3 Concept Questions page 189

1. The derivative of $h(x) = g(f(x))$ is equal to the derivative of g evaluated at $f(x)$ times the derivative of f.

3. $(g \circ f)'(t) = \left[(g \circ f)(t)\right]' = g'(f(t))f'(t)$ describes the rate of change of the revenue as a function of time.

© 2011 Cengage Learning. All Rights Reserved. May not be scanned, copied or duplicated, or posted to a publicly accessible website, in whole or in part.

3.3 The Chain Rule page 189

1. $f(x) = (2x-1)^4$, so $f'(x) = 4(2x-1)^3 \frac{d}{dx}(2x-1) = 4(2x-1)^3(2) = 8(2x-1)^3$.

3. $f(x) = (x^2+2)^5$, so $f'(x) = 5(x^2+2)^4(2x) = 10x(x^2+2)^4$.

5. $f(x) = (2x-x^2)^3$, so $f'(x) = 3(2x-x^2)^2 \frac{d}{dx}(2x-x^2) = 3(2x-x^2)^2(2-2x) = 6x^2(1-x)(2-x)^2$.

7. $f(x) = (2x+1)^{-2}$, so $f'(x) = -2(2x+1)^{-3} \frac{d}{dx}(2x+1) = -2(2x+1)^{-3}(2) = -4(2x+1)^{-3}$.

9. $f(x) = (x^2-4)^{3/2}$, so $f'(x) = \frac{3}{2}(x^2-4)^{1/2} \frac{d}{dx}(x^2-4) = \frac{3}{2}(x^2-4)^{1/2}(2x) = 3x(x^2-4)^{1/2}$.

11. $f(x) = \sqrt{3x-2} = (3x-2)^{1/2}$, so $f'(x) = \frac{1}{2}(3x-2)^{-1/2}(3) = \frac{3}{2}(3x-2)^{-1/2} = \dfrac{3}{2\sqrt{3x-2}}$.

13. $f(x) = \sqrt[3]{1-x^2}$, so

$$f'(x) = \frac{d}{dx}(1-x^2)^{1/3} = \frac{1}{3}(1-x^2)^{-2/3} \frac{d}{dx}(1-x^2) = \frac{1}{3}(1-x^2)^{-2/3}(-2x) = -\frac{2}{3}x(1-x^2)^{-2/3}$$

$$= \frac{-2x}{3(1-x^2)^{2/3}}.$$

15. $f(x) = \dfrac{1}{(2x+3)^3} = (2x+3)^{-3}$, so $f'(x) = -3(2x+3)^{-4}(2) = -6(2x+3)^{-4} = -\dfrac{6}{(2x+3)^4}$.

17. $f(t) = \dfrac{1}{\sqrt{2t-3}}$, so $f'(t) = \frac{d}{dt}(2t-3)^{-1/2} = -\frac{1}{2}(2t-3)^{-3/2}(2) = -(2t-3)^{-3/2} = -\dfrac{1}{(2t-3)^{3/2}}$.

19. $y = \dfrac{1}{(4x^4+x)^{3/2}}$, so $\frac{dy}{dx} = \frac{d}{dx}(4x^4+x)^{-3/2} = -\frac{3}{2}(4x^4+x)^{-5/2}(16x^3+1) = -\frac{3}{2}(16x^3+1)(4x^4+x)^{-5/2}$.

21. $f(x) = (3x^2+2x+1)^{-2}$, so

$$f'(x) = -2(3x^2+2x+1)^{-3} \frac{d}{dx}(3x^2+2x+1) = -2(3x^2+2x+1)^{-3}(6x+2)$$

$$= -4(3x+1)(3x^2+2x+1)^{-3}.$$

23. $f(x) = (x^2+1)^3 - (x^3+1)^2$, so

$$f'(x) = 3(x^2+1)^2 \frac{d}{dx}(x^2+1) - 2(x^3+1) \frac{d}{dx}(x^3+1) = 3(x^2+1)^2(2x) - 2(x^3+1)(3x^2)$$

$$= 6x\left[(x^2+1)^2 - x(x^3+1)\right] = 6x(2x^2-x+1).$$

25. $f(t) = (t^{-1} - t^{-2})^3$, so $f'(t) = 3(t^{-1} - t^{-2})^2 \frac{d}{dt}(t^{-1} - t^{-2}) = 3(t^{-1} - t^{-2})^2(-t^{-2} + 2t^{-3})$.

27. $f(x) = \sqrt{x+1} + \sqrt{x-1} = (x+1)^{1/2} + (x-1)^{1/2}$, so
$$f'(x) = \frac{1}{2}(x+1)^{-1/2}(1) + \frac{1}{2}(x-1)^{-1/2}(1) = \frac{1}{2}\left[(x+1)^{-1/2} + (x-1)^{-1/2}\right].$$

29. $f(x) = 2x^2(3-4x)^4$, so

$$f'(x) = 2x^2(4)(3-4x)^3(-4) + (3-4x)^4(4x) = 4x(3-4x)^3(-8x+3-4x)$$

$$= 4x(3-4x)^3(-12x+3) = (-12x)(4x-1)(3-4x)^3.$$

© 2011 Cengage Learning. All Rights Reserved. May not be scanned, copied or duplicated, or posted to a publicly accessible website, in whole or in part.

31. $f(x) = (x-1)^2 (2x+1)^4$, so

$$f'(x) = (x-1)^2 \frac{d}{dx} (2x+1)^4 + (2x+1)^4 \frac{d}{dx} (x-1)^2 \quad \text{(by the Product Rule)}$$

$$= (x-1)^2 (4)(2x+1)^3 \frac{d}{dx} (2x+1) + (2x+1)^4 (2)(x-1) \frac{d}{dx} (x-1)$$

$$= 8(x-1)^2 (2x+1)^3 + 2(x-1)(2x+1)^4 = 2(x-1)(2x+1)^3 (4x-4+2x+1)$$

$$= 6(x-1)(2x-1)(2x+1)^3.$$

33. $f(x) = \left(\dfrac{x+3}{x-2}\right)^3$, so

$$f'(x) = 3\left(\frac{x+3}{x-2}\right)^2 \frac{d}{dx}\left(\frac{x-3}{x-2}\right) = 3\left(\frac{x+3}{x-2}\right)^2 \left[\frac{(x-2)(1)-(x+3)(1)}{(x-2)^2}\right]$$

$$= 3\left(\frac{x+3}{x-2}\right)^2 \left[-\frac{5}{(x-2)^2}\right] = -\frac{15(x+3)^2}{(x-2)^4}.$$

35. $s(t) = \left(\dfrac{t}{2t+1}\right)^{3/2}$, so

$$s'(t) = \frac{3}{2}\left(\frac{t}{2t+1}\right)^{1/2} \frac{d}{dt}\left(\frac{t}{2t+1}\right) = \frac{3}{2}\left(\frac{t}{2t+1}\right)^{1/2} \left[\frac{(2t+1)(1)-t(2)}{(2t+1)^2}\right]$$

$$= \frac{3}{2}\left(\frac{t}{2t+1}\right)^{1/2} \left[\frac{1}{(2t+1)^2}\right] = \frac{3t^{1/2}}{2(2t+1)^{5/2}}.$$

37. $g(u) = \left(\dfrac{u+1}{3u+2}\right)^{1/2}$, so

$$g'(u) = \frac{1}{2}\left(\frac{u+1}{3u+2}\right)^{-1/2} \frac{d}{du}\left(\frac{u+1}{3u+2}\right) = \frac{1}{2}\left(\frac{u+1}{3u+2}\right)^{-1/2} \left[\frac{(3u+2)(1)-(u+1)(3)}{(3u+2)^2}\right]$$

$$= -\frac{1}{2\sqrt{u+1}\,(3u+2)^{3/2}}.$$

39. $f(x) = \dfrac{x^2}{\left(x^2-1\right)^4}$, so

$$f'(x) = \frac{\left(x^2-1\right)^4 \frac{d}{dx}\left(x^2\right) - \left(x^2\right)\frac{d}{dx}\left(x^2-1\right)^4}{\left[\left(x^2-1\right)^4\right]^2} = \frac{\left(x^2-1\right)^4 (2x) - x^2 (4)\left(x^2-1\right)^3 (2x)}{\left(x^2-1\right)^8}$$

$$= \frac{\left(x^2-1\right)^3 (2x)\left(x^2-1-4x^2\right)}{\left(x^2-1\right)^8} = \frac{(-2x)\left(3x^2+1\right)}{\left(x^2-1\right)^5}.$$

41. $h(x) = \dfrac{\left(3x^2+1\right)^3}{\left(x^2-1\right)^4}$, so

$$h'(x) = \frac{\left(x^2-1\right)^4 (3)\left(3x^2+1\right)^2 (6x) - \left(3x^2+1\right)^3 (4)\left(x^2-1\right)^3 (2x)}{\left(x^2-1\right)^8}$$

$$= \frac{2x\left(x^2-1\right)^3 \left(3x^2+1\right)^2 \left[9\left(x^2-1\right) - 4\left(3x^2+1\right)\right]}{\left(x^2-1\right)^8} = -\frac{2x\left(3x^2+13\right)\left(3x^2+1\right)^2}{\left(x^2-1\right)^5}.$$

© 2011 Cengage Learning. All Rights Reserved. May not be scanned, copied or duplicated, or posted to a publicly accessible website, in whole or in part.

43. $f(x) = \dfrac{\sqrt{2x+1}}{x^2-1}$, so

$$f'(x) = \frac{(x^2-1)\left(\frac{1}{2}\right)(2x+1)^{-1/2}(2) - (2x+1)^{1/2}(2x)}{(x^2-1)^2} = \frac{(2x+1)^{-1/2}\left[(x^2-1) - (2x+1)(2x)\right]}{(x^2-1)^2}$$

$$= -\frac{3x^2+2x+1}{\sqrt{2x+1}\,(x^2-1)^2}.$$

45. $g(t) = \dfrac{(t+1)^{1/2}}{(t^2+1)^{1/2}}$, so

$$g'(t) = \frac{(t^2+1)^{1/2}\dfrac{d}{dt}(t+1)^{1/2} - (t+1)^{1/2}\dfrac{d}{dt}(t^2+1)^{1/2}}{t^2+1}$$

$$= \frac{(t^2+1)^{1/2}\left(\frac{1}{2}\right)(t+1)^{-1/2}(1) - (t+1)^{1/2}\left(\frac{1}{2}\right)(t^2+1)^{-1/2}(2t)}{t^2+1}$$

$$= \frac{\frac{1}{2}(t+1)^{-1/2}(t^2+1)^{-1/2}\left[(t^2+1) - 2t(t+1)\right]}{t^2+1} = -\frac{t^2+2t-1}{2\sqrt{t+1}\,(t^2+1)^{3/2}}.$$

47. $f(x) = (3x+1)^4(x^2-x+1)^3$, so

$$f'(x) = (3x+1)^4\frac{d}{dx}(x^2-x+1)^3 + (x^2-x+1)^3\frac{d}{dx}(3x+1)^4$$

$$= (3x+1)^4 \cdot 3(x^2-x+1)^2(2x-1) + (x^2-x+1)^3 \cdot 4(3x+1)^3 \cdot 3$$

$$= 3(3x+1)^3(x^2-x+1)^2\left[(3x+1)(2x-1) + 4(x^2-x+1)\right]$$

$$= 3(3x+1)^3(x^2-x+1)^2\left(6x^2-3x+2x-1+4x^2-4x+4\right)$$

$$= 3(3x+1)^3(x^2-x+1)^2\left(10x^2-5x+3\right).$$

49. $y = g(u) = u^{4/3}$, so $\dfrac{dy}{du} = \frac{4}{3}u^{1/3}$, and $u = f(x) = 3x^2-1$, so $\dfrac{du}{dx} = 6x$. Thus,

$$\frac{dy}{dx} = \frac{dy}{du}\cdot\frac{du}{dx} = \frac{4}{3}u^{1/3}(6x) = \frac{4}{3}(3x^2-1)^{1/3}6x = 8x(3x^2-1)^{1/3}.$$

51. $\dfrac{dy}{du} = -\frac{2}{3}u^{-5/3} = -\dfrac{2}{3u^{5/3}}$ and $\dfrac{du}{dx} = 6x^2-1$, so $\dfrac{dy}{dx} = \dfrac{dy}{du}\cdot\dfrac{du}{dx} = -\dfrac{2(6x^2-1)}{3u^{5/3}} = -\dfrac{2(6x^2-1)}{3(2x^3-x+1)^{5/3}}.$

53. $\dfrac{dy}{du} = \frac{1}{2}u^{-1/2} - \frac{1}{2}u^{-3/2}$ and $\dfrac{du}{dx} = 3x^2-1$, so

$$\frac{dy}{dx} = \frac{dy}{du}\cdot\frac{du}{dx} = \left[\frac{1}{2\sqrt{x^3-x}} - \frac{1}{2(x^3-x)^{3/2}}\right](3x^2-1) = \frac{(3x^2-1)(x^3-x-1)}{2(x^3-x)^{3/2}}.$$

55. $F(x) = g(f(x))$, so $F'(x) = g'(f(x))f'(x)$. Thus, $F'(2) = g'(3)(-3) = (4)(-3) = -12.$

57. Let $g(x) = x^2+1$. Then $F(x) = f(g(x))$. Next, $F'(x) = f'(g(x))g'(x)$ and
$F'(1) = f'(2)(2x) = (3)(2) = 6.$

59. No. Suppose $h = g(f(x))$. Let $f(x) = x$ and $g(x) = x^2$. Then $h = g(f(x)) = g(x) = x^2$ and
$h'(x) = 2x \neq g'(f'(x)) = g'(1) = 2(1) = 2.$

© 2011 Cengage Learning. All Rights Reserved. May not be scanned, copied or duplicated, or posted to a publicly accessible website, in whole or in part.

61. $f(x) = (1-x)(x^2-1)^2$, so

$f'(x) = (1-x)2(x^2-1)(2x) + (-1)(x^2-1)^2 = (x^2-1)(4x-4x^2-x^2+1) = (x^2-1)(-5x^2+4x+1)$.
Therefore, the slope of the tangent line at $(2, -9)$ is $f'(2) = [(2)^2-1][-5(2)^2+4(2)+1] = -33$. Then an equation of the line is $y + 9 = -33(x-2)$, or $y = -33x + 57$.

63. $f(x) = x\sqrt{2x^2+7}$, so $f'(x) = \sqrt{2x^2+7} + x\left(\frac{1}{2}\right)(2x^2+7)^{-1/2}(4x)$. The slope of the tangent line at $x = 3$ is

$f'(3) = \sqrt{25} + \left(\frac{3}{2}\right)(25)^{-1/2}(12) = \frac{43}{5}$, so an equation is $y - 15 = \frac{43}{5}(x-3)$, or $y = \frac{43}{5}x - \frac{54}{5}$.

65. $N(t) = (60+2t)^{2/3}$, so $N'(t) = \frac{2}{3}(60+2t)^{-1/3}\frac{d}{dt}(60+2t) = \frac{4}{3}(60+2t)^{-1/3}$. The rate of

increase at the end of the second week is $N'(2) = \frac{4}{3}(64)^{-1/3} = \frac{1}{3}$, or $\frac{1}{3}$ million/week. At the end

of the 12th week, $N'(12) = \frac{4}{3}(84)^{-1/3} \approx 0.3$, or 0.3 million/week. The number of viewers in the

2nd week is $N(2) = (60+4)^{2/3} = 16$, or 16 million, and the number of viewers in the 24th week is

$N(24) = (60+48)^{2/3} \approx 22.7$, or approximately 22.7 million.

67. $P(t) = 33.55(t+5)^{0.205}$. $P'(t) = 33.55(0.205)(t+5)^{-0.795}(1) = 6.87775(t+5)^{-0.795}$. The rate of change at

the beginning of 2000 is $P'(20) = 6.87775(25)^{-0.795} \approx 0.5322$, or 0.53%/yr. The percent of these mothers was

$P(20) = 33.55(25)^{0.205} \approx 64.90$, or 64.9%.

69. a. $f(t) = 23.7(0.2t+1)^{1.32}$. The rate of change at any time t is given by

$f'(t) = (23.7)(1.32)(0.2t+1)^{0.32}(0.2) = 6.2568(0.2t+1)^{0.32}$. At the beginning of 2000, the rate of change

is $f'(9) = 6.2568[0.2(9)+1]^{0.32} \approx 8.6985$, or approximately \$8.7 billion/yr.

b. The assets were $f(9) \approx 92.256$, or approximately \$92.3 billion.

71. $C(t) = 0.01(0.2t^2+4t+64)^{2/3}$.

a. $C'(t) = 0.01\left(\frac{2}{3}\right)(0.2t^2+4t+64)^{-1/3}\frac{d}{dt}(0.2t^2+4t+64)$

$= (0.01)(0.667)(0.4t+4)(0.2t^2+4t+64)^{-1/3} = 0.027(0.1t+1)(0.2t^2+4t+64)^{-1/3}$.

b. $C'(5) = 0.027(0.5+1)[0.2(25)+4(5)+64]^{-1/3} \approx 0.009$, or 0.009 parts per million per year.

73. a. $A(t) = 0.03t^3(t-7)^4 + 60.2$, so

$A'(t) = 0.03[3t^2(t-7)^4 + t^3(4)(t-7)^3] = 0.03t^2(t-7)^3[3(t-7)+4t] = 0.21t^2(t-3)(t-7)^3$.

b. $A'(1) = 0.21(-2)(-6)^3 = 90.72$, $A'(3) = 0$, and $A'(4) = 0.21(16)(1)(-3)^3 = -90.72$. The amount of

pollutant is increasing at the rate of 90.72 units/hr at 8 a.m. The rate of change is 0 units/hr at 10 a.m. and

-90.72 units/hr at 11 a.m.

© 2011 Cengage Learning. All Rights Reserved. May not be scanned, copied or duplicated, or posted to a publicly accessible website, in whole or in part.

75. $P(t) = \dfrac{300\sqrt{\frac{1}{2}t^2 + 2t + 25}}{t + 25} = \dfrac{300\left(\frac{1}{2}t^2 + 2t + 25\right)^{1/2}}{t + 25}$, so

$P'(t) = 300\left[\dfrac{(t + 25)\frac{1}{2}\left(\frac{1}{2}t^2 + 2t + 25\right)^{-1/2}(t + 2) - \left(\frac{1}{2}t^2 + 2t + 25\right)^{1/2}(1)}{(t + 25)^2}\right]$

$= 300\left[\dfrac{\left(\frac{1}{2}t^2 + 2t + 25\right)^{-1/2}\left[(t + 25)(t + 2) - 2\left(\frac{1}{2}t^2 + 2t + 25\right)\right]}{2(t + 25)^2}\right] = \dfrac{3450t}{(t + 25)^2\sqrt{\frac{1}{2}t^2 + 2t + 25}}$.

Ten seconds into the run, the athlete's pulse rate is increasing at $P'(10) = \dfrac{3450(10)}{(35)^2\sqrt{50 + 20 + 25}} \approx 2.9$,

or approximately 2.9 beats per minute per minute. Sixty seconds into the run, it is increasing at

$P'(60) = \dfrac{3450(60)}{(85)^2\sqrt{1800 + 120 + 25}} \approx 0.65$, or approximately 0.7 beats per minute per minute. Two minutes into

the run, it is increasing at $P'(120) = \dfrac{3450(120)}{(145)^2\sqrt{7200 + 240 + 25}} \approx 0.23$, or approximately 0.2 beats per minute

per minute. The pulse rate two minutes into the run is given by $P(120) = \dfrac{300\sqrt{7200 + 240 + 25}}{120 + 25} \approx 178.8$, or

approximately 179 beats per minute.

77. The area is given by $A = \pi r^2$. The rate at which the area is increasing is given by dA/dt, that is,

$\dfrac{dA}{dt} = \dfrac{d}{dt}\left(\pi r^2\right) = \dfrac{d}{dt}\left(\pi r^2\right)\dfrac{dr}{dt} = 2\pi r\dfrac{dr}{dt}$. If $r = 40$ and $dr/dt = 2$, then $\dfrac{dA}{dt} = 2\pi(40)(2) = 160\pi$, that is, it is

increasing at the rate of 160π, or approximately 503 ft²/sec.

79. $f(t) = 6.25t^2 + 19.75t + 74.75$. $g(x) = -0.00075x^2 + 67.5$, so $\dfrac{dS}{dt} = g'(x)f'(t) = (-0.0015x)(12.5t + 19.75)$.

When $t = 4$, we have $x = f(4) = 6.25(16) + 19.75(4) + 74.75 = 253.75$ and

$\left.\dfrac{dS}{dt}\right|_{t=4} = (-0.0015)(253.75)[12.5(4) + 19.75] \approx -26.55$; that is, the average speed will be dropping

at the rate of approximately 27 mph per decade. The average speed of traffic flow at that time will be

$S = g(f(4)) = -0.00075(253.75^2) + 67.5 \approx 19.2$, or approximately 19 mph.

81. $N(x) = 1.42x$ and $x(t) = \dfrac{7t^2 + 140t + 700}{3t^2 + 80t + 550}$. The number of construction jobs as a function of time is

$n(t) = N(x(t))$. Using the Chain Rule,

$n'(t) = \dfrac{dN}{dx}\cdot\dfrac{dx}{dt} = 1.42\dfrac{dx}{dt} = (1.42)\left[\dfrac{(3t^2 + 80t + 550)(14t + 140) - (7t^2 + 140t + 700)(6t + 80)}{(3t^2 + 80t + 550)^2}\right]$

$= \dfrac{1.42\left(140t^2 + 3500t + 21000\right)}{(3t^2 + 80t + 550)^2}$.

$n'(12) = \dfrac{1.42\left[140(12)^2 + 3500(12) + 21000\right]}{\left[3(12)^2 + 80(12) + 550\right]^2} \approx 0.0313115$, or approximately 31,312 jobs/year.

© 2011 Cengage Learning. All Rights Reserved. May not be scanned, copied or duplicated, or posted to a publicly accessible website, in whole or in part.

83. $x = f(p) = 10\sqrt{\dfrac{50 - p}{p}}$, so

$$\frac{dx}{dp} = \frac{d}{dp}\left[10\left(\frac{50 - p}{p}\right)^{1/2}\right] = (10)\left(\tfrac{1}{2}\right)\left(\frac{50 - p}{p}\right)^{-1/2}\frac{d}{dp}\left(\frac{50 - p}{p}\right)$$

$$= 5\left(\frac{50 - p}{p}\right)^{-1/2}\cdot\frac{d}{dp}\left(\frac{50}{p} - 1\right) = 5\left(\frac{50 - p}{p}\right)^{-1/2}\left(-\frac{50}{p^2}\right)$$

$$= -\frac{250}{p^2\left(\frac{50 - p}{p}\right)^{1/2}}\quad\frac{dx}{dp}\Bigg|_{p=25} = -\frac{250}{p^2\left(\frac{50 - p}{p}\right)^{1/2}} = -\frac{250}{(625)\left(\frac{25}{25}\right)^{1/2}} = -0.4.$$

Thus, the quantity demanded is falling at the rate of 0.4 (1000) or 400 wristwatches per dollar increase in price.

85. True. This is just the statement of the Chain Rule.

87. True. $\dfrac{d}{dx}\sqrt{f(x)} = \dfrac{d}{dx}\left[f(x)\right]^{1/2} = \tfrac{1}{2}\left[f(x)\right]^{-1/2}f'(x) = \dfrac{f'(x)}{2\sqrt{f(x)}}$.

89. Let $f(x) = x^{1/n}$ so that $\left[f(x)\right]^n = x$. Differentiating both sides with respect to x, we get $n\left[f(x)\right]^{n-1}f'(x) = 1$,

so $f'(x) = \dfrac{1}{n\left[f(x)\right]^{n-1}} = \dfrac{1}{n\left[x^{1/n}\right]^{n-1}} = \dfrac{1}{nx^{1-(1/n)}} = \dfrac{1}{n}x^{(1/n)-1}$, as was to be shown.

3.3   Using Technology   page 193

1. 0.5774

3. 0.9390

5. −4.9498

7. a. 10,146,200/decade

b. 7,810,520/decade

3.4   Problem-Solving Tips

1. **The marginal cost function is the derivative of the cost function.** Similarly, the marginal profit function and the marginal revenue function are the derivatives of the profit function and the revenue function, respectively. The key word here is "marginal": it indicates that we are dealing with the derivative of a function.

2. **The average cost function** is given by $\overline{C}(x) = \dfrac{C(x)}{x}$ and the **marginal average cost function** is given by $\overline{C}'(x)$.

3. Remember that the revenue is increasing on an interval where the demand is inelastic, decreasing on an interval where the demand is elastic, and stationary at the point where the demand is unitary.

3.4   Concept Questions   page 204

1. a. The marginal cost function is the derivative of the cost function.

   b. The average cost function is equal to the total cost function divided by the total number of the commodity produced.

© 2011 Cengage Learning. All Rights Reserved. May not be scanned, copied or duplicated, or posted to a publicly accessible website, in whole or in part.

c. The marginal average cost function is the derivative of the average cost function.

d. The marginal revenue function is the derivative of the revenue function.

e. The marginal profit function is the derivative of the profit function.

3.4 Marginal Functions in Economics   page 204

1. a. $C(x)$ is always increasing because as x, the number of units produced, increases, the amount of money that must be spent on production also increases.

b. This occurs at $x = 4$, a production level of 4000. You can see this by looking at the slopes of the tangent lines for x less than, equal to, and a little larger then $x = 4$.

3. a. The actual cost incurred in the production of the 1001st disc is given by

$$C(1001) - C(1000) = \left[2000 + 2(1001) - 0.0001(1001)^2\right] - \left[2000 + 2(1000) - 0.0001(1000)^2\right]$$

$$= 3901.7999 - 3900 = 1.7999, \text{ or approximately } \$1.80.$$

The actual cost incurred in the production of the 2001st disc is given by

$$C(2001) - C(2000) = \left[2000 + 2(2001) - 0.0001(2001)^2\right] - \left[2000 + 2(2000) - 0.0001(2000)^2\right]$$

$$= 5601.5999 - 5600 = 1.5999, \text{ or approximately } \$1.60.$$

b. The marginal cost is $C'(x) = 2 - 0.0002x$. In particular, $C'(1000) = 2 - 0.0002(1000) = 1.80$ and $C'(2000) = 2 - 0.0002(2000) = 1.60$.

5. a. $\overline{C}(x) = \dfrac{C(x)}{x} = \dfrac{100x + 200{,}000}{x} = 100 + \dfrac{200{,}000}{x}.$

b. $\overline{C}'(x) = \dfrac{d}{dx}(100) + \dfrac{d}{dx}(200{,}000x^{-1}) = -200{,}000x^{-2} = -\dfrac{200{,}000}{x^2}.$

c. $\displaystyle\lim_{x \to \infty} \overline{C}(x) = \lim_{x \to \infty}\left(100 + \dfrac{200{,}000}{x}\right) = 100.$ This says that the average cost approaches $100 per unit if the production level is very high.

7. $\overline{C}(x) = \dfrac{C(x)}{x} = \dfrac{2000 + 2x - 0.0001x^2}{x} = \dfrac{2000}{x} + 2 - 0.0001x,$ so

$\overline{C}'(x) = -\dfrac{2000}{x^2} + 0 - 0.0001 = -\dfrac{2000}{x^2} - 0.0001.$

9. a. $R'(x) = \dfrac{d}{dx}(8000x - 100x^2) = 8000 - 200x.$

b. $R'(39) = 8000 - 200(39) = 200,\ R'(40) = 8000 - 200(40) = 0,$ and $R'(41) = 8000 - 200(41) = -200.$

c. This suggests the total revenue is maximized if the price charged per passenger is $40.

11. a. $P(x) = R(x) - C(x) = (-0.04x^2 + 800x) - (200x + 300{,}000) = -0.04x^2 + 600x - 300{,}000.$

b. $P'(x) = -0.08x + 600.$

c. $P'(5000) = -0.08(5000) + 600 = 200$ and $P'(8000) = -0.08(8000) + 600 = -40.$

© 2011 Cengage Learning. All Rights Reserved. May not be scanned, copied or duplicated, or posted to a publicly accessible website, in whole or in part.

d.

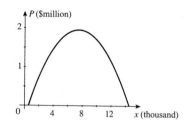

The profit realized by the company increases as production increases, peaking at a production level of 7500 units. Beyond this level of production, the profit begins to fall.

13. a. The revenue function is $R(x) = px = (600 - 0.05x)x = 600x - 0.05x^2$ and the profit function is

$$P(x) = R(x) - C(x) = (600x - 0.05x^2) - (0.000002x^3 - 0.03x^2 + 400x + 80,000)$$

$$= -0.000002x^3 - 0.02x^2 + 200x - 80,000.$$

b. $C'(x) = \dfrac{d}{dx}(0.000002x^3 - 0.03x^2 + 400x + 80,000) = 0.000006x^2 - 0.06x + 400,$

$R'(x) = \dfrac{d}{dx}(600x - 0.05x^2) = 600 - 0.1x,$ and

$P'(x) = \dfrac{d}{dx}(-0.000002x^3 - 0.02x^2 + 200x - 80,000) = -0.000006x^2 - 0.04x + 200.$

c. $C'(2000) = 0.000006(2000)^2 - 0.06(2000) + 400 = 304$, and this says that at a production level of 2000 units, the cost for producing the 2001st unit is $304. $R'(2000) = 600 - 0.1(2000) = 400$, and this says that the revenue realized in selling the 2001st unit is $400. $P'(2000) = R'(2000) - C'(2000) = 400 - 304 = 96$, and this says that the revenue realized in selling the 2001st unit is $96.

d.

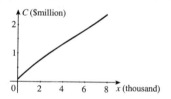 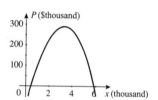

15. $\overline{C}(x) = \dfrac{C(x)}{x} = \dfrac{0.000002x^3 - 0.03x^2 + 400x + 80,000}{x} = 0.000002x^2 - 0.03x + 400 + \dfrac{80,000}{x}.$

a. $\overline{C}'(x) = 0.000004x - 0.03 - \dfrac{80,000}{x^2}.$

b. $\overline{C}'(5000) = 0.000004(5000) - 0.03 - \dfrac{80,000}{5000^2} \approx -0.0132$, and this

says that at a production level of 5000 units, the average cost of production is dropping at the rate of approximately a penny per unit.

c.

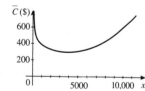

$\overline{C}'(10,000) = 0.000004(10000) - 0.03 - \dfrac{80,000}{10,000^2} \approx 0.0092,$

and this says that, at a production level of 10,000 units, the average cost of production is increasing at the rate of approximately a penny per unit.

17. a. $R(x) = px = \dfrac{50x}{0.01x^2 + 1}.$

b. $R'(x) = \dfrac{(0.01x^2 + 1)50 - 50x(0.02x)}{(0.01x^2 + 1)^2} = \dfrac{50 - 0.5x^2}{(0.01x^2 + 1)^2}.$

© 2011 Cengage Learning. All Rights Reserved. May not be scanned, copied or duplicated, or posted to a publicly accessible website, in whole or in part.

c. $R'(2) = \dfrac{50 - 0.5(4)}{[0.01(4) + 1]^2} \approx 44.379$. This result says that at a sales level of 2000 units, the revenue increases at

the rate of approximately \$44,379 per 1000 units.

19. $C(x) = 0.873x^{1.1} + 20.34$, so $C'(x) = 0.873(1.1)x^{0.1}$. $C'(10) = 0.873(1.1)(10)^{0.1} = 1.21$.

21. The consumption function is given by $C(x) = 0.712x + 95.05$. The marginal propensity to consume is given by
$\frac{dC}{dx} = 0.712$. The marginal propensity to save is given by$\frac{dS}{dx} = 1 - \frac{dC}{dx} = 1 - 0.712 = 0.288$.

23. Here $x = f(p) = -\frac{5}{4}p + 20$ and so $f'(p) = -\frac{5}{4}$. Therefore, $E(p) = -\dfrac{pf'(p)}{f(p)} = -\dfrac{p\left(-\frac{5}{4}\right)}{-\frac{5}{4}p + 20} = \dfrac{5p}{80 - 5p}$.

$E(10) = \dfrac{5(10)}{80 - 5(10)} = \dfrac{50}{30} = \dfrac{5}{3} > 1$, and so the demand is elastic.

25. $f(p) = -\frac{1}{3}p + 20$, so $f'(p) = -\frac{1}{3}$. Then the elasticity of demand is given by $E(p) = -\dfrac{p\left(-\frac{1}{3}\right)}{-\frac{1}{3}p + 20}$, and

$E(30) = -\dfrac{30\left(-\frac{1}{3}\right)}{-\frac{1}{3}(30) + 20} = 1$, and we conclude that the demand is unitary at this price.

27. $x^2 = 169 - p$ and $f(p) = (169 - p)^{1/2}$. Next, $f'(p) = \frac{1}{2}(169 - p)^{-1/2}(-1) = -\frac{1}{2}(169 - p)^{-1/2}$. Then the

elasticity of demand is given by $E(p) = -\dfrac{pf'(p)}{f(p)} = -\dfrac{p\left(-\frac{1}{2}\right)(169 - p)^{-1/2}}{(169 - p)^{1/2}} = \dfrac{\frac{1}{2}p}{169 - p}$. Therefore, when

$p = 29$, $E(p) = \dfrac{\frac{1}{2}(29)}{169 - 29} = \dfrac{14.5}{140} = 0.104$. Because $E(p) < 1$, we conclude that demand is inelastic at this

price.

29. $f(p) = \frac{1}{5}(225 - p^2)$, so $f'(p) = \frac{1}{5}(-2p) = -\frac{2}{5}p$. Then the elasticity of demand is given by

$E(p) = -\dfrac{pf'(p)}{f(p)} = -\dfrac{p\left(-\frac{2}{5}p\right)}{\frac{1}{5}(225 - p^2)} = \dfrac{2p^2}{225 - p^2}$.

a. When $p = 8$, $E(8) = \dfrac{2(64)}{225 - 64} \approx 0.8 < 1$ and the demand is inelastic. When $p = 10$,

$E(10) = \dfrac{2(100)}{225 - 100} = 1.6 > 1$ and the demand is elastic.

b. The demand is unitary when $E = 1$. Solving $\dfrac{2p^2}{225 - p^2} = 1$, we find $2p^2 = 225 - p^2$, $3p^2 = 225$, and

$p \approx 8.66$. So the demand is unitary when $p \approx 8.66$.

c. Because demand is elastic when $p = 10$, lowering the unit price will cause the revenue to increase.

d. Because the demand is inelastic at $p = 8$, a slight increase in the unit price will cause the revenue to increase.

31. $f(p) = \frac{2}{3}(36 - p^2)^{1/2}$. $f'(p) = \frac{2}{3}\left(\frac{1}{2}\right)(36 - p^2)^{-1/2}(-2p) = -\frac{2}{3}p(36 - p^2)^{-1/2}$. Then the elasticity of

demand is given by $E(p) = -\dfrac{pf'(p)}{f(p)} = -\dfrac{-\frac{2}{3}p(36 - p^2)^{-1/2}p}{\frac{2}{3}(36 - p^2)^{1/2}} = \dfrac{p^2}{36 - p^2}$.

a. When $p = 2$, $E(2) = \dfrac{4}{36 - 4} = \dfrac{1}{8} < 1$, and we conclude that the demand is inelastic.

© 2011 Cengage Learning. All Rights Reserved. May not be scanned, copied or duplicated, or posted to a publicly accessible website, in whole or in part.

b. Because the demand is inelastic, the revenue will increase when the rental price is increased.

33. We first solve the demand equation for x in terms of p. Thus, $p = \sqrt{9 - 0.02x}$, and $p^2 = 9 - 0.02x$, or

$x = -50p^2 + 450$. With $f(p) = -50p^2 + 450$, we find $E(p) = -\dfrac{pf'(p)}{f(p)} = -\dfrac{p(-100p)}{-50p^2 + 450} = \dfrac{2p^2}{9 - p^2}$.

Setting $E(p) = 1$ gives $2p^2 = 9 - p^2$, so $p = \sqrt{3}$. So the demand is inelastic in $\left(0, \sqrt{3}\right)$, unitary when $p = \sqrt{3}$,

and elastic in $\left(\sqrt{3}, 3\right)$.

35. True. $\overline{C}'(x) = \dfrac{d}{dx}\left[\dfrac{C(x)}{x}\right] = \dfrac{xC'(x) - C(x)\dfrac{d}{dx}(x)}{x^2} = \dfrac{xC'(x) - C(x)}{x^2}$.

3.5   Problem-Solving Tips

When you work applied problems, keep track of the units of measure used. For example, if velocity is measured in ft/sec, then the units of acceleration will be ft/sec^2. If you are working an applied problem and the units in your answer are not appropriate, it may indicate that you have made an error in your calculations or in the formulation of the problem.

Here are some tips for solving the problems in the exercises that follow:

1. Make sure to simplify expressions before finding higher order derivatives.

2. The velocity of an object moving in a straight path is given by the derivative of the position function for that object. The acceleration of the object is given by the derivative of the velocity function.

3.5   Concept Questions   page 211

1. **a.** The second derivative of f is the derivative of f'.
   b. To find the second derivative of f, we differentiate f'.

3. **a.** $f'(t) > 0$ and $f''(t) > 0$ in (a, b).   **b.** $f'(t) > 0$ and $f''(t) < 0$ in (a, b).
   c. $f'(t) < 0$ and $f''(t) < 0$ in (a, b).   **d.** $f'(t) < 0$ and $f''(t) > 0$ in (a, b).

3.5   Higher-Order Derivatives   page 212

1. $f(x) = 4x^2 - 2x + 1$, so $f'(x) = 8x - 2$ and $f''(x) = 8$.

3. $f(x) = 2x^3 - 3x^2 + 1$, so $f'(x) = 6x^2 - 6x$ and $f''(x) = 12x - 6 = 6(2x - 1)$.

5. $h(t) = t^4 - 2t^3 + 6t^2 - 3t + 10$, so $h'(t) = 4t^3 - 6t^2 + 12t - 3$ and $h''(x) = 12t^2 - 12t + 12 = 12\left(t^2 - t + 1\right)$.

© 2011 Cengage Learning. All Rights Reserved. May not be scanned, copied or duplicated, or posted to a publicly accessible website, in whole or in part.

7. $f(x) = (x^2 + 2)^5$, so $f'(x) = 5(x^2 + 2)^4 (2x) = 10x(x^2 + 2)^4$ and

$\quad f''(x) = 10(x^2 + 2)^4 + 10x(4)(x^2 + 2)^3 (2x) = 10(x^2 + 2)^3 [(x^2 + 2) + 8x^2] = 10(9x^2 + 2)(x^2 + 2)^3$.

9. $g(t) = (2t^2 - 1)^2 (3t^2)$, so

$\quad g'(t) = 2(2t^2 - 1)(4t)(3t^2) - (2t^2 - 1)^2 (6t) = 6t(2t^2 - 1)[4t^2 + (2t^2 - 1)] = 6t(2t^2 - 1)(6t^2 - 1)$

$\qquad = 6t(12t^4 - 8t^2 + 1) = 72t^5 - 48t^3 + 6t$

and $g''(t) = 360t^4 - 144t^2 + 6 = 6(60t^4 - 24t^2 + 1)$.

11. $f(x) = (2x^2 + 2)^{7/2}$, so $f'(x) = \frac{7}{2}(2x^2 + 2)^{5/2}(4x) = 14x(2x^2 + 2)^{5/2}$ and

$\quad f''(x) = 14(2x^2 + 2)^{5/2} + 14x\left(\frac{5}{2}\right)(2x^2 + 2)^{3/2}(4x) = 14(2x^2 + 2)^{3/2}[(2x^2 + 2) + 10x^2]$

$\qquad = 28(6x^2 + 1)(2x^2 + 2)^{3/2}$.

13. $f(x) = x(x^2 + 1)^2$, so

$\quad f'(x) = (x^2 + 1)^2 + x(2)(x^2 + 1)(2x) = (x^2 + 1)[(x^2 + 1) + 4x^2] = (x^2 + 1)(5x^2 + 1)$ and

$\quad f''(x) = 2x(5x^2 + 1) + (x^2 + 1)(10x) = 2x(5x^2 + 1 + 5x^2 + 5) = 4x(5x^2 + 3)$.

15. $f(x) = \dfrac{x}{2x + 1}$, so $f'(x) = \dfrac{(2x + 1)(1) - x(2)}{(2x + 1)^2} = \dfrac{1}{(2x + 1)^2}$ and

$\quad f''(x) = \dfrac{d}{dx}(2x + 1)^{-2} = -2(2x + 1)^{-3}(2) = -\dfrac{4}{(2x + 1)^3}$.

17. $f(s) = \dfrac{s - 1}{s + 1}$, so $f'(s) = \dfrac{(s + 1)(1) - (s - 1)(1)}{(s + 1)^2} = \dfrac{2}{(s + 1)^2}$ and

$\quad f''(s) = 2\dfrac{d}{ds}(s + 1)^{-2} = -4(s + 1)^{-3} = -\dfrac{4}{(s + 1)^3}$.

19. $f(u) = \sqrt{4 - 3u} = (4 - 3u)^{1/2}$, so $f'(u) = \frac{1}{2}(4 - 3u)^{-1/2}(-3) = -\dfrac{3}{2\sqrt{4 - 3u}}$ and

$\quad f''(u) = -\dfrac{3}{2}\cdot\dfrac{d}{du}(4 - 3u)^{-1/2} = -\dfrac{3}{2}\left(-\dfrac{1}{2}\right)(4 - 3u)^{-3/2}(-3) = -\dfrac{9}{4(4 - 3u)^{3/2}}$.

21. $f(x) = 3x^4 - 4x^3$, so $f'(x) = 12x^3 - 12x^2$, $f''(x) = 36x^2 - 24x$, and $f'''(x) = 72x - 24$.

23. $f(x) = \dfrac{1}{x}$, so $f'(x) = \dfrac{d}{dx}(x^{-1}) = -x^{-2} = -\dfrac{1}{x^2}$, $f''(x) = 2x^{-3} = \dfrac{2}{x^3}$, and $f'''(x) = -6x^{-4} = -\dfrac{6}{x^4}$.

25. $g(s) = (3s - 2)^{1/2}$, so $g'(s) = \frac{1}{2}(3s - 2)^{-1/2}(3) = \dfrac{3}{2(3s - 2)^{1/2}}$,

$\quad g''(s) = \frac{3}{2}\left(-\frac{1}{2}\right)(3s - 2)^{-3/2}(3) = -\frac{9}{4}(3s - 2)^{-3/2} = -\dfrac{9}{4(3s - 2)^{3/2}}$, and

$\quad g'''(s) = \frac{27}{8}(3s - 2)^{-5/2}(3) = \frac{81}{8}(3s - 2)^{-5/2} = \dfrac{81}{8(3s - 2)^{5/2}}$.

27. $f(x) = (2x - 3)^4$, so $f'(x) = 4(2x - 3)^3(2) = 8(2x - 3)^3$, $f''(x) = 3(2x - 3)^2(2) = 48(2x - 3)^2$, and

$\quad f'''(x) = 96(2x - 3)(2) = 192(2x - 3)$.

© 2011 Cengage Learning. All Rights Reserved. May not be scanned, copied or duplicated, or posted to a publicly accessible website, in whole or in part.

29. Its velocity at any time t is $v(t) = \frac{d}{dt}(16t^2) = 32t$. The hammer strikes the ground when $16t^2 = 256$ or $t = 4$ (we reject the negative root). Therefore, its velocity at the instant it strikes the ground is $v(4) = 32(4) = 128$ ft/sec. Its acceleration at time t is $a(t) = \frac{d}{dt}(32t) = 32$. In particular, its acceleration at $t = 4$ is 32 ft/sec^2.

31. $N(t) = -0.1t^3 + 1.5t^2 + 100$.

    a. $N'(t) = -0.3t^2 + 3t = 0.3t(10 - t)$. Because $N'(t) > 0$ for $t = 0, 1, 2, \ldots, 7$, it is evident that $N(t)$ (and therefore the crime rate) was increasing from 1998 through 2005.

    b. $N''(t) = -0.6t + 3 = 0.6(5 - t)$. Now $N''(4) = 0.6 > 0$, $N''(5) = 0$, $N''(6) = -0.6 < 0$, and $N''(7) = -1.2 < 0$. This shows that the rate of the rate of change was decreasing beyond $t = 5$ (in the year 2000). This shows that the program was working.

33. $N(t) = 0.00037t^3 - 0.0242t^2 + 0.52t + 5.3$ for $0 \leq t \leq 10$, so $N'(t) = 0.00111t^2 - 0.0484t + 0.52$ and $N''(t) = 0.00222t - 0.0484$. Therefore, $N(8) = 0.00037(8)^3 - 0.0242(8)^2 + 0.52(8) + 5.3 \approx 8.1$, $N'(8) = 0.00111(8)^2 - 0.0484(8) + 0.52 \approx 0.204$, and $N''(8) = 0.00222(8) - 0.0484 \approx -0.031$. We conclude that at the beginning of 1998, there were 8.1 million persons receiving disability benefits, the number was increasing at the rate of 0.2 million/yr, and the rate of the rate of change of the number of persons was decreasing at the rate of 0.03 million persons/yr^2.

35. **a.** $h(t) = \frac{1}{16}t^4 - t^3 + 4t^2$, so $h'(t) = \frac{1}{4}t^3 - 3t^2 + 8t$.

    b. $h'(0) = 0$, or 0 ft/sec. $h'(4) = \frac{1}{4}(64) - 3(16) + 8(4) = 0$, or 0 ft/sec, and $h'(8) = \frac{1}{4}(8)^3 - 3(64) + 8(8) = 0$, or 0 ft/sec.

    c. $h''(t) = \frac{3}{4}t^2 - 6t + 8$.

    d. $h''(0) = 8$ ft/sec^2, $h''(4) = \frac{3}{4}(16) - 6(4) + 8 = -4$ ft/sec^2, and $h''(8) = \frac{3}{4}(64) - 6(8) + 8 = 8$ ft/sec^2.

    e. $h(0) = 0$ ft, $h(4) = \frac{1}{16}(4)^4 - (4)^3 + 4(4)^2 = 16$ ft, and $h(8) = \frac{1}{16}(8)^4 - (8)^3 + 4(8)^2 = 0$ ft.

37. $f(t) = 10.72(0.9t + 10)^{0.3}$, so $f'(t) = 10.72(0.3)(0.9t + 10)^{-0.7}(0.9) = 2.8944(0.9t + 10)^{-0.7}$ and $f''(t) = 2.8944(-0.7)(0.9t + 10)^{-1.7}(0.9) = -1.823472(0.9t + 10)^{-1.7}$. Thus, $f''(10) = -1.823472(19)^{-1.7} \approx -0.01222$, which says that the rate of the rate of change of the population is decreasing at the rate of 0.01%/yr^2.

39. False. If f has derivatives of order two at $x = a$, then $f''(a) = \left[f'(x)\right]'\big|_{x=a}$.

41. True. If $f(x)$ is a polynomial function of degree n, then $f^{(n+1)}(x) = 0$.

43. True. Using the Chain Rule, $h'(x) = f'(2x) \cdot \frac{d}{dx}(2x) = f'(x) \cdot 2 = 2f'(2x)$. Using the Chain Rule again, $h''(x) = 2f''(2x) \cdot 2 = 4f''(2x)$.

45. Consider the function $f(x) = x^{(2n+1)/2} = x^{n+(1/2)}$. We calculate $f'(x) = \left(n + \frac{1}{2}\right)x^{n-(1/2)}$, $f''(x) = \left(n + \frac{1}{2}\right)\left(n - \frac{1}{2}\right)x^{n-(3/2)}$, $\ldots$, $f^{(n)}(x) = \left(n + \frac{1}{2}\right)\left(n - \frac{1}{2}\right)\cdots\frac{3}{2}x^{1/2}$, and $f^{(n+1)}(x) = \left(n + \frac{1}{2}\right)\left(n - \frac{1}{2}\right)\cdots\frac{1}{2}x^{-1/2}$. The first n derivatives exist at $x = 0$, but the $(n + 1)$st derivative fails to be defined there.

© 2011 Cengage Learning. All Rights Reserved. May not be scanned, copied or duplicated, or posted to a publicly accessible website, in whole or in part.

3.5   Using Technology   page 215

1. -18 **3.** 15.2762 **5.** -0.6255 **7.** 0.1973

9. $f''(6) = -68.46214$. This tells us that at the beginning of 1988, the rate of the rate at which banks were failing was decreasing at 68 banks per year per year.

3.6   Problem-Solving Tips

1. If an equation expresses y implicitly as a function of x, then we can use implicit differentiation to find its derivative. We apply the chain rule to find the derivative of any term involving y. (Note that the derivative of any term involving y will include the factor dy/dx.) The terms involving only x are differentiated in the usual manner.

2. Guideline 3 for solving related rates problems, on page 221 of the text, asks you to find an equation giving the relationship between the variables in the related rates problem.. Make sure that you differentiate this equation implicitly with respect to t before substituting the values of the variables (Step 5).

3.6   Concept Questions   page 223

1. a. We differentiate both sides of $F(x, y) = 0$ with respect to x, then solve for dy/dx.
  b. The Chain Rule is used to differentiate any expression involving the dependent variable y.

3. Suppose x and y are two variables that are related by an equation. Furthermore, suppose x and y are both functions of a third variable t. (Normally t represents time.) Then a related rates problem involves finding dx/dt or dy/dt.

3.6   Implicit Differentiation and Related Rates   page 223

1. a. Solving for y in terms of x, we have $y = -\frac{1}{2}x + \frac{5}{2}$. Therefore, $y' = -\frac{1}{2}$.
  b. Next, differentiating $x + 2y = 5$ implicitly, we have $1 + 2y' = 0$, or $y' = -\frac{1}{2}$.

3. a. $xy = 1$, $y = \dfrac{1}{x}$, and $\dfrac{dy}{dx} = -\dfrac{1}{x^2}$.
  b. $x\dfrac{dy}{dx} + y = 0$, so $x\dfrac{dy}{dx} = -y$ and $\dfrac{dy}{dx} = -\dfrac{y}{x} = \dfrac{-1/x}{x} = -\dfrac{1}{x^2}$.

5. $x^3 - x^2 - xy = 4$.

  a. $-xy = 4 - x^3 + x^2$, so $y = -\dfrac{4}{x} + x^2 - x$ and $\dfrac{dy}{dx} = \dfrac{4}{x^2} + 2x - 1$.

  b. $x^3 - x^2 - xy = 4$, so $-x\dfrac{dy}{dx} = -3x^2 + 2x + y$, and therefore
$$\frac{dy}{dx} = 3x - 2 - \frac{y}{x} = 3x - 2 - \frac{1}{x}\left(-\frac{4}{x} + x^2 - x\right) = 3x - 2 + \frac{4}{x^2} - x + 1 = \frac{4}{x^2} + 2x - 1.$$

© 2011 Cengage Learning. All Rights Reserved. May not be scanned, copied or duplicated, or posted to a publicly accessible website, in whole or in part.

7. a. $\dfrac{x}{y} - x^2 = 1$ is equivalent to $\dfrac{x}{y} = x^2 + 1$, or $y = \dfrac{x}{x^2+1}$. Therefore, $y' = \dfrac{(x^2+1) - x(2x)}{(x^2+1)^2} = \dfrac{1-x^2}{(x^2+1)^2}$.

   b. Next, differentiating the equation $x - x^2 y = y$ implicitly, we obtain $1 - 2xy - x^2 y' = y'$, $y'\left(1+x^2\right) = 1 - 2xy$,

   and thus $y' = \dfrac{1-2xy}{\left(1+x^2\right)}$. This may also be written in the form $-2y^2 + \dfrac{y}{x}$. To show that this is equivalent to the

   results obtained earlier, use the earlier value of y to get $y' = \dfrac{1 - 2x\left(\dfrac{x}{x^2+1}\right)}{1+x^2} = \dfrac{x^2+1-2x^2}{\left(1+x^2\right)^2} = \dfrac{1-x^2}{\left(1+x^2\right)^2}$.

9. $x^2 + y^2 = 16$. Differentiating both sides of the equation implicitly, we obtain $2x + 2yy' = 0$, and so $y' = -x/y$.

11. $x^2 - 2y^2 = 16$. Differentiating implicitly with respect to x, we have $2x - 4y\dfrac{dy}{dx} = 0$, and so $\dfrac{dy}{dx} = \dfrac{x}{2y}$.

13. $x^2 - 2xy = 6$. Differentiating both sides of the equation implicitly, we obtain $2x - 2y - 2xy' = 0$ and so

   $y' = \dfrac{x-y}{x} = 1 - \dfrac{y}{x}$.

15. $x^2 y^2 - xy = 8$. Differentiating both sides of the equation implicitly, we obtain $2xy^2 + 2x^2 yy' - y - xy' = 0$,

   $2xy^2 - y + y'\left(2x^2 y - x\right) = 0$, and so $y' = \dfrac{y(1-2xy)}{x(2xy-1)} = -\dfrac{y}{x}$.

17. $x^{1/2} + y^{1/2} = 1$. Differentiating implicitly with respect to x, we have $\frac{1}{2}x^{-1/2} + \frac{1}{2}y^{-1/2}\dfrac{dy}{dx} = 0$. Therefore,

   $\dfrac{dy}{dx} = -\dfrac{x^{-1/2}}{y^{-1/2}} = -\dfrac{\sqrt{y}}{\sqrt{x}}$.

19. $\sqrt{x+y} = x$. Differentiating both sides of the equation implicitly, we obtain $\frac{1}{2}(x+y)^{-1/2}\left(1+y'\right) = 1$,

   $1 + y' = 2(x+y)^{1/2}$, and so $y' = 2\sqrt{x+y} - 1$.

21. $\dfrac{1}{x^2} + \dfrac{1}{y^2} = 1$. Differentiating both sides of the equation implicitly, we obtain $-\dfrac{2}{x^3} - \dfrac{2}{y^3}y' = 0$, or $y' = -\dfrac{y^3}{x^3}$.

23. $\sqrt{xy} = x + y$. Differentiating both sides of the equation implicitly, we obtain $\frac{1}{2}(xy)^{-1/2}\left(xy' + y\right) = 1 + y'$, so

   $xy' + y = 2\sqrt{xy}\left(1+y'\right)$, $y'\left(x - 2\sqrt{xy}\right) = 2\sqrt{xy} - y$, and so $y' = -\dfrac{2\sqrt{xy} - y}{2\sqrt{xy} - x} = \dfrac{2\sqrt{xy} - y}{x - 2\sqrt{xy}}$.

25. $\dfrac{x+y}{x-y} = 3x$, or $x + y = 3x^2 - 3xy$. Differentiating both sides of the equation implicitly, we obtain

   $1 + y' = 6x - 3xy' - 3y$, so $y' + 3xy' = 6x - 3y - 1$ and $y' = \dfrac{6x - 3y - 1}{3x+1}$.

27. $xy^{3/2} = x^2 + y^2$. Differentiating implicitly with respect to x, we obtain $y^{3/2} + x\left(\frac{3}{2}\right)y^{1/2}\dfrac{dy}{dx} = 2x + 2y\dfrac{dy}{dx}$.

   Multiply both sides by 2 to get $2y^{3/2} + 3xy^{1/2}\dfrac{dy}{dx} = 4x + 4y\dfrac{dy}{dx}$. Then $\left(3xy^{1/2} - 4y\right)\dfrac{dy}{dx} = 4x - 2y^{3/2}$, so

   $\dfrac{dy}{dx} = \dfrac{2\left(2x - y^{3/2}\right)}{3xy^{1/2} - 4y}$.

© 2011 Cengage Learning. All Rights Reserved. May not be scanned, copied or duplicated, or posted to a publicly accessible website, in whole or in part.

29. $(x + y)^3 + x^3 + y^3 = 0$. Differentiating implicitly with respect to x, we obtain

$$3(x+y)^2 \left(1 + \frac{dy}{dx}\right) + 3x^2 + 3y^2\frac{dy}{dx} = 0, \ (x+y)^2 + (x+y)^2\frac{dy}{dx} + x^2 + y^2\frac{dy}{dx} = 0,$$

$$\left[(x+y)^2 + y^2\right]\frac{dy}{dx} = -\left[(x+y)^2 + x^2\right], \text{ and thus } \frac{dy}{dx} = -\frac{2x^2 + 2xy + y^2}{x^2 + 2xy + 2y^2}.$$

31. $4x^2 + 9y^2 = 36$. Differentiating the equation implicitly, we obtain $8x + 18yy' = 0$. At the point $(0, 2)$, we have $0 + 36y' = 0$, and the slope of the tangent line is 0. Therefore, an equation of the tangent line is $y = 2$.

33. $x^2y^3 - y^2 + xy - 1 = 0$. Differentiating implicitly with respect to x, we have $2xy^3 + 3x^2y^2\frac{dy}{dx} - 2y\frac{dy}{dx} + y + x\frac{dy}{dx} = 0$.

At $(1, 1)$, $2 + 3\frac{dy}{dx} - 2\frac{dy}{dx} + 1 + \frac{dy}{dx} = 0$, and so $2\frac{dy}{dx} = -3$ and $\frac{dy}{dx} = -\frac{3}{2}$. Using the point-slope form of an equation of a line, we have $y - 1 = -\frac{3}{2}(x - 1)$, and the equation of the tangent line to the graph of the function f at $(1, 1)$ is $y = -\frac{3}{2}x + \frac{5}{2}$.

35. $xy = 1$. Differentiating implicitly, we have $xy' + y = 0$, or $y' = -\frac{y}{x}$. Differentiating implicitly once again, we

have $xy'' + y' + y' = 0$. Therefore, $y'' = -\frac{2y'}{x} = \frac{2\left(\frac{y}{x}\right)}{x} = \frac{2y}{x^2}$.

37. $y^2 - xy = 8$. Differentiating implicitly we have $2yy' - y - xy' = 0$, and so $y' = \frac{y}{2y - x}$. Differentiating

implicitly again, we have $2(y')^2 2yy'' - y' - y' - xy'' = 0$, so $y'' = \frac{2y' - 2(y')^2}{2y - x} = \frac{2y'(1 - y')}{2y - x}$. Then

$$y'' = \frac{2\left(\frac{y}{2y - x}\right)\left(1 - \frac{y}{2y - x}\right)}{2y - x} = \frac{2y(2y - x - y)}{(2y - x)^3} = \frac{2y(y - x)}{(2y - x)^3}.$$

39. a. Differentiating the given equation with respect to t, we obtain $\frac{dV}{dt} = \pi r^2\frac{dh}{dt} + 2\pi rh\frac{dr}{dt} = \pi r\left(r\frac{dh}{dt} + 2h\frac{dr}{dt}\right)$.

b. Substituting $r = 2$, $h = 6$, $\frac{dr}{dt} = 0.1$, and $\frac{dh}{dt} = 0.3$ into the expression for $\frac{dV}{dt}$, we

obtain $\frac{dV}{dt} = \pi(2)[2(0.3) + 2(6)(0.1)] = 3.6\pi$, and so the volume is increasing at the rate of 3.6π in^3/sec.

41. We are given $\frac{dp}{dt} = 2$ and wish to find $\frac{dx}{dt}$ when $x = 9$ and $p = 63$. Differentiating the equation $p + x^2 = 144$ with respect to t, we obtain $\frac{dp}{dt} + 2x\frac{dx}{dt} = 0$. When $x = 9$, $p = 63$, and $\frac{dp}{dt} = 2$, we have $2 + 2(9)\frac{dx}{dt} = 0$, and so and $\frac{dx}{dt} = -\frac{1}{9} \approx -0.111$. Thus, the quantity demanded is decreasing at the rate of approximately 111 tires per week.

43. $100x^2 + 9p^2 = 3600$. Differentiating the given equation implicitly with respect to t, we have $200x\frac{dx}{dt} + 18p\frac{dp}{dt} = 0$. Next, when $p = 14$, the given equation yields $100x^2 + 9(14)^2 = 3600$, so $100x^2 = 1836$, or $x \approx 4.2849$. When $p = 14$, $\frac{dp}{dt} = -0.15$, and $x \approx 4.2849$, we have $200(4.2849)\frac{dx}{dt} + 18(14)(-0.15) = 0$, and so $\frac{dx}{dt} \approx 0.0441$. Thus, the quantity demanded is increasing at the rate of approximately 44 headphones per week.

45. From the results of Problem 44, we have $1250p\frac{dp}{dt} - 2x\frac{dx}{dt} = 0$. When $p = 1.0770$, $x = 25$, and $\frac{dx}{dt} = -1$, we find that $1250(1.077)\frac{dp}{dt} - 2(25)(-1) = 0$, and so $\frac{dp}{dt} = -\frac{50}{1250(1.077)} = -0.037$. We conclude that the price is decreasing at the rate of 3.7 cents per carton.

© 2011 Cengage Learning. All Rights Reserved. May not be scanned, copied or duplicated, or posted to a publicly accessible website, in whole or in part.

47. $p = -0.01x^2 - 0.2x + 8$. Differentiating the given equation implicitly with respect to p, we

have $1 = -0.02x\dfrac{dx}{dp} - 0.2\dfrac{dx}{dp} = -[0.02x + 0.2]\dfrac{dx}{dp}$, so $\dfrac{dx}{dp} = -\dfrac{1}{0.02x + 0.2}$. When $x = 15$,

$p = -0.01\,(15)^2 - 0.2\,(15) + 8 = 2.75$, and so and $\dfrac{dx}{dp} = -\dfrac{1}{0.02\,(15) + 0.2} = -2$. Therefore,

$E(p) = -\dfrac{pf'(p)}{f(p)} = -\dfrac{(2.75)\,(-2)}{15} \approx 0.37 < 1$, and the demand is inelastic.

49. $A = \pi r^2$. Differentiating with respect to t, we obtain $\frac{dA}{dt} = 2\pi r\frac{dr}{dt}$. When the radius of the circle is 40 ft and
increasing at the rate of 2 ft/sec, $\frac{dA}{dt} = 2\pi\,(40)\,(2) = 160\pi$ ft^2/sec.

51. $A = \pi r^2$. Differentiating with respect to t, we obtain $\dfrac{dA}{dt} = 2\pi r\dfrac{dr}{dt}$. When the radius of the circle is 60 ft and

increasing at the rate of $\frac{1}{2}$ ft/s, $\dfrac{dA}{dt} = 2\pi\,(60)\left(\frac{1}{2}\right) = 60\pi$ ft^2/sec. Thus, the area is increasing at the rate of

approximately 188.5 ft^2/sec.

53. Let $(x, 0)$ and $(0, y)$ denote the position of the two cars at time t. Then $y = t^2 + 2t$. Now $D^2 = x^2 + y^2$ so

$2D\dfrac{dD}{dt} = 2x\dfrac{dx}{dt} + 2y\dfrac{dy}{dt}$ and thus $D\dfrac{dD}{dt} = x\dfrac{dx}{dt} + (t^2 + 2t)\,(2t + 2)$. When $t = 4$, we have $x = -20$, $\frac{dx}{dt} = -9$,

and $y = 24$, so $\sqrt{(-20)^2 + (24)^2}\frac{dD}{dt} = (-20)\,(-9) + (24)\,(10)$, and therefore $\frac{dD}{dt} = \frac{420}{\sqrt{976}} \approx 13.44$. That is, the

distance is changing at approximately 13.44 ft/sec.

55. Referring to the diagram, we see that $D^2 = 120^2 + x^2$. Differentiating

this last equation with respect to t, we have $2D\dfrac{dD}{dt} = 2x\dfrac{dx}{dt}$, and so

$\dfrac{dD}{dt} = \dfrac{x\frac{dx}{dt}}{D}$. When $x = 50$ and $\dfrac{dx}{dt} = 20$, $D = \sqrt{120^2 + 50^2} = 130$

and $\dfrac{dD}{dt} = \dfrac{(20)\,(50)}{130} \approx 7.69$, or 7.69 ft/sec.

57. Let V and S denote its volume and surface area. Then we are given that $\frac{dV}{dt} = kS$, where k is the constant

of proportionality. But from $V = \frac{4}{3}\pi r^3$, we find, upon differentiating both sides with respect to t, that

$\frac{dV}{dt} = \frac{d}{dt}\left(\frac{4}{3}\pi r^3\right) = 4\pi r^2\frac{dr}{dt} = kS = k\,(4\pi r^2)$. Therefore, $\frac{dr}{dt} = k$ a constant.

59. We are given that $\frac{dx}{dt} = 264$. Using the Pythagorean Theorem,

$s^2 = x^2 + 1000^2 = x^2 + 1,000,000$. We want to find $\dfrac{ds}{dt}$ when

$s = 1500$. Differentiating both sides of the equation with respect to t,

we have $2s\dfrac{ds}{dt} = 2x\dfrac{dx}{dt}$ and so $\dfrac{ds}{dt} = \dfrac{x\frac{dx}{dt}}{s}$. When $s = 1500$, we have

$1500^2 = x^2 + 1,000,000$, or $x = \sqrt{1,250,000}$. Therefore, $\dfrac{ds}{dt} = \dfrac{\sqrt{1,250,000}\cdot(264)}{1500} \approx 196.8$, that is, the aircraft is

receding from the trawler at the speed of approximately 196.8 ft/sec.

© 2011 Cengage Learning. All Rights Reserved. May not be scanned, copied or duplicated, or posted to a publicly accessible website, in whole or in part.

61. $\dfrac{y}{6} = \dfrac{y+x}{18}$, $18y = 6(y+x)$, so $3y = y+x$, $2y = x$, and $y = \frac{1}{2}x$.

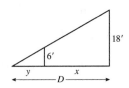

Thus, $D = y + x = \frac{3}{2}x$. Differentiating implicitly, we have

$\frac{dD}{dt} = \frac{3}{2} \cdot \frac{dx}{dt}$, and when $\frac{dx}{dt} = 6$, $\frac{dD}{dt} = \frac{3}{2}(6) = 9$, or 9 ft/sec.

63. Differentiating $x^2 + y^2 = 13^2 = 169$ with respect to t gives

$2x\frac{dx}{dt} + 2y\frac{dy}{dt} = 0$. When $x = 12$, we have $144 + y^2 = 169$, or $y = 5$.

Therefore, with $x = 12$, $y = 5$, and $\frac{dx}{dt} = 8$, we find

$2(12)(8) + 2(5)\frac{dy}{dt} = 0$, or $\frac{dy}{dt} = -19.2$. Thus, the top of the ladder is

sliding down the wall at the rate of 19.2 ft/sec.

65. $P^5 V^7 = C$, so $V^7 = CP^{-5}$ and $7V^6 \dfrac{dV}{dt} = -5CP^{-6} \dfrac{dP}{dt}$. Therefore,

$\dfrac{dV}{dt} = -\dfrac{5C}{7P^6V^6}\dfrac{dP}{dt} = -\dfrac{5P^5V^7}{7P^6V^6}\dfrac{dP}{dt} = -\dfrac{5}{7}\dfrac{V}{P}\dfrac{dP}{dt}$. When $V = 4$ L, $P = 100$ kPa, and $\dfrac{dP}{dt} = -5\,\dfrac{\text{kPa}}{\text{sec}}$, we have

$\dfrac{dV}{dt} = -\dfrac{5}{7} \cdot \dfrac{4}{100}(-5) = \dfrac{1}{7}\left(\dfrac{\text{L}}{\text{kPa}} \cdot \dfrac{\text{kPa}}{\text{s}}\right) = \dfrac{1}{7}\dfrac{\text{L}}{\text{s}}$.

67. False. There are no real numbers x and y such that $x^2 + y^2 = -1$.

69. True. Differentiating both sides of the equation with respect to x, we have $\frac{d}{dx}\left[f(x)g(y)\right] = \frac{d}{dx}(0)$, so

$f(x)g'(y)\dfrac{dy}{dx} + f'(x)g(y) = 0$, and therefore $\frac{dy}{dx} = -\frac{f'(x)g(y)}{f(x)g'(y)}$, provided $f(x) \neq 0$ and $g'(y) \neq 0$.

3.7   **Concept Questions**   page 233

1. The differential of x is dx. The differential of y is $dy = f'(x)\,dx$.

3.7   **Differentials**   page 233

1. $f(x) = 2x^2$ and $dy = 4x\,dx$.

3. $f(x) = x^3 - x$ and $dy = (3x^2 - 1)\,dx$.

5. $f(x) = \sqrt{x+1} = (x+1)^{1/2}$ and $dy = \frac{1}{2}(x+1)^{-1/2}\,dx = \dfrac{dx}{2\sqrt{x+1}}$.

7. $f(x) = 2x^{3/2} + x^{1/2}$ and $dy = \left(3x^{1/2} + \frac{1}{2}x^{-1/2}\right)dx = \frac{1}{2}x^{-1/2}(6x+1)\,dx = \dfrac{6x+1}{2\sqrt{x}}\,dx$.

9. $f(x) = x + \dfrac{2}{x}$ and $dy = \left(1 - \dfrac{2}{x^2}\right)dx = \dfrac{x^2-2}{x^2}dx$.

© 2011 Cengage Learning. All Rights Reserved. May not be scanned, copied or duplicated, or posted to a publicly accessible website, in whole or in part.

11. $f(x) = \dfrac{x-1}{x^2+1}$ and $dy = \dfrac{x^2+1-(x-1)\,2x}{\left(x^2+1\right)^2}\,dx = \dfrac{-x^2+2x+1}{\left(x^2+1\right)^2}\,dx$.

13. $f(x) = \sqrt{3x^2-x} = \left(3x^2-x\right)^{1/2}$ and $dy = \frac{1}{2}\left(3x^2-x\right)^{-1/2}(6x-1)\,dx = \dfrac{6x-1}{2\sqrt{3x^2-x}}\,dx$.

15. $f(x) = x^2 - 1$.

    a. $dy = 2x\,dx$.

    b. $dy \approx 2\,(1)\,(0.02) = 0.04$.

    c. $\Delta y = \left[(1.02)^2 - 1\right] - (1-1) = 0.0404$.

17. $f(x) = \dfrac{1}{x}$.

    a. $dy = -\dfrac{dx}{x^2}$.

    b. $dy \approx -0.05$.

    c. $\Delta y = \frac{1}{-0.95} - \frac{1}{-1} \approx -0.05263$.

19. $y = \sqrt{x}$ and $dy = \dfrac{dx}{2\sqrt{x}}$. Therefore, $\sqrt{10} \approx 3 + \dfrac{1}{2\cdot\sqrt{9}} \approx 3.167$.

21. $y = \sqrt{x}$ and $dy = \dfrac{dx}{2\sqrt{x}}$. Therefore, $\sqrt{49.5} \approx 7 + \dfrac{0.5}{2\cdot 7} \approx 7.0357$.

23. $y = x^{1/3}$ and $dy = \frac{1}{3}x^{-2/3}\,dx$. Therefore, $\sqrt[3]{7.8} \approx 2 - \dfrac{0.2}{3\cdot 4} \approx 1.983$.

25. $y = \sqrt{x}$ and $dy = \dfrac{dx}{2\sqrt{x}}$. Therefore, $\sqrt{0.089} = \frac{1}{10}\sqrt{8.9} \approx \frac{1}{10}\left(3 - \frac{0.1}{2.3}\right) \approx 0.298$.

27. $y = f(x) = \sqrt{x} + \dfrac{1}{\sqrt{x}} = x^{1/2} + x^{-1/2}$. Therefore, $\dfrac{dy}{dx} = \frac{1}{2}x^{-1/2} - \frac{1}{2}x^{-3/2}$, so $dy = \left(\dfrac{1}{2x^{1/2}} - \dfrac{1}{2x^{3/2}}\right)dx$.

Letting $x = 4$ and $dx = 0.02$, we find $\sqrt{4.02} + \dfrac{1}{\sqrt{4.02}} - f(4) = f(4.02) - f(4) = \Delta y \approx dy$, so

$\sqrt{4.02} + \dfrac{1}{\sqrt{4.02}} \approx f(4) + dy \approx 2 + \dfrac{1}{2} + \left(\dfrac{1}{2\cdot 2} - \dfrac{1}{16}\right)(0.02) = 2.50375$.

29. The volume of the cube is given by $V = x^3$. Then $dV = 3x^2\,dx$, and when $x = 12$ and $dx = 0.02$, $dV = 3\,(144)\,(\pm 0.02) = \pm 8.64$. The possible error that might occur in calculating the volume is ± 8.64 cm^3.

31. The volume of the hemisphere is given by $V = \frac{2}{3}\pi r^3$. The amount of rust-proofer needed is

$\Delta V = \frac{2}{3}\pi\,(r+\Delta r)^3 - \frac{2}{3}\pi r^3 \approx dV = \frac{2}{3}\left(3\pi r^2\right)dr$. Thus, with $r = 60$ and $dr = \dfrac{1}{12}\,(0.01)$, we have

$\Delta V \approx 2\pi\,(60^2)\left(\frac{1}{12}\right)(0.01) \approx 18.85$. So we need approximately 18.85 ft^3 of rust-proofer.

33. $dR = \dfrac{d}{dr}\left(k\ell r^{-4}\right)dr = -4k\ell r^{-5}\,dr$. With $\dfrac{dr}{r} = 0.1$, we find $\dfrac{dR}{R} = -\dfrac{4k\ell r^{-5}}{k\ell r^{-4}}\,dr = -4\dfrac{dr}{r} = -4\,(0.1) = -0.4$.

In other words, the resistance will drop by 40%.

35. $f(n) = 4n\sqrt{n-4} = 4n\,(n-4)^{1/2}$, so $df = 4\left[(n-4)^{1/2} + \frac{1}{2}n\,(n-4)^{-1/2}\right]dn$. When $n = 85$ and $dn = 5$,

$df = 4\left(9 + \dfrac{85}{2\cdot 9}\right)5 \approx 274$ seconds.

© 2011 Cengage Learning. All Rights Reserved. May not be scanned, copied or duplicated, or posted to a publicly accessible website, in whole or in part.

37. $N(r) = \dfrac{7}{1 + 0.02r^2}$ and $dN = -\dfrac{0.28r}{(1 + 0.02r^2)^2}\, dr$. To estimate the decrease in the number of housing

starts when the mortgage rate is increased from 6% to 6.5%, we set $r = 6$ and $dr = 0.5$ and compute

$dN = -\dfrac{(0.28)(6)(0.5)}{(1.72)^2} \approx -0.283937$, or 283,937 fewer housing starts.

39. $p = \dfrac{30}{0.02x^2 + 1}$ and $dp = -\dfrac{1.2x}{(0.02x^2 + 1)^2}\, dx$. To estimate the change in the price p when the quantity

demanded changed from 5000 to 5500 units per week (that is, x changes from 5 to 5.5), we compute

$dp = \dfrac{(-1.2)(5)(0.5)}{[0.02(25) + 1]^2} \approx -1.33$, a decrease of $1.33.

41. $P(x) = -0.000032x^3 + 6x - 100$ and $dP = (-0.000096x^2 + 6)\, dx$. To determine the error in the estimate of Trappee's profits corresponding to a maximum error in the forecast of 15 percent [that is, $dx = \pm 0.15\,(200)$], we compute $dP = [(-0.000096)(200)^2 + 6](\pm 30) = (2.16)(30) = \pm 64.80$, or $64,800.

43. $N(x) = \dfrac{500(400 + 20x)^{1/2}}{(5 + 0.2x)^2}$ and

$N'(x) = \dfrac{(5 + 0.2x)^2\, 250\,(400 + 20x)^{-1/2}\,(20) - 500\,(400 + 20x)^{1/2}\,(2)\,(5 + 0.2x)\,(0.2)}{(5 + 0.2x)^4}\, dx$. To estimate the

change in the number of crimes if the level of reinvestment changes from 20 cents to 22 cents per dollar deposited, we compute

$dN = \dfrac{(5 + 4)^2\,(250)\,(800)^{-1/2}\,(20) - 500\,(400 + 400)^{1/2}\,(2)\,(9)\,(0.2)}{(5 + 4)^4}\,(2) \approx \dfrac{(14318.91 - 50911.69)}{9^4}\,(2)$

≈ -11, a decrease of approximately 11 crimes per year.

45. $A = 10,000\left(1 + \dfrac{r}{12}\right)^{120}$.

 a. $dA = 10,000\,(120)\left(1 + \dfrac{r}{12}\right)^{119}\left(\dfrac{1}{12}\right)dr = 100,000\left(1 + \dfrac{r}{12}\right)^{119}dr$.

 b. At 8.1%, it will be worth $100,000\left(1 + \dfrac{0.08}{12}\right)^{119}(0.001)$, or approximately $220.50 more. At 8.2%, it

 will be worth $100,000\left(1 + \dfrac{0.08}{12}\right)^{119}(0.002)$, or approximately $440.99 more. At 8.3%, it will be worth

 $100,000\left(1 + \dfrac{0.08}{12}\right)^{119}(0.003)$, or approximately $661.48 more.

47. True. $dy = f'(x)\,dx = \dfrac{d}{dx}(ax + b)\,dx = a\,dx$. On the other hand,
 $\Delta y = f(x + \Delta x) - f(x) = [a(x + \Delta x) + b] - (ax + b) = a\,\Delta x = a\,dx$.

3.7 Using Technology   page 237

1. $dy = f'(3)\,dx = 757.87\,(0.01) \approx 7.5787$.

3. $dy = f'(1)\,dx = 1.04067285926\,(0.03) \approx 0.031220$.

5. $dy = f'(4)\,(0.1) = -0.198761598\,(0.1) = -0.01988$.

© 2011 Cengage Learning. All Rights Reserved. May not be scanned, copied or duplicated, or posted to a publicly accessible website, in whole or in part.

7. If the interest rate changes from 7% to 7.3% per year, the monthly payment will increase by $dP = f'(0.07)(0.003) \approx 48.3546$, or approximately \$48.35 per month. If the interest rate changes from 7% to 7.4% per year, it will be \$64.47 per month. If the interest rate changes from 7% to 7.5% per year, it will be \$80.58 per month.

9. $dx = f'(40)(2) \approx -0.625$. That is, the quantity demanded will decrease by 625 watches per week.

CHAPTER 3   **Concept Review**   page 238

1. a. 0   **b.** nx^{n-1} **c.** $cf'(x)$.   **d.** $f'(x) \pm g'(x)$

3. a. $g'(f(x)) f'(x)$ **b.** $n[f(x)]^{n-1} f'(x)$

5. a. $-\dfrac{pf'(p)}{f(p)}$ **b.** Elastic, unitary, inelastic

7. $y, dy/dt, a$

9. a. $x_2 - x_1$ **b.** $f(x + \Delta x) - f(x)$

CHAPTER 3   **Review**   page 239

1. $f'(x) = \dfrac{d}{dx}\left(3x^5 - 2x^4 + 3x^2 - 2x + 1\right) = 15x^4 - 8x^3 + 6x - 2.$

3. $g'(x) = \dfrac{d}{dx}\left(-2x^{-3} + 3x^{-1} + 2\right) = 6x^{-4} - 3x^{-2}.$

5. $g'(t) = \dfrac{d}{dt}\left(2t^{-1/2} + 4t^{-3/2} + 2\right) = -t^{-3/2} - 6t^{-5/2}.$

7. $f'(t) = \dfrac{d}{dt}\left(t + 2t^{-1} + 3t^{-2}\right) = 1 - 2t^{-2} - 6t^{-3} = 1 - \dfrac{2}{t^2} - \dfrac{6}{t^3}.$

9. $h'(x) = \dfrac{d}{dx}\left(x^2 - 2x^{-3/2}\right) = 2x + 3x^{-5/2} = 2x + \dfrac{3}{x^{5/2}}.$

11. $g(t) = \dfrac{t^2}{2t^2 + 1}$, so $g'(t) = \dfrac{(2t^2 + 1)\dfrac{d}{dt}(t^2) - t^2\dfrac{d}{dt}(2t^2 + 1)}{(2t^2 + 1)^2} = \dfrac{(2t^2 + 1)(2t) - t^2(4t)}{(2t^2 + 1)^2} = \dfrac{2t}{(2t^2 + 1)^2}.$

© 2011 Cengage Learning. All Rights Reserved. May not be scanned, copied or duplicated, or posted to a publicly accessible website, in whole or in part.

13. $f(x) = \dfrac{\sqrt{x}-1}{\sqrt{x}+1} = \dfrac{x^{1/2}-1}{x^{1/2}+1}$, so

$$f'(x) = \frac{\left(x^{1/2}+1\right)\left(\frac{1}{2}x^{-1/2}\right) - \left(x^{1/2}-1\right)\left(\frac{1}{2}x^{-1/2}\right)}{\left(x^{1/2}+1\right)^2} = \frac{\frac{1}{2} + \frac{1}{2}x^{-1/2} - \frac{1}{2} + \frac{1}{2}x^{-1/2}}{\left(x^{1/2}+1\right)^2} = \frac{x^{-1/2}}{\left(x^{1/2}+1\right)^2}$$

$$= \frac{1}{\sqrt{x}\left(\sqrt{x}+1\right)^2}.$$

15. $f(x) = \dfrac{x^2\left(x^2+1\right)}{x^2-1}$, so

$$f'(x) = \frac{\left(x^2-1\right)\dfrac{d}{dx}\left(x^4+x^2\right) - \left(x^4+x^2\right)\dfrac{d}{dx}\left(x^2-1\right)}{\left(x^2-1\right)^2} = \frac{\left(x^2-1\right)\left(4x^3+2x\right) - \left(x^4+x^2\right)(2x)}{\left(x^2-1\right)^2}$$

$$= \frac{4x^5 + 2x^3 - 4x^3 - 2x - 2x^5 - 2x^3}{\left(x^2-1\right)^2} = \frac{2x^5 - 4x^3 - 2x}{\left(x^2-1\right)^2} = \frac{2x\left(x^4 - 2x^2 - 1\right)}{\left(x^2-1\right)^2}.$$

17. $f(x) = \left(3x^3 - 2\right)^8$, so $f'(x) = 8\left(3x^3 - 2\right)^7\left(9x^2\right) = 72x^2\left(3x^3 - 2\right)^7.$

19. $f'(t) = \dfrac{d}{dt}\left(2t^2 + 1\right)^{1/2} = \frac{1}{2}\left(2t^2 + 1\right)^{-1/2}\dfrac{d}{dt}\left(2t^2 + 1\right) = \frac{1}{2}\left(2t^2 + 1\right)^{-1/2}(4t) = \dfrac{2t}{\sqrt{2t^2 + 1}}.$

21. $s(t) = \left(3t^2 - 2t + 5\right)^{-2}$, so

$$s'(t) = -2\left(3t^2 - 2t + 5\right)^{-3}(6t - 2) = -4\left(3t^2 - 2t + 5\right)^{-3}(3t - 1) = -\frac{4(3t - 1)}{\left(3t^2 - 2t + 5\right)^3}.$$

23. $h(x) = \left(x + \dfrac{1}{x}\right)^2 = \left(x + x^{-1}\right)^2$, so

$$h'(x) = 2\left(x + x^{-1}\right)\left(1 - x^{-2}\right) = 2\left(x + \frac{1}{x}\right)\left(1 - \frac{1}{x^2}\right) = 2\left(\frac{x^2 + 1}{x}\right)\left(\frac{x^2 - 1}{x^2}\right) = \frac{2\left(x^2 + 1\right)\left(x^2 - 1\right)}{x^3}.$$

25. $h'(t) = \left(t^2 + t\right)^4\dfrac{d}{dt}\left(2t^2\right) + 2t^2\dfrac{d}{dt}\left(t^2 + t\right)^4 = \left(t^2 + t\right)^4(4t) + 2t^2 \cdot 4\left(t^2 + t\right)^3(2t + 1)$

$$= 4t\left(t^2 + t\right)^3\left[\left(t^2 + t\right) + 4t^2 + 2t\right] = 4t^2(5t + 3)\left(t^2 + t\right)^3.$$

27. $g(x) = x^{1/2}\left(x^2 - 1\right)^3$, so

$$g'(x) = \frac{d}{dx}\left[x^{1/2}\left(x^2 - 1\right)^3\right] = x^{1/2} \cdot 3\left(x^2 - 1\right)^2(2x) + \left(x^2 - 1\right)^3 \cdot \frac{1}{2}x^{-1/2}$$

$$= \frac{1}{2}x^{-1/2}\left(x^2 - 1\right)^2\left[12x^2 + \left(x^2 - 1\right)\right] = \frac{\left(13x^2 - 1\right)\left(x^2 - 1\right)^2}{2\sqrt{x}}.$$

29. $h(x) = \dfrac{(3x + 2)^{1/2}}{4x - 3}$, so

$$h'(x) = \frac{(4x - 3)\frac{1}{2}(3x + 2)^{-1/2}(3) - (3x + 2)^{1/2}(4)}{(4x - 3)^2} = \frac{\frac{1}{2}(3x + 2)^{-1/2}[3(4x - 3) - 8(3x + 2)]}{(4x - 3)^2}$$

$$= -\frac{12x + 25}{2\sqrt{3x + 2}(4x - 3)^2}.$$

© 2011 Cengage Learning. All Rights Reserved. May not be scanned, copied or duplicated, or posted to a publicly accessible website, in whole or in part.

31. $f(x) = 2x^4 - 3x^3 + 2x^2 + x + 4$, so $f'(x) = \dfrac{d}{dx}(2x^4 - 3x^3 + 2x^2 + x + 4) = 8x^3 - 9x^2 + 4x + 1$ and

$f''(x) = \dfrac{d}{dx}(8x^3 - 9x^2 + 4x + 1) = 24x^2 - 18x + 4 = 2(12x^2 - 9x + 2)$.

33. $h(t) = \dfrac{t}{t^2 + 4}$, so $h'(t) = \dfrac{(t^2 + 4)(1) - t(2t)}{(t^2 + 4)^2} = \dfrac{4 - t^2}{(t^2 + 4)^2}$ and

$h''(t) = \dfrac{(t^2 + 4)^2(-2t) - (4 - t^2)2(t^2 + 4)(2t)}{(t^2 + 4)^4} = \dfrac{-2t(t^2 + 4)[(t^2 + 4) + 2(4 - t^2)]}{(t^2 + 4)^4} = \dfrac{2t(t^2 - 12)}{(t^2 + 4)^3}$.

35. $f'(x) = \dfrac{d}{dx}(2x^2 + 1)^{1/2} = \dfrac{1}{2}(2x^2 + 1)^{-1/2}(4x) = 2x(2x^2 + 1)^{-1/2}$, so

$f''(x) = 2(2x^2 + 1)^{-1/2} + 2x \cdot \left(-\dfrac{1}{2}\right)(2x^2 + 1)^{-3/2}(4x) = 2(2x^2 + 1)^{-3/2}[(2x^2 + 1) - 2x^2] = \dfrac{2}{(2x^2 + 1)^{3/2}}$.

37. $6x^2 - 3y^2 = 9$. Differentiating this equation implicitly, we have $12x - 6y\dfrac{dy}{dx} = 0$ and $-6y\dfrac{dy}{dx} = -12x$. Therefore,

$\dfrac{dy}{dx} = \dfrac{-12x}{-6y} = \dfrac{2x}{y}$.

39. $y^3 + 3x^2 = 3y$. Differentiating this equation implicitly, we have $3y^2y' + 6x = 3y'$, $3y^2y' - 3y' = -6x$, and

$y'(3y^2 - 3) = -6x$. Therefore, $y' = -\dfrac{6x}{3(y^2 - 1)} = -\dfrac{2x}{y^2 - 1}$.

41. $x^2 - 4xy - y^2 = 12$. Differentiating this equation implicitly, we have $2x - 4xy' - 4y - 2yy' = 0$ and

$y'(-4x - 2y) = -2x + 4y$. Therefore, $y' = \dfrac{-2(x - 2y)}{-2(2x + y)} = \dfrac{x - 2y}{2x + y}$.

43. $df = f'(x)\,dx = (2x - 2x^{-3})\,dx = \left(2x - \dfrac{2}{x^3}\right)dx = \dfrac{2(x^4 - 1)}{x^3}\,dx$.

45. a. $df = f'(x)\,dx = \dfrac{d}{dx}(2x^2 + 4)^{1/2}\,dx = \dfrac{1}{2}(2x^2 + 4)^{-1/2}(4x) = \dfrac{2x}{\sqrt{2x^2 + 4}}\,dx$.

b. Setting $x = 4$ and $dx = 0.1$, we find $\Delta f \approx df = \dfrac{2(4)(0.1)}{\sqrt{2(16) + 4}} = \dfrac{0.8}{6} = \dfrac{8}{60} = \dfrac{2}{15}$.

c. $\Delta f = f(4.1) - f(4) = \sqrt{2(4.1)^2 + 4} - \sqrt{2(16) + 4} \approx 0.1335$. From part (b), $\Delta f \approx \dfrac{2}{15} \approx 0.1333$.

47. $f(x) = 2x^3 - 3x^2 - 16x + 3$ and $f'(x) = 6x^2 - 6x - 16$.

a. To find the point(s) on the graph of f where the slope of the tangent line is equal to -4,
we solve $6x^2 - 6x - 16 = -4$, obtaining $6x^2 - 6x - 12 = 0$, $6(x^2 - x - 2) = 0$, and
$6(x - 2)(x + 1) = 0$. Thus, $x = 2$ or $x = -1$. Now $f(2) = 2(2)^3 - 3(2)^2 - 16(2) + 3 = -25$ and
$f(-1) = 2(-1)^3 - 3(-1)^2 - 16(-1) + 3 = 14$, so the points are $(2, -25)$ and $(-1, 14)$.

b. Using the point-slope form of the equation of a line, we find that the equation of the tangent line at $(2, -25)$ is
$y - (-25) = -4(x - 2)$, $y + 25 = -4x + 8$, or $y = -4x - 17$, and the equation of the tangent line at $(-1, 14)$
is $y - 14 = -4(x + 1)$, or $y = -4x + 10$.

© 2011 Cengage Learning. All Rights Reserved. May not be scanned, copied or duplicated, or posted to a publicly accessible website, in whole or in part.

49. $y = \left(4 - x^2\right)^{1/2}$, so $y' = \frac{1}{2}\left(4 - x^2\right)^{-1/2}(-2x) = -\frac{x}{\sqrt{4 - x^2}}$. The slope of the tangent line is obtained by letting

$x = 1$, giving $m = -\frac{1}{\sqrt{3}} = -\frac{\sqrt{3}}{3}$. Therefore, an equation of the tangent line at $x = 1$ is $y - \sqrt{3} = -\frac{\sqrt{3}}{3}(x - 1)$, or

$y = -\frac{\sqrt{3}}{3}x + \frac{4\sqrt{3}}{3}$.

51. $f(x) = (2x - 1)^{-1}$, so $f'(x) = -2(2x - 1)^{-2}$, $f''(x) = 8(2x - 1)^{-3} = \frac{8}{(2x - 1)^3}$, and

$f'''(x) = -48(2x - 1)^4 = -\frac{48}{(2x - 1)^4}$. Because $(2x - 1)^4 = 0$ when $x = \frac{1}{2}$, we see that the domain of f''' is

$\left(-\infty, \frac{1}{2}\right) \cup \left(\frac{1}{2}, \infty\right)$.

53. $x = \frac{25}{\sqrt{p}} - 1$, so $f'(p) = -\frac{25}{2p^{3/2}}$ and $E(p) = -\frac{p\left(-\frac{25}{2p^{3/2}}\right)}{\frac{25}{p^{1/2}} - 1} = \frac{\frac{25}{2p^{1/2}}}{\frac{25 - p^{1/2}}{p^{1/2}}} = \frac{25}{2\left(25 - p^{1/2}\right)}$. If $E(p) = 1$, then

$2\left(25 - p^{1/2}\right) = 25$, so $25 - p^{1/2} = \frac{25}{2}$, $p^{1/2} = \frac{25}{2}$, and $p = \frac{625}{4}$. $E(p) > 1$ and demand is elastic if $p > 156.25$,
$E(p) = 1$ and demand is unitary if $p = 156.25$, and $E(p) < 1$ and demand is inelastic if $p < 156.25$.

55. a. $p = 9\sqrt[3]{1000 - x}$, so $\sqrt[3]{1000 - x} = \frac{p}{9}$, $1000 - x = \frac{p^3}{729}$, and $x = 1000 - \frac{p^3}{729}$. Therefore,

$x = f(p) = \frac{729{,}000 - p^3}{729}$ and $f'(p) = -\frac{3p^2}{729} = -\frac{p^2}{243}$. Then $E(p) = -\frac{p\left(-\frac{p^2}{243}\right)}{\frac{729{,}000 - p^3}{729}} = \frac{3p^3}{729{,}000 - p^3}$.

$E(60) = \frac{3(60)^3}{729{,}000 - 60^3} = \frac{648{,}000}{513{,}000} = \frac{648}{513} > 1$, and so demand is elastic.

b. From part (a), we see that raising the price slightly causes revenue to decrease.

57. a. $P(9) = 24.4(9)^{0.34} \approx 51.5$, or 51.5%.

b. $P(t) = 24.4t^{0.34}$, so $P'(t) = \frac{d}{dt}\left(24.4t^{0.34}\right) = (0.34)(24.4)t^{-0.66} = 8.296t^{-0.66}$.

$P'(9) = 8.296 \cdot 9^{-0.66} \approx 1.946$, or approximately 1.95%/year.

59. a. $S(0) = 3.1$, or $\$3.1$ billion. $S(5) = 0.14(5)^2 + 0.68(5) + 3.1 = 10$, or $\$10$ billion.

b. $S'(t) = 0.28t + 0.68$, so $S'(0) = 0.28(0) + 0.68 = 0.68$, or $\$0.68$ billion, and $S'(5) = 0.28(5) + 0.68 = 2.08$,
or $\$2.08$ billion/yr.

61. a. The population after 3 years is given by $P(3) = 30 - \frac{20}{2(3) + 3} \approx 27.7778$, or approximately $27{,}778$. The

current population is $P(0) = 30 - \frac{20}{3} \approx 23.333$, or approximately $23{,}333$. So the population will have changed
by $27{,}778 - 23{,}333 = 4445$; that is, it would have increased by 4445.

b. $P'(t) = \frac{d}{dt}\left[30 - 20(2t + 3)^{-1}\right] = \frac{40}{(2t + 3)^2}$, so the rate of change after 3 years is

$P'(3) = \frac{40}{[2(3) + 3]^2} \approx 0.4938$; that is, it will be increasing at the rate of approximately 494 people/yr.

63. $N(x) = 1000(1 + 2x)^{1/2}$, so $N'(x) = 1000\left(\frac{1}{2}\right)(1 + 2x)^{-1/2}(2) = \frac{1000}{\sqrt{1 + 2x}}$. The rate of increase at the end of

the twelfth week is $N'(12) = \frac{1000}{\sqrt{25}} = 200$, or 200 subscribers/week.

© 2011 Cengage Learning. All Rights Reserved. May not be scanned, copied or duplicated, or posted to a publicly accessible website, in whole or in part.

65. He can expect to live $f(100) = 46.9[1 + 1.09(100)]^{0.1} \approx 75.0433$, or approximately 75.04 years.

$f'(t) = 46.9(0.1)(1.1.09t)^{-0.9}(1.09) = 5.1121(1 + 1.09t)^{-0.9}$, so the required rate of change is

$f'(100) = 5.1121[1 + 1.09(100)]^{-0.9} \approx 0.074$, or approximately 0.07 yr/yr.

67. $p'(x) = \dfrac{d}{dt}\left[\dfrac{1}{10}x^{3/2} + 10\right] = \dfrac{3}{20}x^{1/2} = \dfrac{3}{20}\sqrt{x}$, so $p'(40) = \dfrac{3}{20}\sqrt{40} \approx 0.9487$, or \$0.9487. When the number of

units is 40,000, the price will increase \$0.9487 for each 1000 radios demanded.

69. a. The actual cost incurred in the manufacturing of the 301st MP3 player is

$$C(301) - C(300) = \left[0.0001(301)^3 - 0.02(301)^2 + 24(301) + 2000\right]$$
$$- \left[0.0001(300)^3 - 0.02(300)^2 + 24(300) + 2000\right]$$
$$\approx 39.07, \text{ or approximately } \$39.07.$$

b. The marginal cost is $C'(300) = \left(0.0003x^2 - 0.04x + 24\right)\big|_{x=300} \approx 39$, or approximately \$39.

71. a. $R(x) = px = (2000 - 0.04x)x = 2000x - 0.04x^2$, so

$$P(x) = R(x) - C(x) = \left(2000x - 0.04x^2\right) - \left(0.000002x^3 - 0.02x^2 + 1000x + 120{,}000\right)$$
$$= -0.000002x^3 - 0.02x^2 + 1000x - 120{,}000.$$

Therefore,

$$\overline{C}(x) = \dfrac{C(x)}{x} = \dfrac{0.000002x^3 - 0.02x^2 + 1000x + 120{,}000}{x} = 0.000002x^2 - 0.02x + 1000 + \dfrac{120{,}000}{x}.$$

b. $C'(x) = \dfrac{d}{dx}\left(0.000002x^3 - 0.02x^2 + 1000x + 120{,}000\right) = 0.000006x^2 - 0.04x + 1000$,

$R'(x) = \dfrac{d}{dx}\left(2000x - 0.04x^2\right) = 2000 - 0.08x$,

$P'(x) = \dfrac{d}{dx}\left(-0.000002x^3 - 0.02x^2 + 1000x - 120{,}000\right) = -0.000006x^2 - 0.04x + 1000$, and

$\overline{C}'(x) = \dfrac{d}{dx}\left(0.000002x^2 - 0.02x + 1000 + 120{,}000x^{-1}\right) = 0.000004x - 0.02 - 120{,}000x^{-2}$.

c. $C'(3000) = 0.000006(3000)^2 - 0.04(3000) + 1000 = 934$, $R'(3000) = 2000 - 0.08(3000) = 1760$, and

$P'(3000) = -0.000006(3000)^2 - 0.04(3000) + 1000 = 826$.

d. $\overline{C}'(5000) = 0.000004(5000) - 0.02 - 120{,}000(5000)^{-2} = -0.0048$, and

$\overline{C}'(8000) = 0.000004(8000) - 0.02 - 120{,}000(8000)^{-2} \approx 0.0101$. At a production level of 5000 machines, the

average cost of each additional unit is decreasing at a rate of 0.48 cents. At a production level of 8000 machines,

the average cost of each additional unit is increasing at a rate of approximately 1 cent per unit.

73. $G'(t) = \dfrac{d}{dt}\left(-0.3t^3 + 1.2t^2 + 500\right) = -0.9t^2 + 2.4t$, so $G'(2) = -0.9(4) + 2.4(2) = 1.2$. Thus, the GDP is

growing at the rate of \$1.2 billion/year. $G''(2) = (-1.8t + 2.4)|_{t=2} = -1.2$, so the rate of rate of change of the

GDP is decreasing at the rate of \$1.2 billion/yr/yr.

© 2011 Cengage Learning. All Rights Reserved. May not be scanned, copied or duplicated, or posted to a publicly accessible website, in whole or in part.

1. $f'(x) = 2(3x^2) - 3\left(\frac{1}{3}x^{-2/3}\right) + 5\left(-\frac{2}{3}x^{-5/3}\right) = 6x^2 - x^{-2/3} - \frac{10}{3}x^{-5/3}.$

2. $g'(x) = \dfrac{d}{dx}\left[x\left(2x^2 - 1\right)^{1/2}\right] = \left(2x^2 - 1\right)^{1/2} + x\left(\frac{1}{2}\right)\left(2x^2 - 1\right)^{-1/2}\dfrac{d}{dx}\left(2x^2 - 1\right)$

$$= \left(2x^2 - 1\right)^{1/2} + \frac{1}{2}x\left(2x^2 - 1\right)^{-1/2}(4x) = \left(2x^2 - 1\right)^{-1/2}\left[\left(2x^2 - 1\right) + 2x^2\right] = \frac{4x^2 - 1}{\sqrt{2x^2 - 1}}.$$

3. $\dfrac{dy}{dx} = \dfrac{\left(x^2 + x + 1\right)(2) - (2x + 1)(2x + 1)}{\left(x^2 + x + 1\right)^2} = \dfrac{2x^2 + 2x + 2 - \left(4x^2 + 4x + 1\right)}{\left(x^2 + x + 1\right)^2} = -\dfrac{2x^2 + 2x - 1}{\left(x^2 + x + 1\right)^2}.$

4. $f'(x) = \dfrac{d}{dx}(x + 1)^{-1/2} = -\frac{1}{2}(x + 1)^{-3/2} = -\dfrac{1}{2(x + 1)^{3/2}}$, so

$f''(x) = -\frac{1}{2}\left(-\frac{3}{2}\right)(x + 1)^{-5/2} = \frac{3}{4}(x + 1)^{-5/2} = \dfrac{3}{4(x + 1)^{5/2}}$ and

$f'''(x) = \frac{3}{4}\left(-\frac{5}{2}\right)(x + 1)^{-7/2} = -\frac{15}{8}(x + 1)^{-7/2} = -\dfrac{15}{8(x + 1)^{7/2}}.$

5. Differentiating both sides of the equation implicitly with respect to x gives $y^2 + x\left(2yy'\right) - 2xy - x^2y' + 3x^2 = 0$,

so $\left(2xy - x^2\right)y' + \left(y^2 - 2xy + 3x^2\right) = 0$ and $y' = \dfrac{-y^2 + 2xy - 3x^2}{2xy - x^2} = \dfrac{-y^2 + 2xy - 3x^2}{x(2y - x)}.$

6. a. $dy = \dfrac{d}{dx}\left[x\left(x^2 + 5\right)^{1/2}\right]dx = \left[x\left(\frac{1}{2}\right)\left(x^2 + 5\right)^{-1/2}(2x)\right]dx + \left[\left(x^2 + 5\right)^{1/2}(1)\right]dx$

$$= \left(x^2 + 5\right)^{-1/2}\left[\left(x^2 + 5\right) + x^2\right]dx = \dfrac{2x^2 + 5}{\sqrt{x^2 + 5}}dx.$$

b. Here $dx = \Delta x = 2.01 - 2 = 0.01$. Therefore, $\Delta y \approx dy = \dfrac{2(4) + 5}{\sqrt{4 + 5}}(0.01) = \dfrac{0.13}{3} \approx 0.043.$

© 2011 Cengage Learning. All Rights Reserved. May not be scanned, copied or duplicated, or posted to a publicly accessible website, in whole or in part.

4 APPLICATIONS OF THE DERIVATIVE

Problem-Solving Tips

1. The critical number of a function f is any number x in the domain of f such that $f'(x) = 0$ or $f'(x)$ does not exist. Note that the definition requires that x be in the domain of f. For example, consider the function $f(x) = x + \frac{1}{x}$ in Example 8 on page 253 of the text. Even though f' is discontinuous at $x = 0$, this value does not qualify as a critical number because it does not lie in the domain of f.

2. Note that when you use test values to find the sign of a derivative over an interval, you don't need to evaluate the derivative at a test value. You need only find the **sign** of the derivative at that test value. For example, to find the sign of $f'(x) = \dfrac{(x+1)(x-1)}{x^2}$ in the interval $(0, 1)$ using the test value $x = \frac{1}{2}$, we simply note that the numerator is the product of a positive number and a negative number, so it is negative. Because x^2 is always positive, the denominator is always positive. The quotient of a negative number and a positive number is negative, so f' is negative over the interval $(0, 1)$.

4.1 Concept Questions   page 255

1. a. f is increasing on I if whenever x_1 and x_2 are in I with $x_1 < x_2$, then $f(x_1) < f(x_2)$.
   b. f is decreasing on I if whenever x_1 and x_2 are in I with $x_1 < x_2$, then $f(x_1) > f(x_2)$.

3. a. f has a relative maximum at $x = a$ if there is an open interval I containing a such that $f(x) \leq f(a)$ for all x in I.
   b. f has a relative minimum at $x = a$ if there is an open interval I containing a such that $f(x) \geq f(a)$ for all x in I.

5. See page 252 of the text.

4.1 Applications of the First Derivative   page 255

1. f is decreasing on $(-\infty, 0)$ and increasing on $(0, \infty)$.

3. f is increasing on $(-\infty, -1) \cup (1, \infty)$, and decreasing on $(-1, 1)$.

5. f is increasing on $(0, 2)$ and decreasing on $(-\infty, 0) \cup (2, \infty)$.

7. f is decreasing on $(-\infty, -1) \cup (1, \infty)$ and increasing on $(-1, 1)$.

© 2011 Cengage Learning. All Rights Reserved. May not be scanned, copied or duplicated, or posted to a publicly accessible website, in whole or in part.

9. Increasing on $(20.2, 20.6) \cup (21.7, 21.8)$, constant on $(19.6, 20.2) \cup (20.6, 21.1)$, and decreasing on $(21.1, 21.7) \cup (21.8, 22.7)$.

11. a. Positive   **b.** Positive   **c.** Zero   **d.** Zero

 e. Negative   **f.** Negative   **g.** Positive

13. $f(x) = 3x + 5$, so $f'(x) = 3 > 0$ for all x. Thus, f is increasing on $(-\infty, \infty)$.

15. $f(x) = x^2 - 3x$, so $f'(x) = 2x - 3$. f' is continuous everywhere and is equal to zero when $x = \frac{3}{2}$. From the sign diagram, we see that f is decreasing on $\left(-\infty, \frac{3}{2}\right)$ and increasing on $\left(\frac{3}{2}, \infty\right)$.

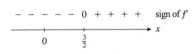

17. $g(x) = x - x^3$, so $g'(x) = 1 - 3x^2$ is continuous everywhere and is equal to zero when $1 - 3x^2 = 0$, or $x = \pm\frac{\sqrt{3}}{3}$. From the sign diagram, we see that f is decreasing on $\left(-\infty, -\frac{\sqrt{3}}{3}\right) \cup \left(\frac{\sqrt{3}}{3}, \infty\right)$ and increasing on $\left(-\frac{\sqrt{3}}{3}, \frac{\sqrt{3}}{3}\right)$.

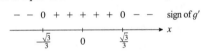

19. $g(x) = x^3 + 3x^2 + 1$, so $g'(x) = 3x^2 + 6x = 3x(x + 2)$. From the sign diagram, we see that g is increasing on $(-\infty, -2) \cup (0, \infty)$ and decreasing on $(-2, 0)$.

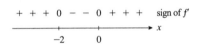

21. $f(x) = \frac{1}{3}x^3 - 3x^2 + 9x + 20$, so $f'(x) = x^2 - 6x + 9 = (x - 3)^2 > 0$ for all x except $x = 3$, at which point $f'(3) = 0$. Therefore, f is increasing on $(-\infty, 3) \cup (3, \infty)$.

23. $h(x) = x^4 - 4x^3 + 10$, so $h'(x) = 4x^3 - 12x^2 = 4x^2(x - 3) = 0$ if $x = 0$ or 3. From the sign diagram of h', we see that h is increasing on $(3, \infty)$ and decreasing on $(-\infty, 0) \cup (0, 3)$.

25. $f(x) = \frac{1}{x - 2} = (x - 2)^{-1}$, so $f'(x) = -1(x - 2)^{-2}(1) = -\frac{1}{(x - 2)^2}$ is discontinuous at $x = 2$ and is continuous and nonzero everywhere else. From the sign diagram, we see that f is decreasing on $(-\infty, 2) \cup (2, \infty)$.

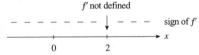

27. $h(t) = \frac{t}{t - 1}$, so $h'(t) = \frac{(t - 1)(1) - t(1)}{(t - 1)^2} = -\frac{1}{(t - 1)^2}$. From the sign diagram, we see that $h'(t) < 0$ whenever it is defined. We conclude that h is decreasing on $(-\infty, 1) \cup (1, \infty)$.

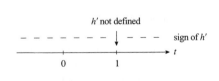

© 2011 Cengage Learning. All Rights Reserved. May not be scanned, copied or duplicated, or posted to a publicly accessible website, in whole or in part.

29. $f(x) = x^{3/5}$, so $f'(x) = \frac{3}{5}x^{-2/5} = \frac{3}{5x^{2/5}}$. Observe that $f'(x)$ is not defined at $x = 0$, but is positive everywhere else. Therefore, f is increasing on $(-\infty, 0) \cup (0, \infty)$.

31. $f(x) = \sqrt{x+1}$, so $f'(x) = \frac{d}{dx}(x+1)^{1/2} = \frac{1}{2}(x+1)^{-1/2} = \frac{1}{2\sqrt{x+1}}$, and we see that $f'(x) > 0$ if $x > -1$. Therefore, f is increasing on $(-1, \infty)$.

33. $f(x) = \sqrt{16 - x^2} = (16 - x^2)^{1/2}$, so

$f'(x) = \frac{1}{2}(16 - x^2)^{-1/2}(-2x) = -\frac{x}{\sqrt{16 - x^2}}$. Because

the domain of f is $[-4, 4]$, we consider the sign diagram
for f' on this interval. We see that f is increasing on
$(-4, 0)$ and decreasing on $(0, 4)$.

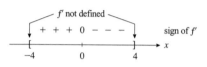

35. $f'(x) = \frac{d}{dx}(x - x^{-1}) = 1 + \frac{1}{x^2} = \frac{x^2 + 1}{x^2} > 0$ for all $x \neq 0$. Therefore, f is increasing on $(-\infty, 0) \cup (0, \infty)$.

37. f has a relative maximum of $f(0) = 1$ and relative minima of $f(-1) = 0$ and $f(1) = 0$.

39. f has a relative maximum of $f(-1) = 2$ and a relative minimum of $f(1) = -2$.

41. f has a relative maximum of $f(1) = 3$ and a relative minimum of $f(2) = 2$.

43. f has a relative minimum at $(0, 2)$.

45. a   **47.** d

49. $f(x) = x^2 - 4x$, so $f'(x) = 2x - 4 = 2(x - 2)$ has a
critical point at $x = 2$. From the sign diagram, we see that
$f(2) = -4$ is a relative minimum by the First Derivative
Test.

51. $h(t) = -t^2 + 6t + 6$, so $h'(t) = -2t + 6 = -2(t - 3) = 0$
if $t = 3$, a critical number. The sign diagram and the First
Derivative Test imply that h has a relative maximum at 3
with value $h(3) = -9 + 18 + 6 = 15$.

53. $f(x) = x^{5/3}$, so $f'(x) = \frac{5}{3}x^{2/3}$, $x = 0$ as the critical
number of f. From the sign diagram, we see that f' does
not change sign as we move across $x = 0$, and conclude
that f has no relative extremum.

55. $g(x) = x^3 - 3x^2 + 4$, so $g'(x) = 3x^2 - 6x = 3x(x - 2) = 0$ if $x = 0$ or 2. From the sign diagram, we see that
the critical number $x = 0$ gives a relative maximum,
whereas $x = 2$ gives a relative minimum. The values are
$g(0) = 4$ and $g(2) = 8 - 12 + 4 = 0$.

© 2011 Cengage Learning. All Rights Reserved. May not be scanned, copied or duplicated, or posted to a publicly accessible website, in whole or in part.

57. $f(x) = \frac{1}{2}x^4 - x^2$, so $f'(x) = 2x^3 - 2x = 2x(x^2 - 1) = 2x(x+1)(x-1)$ is continuous everywhere and has

zeros at $x = -1, 0$, and 1, the critical numbers of f. Using
the First Derivative Test and the sign diagram of f', we see
that $f(-1) = -\frac{1}{2}$ and $f(1) = -\frac{1}{2}$ are relative minima of
f and $f(0) = 0$ is a relative maximum of f.

```
    - - - 0 + + 0 - - - 0 + + +    sign of f′
  +-----------+-----------+---------→ x
          -1          0          1
```

59. $F(x) = \frac{1}{3}x^3 - x^2 - 3x + 4$, so $F'(x) = x^2 - 2x - 3 = (x-3)(x+1) = 0$ gives $x = -1$ and $x = 3$ as critical

numbers. From the sign diagram, we see that $x = -1$ gives a relative maximum and $x = 3$ gives a relative

minimum. The values are $F(-1) = -\frac{1}{3} - 1 + 3 + 4 = \frac{17}{3}$ and $F(3) = 9 - 9 - 9 + 4 = -5$.

61. $g(x) = x^4 - 4x^3 + 8$. Setting
$g'(x) = 4x^3 - 12x^2 = 4x^2(x-3) = 0$ gives $x = 0$ and
$x = 3$ as critical numbers. From the sign diagram, we see
that $x = 3$ gives a relative minimum. Its value is
$g(3) = 3^4 - 4(3)^3 + 8 = -19$.

```
    - - - 0 - - - 0 + + +    sign of g′
  +-----------+-----------+---------→ x
          0           3
```

63. $g'(x) = \dfrac{d}{dx}\left(1 + \dfrac{1}{x}\right) = -\dfrac{1}{x^2}$. Observe that g' is nonzero for all values of x. Furthermore, g' is undefined at

$x = 0$, but $x = 0$ is not in the domain of g. Therefore, g has no critical number and hence no relative extremum.

65. $f(x) = x + \dfrac{9}{x} + 2$, so $f'(x) = 1 - \dfrac{9}{x^2} = \dfrac{x^2 - 9}{x^2} = \dfrac{(x+3)(x-3)}{x^2} = 0$ gives $x = -3$ and $x = 3$ as critical

numbers. From the sign diagram, we see that $(-3, -4)$ is a
relative maximum and $(3, 8)$ is a relative minimum.

```
                    f′ not defined
    + + + 0 - - -  |  - - 0 + + +    sign of f′
  +-----------+-----------+---------→ x
          -3          0          3
```

67. $f(x) = \dfrac{x}{1+x^2}$, so $f'(x) = \dfrac{(1+x^2)(1) - x(2x)}{(1+x^2)^2} = \dfrac{1-x^2}{(1+x^2)^2} = \dfrac{(1-x)(1+x)}{(1+x^2)^2} = 0$ if $x = \pm 1$, and these are

critical numbers of f. From the sign diagram of f', we see
that f has a relative minimum at $\left(-1, -\frac{1}{2}\right)$ and a relative
maximum at $\left(1, \frac{1}{2}\right)$.

```
    - - - 0 + + + + + 0 - - -    sign of f′
  +-----------+-----------+---------→ t
          -1          0          1
```

69. $f(x) = (x-1)^{2/3}$, so

$f'(x) = \frac{2}{3}(x-1)^{-1/3} = \dfrac{2}{3(x-1)^{1/3}}$. $f'(x)$ is

discontinuous at $x = 1$. From the sign diagram for f', we
conclude that $f(1) = 0$ is a relative minimum.

```
                    f′ not defined
    - - - - - - - - - -  |  + + +    sign of f′
  +-----------+-----------+---------→ x
          0           1
```

© 2011 Cengage Learning. All Rights Reserved. May not be scanned, copied or duplicated, or posted to a publicly accessible website, in whole or in part.

71. $h(t) = -16t^2 + 64t + 80$, so

$h'(t) = -32t + 64 = -32(t - 2)$. The sign diagram

shows us that the stone is rising in the time interval $(0, 2)$

and falling when $t > 2$. It hits the ground when

$h(t) = -16t^2 + 64t + 80 = 0$, that is,

$t^2 - 4t - 5 = (t - 5)(t + 1) = 0$ or $t = 5$. (We reject the

root $t = -1$.)

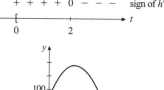

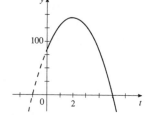

73. $P'(x) = \dfrac{d}{dx}\left(0.0726x^2 + 0.7902x + 4.9623\right) = 0.1452x + 0.7902$. Because $P'(x) > 0$ on $(0, 25)$, we see that P

is increasing on the interval in question. Our result tells us that the proportion of the population afflicted with

Alzheimer's disease increases with age for those that are 65 and over.

75. $h(t) = -\frac{1}{3}t^3 + 16t^2 + 33t + 10$, so $h'(t) = -t^2 + 32t + 33 = -(t + 1)(t - 33)$. The sign diagram for h' shows

that the rocket is ascending on the time interval $(0, 33)$ and

descending on $(33, T)$ for some positive number T. The

parachute is deployed 33 seconds after liftoff.

$$+ + + + 0\ -\ -\ -\quad \text{sign of } h'$$

$$\underset{0\qquad\qquad 33}{\xrightarrow{\hspace{3cm}} t}$$

77. $f(t) = 20t - 40\sqrt{t} + 50 = 20t - 40t^{1/2} + 50$, so $f'(t) = 20 - 40\left(\frac{1}{2}t^{-1/2}\right) = 20\left(1 - \dfrac{1}{\sqrt{t}}\right) = \dfrac{20(\sqrt{t} - 1)}{\sqrt{t}}$.

Thus, f' is continuous on $(0, 4)$ and is equal to zero at $t = 1$. From the sign diagram, we see that f is decreasing on

$(0, 1)$ and increasing on $(1, 4)$. We conclude that the

average speed decreases from 6 a.m. to 7 a.m. and then

picks up from 7 a.m. to 10 a.m.

$$-\ 0 + + + + + +\quad \text{sign of } f'$$

$$\underset{0\quad 1\qquad\qquad 4}{\xrightarrow{\hspace{3cm}} t}$$

79. a. $f'(t) = \dfrac{d}{dt}\left(-0.05t^3 + 0.56t^2 + 5.47t + 7.5\right) = -0.15t^2 + 1.12t + 5.47$. Setting $f'(t) = 0$ gives

$-0.15t^2 + 1.12t + 5.47 = 0$. Using the quadratic formula, we find $t = \dfrac{-1.12 \pm \sqrt{(1.12)^2 - 4(-0.15)(5.47)}}{-0.3}$;

that is, $t = -3.37$ or 10.83. Because f' is continuous, the only critical numbers of f are $t \approx -3.4$ and $t \approx 10.8$,

both of which lie outside the interval of interest. Nevertheless, this result can be used to tell us that f' does not

change sign in the interval $(-3.4, 10.8)$. Using $t = 0$ as the test number, we see that $f'(0) = 5.47 > 0$ and so we

see that f is increasing on $(-3.4, 10.8)$ and, in particular, in the interval $(0, 6)$. Thus, we conclude that f is

increasing on $(0, 6)$.

b. The result of part (a) tells us that sales in the web hosting industry will be increasing from 1999 through 2005.

81. $S'(t) = \dfrac{d}{dt}\left(0.46t^3 - 2.22t^2 + 6.21t + 17.25\right) = 1.38t^2 - 4.44t + 6.21$. Observe that S' is continuous everywhere.

Setting $S'(t) = 0$ and solving, we find $t = \dfrac{4.44\sqrt{4.44^2 - 4(1.38)(6.21)}}{2(1.38)}$. The discriminant is $-14.5656 < 0$,

which shows that the equation has no real roots. Because $S'(0) = 6.21 > 0$, we conclude that $S'(t) > 0$ for all t, in

particular, for t in the interval $(0, 4)$. This shows that S is increasing on $(0, 4)$.

© 2011 Cengage Learning. All Rights Reserved. May not be scanned, copied or duplicated, or posted to a publicly accessible website, in whole or in part.

83. $S'(t) = -6.945t^2 + 68.65t + 1.32$. Setting $S'(t) = 0$ and solving the resulting equation, we obtain

$t = \dfrac{-68.65 \pm \sqrt{(68.65)^2 - 4(-6.945)(1.32)}}{2(-6.945)} \approx -0.02$ or 9.90. From the sign diagram for S', we see that S' is

increasing on the interval $(0, 5)$. We conclude that U.S. telephone company spending was projected to increase from 2001 through 2006.

85. $C(t) = \dfrac{t^2}{2t^3 + 1}$, so $C'(t) = \dfrac{(2t^3 + 1)(2t) - t^2(6t^2)}{(2t^3 + 1)^2} = \dfrac{2t - 2t^4}{(2t^3 + 1)^2} = \dfrac{2t(1 - t^3)}{(2t^3 + 1)^2}$. From the sign diagram of C',

we see that the drug concentration is increasing on $(0, 1)$ and decreasing on $(1, 4)$.

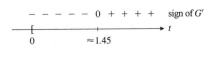

87. $A(t) = \dfrac{136}{1 + 0.25(t - 4.5)^2} + 28$, so

$A'(t) = 136 \dfrac{d}{dt}\left[1 + 0.25(t - 4.5)^2\right]^{-1} = -136\left[1 + 0.25(t - 4.5)^2\right]^{-2} 2(0.25)(t - 4.5) = -\dfrac{68(t - 4.5)}{\left[1 + 0.25(t - 4.5)^2\right]^2}$.

Observe that $A'(t) > 0$ if $t < 4.5$ and $A'(t) < 0$ if $t > 4.5$, so the pollution is increasing from 7 a.m. to 11:30 a.m. and decreasing from 11:30 a.m. to 6 p.m.

89. a. $G(t) = (D - S)(t) = D(t) - S(t) = \left(0.0007t^2 + 0.0265t + 2\right) - \left(-0.0014t^2 + 0.0326t + 1.9\right)$

$= 0.0021t^2 - 0.0061t + 0.1$.

b. $G'(t) = 0.0042t - 0.0061 = 0$ implies $t \approx 1.45$. From the sign diagram of G', we see that G is decreasing on $(0, 1.5)$ and increasing on $(1.5, 15)$. This shows that the gap between the demand and supply of nurses was increasing from 2000 through the middle of 2001 but starts widening from the middle of 2001 through 2015.

c. The relative minimum of G occurs at $t = 1.5$ and is $f(1.5) \approx 0.0956$. This says that the smallest shortage is approximately 96,000.

91. True. Let $a < x_1 < x_2 < b$. Then $f(x_2) > f(x_1)$ and $g(x_2) > g(x_1)$. Therefore, $(f + g)(x_2) = f(x_2) + g(x_2) > f(x_1) + g(x_1) = (f + g)(x_1)$, and so $f + g$ is increasing on (a, b).

93. True. Let $a < x_1 < x_2 < b$. Then $f(x_1) < f(x_2)$ and $g(x_1) < g(x_2)$. We find

$(fg)(x_2) - (fg)(x_1) = f(x_2)g(x_2) - f(x_1)g(x_1)$

$= f(x_2)g(x_2) - f(x_2)g(x_1) + f(x_2)g(x_1) - f(x_1)g(x_1)$

$= f(x_2)\left[g(x_2) - g(x_1)\right] + g(x_1)\left[f(x_2) - f(x_1)\right] > 0$,

so $(fg)(x_2) > (fg)(x_1)$ and fg is increasing on (a, b).

95. False. Let $f(x) = |x|$. Then f has a relative minimum at $x = 0$, but $f'(0)$ does not exist.

97. $f'(x) = 3x^2 + 1$ is continuous on $(-\infty, \infty)$ and is always greater than or equal to 1, so f has no critical number in $(-\infty, \infty)$. Therefore, f has no relative extremum on $(-\infty, \infty)$.

© 2011 Cengage Learning. All Rights Reserved. May not be scanned, copied or duplicated, or posted to a publicly accessible website, in whole or in part.

99. We require that $f'(-1) = 0$; that is, $f'(-1) = (3ax^2 + 12x + b)\big|_{x=-1} = 3a - 12 + b = 0$, and

therefore $f'(2) = 0$, or $f'(2) = (3ax^2 + 12x + b)\big|_{x=2} = 12a + 24 + b = 0$. Solving the system

$$\begin{cases} 3a + b = 12 \\ 12a + b = -24 \end{cases} \text{ simultaneously gives } a = -4 \text{ and } b = 24.$$

101. a. $f'(x) = -2x$ if $x \neq 0$, $f'(-1) = 2$, and $f'(1) = -2$, so $f'(x)$ changes sign from positive to negative as we move across $x = 0$.

b. f does not have a relative maximum at $x = 0$ because $f(0) = 2$ but a neighborhood of $x = 0$, for example $\left(-\frac{1}{2}, \frac{1}{2}\right)$, contains numbers with values larger than 2. This does not contradict the First Derivative Test because f is not continuous at $x = 0$.

103. $f(x) = ax^2 + bx + c$. Setting $f'(x) = 2ax + b = 2a\left(x + \frac{b}{2a}\right) = 0$ gives $x = -\frac{b}{2a}$ as the only critical number of f. If $a < 0$, we the sign diagram shows that $x = -\frac{b}{2a}$ gives a relative maximum. Similarly, it can be shown that if $a > 0$, then $x = -\frac{b}{2a}$ gives a relative minimum.

```
+ + + + + 0 - - - -    sign of f'
─────────┼────┼──────→ x
         0   -b/2a
```

105. a. $f'(x) = 3x^2 + 1$, and so $f'(x) > 1$ on the interval $(0, 1)$. Therefore, f is increasing on $(0, 1)$.

b. $f(0) = -1$ and $f(1) = 1 + 1 - 1 = 1$. Thus, the Intermediate Value Theorem guarantees that there is at least one root of $f(x) = 0$ in $(0, 1)$. Because f is increasing on $(0, 1)$, the graph of f can cross the x-axis at only one point in $(0, 1)$, and so $f(x) = 0$ has exactly one root.

4.1   Using Technology   page 263

1. a. f is decreasing on $(-\infty, -0.2934)$ and increasing on $(-0.2934, \infty)$.

b. Relative minimum $f(-0.2934) = -2.5435$.

3. a. f is increasing on $(-\infty, -1.6144) \cup (0.2390, \infty)$ and decreasing on $(-1.6144, 0.2390)$.

b. Relative maximum $f(-1.6144) = 26.7991$, relative minimum $f(0.2390) = 1.6733$.

5. a. f is decreasing on $(-\infty, -1) \cup (0.33, \infty)$ and increasing on $(-1, 0.33)$.

b. Relative maximum $f(0.33) = 1.11$, relative minimum $f(-1) = -0.63$.

7. a. f is decreasing on $(-1, -0.71)$ and increasing on $(-0.71, 1)$.

b. f has a relative minimum at $(-0.71, -1.41)$.

9. a.

b. Increasing on $(0, 3.6676)$, decreasing on $(3.6676, 6)$.

© 2011 Cengage Learning. All Rights Reserved. May not be scanned, copied or duplicated, or posted to a publicly accessible website, in whole or in part.

11. The PSI is increasing on the interval $(0, 4.5)$ and decreasing on $(4.5, 11)$. It is highest when $t = 4.5$ (at 11:30 a.m.) and has value 164.

4.2   Problem-Solving Tips

1. If $f''(x) > 0$, then the graph of f "holds water" and we say the graph of f is concave upward.

If $f''(x) < 0$, then the graph of f "loses water" and we say the graph of f is concave downward.

2. To find the inflection points of a function f, determine the number(s) in the domain of f for which $f''(x) = 0$ or $f''(x)$ does not exist. Note that each of these numbers c provides us with a candidate $(c, f(c))$ for an inflection point of f.

3. Note that the second derivative test is not valid when $f''(c) = 0$ or $f''(c)$ does not exist. In these cases you need to use the first derivative test to determine the relative extrema.

4.2   Concept Questions   page 274

1. a. f is concave upward on (a, b) if f' is increasing on (a, b). f is concave downward on (a, b) if f' is decreasing on (a, b).

   b. For the procedure for determining where f is concave upward and where f is concave downward, see page 265 of the text.

3. The second derivative test is stated in the text on page 272 of the text. In general, if f'' is easy to compute, then use the Second Derivative Test. However, keep in mind that (1) in order to use this test f'' must exist, (2) the test is inconclusive if $f''(c) = 0$, and (3) the test is inconvenient to use if f'' is difficult to compute.

4.2   Applications of the Second Derivative   page 274

1. f is concave downward on $(-\infty, 0)$ and concave upward on $(0, \infty)$. f has an inflection point at $(0, 0)$.

3. f is concave downward on $(-\infty, 0) \cup (0, \infty)$.

5. f is concave upward on $(-\infty, 0) \cup (1, \infty)$ and concave downward on $(0, 1)$. $(0, 0)$ and $(1, -1)$ are inflection points of f.

7. f is concave downward on $(-\infty, -2) \cup (-2, 2) \cup (2, \infty)$.

9. a. f is concave upward on $(0, 2) \cup (4, 6) \cup (7, 9) \cup (9, 12)$ and concave downward on $(2, 4) \cup (6, 7)$.

   b. f has inflection points at $\left(2, \frac{5}{2}\right)$, $(4, 2)$, $(6, 2)$, and $(7, 3)$.

© 2011 Cengage Learning. All Rights Reserved. May not be scanned, copied or duplicated, or posted to a publicly accessible website, in whole or in part.

11. a

13. b

15. a. $D_1'(t) > 0$, $D_2'(t) > 0$, $D_1''(t) > 0$, and $D_2''(t) < 0$ on $(0, 12)$.

b. With or without the proposed promotional campaign, the deposits will increase, but with the promotion, the deposits will increase at an increasing rate whereas without the promotion, the deposits will increase at a decreasing rate.

17. The significance of the inflection point Q is that the restoration process is working at its peak at the time t_0 corresponding to its t-coordinate.

19. $f(x) = 4x^2 - 12x + 7$, so $f'(x) = 8x - 12$ and $f''(x) = 8$. Thus, $f''(x) > 0$ everywhere, and so f is concave upward everywhere.

21. $f(x) = \dfrac{1}{x^4} = x^{-4}$, so $f'(x) = -\dfrac{4}{x^5}$ and $f''(x) = \dfrac{20}{x^6} > 0$ for all values of x in $(-\infty, 0)$ and $(0, \infty)$, and so f is concave upward on its domain.

23. $f(x) = 2x^2 - 3x + 4$, so $f'(x) = 4x - 3$ and $f''(x) = 4x > 0$ for all values of x. Thus, f is concave upward on $(-\infty, \infty)$.

25. $f(x) = x^3 - 1$, so $f'(x) = 3x^2$ and $f''(x) = 6x$. From the sign diagram of f'', we see that f is concave downward on $(-\infty, 0)$ and concave upward on $(0, \infty)$.

27. $f(x) = x^4 - 6x^3 + 2x + 8$, so $f'(x) = 4x^3 - 18x^2 + 2$ and $f''(x) = 12x^2 - 36x = 12x(x - 3)$. The sign diagram of f'' shows that f is concave upward on $(-\infty, 0) \cup (3, \infty)$ and concave downward on $(0, 3)$.

29. $f(x) = x^{4/7}$, so $f'(x) = \frac{4}{7}x^{-3/7}$ and $f''(x) = -\frac{12}{49}x^{-10/7} = -\dfrac{12}{49x^{10/7}}$. Observe that $f''(x) < 0$ for all $x \neq 0$, so f is concave downward on $(-\infty, 0) \cup (0, \infty)$.

31. $f(x) = (4 - x)^{1/2}$, so $f'(x) = \frac{1}{2}(4 - x)^{-1/2}(-1) = \frac{1}{2}(4 - x)^{-1/2}$ and $f''(x) = \frac{1}{4}(4 - x)^{-3/2}(-1) = -\dfrac{1}{4(4 - x)^{3/2}} < 0$ whenever it is defined, so f is concave downward on $(-\infty, 4)$.

33. $f'(x) = \dfrac{d}{dx}(x - 2)^{-1} = -(x - 2)^{-2}$ and $f''(x) = 2(x - 2)^{-3} = \dfrac{2}{(x - 2)^3}$. The sign diagram of f'' shows that f is concave downward on $(-\infty, 2)$ and concave upward on $(2, \infty)$.

© 2011 Cengage Learning. All Rights Reserved. May not be scanned, copied or duplicated, or posted to a publicly accessible website, in whole or in part.

35. $f'(x) = \dfrac{d}{dx}(2+x^2)^{-1} = -(2+x^2)^{-2}(2x) = -2x(2+x^2)^{-2}$ and

$f''(x) = -2(2+x^2)^{-2} - 2x(-2)(2+x^2)^{-3}(2x) = 2(2+x^2)^{-3}[-(2+x^2)+4x^2] = \dfrac{2(3x^2-2)}{(2+x^2)^3} = 0$ if

$x = \pm\sqrt{\frac{2}{3}} = \frac{\sqrt{6}}{3}$. From the sign diagram of f'', we see that

f is concave upward on $\left(-\infty, -\frac{\sqrt{6}}{3}\right) \cup \left(\frac{\sqrt{6}}{3}, \infty\right)$ and

concave downward on $\left(-\frac{\sqrt{6}}{3}, \frac{\sqrt{6}}{3}\right)$.

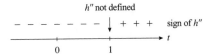

37. $h(t) = \dfrac{t^2}{t-1}$, so $h'(t) = \dfrac{(t-1)(2t)-t^2(1)}{(t-1)^2} = \dfrac{t^2-2t}{(t-1)^2}$ and

$h''(t) = \dfrac{(t-1)^2(2t-2)-(t^2-2t)2(t-1)}{(t-1)^4} = \dfrac{(t-1)(2t^2-4t+2-2t^2+4t)}{(t-1)^4} = \dfrac{2}{(t-1)^3}$.

The sign diagram of h'' shows that h is concave downward

on $(-\infty, 1)$ and concave upward on $(1, \infty)$.

39. $g(x) = x + \dfrac{1}{x^2}$, so $g'(x) = 1 - 2x^{-3}$ and $g''(x) = 6x^{-4} = \dfrac{6}{x^4} > 0$ whenever $x \neq 0$. Therefore, g is concave

upward on $(-\infty, 0) \cup (0, \infty)$.

41. $g(t) = (2t-4)^{1/3}$, so $g'(t) = \frac{1}{3}(2t-4)^{-2/3}(2) = \frac{2}{3}(2t-4)^{-2/3}$ and

$g''(t) = -\frac{4}{9}(2t-4)^{-5/3} = -\dfrac{4}{9(2t-4)^{5/3}}$. The sign

diagram of g'' shows that g is concave upward on $(-\infty, 2)$

and concave downward on $(2, \infty)$.

43. $f(x) = x^3 - 2$, so $f'(x) = 3x^2$ and $f''(x) = 6x$. $f''(x)$

is continuous everywhere and has a zero at $x = 0$. From the

sign diagram of f'', we conclude that $(0, -2)$ is an

inflection point of f.

45. $f(x) = 6x^3 - 18x^2 + 12x - 15$, so

$f'(x) = 18x^2 - 36x + 12$ and

$f''(x) = 36x - 36 = 36(x-1) = 0$ if $x = 1$. The sign

diagram of f'' shows that f has an inflection point at

$(1, -15)$.

47. $f(x) = 3x^4 - 4x^3 + 1$, so $f'(x) = 12x^3 - 12x^2$ and $f''(x) = 36x^2 - 24x = 12x(3x-2) = 0$ if $x = 0$ or $\frac{2}{3}$.

These are candidates for inflection points. The sign diagram

of f'' shows that $(0, 1)$ and $\left(\frac{2}{3}, \frac{11}{27}\right)$ are inflection points of

f.

© 2011 Cengage Learning. All Rights Reserved. May not be scanned, copied or duplicated, or posted to a publicly accessible website, in whole or in part.

49. $g(t) = t^{1/3}$, so $g'(t) = \frac{1}{3}t^{-2/3}$ and $g''(t) = -\frac{2}{9}t^{-5/3} = -\frac{2}{9t^{5/3}}$. Observe that $t = 0$ is in the domain of g. Next, since $g''(t) > 0$ if $t < 0$ and $g''(t) < 0$ if $t > 0$, we see that $(0, 0)$ is an inflection point of g.

51. $f(x) = (x - 1)^3 + 2$, so $f'(x) = 3(x - 1)^2$ and $f''(x) = 6(x - 1)$. Observe that $f''(x) < 0$ if $x < 1$ and $f''(x) > 0$ if $x > 1$ and so $(1, 2)$ is an inflection point of f.

53. $f(x) = \dfrac{2}{1 + x^2} = 2(1 + x^2)^{-1}$, so $f'(x) = -2(1 + x^2)^{-2}(2x) = -4x(1 + x^2)^{-2}$ and

$f''(x) = -4(1 + x^2)^{-2} - 4x(-2)(1 + x^2)^{-3}(2x) = 4(1 + x^2)^{-3}\left[-(1 + x^2) + 4x^2\right] = \dfrac{4(3x^2 - 1)}{(1 + x^2)^3}$, which is

continuous everywhere and has zeros at $x = \pm\frac{\sqrt{3}}{3}$. From the sign diagram of f'', we conclude that $\left(-\frac{\sqrt{3}}{3}, \frac{3}{2}\right)$ and $\left(\frac{\sqrt{3}}{3}, \frac{3}{2}\right)$ are inflection points of f.

$$+\ +\ +\ 0\ -\ -\ -\ -\ -\ -\ 0\ +\ +\ +\quad \text{sign of } f''$$

55. $f(x) = -x^2 + 2x + 4$, so $f'(x) = -2x + 2$. The critical number of f is $x = 1$. Because $f''(x) = -2$ and $f''(1) = -2 < 0$, we conclude that $f(1) = 5$ is a relative maximum of f.

57. $f(x) = 2x^3 + 1$, so $f'(x) = 6x^2 = 0$ if $x = 0$ and this is a critical number of f. Next, $f''(x) = 12x$, and so $f''(0) = 0$. Thus, the Second Derivative Test fails. But the First Derivative Test shows that $(0, 0)$ is not a relative extremum.

59. $f(x) = \frac{1}{3}x^3 - 2x^2 - 5x - 10$, so $f'(x) = x^2 - 4x - 5 = (x - 5)(x + 1)$ and this gives $x = -1$ and $x = 5$ as critical numbers of f. Next, $f''(x) = 2x - 4$. Because $f''(-1) = -6 < 0$, we see that $\left(-1, -\frac{22}{3}\right)$ is a relative maximum of f. Next, $f''(5) = 6 > 0$ and this shows that $\left(5, -\frac{130}{3}\right)$ is a relative minimum of f.

61. $g(t) = t + \dfrac{9}{t}$, so $g'(t) = 1 - \dfrac{9}{t^2} = \dfrac{t^2 - 9}{t^2} = \dfrac{(t + 3)(t - 3)}{t^2}$, showing that $t = \pm 3$ are critical numbers of g. Now, $g''(t) = 18t^{-3} = \dfrac{18}{t^3}$. Because $g''(-3) = -\frac{18}{27} < 0$, the Second Derivative Test implies that g has a relative maximum at $(-3, -6)$. Also, $g''(3) = \frac{18}{27} > 0$ and so g has a relative minimum at $(3, 6)$.

63. $f(x) = \dfrac{x}{1 - x}$, so $f'(x) = \dfrac{(1 - x)(1) - x(-1)}{(1 - x)^2} = \dfrac{1}{(1 - x)^2}$ is never zero. Thus, there is no critical number and f has no relative extremum.

65. $f(t) = t^2 - \dfrac{16}{t}$, so $f'(t) = 2t + \dfrac{16}{t^2} = \dfrac{2t^3 + 16}{t^2} = \dfrac{2(t^3 + 8)}{t^2}$. Setting $f'(t) = 0$ gives $t = -2$ as a critical number. Next, we compute $f''(t) = \dfrac{d}{dt}(2t + 16t^{-2}) = 2 - 32t^{-3} = 2 - \dfrac{32}{t^3}$. Because $f''(-2) = 2 - \dfrac{32}{(-8)} = 6 > 0$, we see that $(-2, 12)$ is a relative minimum.

© 2011 Cengage Learning. All Rights Reserved. May not be scanned, copied or duplicated, or posted to a publicly accessible website, in whole or in part.

67. $g(s) = \dfrac{s}{1+s^2}$, so $g'(s) = \dfrac{(1+s^2)(1) - s(2s)}{(1+s^2)^2} = \dfrac{1-s^2}{(1+s^2)^2} = 0$

gives $s = -1$ and $s = 1$ as critical numbers of g. Next, we compute

$$g''(s) = \frac{(1+s^2)^2(-2s) - (1-s^2)2(1+s^2)(2s)}{(1+s^2)^4} = \frac{2s(1+s^2)(-1-s^2-2+2s^2)}{(1+s^2)^4} = \frac{2s(s^2-3)}{(1+s^2)^3}. \text{ Now}$$

$g''(-1) = \frac{1}{2} > 0$, and so $g(-1) = -\frac{1}{2}$ is a relative minimum of g. Next, $g''(1) = -\frac{1}{2} < 0$ and so $g(1) = \frac{1}{2}$ is a

relative maximum of g.

69. $f(x) = \dfrac{x^4}{x-1}$, so $f'(x) = \dfrac{(x-1)(4x^3) - x^4(1)}{(x-1)^2} = \dfrac{4x^4 - 4x^3 - x^4}{(x-1)^2} = \dfrac{3x^4 - 4x^3}{(x-1)^2} = \dfrac{x^3(3x-4)}{(x-1)^2}. \text{ Thus, } x = 0$

and $x = \frac{4}{3}$ are critical numbers of f. Next,

$$f''(x) = \frac{(x-1)^2(12x^3 - 12x^2) - (3x^4 - 4x^3)(2)(x-1)}{(x-1)^4}$$

$$= \frac{(x-1)(12x^4 - 12x^3 - 12x^3 + 12x^2 - 6x^4 + 8x^3)}{(x-1)^4}$$

$$= \frac{6x^4 - 16x^3 + 12x^2}{(x-1)^3} = \frac{2x^2(3x^2 - 8x + 6)}{(x-1)^3}.$$

Because $f''\left(\frac{4}{3}\right) > 0$, we see that $f\left(\frac{4}{3}\right) = \frac{256}{27}$ is a relative

minimum. Because $f''(0) = 0$, the Second Derivative Test

fails. Using the sign diagram for f' and the First Derivative

Test, we see that $f(0) = 0$ is a relative maximum.

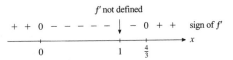

71.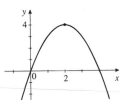

73.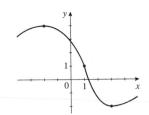

75.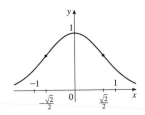

77. a. $N'(t)$ is positive because N is increasing on $(0, 12)$.

   b. $N''(t) < 0$ on $(0, 6)$ and $N''(t) > 0$ on $(6, 12)$.

   c. The rate of growth of the number of help-wanted advertisements was decreasing over the first six months of the year and increasing over the last six months.

79. $f(t)$ increases at an increasing rate until the water level reaches the middle of the vase (this corresponds to the inflection point of f). At this point, $f(t)$ is increasing at the fastest rate. Though $f(t)$ still increases until the vase is filled, it does so at a decreasing rate.

81. a. $f'(x) = \dfrac{d}{dx}\left(0.43x^{0.43}\right) = \left(0.43^2\right)x^{-0.57} = \dfrac{0.1849}{x^{0.57}}$ is positive if $x \geq 1$. This shows that f is increasing for

   $x \geq 1$, and this implies that the average state cigarette tax was increasing during the period in question.

© 2011 Cengage Learning. All Rights Reserved. May not be scanned, copied or duplicated, or posted to a publicly accessible website, in whole or in part.

b. $f''(x) = \dfrac{d}{dx}\left(0.1849x^{-0.57}\right) = (0.1849)(-0.57)x^{-1.57} = -\dfrac{0.105393}{x^{1.57}}$ is negative if $x \geq 1$. Thus, the rate of the increase of the cigarette tax is decreasing over the period in question.

83. a. $S'(t) = 0.39t + 0.32 > 0$ on $[0, 7]$, so sales were increasing through the years in question.

b. $S''(t) = 0.39 > 0$ on $[0, 7]$, so sales continued to accelerate through the years.

85. We wish to find the inflection point of the function $N(t) = -t^3 + 6t^2 + 15t$. Now $N'(t) = -3t^2 + 12t + 15$ and $N''(t) = -6t + 12 = -6(t - 2)$, giving $t = 2$ as the only candidate for an inflection point of N.

From the sign diagram for N'', we conclude that $t = 2$ gives an inflection point of N. Therefore, the average worker is performing at peak efficiency at 10 a.m.

87. $S'(t) = -5.64t^2 + 60.66t - 76.14$ and $S''(t) = -11.28t + 60.66 = 0$ if $t \approx 5.38$. Because $S''(t) > 0$ on $(0, 5)$, the graph of S is concave upward on $(0, 5)$. This says that the rate of business spending on technology is increasing from 2000 through 2005.

89. a. $R'(x) = -0.009x^2 + 2.7x + 2$; $R''(x) = -0.018x + 2.7$. Setting $R''(x) = 0$ gives $x = 150$. Because $R''(x) > 0$ if $x < 150$ and $R''(x) < 0$ if $x > 150$, we see that the graph of R is concave upward on $(0, 150)$ and concave downward on $(150, 400)$, so $x = 150$ gives an inflection point of R. Thus, the inflection point is $(150, 28550)$.

b. $R''(140) = 0.18$ and $R''(160) = -0.18$ This shows that at $x = 140$, a slight increase in x (spending) results in increased revenue. At $x = 160$, the opposite conclusion holds. So it would be more beneficial to increase the expenditure when it is \$140,000 than when it is \$160,000.

91. a. $A'(t) = 0.92(0.61)(t + 1)^{-0.39} = \dfrac{0.5612}{(t + 1)^{0.39}} > 0$ on $(0, 4)$, so A is increasing on $(0, 4)$. this tells us that the spending is increasing over the years in question.

b. $A''(t) = (0.5612)(-0.39)(t + 1)^{-1.39} = -\dfrac{0.218868}{(t + 1)^{1.39}} < 0$ on $(0, 4)$, so A'' is concave downward on $(0, 4)$.

This tells us that the spending is increasing but at a decreasing rate.

93. $S'(t) = -5.418t^2 + 20.476t + 93.35$ and $S''(t) = -10.836t + 20.476 = 0$ if $t \approx 1.9$. From the sign diagram and the fact that $S(1.9) = 784.9$, we see that the inflection point is approximately $(1.9, 784.9)$. The rate of annual spending slowed down near the end of 2000.

95. a. $R(t) = 24.975t^3 - 49.81t^2 + 41.25t + 0.2$, so $R'(t) = 74.925t^2 - 99.62t + 41.25$ and $R''(t) = 149.85t - 99.62$.

b. In solving the equation $R'(t) = 0$, we see that the discriminant is $(-99.62)^2 - 4(74.925)(41.25) = -2438.4806 < 0$, and so R' has no zero. Because $R'(0) = 41.25 > 0$, we see that $R'(t) > 0$ in $(0, 4)$. This shows that the revenue was increasing increasing from 1999 through 2003.

© 2011 Cengage Learning. All Rights Reserved. May not be scanned, copied or duplicated, or posted to a publicly accessible website, in whole or in part.

c. $R''(t) = 0$ implies that $t = 0.66$. From the sign diagram of R'', we conclude that the rate was increasing least rapidly around August of 1999.

97. a. The number of measles deaths in 1999 is given by $N(0) = 506$, or 506,000. The number of measles deaths in 2005 is given by $N(6) = -2.42(6^3) + 24.5(6^2) - 123.3(6) + 506 \approx 125.48$, or approximately 125,480.

b. $N'(t) = \dfrac{d}{dt}\left(-2.42t^3 + 24.5t^2 - 123.3t + 506\right) = -7.26t^2 + 49t - 123.3$. Because

$(49)^2 - 4(-7.26)(-123.3) = -1179.6 < 0$, we see that $N'(t)$ has no zero. Because $N'(0) = 123.3 < 0$, we conclude that $N'(t) < 0$ on $(0, 6)$. This shows that N is decreasing on $(0, 6)$, so the number of measles deaths was dropping from 1999 through 2005.

c. $N''(t) = -14.52t + 49 = 0$ implies that $t \approx 3.37$, so the number of measles deaths was decreasing most rapidly in April 2002. The rate is given by $N'(t) = -7.26(3.37)^2 + 49(3.37) - 123.3 \approx -40.62$, or approximately -41 deaths/yr.

99. $A(t) = 1.0974t^3 - 0.0915t^4$, so $A'(t) = 3.2922t^2 - 0.366t^3$ and $A''(t) = 6.5844t - 1.098t^2$. Setting $A'(t) = 0$, we obtain $t^2(3.2922 - 0.366t) = 0$, and this gives $t = 0$ or $t \approx 8.995 \approx 9$. Using the Second Derivative Test, we find $A''(9) = 6.5844(9) - 1.098(81) = -29.6784 < 0$, and this tells us that $t \approx 9$ gives rise to a relative maximum of A. Our analysis tells us that on that May day, the level of ozone peaked at approximately 4 p.m.

101. a. $N'(t) = \dfrac{d}{dt}\left(-0.9307t^3 + 74.04t^2 + 46.8667t + 3967\right) = -2.7921t^2 + 148.08t + 46.8667$. N' is continuous everywhere and has zeros at $t = \dfrac{-148.08 \pm \sqrt{(148.08)^2 - 4(-0.9307)(46.8667)}}{2(-2.7921)}$, that is, at $t \approx -0.31$ or 53.35. Both these numbers lie outside the interval of interest. Picking $t = 0$ for a test number, we see that $N'(0) = 46.86667 > 0$ and conclude that N is increasing on $(0, 16)$. This shows that the number of participants is increasing over the years in question.

b. $N''(t) = \dfrac{d}{dt}\left(-2.7921t^2 + 148.08t + 46.86667\right) = -5.5842t + 148.08 = 0$ if $t \approx 26.518$. Thus, $N''(t)$ does not change sign in the interval $(0, 16)$. Because $N''(0) = 148.08 > 0$, we see that $N'(t)$ is increasing on $(0, 16)$ and the desired conclusion follows.

103. True. If f' is increasing on (a, b), then $-f'$ is decreasing on (a, b), and so if the graph of f is concave upward on (a, b), the graph of $-f$ must be concave downward on (a, b).

105. True. The given conditions imply that $f''(0) < 0$ and the Second Derivative Test gives the desired conclusion.

107. $f(x) = ax^2 + bx + c$, so $f'(x) = 2ax + b$ and $f''(x) = 2a$. Thus, $f''(x) > 0$ if $a > 0$, and the parabola opens upward. If $a < 0$, then $f''(x) < 0$ and the parabola opens downward.

4.2　Using Technology　page 283

1. a. f is concave upward on $(-\infty, 0) \cup (1.1667, \infty)$ and concave downward on $(0, 1.1667)$.

b. f has inflection points at $(1.1667, 1.1153)$ and $(0, 2)$.

© 2011 Cengage Learning. All Rights Reserved. May not be scanned, copied or duplicated, or posted to a publicly accessible website, in whole or in part.

3. a. f is concave downward on $(-\infty, 0)$ and concave upward on $(0, \infty)$.

   b. f has an inflection point at $(0, 2)$.

5. a. f is concave downward on $(-\infty, 0)$ and concave upward on $(0, \infty)$.

   b. f has an inflection point at $(0, 0)$.

7. a. f is concave downward on $(-\infty, -2.4495) \cup (0, 2.4495)$ and concave upward on $(-2.4495, 0) \cup (2.4495, \infty)$.

   b. f has inflection points at $(-2.4495, -0.3402)$ and $(2.4495, 0.3402)$.

9. a.

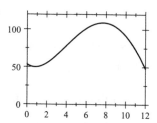

11. a.

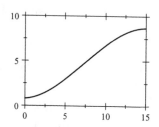

b. The inflection point is approximately $(3.9024, 77.0919)$. This means that the rate of increase of houses' length of time on the market became negative in late 1998, indicating an improving housing market.

b. The inflection point occurs at $t \approx 7.36$, so the number of surgeries is increasing at the fastest rate in approximately August 1993.

4.3   Problem-Solving Tips

1. **To find the horizontal asymptotes of a function** f, find the limit of f as $x \to \infty$ and as $x \to -\infty$. If the limit is equal to a real number b, then $y = b$ is a horizontal asymptote of f.

2. **To find the vertical asymptotes of a rational function** $f(x) = P(x)/Q(x)$, determine the values a for which $Q(a) = 0$. If $Q(a) = 0$ but $P(a) \neq 0$, then the line $x = a$ is a vertical asymptote of f.

3. If a line $x = a$ is a vertical asymptote of the graph of a rational function f, then the denominator of $f(x)$ is equal to zero at $x = a$. However, if both numerator and denominator of $f(x)$ are equal to zero, then $x = a$ is not necessarily a vertical asymptote.

4.3   Concept Questions   page 291

1. a. See the definition on page 286 of the text.

b. See the definition on page 285 of the text.

3. See the procedure given on page 285 of the text.

© 2011 Cengage Learning. All Rights Reserved. May not be scanned, copied or duplicated, or posted to a publicly accessible website, in whole or in part.

| **4.3** | **Curve Sketching**   page 291 |

1. $y = 0$ is a horizontal asymptote.

3. $y = 0$ is a horizontal asymptote and $x = 0$ is a vertical asymptote.

5. $y = 0$ is a horizontal asymptote and $x = -1$ and $x = 1$ are vertical asymptotes.

7. $y = 3$ is a horizontal asymptote and $x = 0$ is a vertical asymptote.

9. $y = 1$ and $y = -1$ are horizontal asymptotes.

11. $\lim\limits_{x \to \infty} \dfrac{1}{x} = 0$, and so $y = 0$ is a horizontal asymptote. Next, since the numerator of the rational expression is not equal to zero and the denominator is zero at $x = 0$, we see that $x = 0$ is a vertical asymptote.

13. $f(x) = -\dfrac{2}{x^2}$, so $\lim\limits_{x \to \infty} f(x) = \lim\limits_{x \to \infty} \left(-\dfrac{2}{x^2}\right) = 0$. Thus, $y = 0$ is a horizontal asymptote. Next, the denominator of $f(x)$ is equal to zero at $x = 0$. Because the numerator of $f(x)$ is not equal to zero at $x = 0$, we see that $x = 0$ is a vertical asymptote.

15. $\lim\limits_{x \to \infty} \dfrac{x-1}{x+1} = \lim\limits_{x \to \infty} \dfrac{1-\frac{1}{x}}{1+\frac{1}{x}} = 1$, and so $y = 1$ is a horizontal asymptote. Next, the denominator is equal to zero at $x = -1$ and the numerator is not equal to zero at this number, so $x = -1$ is a vertical asymptote.

17. $h(x) = x^3 - 3x^2 + x + 1$. $h(x)$ is a polynomial function, and therefore it does not have any horizontal or vertical asymptotes.

19. $\lim\limits_{t \to \infty} \dfrac{t^2}{t^2 - 9} = \lim\limits_{t \to \infty} \dfrac{1}{1 - \frac{9}{t^2}} = 1$, and so $y = 1$ is a horizontal asymptote. Next, observe that the denominator of the rational expression $t^2 - 9 = (t + 3)(t - 3) = 0$ if $t = -3$ or $t = 3$. But the numerator is not equal to zero at these numbers, so $t = -3$ and $t = 3$ are vertical asymptotes.

21. $\lim\limits_{x \to \infty} \dfrac{3x}{x^2 - x - 6} = \lim\limits_{x \to \infty} \dfrac{\frac{3}{x}}{1 - \frac{1}{x} - \frac{6}{x^2}} = 0$ and so $y = 0$ is a horizontal asymptote. Next, observe that the denominator $x^2 - x - 6 = (x - 3)(x + 2) = 0$ if $x = -2$ or $x = 3$. But the numerator $3x$ is not equal to zero at these numbers, so $x = -2$ and $x = 3$ are vertical asymptotes.

23. $\lim\limits_{t \to \infty} \left[2 + \dfrac{5}{(t-2)^2}\right] = 2$, and so $y = 2$ is a horizontal asymptote. Next observe that

$\lim\limits_{t \to 2^+} g(t) = \lim\limits_{t \to 2^-} \left[2 + \dfrac{5}{(t-2)^2}\right] = \infty$, and so $t = 2$ is a vertical asymptote.

25. $\lim\limits_{x \to \infty} \dfrac{x^2 - 2}{x^2 - 4} = \lim\limits_{x \to \infty} \dfrac{1 - \frac{2}{x^2}}{1 - \frac{4}{x^2}} = 1$, and so $y = 1$ is a horizontal asymptote. Next, observe that the denominator $x^2 - 4 = (x + 2)(x - 2) = 0$ if $x = -2$ or 2. Because the numerator $x^2 - 2$ is not equal to zero at these numbers, the lines $x = -2$ and $x = 2$ are vertical asymptotes.

© 2011 Cengage Learning. All Rights Reserved. May not be scanned, copied or duplicated, or posted to a publicly accessible website, in whole or in part.

27. $g(x) = \dfrac{x^3 - x}{x(x+1)}$. Rewrite $g(x)$ as $\dfrac{x^2 - 1}{x+1}$ for $x \neq 0$, and note that $\displaystyle\lim_{x \to -\infty} g(x) = \lim_{x \to -\infty} \dfrac{x - \frac{1}{x}}{1 + \frac{1}{x}} = -\infty$ and

$\displaystyle\lim_{x \to \infty} g(x) = \infty$. Therefore, there is no horizontal asymptote. Next, note that the denominator of $g(x)$ is equal to

zero at $x = 0$ and $x = -1$. However, since the numerator of $g(x)$ is also equal to zero when $x = 0$, we see that

$x = 0$ is not a vertical asymptote. Also, the numerator of $g(x)$ is equal to zero when $x = -1$, so $x = -1$ is not a

vertical asymptote.

29. f is the derivative function of the function g. Observe that at a relative maximum or minimum of g, $f(x) = 0$.

31.

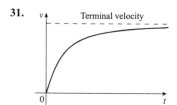

33.

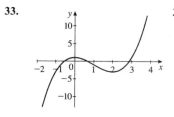

35.

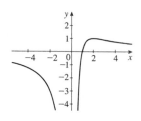

37. $g(x) = 4 - 3x - 2x^3$. We first gather the following information on f.

1. The domain of f is $(-\infty, \infty)$.

2. Setting $x = 0$ gives $y = 4$ as the y-intercept. Setting $y = g(x) = 0$ gives a cubic equation which is not easily solved, and we will not attempt to find the x-intercepts.

3. $\displaystyle\lim_{x \to -\infty} g(x) = \infty$ and $\displaystyle\lim_{x \to \infty} g(x) = -\infty$.

4. The graph of g has no asymptote.

5. $g'(x) = -3 - 6x^2 = -3(2x^2 + 1) < 0$ for all values of x and so g is decreasing on $(-\infty, \infty)$.

6. The results of step 5 show that g has no critical number and hence no relative extremum.

7. $g''(x) = -12x$. Because $g''(x) > 0$ for $x < 0$ and $g''(x) < 0$ for $x > 0$, we see that g is concave upward on $(-\infty, 0)$ and concave downward on $(0, \infty)$.

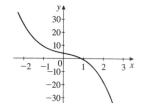

8. From the results of step 7, we see that $(0, 4)$ is an inflection point of g.

39. $h(x) = x^3 - 3x + 1$. We first gather the following information on h.

1. The domain of h is $(-\infty, \infty)$.

2. Setting $x = 0$ gives 1 as the y-intercept. We will not find the x-intercept.

3. $\displaystyle\lim_{x \to -\infty} (x^3 - 3x + 1) = -\infty$ and $\displaystyle\lim_{x \to \infty} (x^3 - 3x + 1) = \infty$.

4. There is no asymptote because $h(x)$ is a polynomial.

5. $h'(x) = 3x^2 - 3 = 3(x + 1)(x - 1)$, and we see that $x = -1$ and $x = 1$ are critical numbers. From the sign diagram, we see that h is increasing on $(-\infty, -1) \cup (1, \infty)$ and decreasing on $(-1, 1)$.

$$+\ +\ 0\ -\ -\ -\ -\ -\ 0\ +\ + \quad \text{sign of } h'$$

$$\xrightarrow{\hspace{2cm}} x$$

$$-1 \qquad 0 \qquad 1$$

© 2011 Cengage Learning. All Rights Reserved. May not be scanned, copied or duplicated, or posted to a publicly accessible website, in whole or in part.

6. The results of step 5 show that $(-1, 3)$ is a relative maximum and $(1, -1)$ is a relative minimum.

7. $h''(x) = 6x$, so $h''(x) < 0$ if $x < 0$ and $h''(x) > 0$ if $x > 0$. Thus, the graph of h is concave downward on $(-\infty, 0)$ and concave upward on $(0, \infty)$.

8. The results of step 7 show that $(0, 1)$ is an inflection point of h.

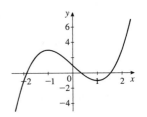

41. $f(x) = -2x^3 + 3x^2 + 12x + 2$. We first gather the following information on f.

1. The domain of f is $(-\infty, \infty)$.

2. Setting $x = 0$ gives 2 as the y-intercept.

3. $\lim\limits_{x \to -\infty} \left(-2x^3 + 3x^2 + 12x + 2\right) = \infty$ and $\lim\limits_{x \to \infty} \left(-2x^3 + 3x^2 + 12x + 2\right) = -\infty$

4. There is no asymptote because $f(x)$ is a polynomial function.

5. $f'(x) = -6x^2 + 6x + 12 = -6\left(x^2 - x - 2\right) = -6(x - 2)(x + 1) = 0$ if $x = -1$ or $x = 2$, the critical numbers of f. From the sign diagram, we see that f is decreasing on $(-\infty, -1) \cup (2, \infty)$ and increasing on $(-1, 2)$.

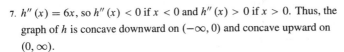

6. The results of step 5 show that $(-1, -5)$ is a relative minimum and $(2, 22)$ is a relative maximum.

7. $f''(x) = -12x + 6 = 0$ if $x = \frac{1}{2}$. The sign diagram of f'' shows that the graph of f is concave upward on $\left(-\infty, \frac{1}{2}\right)$ and concave downward on $\left(\frac{1}{2}, \infty\right)$.

8. The results of step 7 show that $\left(\frac{1}{2}, \frac{17}{2}\right)$ is an inflection point.

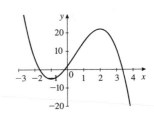

43. $h(x) = \frac{3}{2}x^4 - 2x^3 - 6x^2 + 8$. We first gather the following information on h.

1. The domain of h is $(-\infty, \infty)$.

2. Setting $x = 0$ gives 8 as the y-intercept.

3. $\lim\limits_{x \to -\infty} h(x) = \lim\limits_{x \to \infty} h(x) = \infty$

4. There is no asymptote.

5. $h'(x) = 6x^3 - 6x^2 - 12x = 6x\left(x^2 - x - 2\right) = 6x(x - 2)(x + 1) = 0$ if $x = -1$, 0, or 2, and these are the critical numbers of h. The sign diagram of h' shows that h is increasing on $(-1, 0) \cup (2, \infty)$ and decreasing on $(-\infty, -1) \cup (0, 2)$.

© 2011 Cengage Learning. All Rights Reserved. May not be scanned, copied or duplicated, or posted to a publicly accessible website, in whole or in part.

6. The results of step 5 show that $\left(-1, \frac{11}{2}\right)$ and $(2, -8)$ are relative minima of h and $(0, 8)$ is a relative maximum of h.

7. $h''(x) = 18x^2 - 12x - 12 = 6\left(3x^2 - 2x - 2\right)$. The zeros of h'' are $x = \frac{2 \pm \sqrt{4+24}}{6} \approx -0.5$ or 1.2.

The sign diagram of h'' shows that the graph of h is concave upward on $(-\infty, -0.5) \cup (1.2, \infty)$ and concave downward on $(-0.5, 1.2)$.

8. The results of step 7 also show that $(-0.5, 6.8)$ and $(1.2, -1)$ are inflection points.

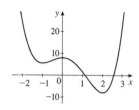

45. $f(t) = \sqrt{t^2 - 4}$. We first gather the following information on f.

1. The domain of f is found by solving $t^2 - 4 \geq 0$ to obtain $(-\infty, -2] \cup [2, \infty)$.

2. Because $t \neq 0$, there is no y-intercept. Next, setting $y = f(t) = 0$ gives the t-intercepts as -2 and 2.

3. $\lim\limits_{t \to -\infty} f(t) = \lim\limits_{t \to \infty} f(t) = \infty$.

4. There is no asymptote.

5. $f'(t) = \frac{1}{2}\left(t^2 - 4\right)^{-1/2}(2t) = t\left(t^2 - 4\right)^{-1/2} = \frac{t}{\sqrt{t^2 - 4}}$. Setting $f'(t) = 0$ gives $t = 0$. But $t = 0$ is not in the

domain of f and so there is no critical number. From the sign diagram for f', we see that f is increasing on $(2, \infty)$ and decreasing on $(-\infty, -2)$.

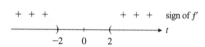

6. From the results of step 5 we see that there is no relative extremum.

7. $f''(t) = \left(t^2 - 4\right)^{-1/2} + t\left(-\frac{1}{2}\right)\left(t^2 - 4\right)^{-3/2}(2t)$

$= \left(t^2 - 4\right)^{-3/2}\left(t^2 - 4 - t^2\right) = -\frac{4}{\left(t^2 - 4\right)^{3/2}}$.

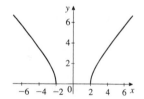

8. Because $f''(t) < 0$ for all t in the domain of f, we see that f is concave downward everywhere. From the results of step 7, we see that there is no inflection point.

47. $g(x) = \frac{1}{2}x - \sqrt{x}$. We first gather the following information on g.

1. The domain of g is $[0, \infty)$.

2. The y-intercept is 0. To find the x-intercept(s), set $y = 0$, giving $\frac{1}{2}x - \sqrt{x} = 0$, $x = 2\sqrt{x}$, $x^2 = 4x$, $x(x - 4) = 0$, and so $x = 0$ or $x = 4$.

3. $\lim\limits_{x \to \infty}\left(\frac{1}{2}x - \sqrt{x}\right) = \lim\limits_{x \to \infty} \frac{1}{2}x\left(1 - \frac{2}{\sqrt{x}}\right) = \infty$.

4. There is no asymptote.

© 2011 Cengage Learning. All Rights Reserved. May not be scanned, copied or duplicated, or posted to a publicly accessible website, in whole or in part.

5. $g'(x) = \frac{1}{2} - \frac{1}{2}x^{-1/2} = \frac{1}{2}x^{-1/2}\left(x^{1/2} - 1\right) = \frac{\sqrt{x} - 1}{2\sqrt{x}}$, which

is zero when $x = 1$. From the sign diagram for g', we see that
g is decreasing on $(0, 1)$ and increasing on $(1, \infty)$.

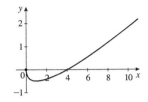

6. From the results of part 5, we see that $g(1) = -\frac{1}{2}$ is a relative minimum.

7. $g''(x) = \left(-\frac{1}{2}\right)\left(-\frac{1}{2}\right)x^{-3/2} = \frac{1}{4x^{3/2}} > 0$ for $x > 0$, and so g is concave

upward on $(0, \infty)$.

8. There is no inflection point.

49. $g(x) = \frac{2}{x - 1}$. We first gather the following information on g.

1. The domain of g is $(-\infty, 1) \cup (1, \infty)$.

2. Setting $x = 0$ gives -2 as the y-intercept. There is no x-intercept because $\frac{2}{x - 1} \neq 0$ for all x.

3. $\lim\limits_{x \to -\infty} \frac{2}{x - 1} = 0$ and $\lim\limits_{x \to \infty} \frac{2}{x - 1} = 0$.

4. The results of step 3 show that $y = 0$ is a horizontal asymptote. Furthermore, the denominator of $g(x)$ is equal to
zero at $x = 1$ but the numerator is not equal to zero there. Therefore, $x = 1$ is a vertical asymptote.

5. $g'(x) = -2(x - 1)^{-2} = -\frac{2}{(x - 1)^2} < 0$ for all $x \neq 1$ and so g is decreasing on $(-\infty, 1) \cup (1, \infty)$.

6. Because g has no critical number, there is no relative extremum.

7. $g''(x) = \frac{4}{(x - 1)^3}$ and so $g''(x) < 0$ if $x < 1$ and $g''(x) > 0$ if $x > 1$.

Therefore, the graph of g is concave downward on $(-\infty, 1)$ and concave
upward on $(1, \infty)$.

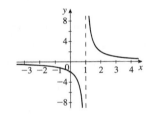

8. Because $g''(x) \neq 0$, there is no inflection point.

51. $h(x) = \frac{x + 2}{x - 2}$. We first gather the following information on h.

1. The domain of h is $(-\infty, 2) \cup (2, \infty)$.

2. Setting $x = 0$ gives $y = -1$ as the y-intercept. Next, setting $y = 0$ gives $x = -2$ as the x-intercept.

3. $\lim\limits_{x \to \infty} h(x) = \lim\limits_{x \to -\infty} \frac{1 + \frac{2}{x}}{1 - \frac{2}{x}} = \lim\limits_{x \to -\infty} h(x) = 1$.

4. Setting $x - 2 = 0$ gives $x = 2$. Furthermore, $\lim\limits_{x \to 2^+} \frac{x + 2}{x - 2} = \infty$, so $x = 2$ is a vertical asymptote of h. Also, from
the results of step 3, we see that $y = 1$ is a horizontal asymptote of h.

5. $h'(x) = \frac{(x - 2)(1) - (x + 2)(1)}{(x - 2)^2} = -\frac{4}{(x - 2)^2}$. We see that

h has no critical number. (Note that $x = 2$ is not in the domain
of h.) The sign diagram of h' shows that h is decreasing on
$(-\infty, 2) \cup (2, \infty)$.

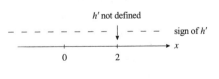

© 2011 Cengage Learning. All Rights Reserved. May not be scanned, copied or duplicated, or posted to a publicly accessible website, in whole or in part.

6. From the results of step 5, we see that there is no relative extremum.

7. $h''(x) = \dfrac{8}{(x-2)^3}$. Note that $x = 2$ is not a candidate for an inflection

   point because $h(2)$ is not defined. Because $h''(x) < 0$ for $x < 2$ and
   $h''(x) > 0$ for $x > 2$, we see that h is concave downward on $(-\infty, 2)$ and
   concave upward on $(2, \infty)$.

8. From the results of step 7, we see that there is no inflection point.

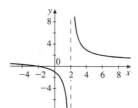

53. $f(t) = \dfrac{t^2}{1+t^2}$. We first gather the following information on f.

1. The domain of f is $(-\infty, \infty)$.

2. Setting $t = 0$ gives the y-intercept as 0. Similarly, setting $y = 0$ gives the t-intercept as 0.

3. $\displaystyle\lim_{t\to-\infty} \dfrac{t^2}{1+t^2} = \lim_{t\to\infty} \dfrac{t^2}{1+t^2} = 1$.

4. The results of step 3 show that $y = 1$ is a horizontal asymptote. There is no vertical asymptote since the
   denominator is never zero.

5. $f'(t) = \dfrac{(1+t^2)(2t) - t^2(2t)}{(1+t^2)^2} = \dfrac{2t}{(1+t^2)^2} = 0$, if $t = 0$, the only critical number of f. Because $f'(t) < 0$ if

   $t < 0$ and $f'(t) > 0$ if $t > 0$, we see that f is decreasing on $(-\infty, 0)$ and increasing on $(0, \infty)$.

6. The results of step 5 show that $(0, 0)$ is a relative minimum.

7. $f''(t) = \dfrac{(1+t^2)^2(2) - 2t(2)(1+t^2)(2t)}{(1+t^2)^4} = \dfrac{2(1+t^2)\left[(1+t^2) - 4t^2\right]}{(1+t^2)^4} = \dfrac{2(1 - 3t^2)}{(1+t^2)^3} = 0$ if $t = \pm\dfrac{\sqrt{3}}{3}$.

   The sign diagram of f'' shows that f is concave downward

   on $\left(-\infty, -\dfrac{\sqrt{3}}{3}\right) \cup \left(\dfrac{\sqrt{3}}{3}, \infty\right)$ and concave upward on

   $\left(-\dfrac{\sqrt{3}}{3}, \dfrac{\sqrt{3}}{3}\right)$.

   $$- - - 0 + + + + + + 0 - - - - \quad \text{sign of } f''$$

8. The results of step 7 show that $\left(-\dfrac{\sqrt{3}}{3}, \dfrac{1}{4}\right)$ and $\left(\dfrac{\sqrt{3}}{3}, \dfrac{1}{4}\right)$ are
   inflection points.

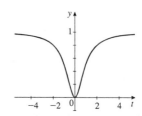

55. $g(t) = -\dfrac{t^2 - 2}{t - 1}$. We first gather the following information on g.

1. The domain of g is $(-\infty, 1) \cup (1, \infty)$

2. Setting $t = 0$ gives -2 as the y-intercept.

3. $\displaystyle\lim_{t\to-\infty}\left(-\dfrac{t^2-2}{t-1}\right) = \infty$ and $\displaystyle\lim_{t\to\infty}\left(-\dfrac{t^2-2}{t-1}\right) = -\infty$.

4. There is no horizontal asymptotes. The denominator is equal to zero at $t = 1$ at which number the numerator is
   not equal to zero. Therefore $t = 1$ is a vertical asymptote.

© 2011 Cengage Learning. All Rights Reserved. May not be scanned, copied or duplicated, or posted to a publicly accessible website, in whole or in part.

5. $g'(t) = -\dfrac{(t-1)(2t) - (t^2-2)(1)}{(t-1)^2}$

$\qquad = -\dfrac{t^2 - 2t + 2}{(t-1)^2} \neq 0$ for all values of t.

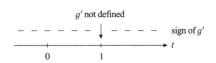

The sign diagram of g' shows that g is decreasing on
$(-\infty, 1) \cup (1, \infty)$.

6. Because there is no critical number, g has no relative extremum.

7. $g''(t) = -\dfrac{(t-1)^2(2t-2) - (t^2-2t+2)(2)(t-1)}{(t-1)^4}$

$\qquad = \dfrac{-2(t-1)(t^2-2t+1-t^2+2t-2)}{(t-1)^4} = \dfrac{2}{(t-1)^3}$.

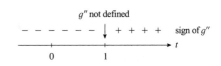

The sign diagram of g'' shows that the graph of g is concave
upward on $(1, \infty)$ and concave downward on $(-\infty, 1)$.

8. There is no inflection point because $g''(x) \neq 0$ for all x.

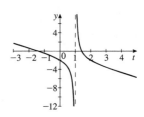

57. $g(t) = \dfrac{t^2}{t^2 - 1}$. We first gather the following information on g.

1. Because $t^2 - 1 = 0$ if $t = \pm 1$, we see that the domain of g is $(-\infty, -1) \cup (-1, 1) \cup (1, \infty)$.

2. Setting $t = 0$ gives 0 as the y-intercept. Setting $y = 0$ gives 0 as the t-intercept.

3. $\displaystyle\lim_{t \to -\infty} g(t) = \lim_{t \to \infty} g(t) = 1$.

4. The results of step 3 show that $y = 1$ is a horizontal asymptote. Because the denominator (but not the numerator)
is zero at $t = \pm 1$, we see that $t = \pm 1$ are vertical asymptotes.

5. $g'(t) = \dfrac{(t^2-1)(2t) - (t^2)(2t)}{(t^2-1)^2} = -\dfrac{2t}{(t^2-1)^2} = 0$ if

$t = 0$. From the sign diagram of g', we see that g is
increasing on $(-\infty, -1) \cup (-1, 0)$ and decreasing on
$(0, 1) \cup (1, \infty)$.

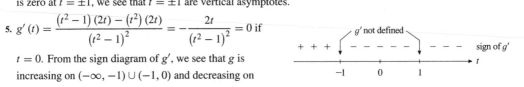

6. From the results of step 5, we see that g has a relative maximum at $t = 0$.

7. $g''(t) = \dfrac{(t^2-1)^2(-2) - (-2t)(2)(t^2-1)(2t)}{(t^2-1)^4} = \dfrac{2(t^2-1)^2[-(t^2-1) + 4t^2]}{(t^2-1)^3} = \dfrac{2(-t^2+1+4t^2)}{(t^2-1)^3}$

$\qquad = \dfrac{2(3t^2+1)}{(t^2-1)^3}$.

From the sign diagram, we see that the graph of g is concave
upward on $(-\infty, -1) \cup (1, \infty)$ and concave downward on
$(-1, 1)$.

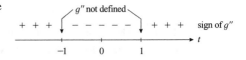

© 2011 Cengage Learning. All Rights Reserved. May not be scanned, copied or duplicated, or posted to a publicly accessible website, in whole or in part.

8. Because g is undefined at ± 1, the graph of g has no inflection
   point.

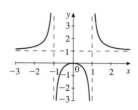

59. $h(x) = (x-1)^{2/3} + 1$. We begin by obtaining the following information on h.

1. The domain of h is $(-\infty, \infty)$.

2. Setting $x = 0$ gives 2 as the y-intercept; since $h(x) \neq 0$ there is no x-intercept.

3. $\lim\limits_{x \to \infty} \left[(x-1)^{2/3} + 1\right] = \infty$ and $\lim\limits_{x \to -\infty} \left[(x-1)^{2/3} + 1\right] = \infty$.

4. There is no asymptote.

5. $h'(x) = \frac{2}{3}(x-1)^{-1/3}$ and is positive if $x > 1$ and negative if $x < 1$. Thus, h is increasing on $(1, \infty)$ and
   decreasing on $(-\infty, 1)$.

6. From step 5, we see that h has a relative minimum at $(1, 1)$.

7. $h''(x) = \frac{2}{3}\left(-\frac{1}{3}\right)(x-1)^{-4/3} = -\frac{2}{9}(x-1)^{-4/3} = -\dfrac{2}{(x-1)^{4/3}}$.

   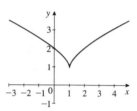

   Because $h''(x) < 0$ on $(-\infty, 1) \cup (1, \infty)$, we see that h is concave
   downward on $(-\infty, 1) \cup (1, \infty)$. Note that $h''(x)$ is not defined at $x = 1$.

8. From the results of step 7, we see h has no inflection point.

61. a. The denominator of $C(x)$ is equal to zero if $x = 100$. Also, $\lim\limits_{x \to 100^-} \dfrac{0.5x}{100 - x} = \infty$ and $\lim\limits_{x \to 100^+} \dfrac{0.5x}{100 - x} = -\infty$.
   Therefore, $x = 100$ is a vertical asymptote of C.

   b. No, because the denominator is equal to zero in that case.

63. a. Because $\lim\limits_{t \to \infty} C(t) = \lim\limits_{t \to \infty} \dfrac{0.2t}{t^2 + 1} = \lim\limits_{t \to \infty} \left(\dfrac{0.2}{t + \frac{1}{t^2}}\right) = 0$, $y = 0$ is a horizontal asymptote.

   b. Our results reveal that as time passes, the concentration of the drug decreases and approaches zero.

65. $G(t) = -0.2t^3 + 2.4t^2 + 60$. We first gather the following information on G.

1. The domain of G is restricted to $[0, 8]$.

2. Setting $t = 0$ gives 60 as the y-intercept.

   Step 3 is unnecessary in this case because of the restricted domain.

4. There is no asymptote because G is a polynomial function.

5. $G'(t) = -0.6t^2 + 4.8t = -0.6t(t-8) = 0$ if $t = 0$ or $t = 8$, critical numbers of G. But $G'(t) > 0$ on $(0, 8)$, so
   G is increasing on its domain.

6. The results of step 5 tell us that there is no relative extremum.

7. $G''(t) = -1.2t + 4.8 = -1.2(t-4)$. The sign diagram of
   G'' shows that G is concave upward on $(0, 4)$ and concave
   downward on $(4, 8)$.

   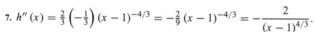

© 2011 Cengage Learning. All Rights Reserved. May not be scanned, copied or duplicated, or posted to a publicly accessible website, in whole or in part.

8. The results of step 7 show that $(4, 85.6)$ is an inflection point.

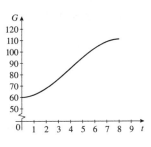

67. $N(t) = -\frac{1}{2}t^3 + 3t^2 + 10t,\ 0 \le t \le 4$. We first gather the following information on N.

1. The domain of N is restricted to $[0, 4]$.

2. The y-intercept is 0.

Step 3 does not apply because the domain of $N(t)$ is $[0, 4]$.

4. There is no asymptote.

5. $N'(t) = -\frac{3}{2}t^2 + 6t + 10 = -\frac{1}{2}\left(3t^2 - 12t - 20\right)$ is never zero. Therefore, N is increasing on $(0, 4)$.

6. There is no relative extremum in $(0, 4)$.

7. $N''(t) = -3t + 6 = -3(t - 2) = 0$ at $t = 2$. From the sign diagram of N'', we see that N is concave upward on $(0, 2)$ and concave downward on $(2, 4)$.

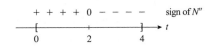

8. The point $(2, 28)$ is an inflection point.

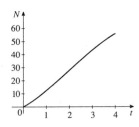

69. $T(x) = \dfrac{120x^2}{x^2 + 4}$. We first gather the following information on T.

1. The domain of T is $[0, \infty)$.

2. Setting $x = 0$ gives 0 as the y-intercept.

3. $\displaystyle\lim_{x \to \infty} \frac{120x^2}{x^2 + 4} = 120$.

4. The results of step 3 show that $y = 120$ is a horizontal asymptote.

5. $T'(x) = 120 \left[\dfrac{(x^2 + 4)\, 2x - x^2\, (2x)}{(x^2 + 4)^2} \right] = \dfrac{960x}{(x^2 + 4)^2}$. Because $T'(x) > 0$ if $x > 0$, we see that T is increasing on $(0, \infty)$.

6. There is no relative extremum.

© 2011 Cengage Learning. All Rights Reserved. May not be scanned, copied or duplicated, or posted to a publicly accessible website, in whole or in part.

7. $T''(x) = 960 \left[\dfrac{(x^2+4)^2 - x(2)(x^2+4)(2x)}{(x^2+4)^4} \right] = \dfrac{960(x^2+4)[(x^2+4)-4x^2]}{(x^2+4)^4} = \dfrac{960(4-3x^2)}{(x^2+4)^3}.$

The sign diagram for T'' shows that T is concave downward on $\left(\frac{2\sqrt{3}}{3}, \infty \right)$ and concave upward on $\left(0, \frac{2\sqrt{3}}{3} \right)$.

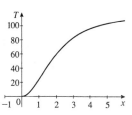

8. We see from the results of step 7 that $\left(\frac{2\sqrt{3}}{3}, 30 \right)$ is an inflection point.

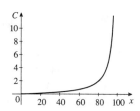

71. $C(x) = \dfrac{0.5x}{100-x}$. We first gather the following information on C.

1. The domain of C is $[0, 100)$.

2. Setting $x = 0$ gives the y-intercept as 0.

Because of the restricted domain, we omit steps 3 and 4.

5. $C'(x) = 0.5 \left[\dfrac{(100-x)(1) - x(-1)}{(100-x)^2} \right] = \dfrac{50}{(100-x)^2} > 0$ for all $x \neq 100$. Therefore C is increasing on $(0, 100)$.

6. There is no relative extremum.

7. $C''(x) = -\dfrac{100}{(100-x)^3}$, so $C''(x) > 0$ if $x < 100$ and the graph of C is concave upward on $(0, 100)$.

8. There is no inflection point.

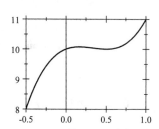

4.3   Using Technology   page 297

1.

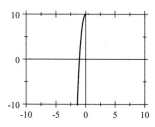 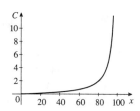

$f(x) = 4x^3 - 4x^2 + x + 10$, so $f'(x) = 12x^2 - 8x + 1 = (6x-1)(2x-1) = 0$ if $x = \frac{1}{6}$ or $x = \frac{1}{2}$. The second graph shows that f has a maximum at $x = \frac{1}{6}$ and a minimum at $x = \frac{1}{2}$.

© 2011 Cengage Learning. All Rights Reserved. May not be scanned, copied or duplicated, or posted to a publicly accessible website, in whole or in part.

3.

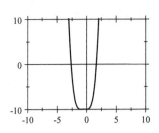
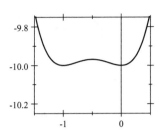

$f(x) = \frac{1}{2}x^4 + x^3 + \frac{1}{2}x^2 - 10$, so $f'(x) = 2x^3 + 3x^2 + x = x(x+1)(2x+1) = 0$ if $x = -1, -\frac{1}{2}$, or 0. The second graph shows that x has minima at $x = -1$ and $x = 0$ and a maximum at $x = -\frac{1}{2}$.

5. -0.9733, 2.3165, and 4.6569.

7. 1.5142

9.

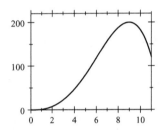

4.4   **Problem-Solving Tips**

1. **To determine the absolute maximum and absolute minimum** of a continuous function f on a closed interval $[a, b]$, find the critical numbers of f that lie in (a, b). Then compute the value of f at each critical number of f and compute $f(a)$ and $f(b)$. The largest and smallest of these values are the absolute maximum and absolute minimum values of f, respectively.

2. Note that the procedure in Tip 1 holds only for a *continuous* function f over a *closed* interval $[a, b]$.

4.4   **Concept Questions**   page 305

1. a. A function f has an absolute maximum at a if $f(x) \leq f(a)$ for all x in the domain of f.
    b. A function f has an absolute minimum at a if $f(x) \geq f(a)$ for all x in the domain of f.

4.4   **Optimization I**   page 305

1. f has no absolute extremum.

3. f has an absolute minimum at $(0, 0)$.

© 2011 Cengage Learning. All Rights Reserved. May not be scanned, copied or duplicated, or posted to a publicly accessible website, in whole or in part.

5. f has an absolute minimum at $(0, -2)$ and an absolute maximum at $(1, 3)$.

7. f has an absolute minimum at $\left(\frac{3}{2}, -\frac{27}{16}\right)$ and an absolute maximum at $(-1, 3)$.

9. The graph of $f(x) = 2x^2 + 3x - 4$ is a parabola that opens upward. Therefore, the vertex of the parabola is the absolute minimum of f. To find the vertex, we solve the equation $f'(x) = 4x + 3 = 0$, finding $x = -\frac{3}{4}$. We conclude that the absolute minimum value is $f\left(-\frac{3}{4}\right) = -\frac{41}{8}$.

11. Because $\lim\limits_{x \to -\infty} x^{1/3} = -\infty$ and $\lim\limits_{x \to \infty} x^{1/3} = \infty$, we see that h is unbounded. Therefore, it has no absolute extremum.

13. $f(x) = \dfrac{1}{1 + x^2}$. Using the techniques of graphing, we sketch the graph of f (see Figure 40 on page 269 of the text). The absolute maximum of f is $f(0) = 1$. Alternatively, observe that $1 + x^2 \geq 1$ for all real values of x. Therefore, $f(x) \leq 1$ for all x, and we see that the absolute maximum is attained when $x = 0$.

15. $f(x) = x^2 - 2x - 3$ and $f'(x) = 2x - 2 = 0$, so $x = 1$ is a critical number. From the table, we conclude that the absolute maximum value is $f(-2) = 5$ and the absolute minimum value is $f(1) = -4$.

x	-2	1	3
$f(x)$	5	-4	0

17. $f(x) = -x^2 + 4x + 6$; The function f is continuous and defined on the closed interval $[0, 5]$. $f'(x) = -2x + 4$, and so $x = 2$ is a critical number. From the table, we conclude that $f(2) = 10$ is the absolute maximum value and $f(5) = 1$ is the absolute minimum value.

x	0	2	5
$f(x)$	6	10	1

19. The function $f(x) = x^3 + 3x^2 - 1$ is continuous and defined on the closed interval $[-3, 2]$ and differentiable in $(-3, 2)$. The critical numbers of f are found by solving $f'(x) = 3x^2 + 6x = 3x(x + 2) = 0$, giving $x = -2$ and $x = 0$. From the table, we see that the absolute maximum value of f is $f(2) = 19$ and the absolute minimum value is $f(-3) = f(0) = -1$.

x	-3	-2	0	2
$f(x)$	-1	3	-1	19

21. The function $g(x) = 3x^4 + 4x^3$ is continuous on the closed interval $[-2, 1]$ and differentiable in $(-2, 1)$. The critical numbers of g are found by solving $g'(x) = 12x^3 + 12x^2 = 12x^2(x + 1) = 0$, giving $x = 0$ and $x = -1$. From the table, we see that $g(-2) = 16$ is the absolute maximum value of g and $g(-1) = -1$ is the absolute minimum value of g.

x	-2	-1	0	1
$g(x)$	16	-1	0	7

23. $f(x) = \dfrac{x + 1}{x - 1}$ on $[2, 4]$. Next, we compute $f'(x) = \dfrac{(x - 1)(1) - (x + 1)(1)}{(x - 1)^2} = -\dfrac{2}{(x - 1)^2}$. Because there is no critical number ($x = 1$ is not in the domain of f), we need only test the endpoints. We conclude that $f(4) = \frac{5}{3}$ is the absolute minimum value and $f(2) = 3$ is the absolute maximum value.

© 2011 Cengage Learning. All Rights Reserved. May not be scanned, copied or duplicated, or posted to a publicly accessible website, in whole or in part.

25. $f(x) = 4x + \dfrac{1}{x}$ is continuous on $[1, 3]$ and differentiable in $(1, 3)$. To find the critical numbers of f, we solve

$f'(x) = 4 - \dfrac{1}{x^2} = 0$, obtaining $x = \pm\frac{1}{2}$. Because these critical numbers lie outside the interval $[1, 3]$, they are not

candidates for the absolute extrema of f. Evaluating f at the endpoints of the interval $[1, 3]$, we find that the

absolute maximum value of f is $f(3) = \frac{37}{3}$, and the absolute minimum value of f is $f(1) = 5$.

27. $f(x) = \frac{1}{2}x^2 - 2\sqrt{x} = \frac{1}{2}x^2 - 2x^{1/2}$. To find the critical

numbers of f, we solve $f'(x) = x - x^{-1/2} = 0$, or

$x^{3/2} - 1 = 0$, obtaining $x = 1$. From the table, we conclude

that $f(3) \approx 1.04$ is the absolute maximum value and

$f(1) = -\frac{3}{2}$ is the absolute minimum value.

x	0	1	3
$f(x)$	0	$-\frac{3}{2}$	$\frac{9}{2} - 2\sqrt{3} \approx 1.04$

29.

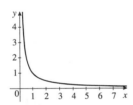

From the graph of $f(x) = \dfrac{1}{x}$ for $x > 0$, we

conclude that f has no absolute extremum.

31. $f(x) = 3x^{2/3} - 2x$. The function f is continuous on $[0, 3]$

and differentiable on $(0, 3)$. To find the critical numbers of f,

we solve $f'(x) = 2x^{-1/3} - 2 = 0$, obtaining $x = 1$ as the

critical number. From the table, we conclude that the absolute

maximum value is $f(1) = 1$ and the absolute minimum value

is $f(0) = 0$.

x	0	1	3
$f(x)$	0	1	$3^{5/3} - 6 \approx 0.24$

33. $f(x) = x^{2/3}\left(x^2 - 4\right)$, so $f'(x) = x^{2/3}(2x) + \frac{2}{3}x^{-1/3}\left(x^2 - 4\right) = \frac{2}{3}x^{-1/3}\left[3x^2 + \left(x^2 - 4\right)\right] = \dfrac{8\left(x^2 - 1\right)}{3x^{1/3}} = 0$.

Observe that f' is not defined at $x = 0$. Furthermore, $f'(x) = 0$ at $x \pm 1$. So the critical numbers of f are -1 and 0,

and 1. From the table, we see that f has absolute minima at

$(-1, -3)$ and $(1, -3)$ and absolute maxima at $(0, 0)$ and $(2, 0)$.

x	-1	0	1	2
$f(x)$	-3	0	-3	0

© 2011 Cengage Learning. All Rights Reserved. May not be scanned, copied or duplicated, or posted to a publicly accessible website, in whole or in part.

35. $f(x) = \dfrac{x}{x^2 + 2}$. To find the critical numbers of f, we solve $f'(x) = \dfrac{(x^2 + 2) - x(2x)}{(x^2 + 2)^2} = \dfrac{2 - x^2}{(x^2 + 2)^2} = 0$,

obtaining $x = \pm\sqrt{2}$. Because $x = -\sqrt{2}$ lies outside $[-1, 2]$, $x = \sqrt{2}$ is the only critical number in the given

interval.

From the table, we conclude that $f\left(\sqrt{2}\right) = \frac{\sqrt{2}}{4} \approx 0.35$ is the

absolute maximum value and $f(-1) = -\frac{1}{3}$ is the absolute

minimum value.

x	-1	$\sqrt{2}$	2
$f(x)$	$-\frac{1}{3}$	$\frac{\sqrt{2}}{4} \approx 0.35$	$\frac{1}{3}$

37. The function $f(x) = \dfrac{x}{\sqrt{x^2 + 1}} = \dfrac{x}{(x^2 + 1)^{1/2}}$ is continuous on the closed interval

$[-1, 1]$ and differentiable on $(-1, 1)$. To find the critical numbers of f, we first compute

$f'(x) = \dfrac{(x^2 + 1)^{1/2}(1) - x\left(\frac{1}{2}\right)(x^2 + 1)^{-1/2}(2x)}{\left[(x^2 + 1)^{1/2}\right]^2} = \dfrac{(x^2 + 1)^{-1/2}\left[x^2 + 1 - x^2\right]}{x^2 + 1} = \dfrac{1}{(x^2 + 1)^{3/2}}$, which is

never equal to zero. We compute $f(x)$ at the endpoints, and conclude that $f(-1) = -\frac{\sqrt{2}}{2}$ is the absolute minimum

value and $f(1) = \frac{\sqrt{2}}{2}$ is the absolute maximum value.

39. $h(t) = -16t^2 + 64t + 80$. To find the maximum value of h, we solve $h'(t) = -32t + 64 = -32(t - 2) = 0$,

giving $t = 2$ as the critical number of h. Furthermore, this value of t gives rise to the absolute maximum

value of h since the graph of h is a parabola that opens downward. The maximum height is given by

$h(2) = -16(4) + 64(2) + 80 = 144$, or 144 feet.

41. $P'(t) = 0.027t - 1.126 = 0$ implies $t \approx 41.7$, a critical number of P. $P''(t) = 0.027$ and $P''(41.7) = 0.027 > 0$,

so $t = 41.7$ gives a minimum of P. This occurred around September of 1991. $P(41.7) \approx 17.72$, or 17.72%.

43. $N(t) = 0.81t - 1.14\sqrt{t} + 1.53$, so $N'(t) = 0.81 - 1.14\left(\frac{1}{2}t^{-1/2}\right) = 0.81 - \dfrac{0.57}{t^{1/2}}$. Setting $N'(t) = 0$ gives

$t^{1/2} = \dfrac{0.57}{0.81}$, so $t = 0.4952$ is a critical number of N. Evaluating $N(t)$ at the endpoints $t = 0$ and $t = 6$ as well as

at the critical number, we see that the absolute maximum of N occurs at $t = 6$ and the absolute minimum occurs at

$t \approx 0.5$. Our results tell us that the number of nonfarm

full-time self-employed women over the time interval from

1963 to 1993 reached its maximum of approximately

3.6 million in 1993.

t	0	0.4952	6
$N(t)$	1.53	1.13	3.60

45. $P(x) = -0.000002x^3 + 6x - 400$, so $P'(x) = -0.000006x^2 + 6 = 0$ if $x = \pm 1000$. We reject the negative

root. Next, we compute $P''(x) = -0.000012x$. Because $P''(1000) = -0.012 < 0$, the Second Derivative

Test shows that $x = 1000$ gives a relative maximum of f. From physical considerations, or from a sketch

of the graph of f, we see that the maximum profit is realized if 1000 cases are produced per day. That profit is

$P(1000) = -0.000002(1000)^3 + 6(1000) - 400$, or $\$3600$/day.

© 2011 Cengage Learning. All Rights Reserved. May not be scanned, copied or duplicated, or posted to a publicly accessible website, in whole or in part.

47. The revenue is $R(x) = px = -0.0004x^2 + 10x$ and the profit is
$P(x) = R(x) - C(x) = -0.0004x^2 + 10x - (400 + 4x + 0.0001x^2) = -0.0005x^2 + 6x - 400$.
$P'(x) = -0.001x + 6 = 0$ if $x = 6000$, a critical number. Because $P''(x) = -0.001 < 0$ for all x, we see that the graph of P is a parabola that opens downward. Therefore, a level of production of 6000 rackets/day will yield a maximum profit.

49. The cost function is $C(x) = V(x) + 20,000 = 0.000001x^3 - 0.01x^2 + 50x + 20,000$, so the profit function is
$P(x) = R(x) - C(x) = -0.02x^2 + 150x - 0.000001x^3 + 0.01x^2 - 50x + 20,000$

$$= -0.000001x^3 - 0.01x^2 + 100x - 20,000.$$

We want to maximize P on $[0, 7000]$. $P'(x) = -0.000003x^2 - 0.02x + 100$. Setting $P'(x) = 0$ gives
$3x^2 + 20,000x - 100,000,000 = 0$, so or $x = \dfrac{-20,000 \pm \sqrt{20,000^2 + 1,200,000,000}}{6} = -10,000$ or $3,333.33$.

Thus, $x = 3333.33$ is a critical number in the interval $[0, 7500]$. From the table, we see that a level of production of 3,333 pagers per week will yield a maximum profit of $165,185.20 per week.

x	0	3333.33	7500
$P(x)$	−20,000	165,185.2	−254,375

51. **a.** $\overline{C}(x) = \dfrac{C(x)}{x} = 0.0025x + 80 + \dfrac{10,000}{x}$.

b. $\overline{C}'(x) = 0.0025 - \dfrac{10,000}{x^2} = 0$ if $0.0025x^2 = 10,000$, or $x = 2000$. Because $\overline{C}''(x) = \dfrac{20,000}{x^3}$, we see that $\overline{C}''(x) > 0$ for $x > 0$ and so $\overline{C}$ is concave upward on $(0, \infty)$. Therefore, $x = 2000$ yields a minimum.

c. We solve $\overline{C}(x) = C'(x)$: $0.0025x + 80 + \dfrac{10,000}{x} = 0.005x + 80$, so $0.0025x^2 = 10,000$ and $x = 2000$.

d. It appears that we can solve the problem in two ways.

53. The demand equation is $p = \sqrt{800 - x} = (800 - x)^{1/2}$, so the revenue function is
$R(x) = xp = x(800 - x)^{1/2}$. To find the maximum of R, we compute $R'(x) = \frac{1}{2}(800 - x)^{-1/2}(-1)(x) + (800 - x)^{1/2} = \frac{1}{2}(800 - x)^{-1/2}[-x + 2(800 - x)] = \frac{1}{2}(800 - x)^{-1/2}(1600 - 3x)$. Next, $R'(x) = 0$ implies $x = 800$ or $x = \frac{1600}{3}$, the critical numbers of R. From the table, we conclude that $R\left(\frac{1600}{3}\right) = 8709$ is the absolute maximum value. Therefore, the revenue is maximized by producing $\frac{1600}{3} \approx 533$ dresses.

x	0	800	$\frac{1600}{3}$
$R(x)$	0	0	8709

© 2011 Cengage Learning. All Rights Reserved. May not be scanned, copied or duplicated, or posted to a publicly accessible website, in whole or in part.

55. $f(t) = 100 \left(\dfrac{t^2 - 4t + 4}{t^2 + 4} \right)$.

a. $f'(t) = 100 \left[\dfrac{(t^2 + 4)(2t - 4) - (t^2 - 4t + 4)(2t)}{(t^2 + 4)^2} \right] = \dfrac{400(t^2 - 4)}{(t^2 + 4)^2} = \dfrac{400(t - 2)(t + 2)}{(t^2 + 4)^2}$.

From the sign diagram for f', we see that $t = 2$ gives a
relative minimum, and we conclude that the oxygen content
is the lowest 2 days after the organic waste has been
dumped into the pond.

$$- - - - - \; 0 \; + + + + \quad \text{sign of } f'$$
$$\underset{\;0}{\vdash} \qquad \underset{\;2}{\dashv} \qquad \longrightarrow t$$

b. $f''(t) = 400 \left[\dfrac{(t^2 + 4)^2 (2t) - (t^2 - 4) \, 2 \, (t^2 + 4)(2t)}{(t + 4)^4} \right] = 400 \left[\dfrac{(2t)(t^2 + 4)(t^2 + 4 - 2t^2 + 8)}{(t^2 + 4)^4} \right]$

$= -\dfrac{800t \, (t^2 - 12)}{(t^2 + 4)^3}$.

$f''(t) = 0$ when $t = 0$ and $t = \pm 2\sqrt{3}$. We reject $t = 0$ and $t = -2\sqrt{3}$. From the sign diagram for f'', we see
that $f' \left(2\sqrt{3} \right)$ gives an inflection point of f and we
conclude that this is an absolute maximum. Therefore, the
rate of oxygen regeneration is greatest 3.5 days after the
organic waste has been dumped into the pond.

$$0 \; + + + + \; 0 \; - - - \quad \text{sign of } f''$$
$$\underset{\;0}{\vdash} \qquad \underset{2\sqrt{3}}{\dashv} \qquad \longrightarrow t$$

57. We compute $\overline{R}'(x) = \dfrac{x R'(x) - R(x)}{x^2}$. Setting $\overline{R}'(x) = 0$ gives $x R'(x) - R(x) = 0$, or

$R'(x) = \dfrac{R(x)}{x} = \overline{R}(x)$, so a critical number of $\overline{R}$ occurs when $\overline{R}(x) = R'(x)$. Next, we compute

$\overline{R}''(x) = \dfrac{x^2 \left[R'(x) + x R''(x) - R'(x) \right] - \left[x R'(x) - R(x) \right](2x)}{x^4} = \dfrac{R''(x)}{x} < 0$. Thus, by the Second

Derivative Test, the critical number does give the maximum revenue.

59. The growth rate is $G'(t) = -0.6t^2 + 4.8t$. To find the
maximum growth rate, we compute $G''(t) = -1.2t + 4.8$.
Setting $G''(t) = 0$ gives $t = 4$ as a critical number. From the
table, we see that G is maximal at $t = 4$; that is, the growth
rate is greatest in 2004.

t	0	4	8
$G'(t)$	0	9.6	0

61. $P'(t) = 0.13089t^2 - 0.534t - 1.59 = 0$ gives $t = \dfrac{0.534 \pm \sqrt{(0.534)^2 - 4(0.13089)(-1.59)}}{2(0.13089)} \approx -2$ or 6.08. We

reject the negative root. Because $P''(t) = 0.26178t - 0.534$ and $P''(6.08) \approx 1.06 > 0$, we conclude that $t = 6.08$
gives a minimum of P, and this number corresponds to approximately early 1970.

© 2011 Cengage Learning. All Rights Reserved. May not be scanned, copied or duplicated, or posted to a publicly accessible website, in whole or in part.

63. $A'(t) = \dfrac{d}{dt}\left(-0.00005t^3 - 0.000826t^2 + 0.0153t + 4.55\right) = -0.00015t^2 - 0.001652t + 0.0153.$ Using the

quadratic formula to solve $f'(t) = 0$, we obtain $t \approx -17.01$,
or 5.997. From the table, we see that A has an absolute
maximum when $t \approx 6$, so the cortex of children of average
intelligence reaches a maximum thickness around the time the
children are 6 years old.

t	5	6	19
$A(t)$	4.5996	4.601	4.200

65. $R'(t) = -2.133t^2 + 7.52t + 0.2 = 0$ implies $t = \dfrac{-7.52 \pm \sqrt{7.52^2 - 4(-2.133)(0.2)}}{2(-2.133)} \approx -0.026$ or 3.55. The

root -0.026 lies outside the interval $[0, 5]$ and is rejected. $R''(t) = -4.266t + 7.52$ and $R''(3.55) \approx -7.62 > 0$,
so $t = 3.55$ gives a relative maximum. This is around the middle of the year 2000. The highest office space rent
was $R(3.55) \approx 52.79$, or approximately $\$52.79/\text{ft}^2$.

67. a. On $[0, 3]$, $f(t) = 0.6t^2 + 2.4t + 7.6$, so $f'(t) = 1.2t + 2.4 = 0$ implies $t = -2$ which lies outside the interval
$[0, 3]$. (We evaluate f at each relevant point below.)
On $[3, 5]$, $f(t) = 3t^2 + 18.8t - 63.2$, so $f'(t) = 6t + 18.8 = 0$ implies $t = -3.13$ which lies outside the
interval $[3, 5]$.
On $[5, 8]$, $f(t) = -3.3167t^3 + 80.1t^2 - 642.583t + 1730.8025$, so $f'(t) = -9.9501t^2 + 160.2t - 642.583 = 0$

implies $t = \dfrac{-160.2 \pm \sqrt{160.2^2 - 4(-9.9501)(642.583)}}{2(-9.9501)} \approx 7.58$ or 8.52. Only the critical number $t = 7.58$ lies

inside the interval $[5, 8]$.
From the table, we see that the investment peaked
when $t = 5$, that is, in the year 2000. The amount
invested was \$105.8 billion.

t	0	3	5	7.58	8
$f(t)$	7.6	20.2	105.8	17.8	18.4

b. Investment was lowest (at \$7.6 billion) when $t = 0$.

69. $R = D^2\left(\dfrac{k}{2} - \dfrac{D}{3}\right) = \dfrac{kD^2}{2} - \dfrac{D^3}{3}$, so $\dfrac{dR}{dD} = \dfrac{2kD}{2} - \dfrac{3D^2}{3} = kD - D^2 = D(k - D)$. Setting $\dfrac{dR}{dD} = 0$, we have

$D = 0$ or $k = D$. We consider only $k = D$ because $D > 0$. If $k > 0$, $\dfrac{dR}{dD} > 0$ and if $k < 0$, $\dfrac{dR}{dD} < 0$. Therefore
$k = D$ gives a relative maximum. The nature of the problem suggests that $k = D$ gives the absolute maximum of
R. We can also verify this by graphing R.

71. Setting $P' = 0$ gives $P' = \dfrac{d}{dR}\left[\dfrac{E^2 R}{(R+r)^2}\right] = E^2\left[\dfrac{(R+r)^2 - R(2)(R+r)}{(R+r)^4}\right] = \dfrac{E^2(r - R)}{(R+r)^3} = 0$. Therefore,

$R = r$ is a critical number of P. Because $P'' = E^2 \dfrac{(R+r)^3(-1) - (r - R)(3)(R+r)^2}{(R+r)^6} = \dfrac{2E^2(R - 2r)}{(R+r)^4}$ and

$P''(R) = \dfrac{-2E^2 r}{(2r)^4} = -\dfrac{E^2}{8r^3} < 0$, the Second Derivative Test and physical considerations both imply that $R = r$

gives a relative maximum value of P. The maximum power is $P = \dfrac{E^2 r}{(2r)^2} = \dfrac{E^2}{4r}$ watts.

© 2011 Cengage Learning. All Rights Reserved. May not be scanned, copied or duplicated, or posted to a publicly accessible website, in whole or in part.

73. $R'(x) = \dfrac{d}{dx}[kx(Q-x)] = k\dfrac{d}{dx}(Qx - x^2) = k(Q - 2x)$ is continuous everywhere and has a zero at $\frac{1}{2}Q$; this is the only critical number of R in $(0, Q)$. $R(0) = 0$, $R\left(\frac{1}{2}Q\right) = \frac{1}{4}kQ^2$, and $R(Q) = 0$, so the absolute maximum value of R is $R\left(\frac{1}{2}Q\right) = \frac{1}{4}kQ^2$, showing that the rate of chemical reaction is greatest when exactly half of the original substrate has been transformed.

75. False. Let $f(x) = \begin{cases} |x| & \text{if } x \neq 0 \\ 1 & \text{if } x = 0 \end{cases}$ on $[-1, 1]$.

77. False. Let $f(x) = \begin{cases} -x & \text{if } -1 \leq x < 0 \\ \frac{1}{2} & \text{if } 0 \leq x < 1 \end{cases}$ Then f is discontinuous at $x = 0$, but f has an absolute maximum value of 1, attained at $x = -1$.

79. Because $f(x) = c$ for all x, the function f satisfies $f(x) \leq c$ for all x and so f has absolute maxima at all values of x. Similarly, f has absolute minima at all values of x.

81. a. f is not continuous at $x = 0$ because $\lim\limits_{x \to 0} f(x)$ does not exist.

b. $\lim\limits_{x \to 0} f(x) = \lim\limits_{x \to 0} \dfrac{1}{x} = -\infty$ and $\lim\limits_{x \to 0} f(x) = \lim\limits_{x \to 0} \dfrac{1}{x} = \infty$.

c.

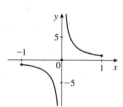

4.4 Using Technology page 312

1. Absolute maximum value 145.9, absolute minimum value -4.3834.

3. Absolute maximum value 16, absolute minimum value -0.1257.

5. Absolute maximum value 2.8889, absolute minimum value 0.

© 2011 Cengage Learning. All Rights Reserved. May not be scanned, copied or duplicated, or posted to a publicly accessible website, in whole or in part.

7. a.

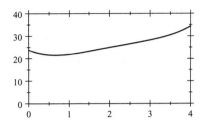

b. The lowest percentage was 21.51%.

9. a. $N(t) = 1.2576t^4 - 26.357t^3 + 127.98t^2 + 82.3t + 43$, so
$N'(t) = 5.0304t^3 - 79.071t^2 + 255.96t + 82.3$.

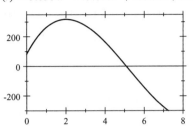

From the graph, we see that $N'(t)$ has a maximum when $t \approx 2$, on February 8.

b. From the graph in part (a), the maximum number of sickouts occurred when $N'(t) = 0$, that is, when $t = 5$. We calculate $N(5) \approx 1145$ canceled flights.

4.5   Problem-Solving Tips

Follow the guidelines given on page 312 of the text to solve the optimization problems that follow. Remember, Theorem 3 in Section 4.4 provides us with a method of computing the absolute extrema of a continuous function over a closed interval $[a, b]$. If the problem involves a function that is to be optimized over an interval that is not closed, then use the graphical method to find the optimal values of f. You might review Example 4 on page 316 of the text to make sure you understand how to use the graphical method.

4.5   Concept Questions   page 319

1. We could solve the problem by sketching the graph of f and checking to see if there is an absolute extremum.

4.5   Optimization II   page 319

1. Let x and y denote the lengths of two adjacent sides of the rectangle. We want to maximize $A = xy$. But the perimeter is $2x + 2y$ and this is equal to 100, so $2x + 2y = 100$, and therefore $y = 50 - x$. Thus, $A = f(x) = x(50 - x) = -x^2 + 50x$, $0 \le x \le 50$. We allow the "degenerate" cases $x = 0$ and $x = 50$. $A' = -2x + 50 = 0$ implies that $x = 25$ is a critical number of f. $A(0) = 0$, $A(25) = 625$, and $A(50) = 0$, so we see that A is maximized for $x = 25$. The required dimensions are 25 ft by 25 ft.

© 2011 Cengage Learning. All Rights Reserved. May not be scanned, copied or duplicated, or posted to a publicly accessible website, in whole or in part.

3. We have $2x + y = 3000$ and we want to maximize the function

$A = f(x) = xy = x(3000 - 2x) = 3000x - 2x^2$ on the interval

$[0, 1500]$. The critical number of A is obtained by solving

$f'(x) = 3000 - 4x = 0$, giving $x = 750$. From the table of values,

we conclude that $x = 750$ yields the absolute maximum value of A.

Thus, the required dimensions are 750×1500 yards. The

maximum area is $1,125,000$ yd^2.

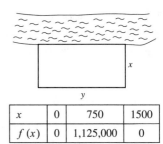

x	0	750	1500
$f(x)$	0	1,125,000	0

5. Let x denote the length of the side made of wood and y the length of the side made of steel. The cost of construction

is $C = 6(2x) + 3y$, but $xy = 800$, so $y = \dfrac{800}{x}$. Therefore, $C = f(x) = 12x + 3\left(\dfrac{800}{x}\right) = 12x + \dfrac{2400}{x}$. To

minimize C, we compute $f'(x) = 12 - \dfrac{2400}{x^2} = \dfrac{12x^2 - 2400}{x^2} = \dfrac{12(x^2 - 200)}{x^2}$. Setting $f'(x) = 0$ gives

$x = \pm\sqrt{200}$ as critical numbers of f. The sign diagram of f'

shows that $x = \pm\sqrt{200}$ gives a relative minimum of f.

$f''(x) = \dfrac{4800}{x^3} > 0$ if $x > 0$, and so f is concave upward for

$x > 0$. Therefore, $x = \sqrt{200} = 10\sqrt{2}$ yields the absolute minimum. Thus, the dimensions of the enclosure should

be $10\sqrt{2}$ ft $\times 40\sqrt{2}$ ft, or 14.1 ft $\times 56.6$ ft.

$$- \, - \, - \; 0 \; + \; + \; + \quad \text{sign of } f'$$

7. Let the dimensions of each square that is cut out be $x'' \times x''$. Then

the dimensions of the box are $(8 - 2x)''$ by $(8 - 2x)''$ by x'', and

its volume is be $V = f(x) = x(8 - 2x)^2$. We want to maximize f

on $[0, 4]$.

$f'(x) = (8 - 2x)^2 + x(2)(8 - 2x)(-2)$ (by the Product Rule)

$\qquad = (8 - 2x)[(8 - 2x) - 4x]$

$\qquad = (8 - 2x)(8 - 6x)$

$\qquad = 0$ if $x = 4$ or $\frac{4}{3}$.

The latter is a critical number in $(0, 4)$. From the table, we see that

$x = \frac{4}{3}$ yields an absolute maximum for f, so the dimensions of the

box should be $\frac{16''}{3} \times \frac{16''}{3} \times \frac{4''}{3}$.

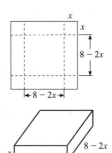

x	0	$\frac{4}{3}$	4
$f(x)$	0	$\frac{1024}{27}$	0

© 2011 Cengage Learning. All Rights Reserved. May not be scanned, copied or duplicated, or posted to a publicly accessible website, in whole or in part.

9. Let x denote the length of the sides of the box and y denote its height. Referring to the figure, we see that the volume of the box is given by $x^2 y = 128$. The amount of material used is given by

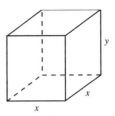

$$S = f(x) = 2x^2 + 4xy = 2x^2 + 4x\left(\frac{128}{x^2}\right) = 2x^2 + \frac{512}{x} \text{ in}^2. \text{ We}$$

want to minimize f subject to the condition that $x > 0$.

Now $f'(x) = 4x - \dfrac{512}{x^2} = \dfrac{4x^3 - 512}{x^2} = \dfrac{4(x^3 - 128)}{x^2}$. Setting $f'(x) = 0$ yields $x = 5.04$, a critical number of f.

Next, $f''(x) = 4 + \dfrac{1024}{x^3} > 0$ for all $x > 0$. Thus, the graph of f is concave upward, and so $x = 5.04$ yields an absolute minimum of f. The required dimensions are $5.04'' \times 5.04'' \times 5.04''$.

11. The length plus the girth of the box is $4x + h = 108$ and $h = 108 - 4x$. Then $V = x^2 h = x^2(108 - 4x) = 108x^2 - 4x^3$ and $V' = 216x - 12x^2$. We want to maximize V on the interval $[0, 27]$. Setting $V'(x) = 0$ and solving for x, we obtain $x = 18$ and $x = 0$. We calculate $V(0) = 0$, $V(18) = 11{,}664$, and $V(27) = 0$. Thus, the dimensions of the box are $18'' \times 18'' \times 36''$ and its maximum volume is approximately $11{,}664 \text{ in}^3$.

13. We take $2\pi r + \ell = 108$. We want to maximize $V = \pi r^2 \ell = \pi r^2(-2\pi r + 108) = -2\pi^2 r^3 + 108\pi r^2$ subject to the condition that $0 \le r \le \frac{54}{\pi}$. Now $V'(r) = -6\pi^2 r^2 + 216\pi r = -6\pi r(\pi r - 36) = 0$ implies that $r = 0$ and $r = \frac{36}{\pi}$ are critical numbers of V. From the table, we conclude that the maximum volume occurs when $r = \frac{36}{\pi} \approx 11.5$ inches and $\ell = 108 - 2\pi\left(\frac{36}{\pi}\right) = 36$ inches and the volume of the parcel is $46{,}656/\pi \text{ in}^3$.

r	0	$\frac{36}{\pi}$	$\frac{54}{\pi}$
V	0	$\frac{46{,}656}{\pi}$	0

15. Let y denote the height and x the width of the cabinet. Then $y = \frac{3}{2}x$. Because the volume is to be 2.4 ft^3, we have $xyd = 2.4$, where d is the depth of the cabinet. Thus, $x\left(\frac{3}{2}x\right)d = 2.4$, so $d = \dfrac{2.4(2)}{3x^2} = \dfrac{1.6}{x^2}$. The cost for constructing the cabinet is

$$C = 40(2xd + 2yd) + 20(2xy) = 80\left[\frac{1.6}{x} + \left(\frac{3x}{2}\right)\left(\frac{1.6}{x^2}\right)\right] + 40x\left(\frac{3x}{2}\right) = \frac{320}{x} + 60x^2, \text{ so}$$

$$C'(x) = -\frac{320}{x^2} + 120x = \frac{120x^3 - 320}{x^2} = 0 \text{ if } x = \sqrt[3]{\frac{8}{3}} = \frac{2}{\sqrt[3]{3}} = \frac{2}{3}\sqrt[3]{9}. \text{ Therefore, } x = \frac{2}{3}\sqrt[3]{9} \text{ is a critical number}$$

of C. The sign diagram shows that $x = \frac{2}{3}\sqrt[3]{9}$ gives a relative minimum. Next, $C''(x) = \dfrac{640}{x^3} + 120 > 0$ for

all $x > 0$, telling us that the graph of C is concave upward, so $x = \frac{2}{3}\sqrt[3]{9}$ yields an absolute minimum. The required dimensions are $\frac{2}{3}\sqrt[3]{9} \times \sqrt[3]{9} \times \frac{2}{3}\sqrt[3]{9}$.

$$\begin{array}{c} \quad\quad - - - - 0 + + + + + \quad \text{sign of } C' \\ \hline \quad\quad\quad\quad\quad\quad\quad\quad\quad\quad\quad\quad\quad\quad x \\ 0 \quad\quad\quad\quad \frac{2\sqrt[3]{9}}{3} \end{array}$$

17. We want to maximize the function $R(x) = (200 + x)(300 - x) = -x^2 + 100x + 60{,}000$. Now $R'(x) = -2x + 100 = 0$ gives $x = 50$, and this is a critical number of R. Because $R''(x) = -2 < 0$, we see that $x = 50$ gives an absolute maximum of R. Therefore, the number of passengers should be 250. The fare will then be \$250/passenger and the revenue will be \$62,500.

© 2011 Cengage Learning. All Rights Reserved. May not be scanned, copied or duplicated, or posted to a publicly accessible website, in whole or in part.

19. Let x denote the number of people beyond 20 who sign up for the cruise. Then the revenue is
$R(x) = (20 + x)(600 - 4x) = -4x^2 + 520x + 12{,}000$. We want to maximize R on the closed bounded interval
$[0, 70]$. $R'(x) = -8x + 520 = 0$ implies $x = 65$, a critical number of R. Evaluating R at this critical number and
the endpoints, we see that R is maximized if $x = 65$. Therefore,
85 passengers will result in a maximum revenue of \$28,900. The
fare in this case is \$340/passenger.

x	0	65	70
$R(x)$	12,000	28,900	28,800

21. The fuel cost is $x/600$ dollars per mile and the labor cost is $18/x$ dollars per mile. Therefore, the total cost is
$C(x) = \dfrac{18}{x} + \dfrac{3x}{600}$. We calculate $C'(x) = -\dfrac{18}{x^2} + \dfrac{3}{600} = 0$, giving $-\dfrac{18}{x^2} = -\dfrac{3}{600}$, $3x^2 = 18(600)$, $x^2 = 3600$,
and so $x = 60$. Next, $C''(x) = \dfrac{48}{x^3} > 0$ for all $x > 0$ so C is concave upward. Therefore, $x = 60$ gives the absolute
minimum. The most economical speed is 60 mph.

23. We want to maximize $S = kh^2 w$. But $h^2 + w^2 = 24^2$, or $h^2 = 576 - w^2$, so
$S = f(w) = kw(576 - w^2) = k(576w - w^3)$. Now, setting $f'(w) = k(576 - 3w^2) = 0$ gives
$w = \pm\sqrt{192} \approx \pm 13.86$. Only the positive root is a critical number of interest. Next, we find $f''(w) = -6kw$,
and in particular, $f''\left(\sqrt{192}\right) = -6\sqrt{192}k < 0$, so that $w \approx 13.86$ gives a relative maximum of f. Because
$f''(w) < 0$ for $w > 0$, we see that the graph of f is concave downward on $(0, \infty)$, and so $w = \sqrt{192}$ gives an
absolute maximum of f. We find $h^2 = 576 - 192 = 384$ and so $h \approx 19.60$, so the width and height of the log
should be approximately 13.86 inches and 19.60 inches, respectively.

25. We want to minimize $C(x) = 1.50(10{,}000 - x) + 2.50\sqrt{3000^2 + x^2}$ subject to $0 \le x \le 10{,}000$. Now
$C'(x) = -1.50 + 2.5\left(\frac{1}{2}\right)(9{,}000{,}000 + x^2)^{-1/2}(2x) = -1.50 + \dfrac{2.50x}{\sqrt{9{,}000{,}000 + x^2}} = 0$ if
$2.5x = 1.50\sqrt{9{,}000{,}000 + x^2}$, or $6.25x^2 = 2.25(9{,}000{,}000 + x^2)$,
or $4x^2 = 20{,}250{,}000$, giving $x = 2250$. From the table, we see that
$x = 2250$, or 2250 ft, gives the absolute minimum.

x	0	2250	10,000
$f(x)$	22,500	21,000	26,101

27. The time of flight is $T = f(x) = \dfrac{12 - x}{6} + \dfrac{\sqrt{x^2 + 9}}{4}$, so
$f'(x) = -\dfrac{1}{6} + \dfrac{1}{4}\left(\dfrac{1}{2}\right)(x^2 + 9)^{-1/2}(2x) = -\dfrac{1}{6} + \dfrac{x}{4\sqrt{x^2 + 9}} = \dfrac{3x - 2\sqrt{x^2 + 9}}{12\sqrt{x^2 + 9}}$. Setting $f'(x) = 0$ gives
$3x = 2\sqrt{x^2 + 9}$, $9x^2 = 4(x^2 + 9)$, and $5x^2 = 36$. Therefore, $x = \pm\frac{6}{\sqrt{5}} = \pm\frac{6\sqrt{5}}{5}$. Only the critical number
$x = \frac{6\sqrt{5}}{5}$ is of interest. The nature of the problem suggests $x \approx 2.68$ gives an absolute minimum for T.

© 2011 Cengage Learning. All Rights Reserved. May not be scanned, copied or duplicated, or posted to a publicly accessible website, in whole or in part.

29. The area enclosed by the rectangular region of the racetrack is $A = (\ell)(2r) = 2r\ell$. The length of the racetrack is $2\pi r + 2\ell$, and is equal to 1760. That is, $2(\pi r + \ell) = 1760$, and $\pi r + \ell = 880$. Therefore, we want to maximize $A = f(r) = 2r(880 - \pi r) = 1760r - 2\pi r^2$. The restriction on r is $0 \le r \le \frac{880}{\pi}$. To maximize A, we compute $f'(r) = 1760 - 4\pi r$. Setting $f'(r) = 0$ gives $r = \frac{1760}{4\pi} = \frac{440}{\pi} \approx 140$. Because $f(0) = f\left(\frac{880}{\pi}\right) = 0$, we see that the maximum rectangular area is enclosed if we take $r = \frac{440}{\pi}$ and $\ell = 880 - \pi\left(\frac{440}{\pi}\right) = 440$. So $r = 140$ and $\ell = 440$. The total area enclosed is

$$2r\ell + \pi r^2 = 2\left(\frac{440}{\pi}\right)(440) + \pi\left(\frac{440}{\pi}\right)^2 = \frac{2(440)^2}{\pi} + \frac{440^2}{\pi} = \frac{580,800}{\pi} \approx 184,874 \text{ ft}^2.$$

31. Let x denote the number of bottles in each order. We want to minimize

$$C(x) = 200\left(\frac{2,000,000}{x}\right) + \frac{x}{2}(0.40) = \frac{400,000,000}{x} + 0.2x. \text{ We compute } C'(x) = -\frac{400,000,000}{x^2} + 0.2.$$

Setting $C'(x) = 0$ gives $x^2 = \frac{400,000,000}{0.2} = 2,000,000,000$, or $x = 44,721$, a critical number of C.

$C''(x) = \dfrac{800,000,000}{x^3} > 0$ for all $x > 0$, and we see that the graph of C is concave upward and so $x = 44,721$ gives an absolute minimum of C. Therefore, there should be $2,000,000/x \approx 45$ orders per year (since we can not have fractions of an order.) Each order should be for $2,000,000/45 \approx 44,445$ bottles.

33. a. Because the sales are assumed to be steady and D units are expected to be sold per year, the number of orders per year is D/x. Because is costs $\$K$ per order, the ordering cost is KD/x. The purchasing cost is PD (cost per item times number purchased). Finally, the holding cost is $\frac{1}{2}xh$ (the average number on hand times holding cost per item). Therefore, $C(x) = \dfrac{KD}{x} + pD + \dfrac{hx}{2}$.

b. $C'(x) = -\dfrac{KD}{x^2} + \dfrac{h}{2} = 0$ implies $\dfrac{KD}{x^2} = \dfrac{h}{2}$, so $x^2 = \dfrac{2KD}{h}$ and $x = \pm\sqrt{\dfrac{2KD}{h}}$. We reject the negative root. So $x = \sqrt{\dfrac{2KD}{h}}$ is the only critical number. Next, $C''(x) = \dfrac{2KD}{x^3} > 0$ for $x > 0$, so $C''\left(\sqrt{\dfrac{2KD}{h}}\right) > 0$ and the Second Derivative Test shows that $x = \sqrt{\dfrac{2KD}{h}}$ does give a relative minimum. Because C is concave upward, this is also the absolute minimum.

CHAPTER 4   **Concept Review**   page 324

1. a. $f(x_1) < f(x_2)$

b. $f(x_1) > f(x_2)$

3. a. $f(x) \le f(c)$

b. $f(x) \ge f(c)$

5. a. $f'(x)$ **b.** > 0 **c.** concavity   **d.** relative maximum; relative extremum

7. $0, 0$

9. a. $f(x) \le f(c)$, absolute maximum value

b. $f(x) \ge f(c)$, open interval

© 2011 Cengage Learning. All Rights Reserved. May not be scanned, copied or duplicated, or posted to a publicly accessible website, in whole or in part.

CHAPTER 4 Review page 324

1. a. $f(x) = \frac{1}{3}x^3 - x^2 + x - 6$, so $f'(x) = x^2 - 2x + 1 = (x-1)^2$. $f'(x) = 0$ gives $x = 1$, the critical number of f. Now $f'(x) > 0$ for all $x \neq 1$. Thus, f is increasing on $(-\infty, 1) \cup (1, \infty)$.

b. Because $f'(x)$ does not change sign as we move across the critical number $x = 1$, the First Derivative Test implies that $x = 1$ does not give a relative extremum of f.

c. $f''(x) = 2(x-1)$. Setting $f''(x) = 0$ gives $x = 1$ as a candidate for an inflection point of f. Because $f''(x) < 0$ for $x < 1$, and $f''(x) > 0$ for $x > 1$, we see that f is concave downward on $(-\infty, 1)$ and concave upward on $(1, \infty)$.

d. The results of part (c) imply that $\left(1, -\frac{17}{3}\right)$ is an inflection point.

3. a. $f(x) = x^4 - 2x^2$, so

$f'(x) = 4x^3 - 4x = 4x(x^2 - 1) = 4x(x+1)(x-1)$. The sign diagram of f' shows that f is decreasing on $(-\infty, -1) \cup (0, 1)$ and increasing on $(-1, 0) \cup (1, \infty)$.

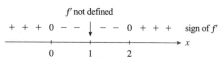

b. The results of part (a) and the First Derivative Test show that $(-1, -1)$ and $(1, -1)$ are relative minima and $(0, 0)$ is a relative maximum.

c. $f''(x) = 12x^2 - 4 = 4(3x^2 - 1) = 0$ if $x = \pm\frac{\sqrt{3}}{3}$. The sign diagram shows that f is concave upward on

$\left(-\infty, -\frac{\sqrt{3}}{3}\right) \cup \left(\frac{\sqrt{3}}{3}, \infty\right)$ and concave downward on

$\left(-\frac{\sqrt{3}}{3}, \frac{\sqrt{3}}{3}\right)$.

d. The results of part (c) show that $\left(-\frac{\sqrt{3}}{3}, -\frac{5}{9}\right)$ and $\left(\frac{\sqrt{3}}{3}, -\frac{5}{9}\right)$ are inflection points.

5. a. $f(x) = \frac{x^2}{x-1}$, so $f'(x) = \frac{(x-1)(2x) - x^2(1)}{(x-1)^2} = \frac{x^2 - 2x}{(x-1)^2} = \frac{x(x-2)}{(x-1)^2}$.

The sign diagram of f' shows that f is increasing on $(-\infty, 0) \cup (2, \infty)$ and decreasing on $(0, 1) \cup (1, 2)$.

b. The results of part (a) show that $(0, 0)$ is a relative maximum and $(2, 4)$ is a relative minimum.

c. $f''(x) = \frac{(x-1)^2(2x-2) - x(x-2)2(x-1)}{(x-1)^4} = \frac{2(x-1)[(x-1)^2 - x(x-2)]}{(x-1)^4} = \frac{2}{(x-1)^3}$. Because $f''(x) < 0$ if $x < 1$ and $f''(x) > 0$ if $x > 1$, we see that f is concave downward on $(-\infty, 1)$ and concave upward on $(1, \infty)$.

d. Because $x = 1$ is not in the domain of f, there is no inflection point.

7. a. $f(x) = (1-x)^{1/3}$, so

$f'(x) = -\frac{1}{3}(1-x)^{-2/3} = -\frac{1}{3(1-x)^{2/3}}$. The sign diagram for f' shows that f is decreasing on $(-\infty, 1) \cup (1, \infty)$.

b. There is no relative extremum.

© 2011 Cengage Learning. All Rights Reserved. May not be scanned, copied or duplicated, or posted to a publicly accessible website, in whole or in part.

c. $f''(x) = -\frac{2}{9}(1-x)^{-5/3} = -\dfrac{2}{9(1-x)^{5/3}}$. The sign diagram

for f'' shows that f is concave downward on $(-\infty, 1)$ and
concave upward on $(1, \infty)$.

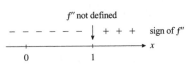

d. $x = 1$ is a candidate for an inflection point of f. Referring to the sign diagram for f'', we see that $(1, 0)$ is an
inflection point.

9. a. $f(x) = \dfrac{2x}{x+1}$, so $f'(x) = \dfrac{(x+1)(2) - 2x(1)}{(x+1)^2} = \dfrac{2}{(x+1)^2} > 0$ if $x \neq -1$. Therefore f is increasing on
$(-\infty, -1) \cup (-1, \infty)$.

b. Because there is no critical number, f has no relative extremum.

c. $f''(x) = -4(x+1)^{-3} = -\dfrac{4}{(x+1)^3}$. Because $f''(x) > 0$ if $x < -1$ and $f''(x) < 0$ if $x > -1$, we see that f
is concave upward on $(-\infty, -1)$ and concave downward on $(-1, \infty)$.

d. There is no inflection point because $f''(x) \neq 0$ for all x in the domain of f.

11. $f(x) = x^2 - 5x + 5$.

1. The domain of f is $(-\infty, \infty)$.

2. Setting $x = 0$ gives 5 as the y-intercept.

3. $\lim\limits_{x \to -\infty} (x^2 - 5x + 5) = \lim\limits_{x \to \infty} (x^2 - 5x + 5) = \infty$.

4. There is no asymptote because f is a quadratic function.

5. $f'(x) = 2x - 5 = 0$ if $x = \frac{5}{2}$. The sign diagram shows that f
is increasing on $\left(\frac{5}{2}, \infty\right)$ and decreasing on $\left(-\infty, \frac{5}{2}\right)$.

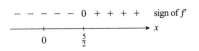

6. The First Derivative Test implies that $\left(\frac{5}{2}, -\frac{5}{4}\right)$ is a relative

   minimum.

7. $f''(x) = 2 > 0$ and so f is concave upward on $(-\infty, \infty)$.

8. There is no inflection point.

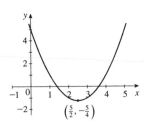

13. $g(x) = 2x^3 - 6x^2 + 6x + 1$.

1. The domain of g is $(-\infty, \infty)$.

2. Setting $x = 0$ gives 1 as the y-intercept.

3. $\lim\limits_{x \to -\infty} g(x) = -\infty$ and $\lim\limits_{x \to \infty} g(x) = \infty$.

4. There is no vertical or horizontal asymptote.

5. $g'(x) = 6x^2 - 12x + 6 = 6(x^2 - 2x + 1) = 6(x-1)^2$. Because $g'(x) > 0$ for all $x \neq 1$, we see that g is
increasing on $(-\infty, 1) \cup (1, \infty)$.

© 2011 Cengage Learning. All Rights Reserved. May not be scanned, copied or duplicated, or posted to a publicly accessible website, in whole or in part.

6. $g'(x)$ does not change sign as we move across the critical number $x = 1$, so there is no extremum.

7. $g''(x) = 12x - 12 = 12(x - 1)$. Because $g''(x) < 0$ if $x < 1$ and $g''(x) > 0$ if $x > 1$, we see that g is concave upward on $(1, \infty)$ and concave downward on $(-\infty, 1)$.

8. The point $x = 1$ gives rise to the inflection point $(1, 3)$.

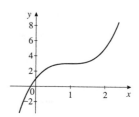

15. $h(x) = x\sqrt{x - 2}$.

1. The domain of h is $[2, \infty)$.

2. There is no y-intercept. Setting $y = 0$ gives 2 as the x-intercept.

3. $\lim_{x \to \infty} x\sqrt{x - 2} = \infty$.

4. There is no asymptote.

5. $h'(x) = (x - 2)^{1/2} + x\left(\frac{1}{2}\right)(x - 2)^{-1/2} = \frac{1}{2}(x - 2)^{-1/2}[2(x - 2) + x] = \frac{3x - 4}{2\sqrt{x - 2}} > 0$ on $[2, \infty)$, and so h is increasing on $[2, \infty)$.

6. Because h has no critical number in $(2, \infty)$, there is no relative extremum.

7. $h''(x) = \frac{1}{2}\left[\frac{(x - 2)^{1/2}(3) - (3x - 4)\frac{1}{2}(x - 2)^{-1/2}}{x - 2}\right] = \frac{(x - 2)^{-1/2}[6(x - 2) - (3x - 4)]}{4(x - 2)} = \frac{3x - 8}{4(x - 2)^{3/2}}$.

The sign diagram for h'' shows that h is concave downward on $\left(2, \frac{8}{3}\right)$ and concave upward on $\left(\frac{8}{3}, \infty\right)$.

8. The results of step 7 tell us that $\left(\frac{8}{3}, \frac{8\sqrt{6}}{9}\right)$ is an inflection point.

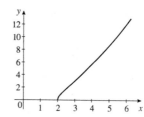

17. $f(x) = \frac{x - 2}{x + 2}$.

1. The domain of f is $(-\infty, -2) \cup (-2, \infty)$.

2. Setting $x = 0$ gives -1 as the y-intercept. Setting $y = 0$ gives 2 as the x-intercept.

3. $\lim_{x \to -\infty} \frac{x - 2}{x + 2} = \lim_{x \to \infty} \frac{x - 2}{x + 2} = 1$.

4. The results of step 3 tell us that $y = 1$ is a horizontal asymptote. Next, observe that the denominator of $f(x)$ is equal to zero at $x = -2$, but its numerator is not equal to zero there. Therefore, $x = -2$ is a vertical asymptote.

5. $f'(x) = \frac{(x + 2)(1) - (x - 2)(1)}{(x + 2)^2} = \frac{4}{(x + 2)^2}$. The sign diagram of f' tells us that f is increasing on $(-\infty, -2) \cup (-2, \infty)$.

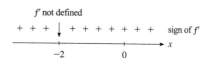

© 2011 Cengage Learning. All Rights Reserved. May not be scanned, copied or duplicated, or posted to a publicly accessible website, in whole or in part.

6. The results of step 5 tell us that there is no relative extremum.

7. $f''(x) = -\dfrac{8}{(x+2)^3}$. The sign diagram of f'' shows that f is concave upward on $(-\infty, -2)$ and concave downward on $(-2, \infty)$.

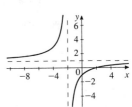

8. There is no inflection point.

19. $\lim\limits_{x \to -\infty} \dfrac{1}{2x+3} = \lim\limits_{x \to \infty} \dfrac{1}{2x+3} = 0$ and so $y = 0$ is a horizontal asymptote. Because the denominator is equal to zero at $x = -\frac{3}{2}$ but the numerator is not equal to zero there, we see that $x = -\frac{3}{2}$ is a vertical asymptote.

21. $\lim\limits_{x \to -\infty} \dfrac{5x}{x^2 - 2x - 8} = \lim\limits_{x \to \infty} \dfrac{5x}{x^2 - 2x - 8} = 0$, so $y = 0$ is a horizontal asymptote. Next, note that the denominator is zero if $x^2 - 2x - 8 = (x-4)(x+2) = 0$, that is, if $x = -2$ or $x = 4$. Because the numerator is not equal to zero at these points, we see that $x = -2$ and $x = 4$ are vertical asymptotes.

23. $f(x) = 2x^2 + 3x - 2$, so $f'(x) = 4x + 3$. Setting $f'(x) = 0$ gives $x = -\frac{3}{4}$ as a critical number of f. Next, $f''(x) = 4 > 0$ for all x, so f is concave upward on $(-\infty, \infty)$. Therefore, $f\left(-\frac{3}{4}\right) = -\frac{25}{8}$ is an absolute minimum of f. There is no absolute maximum.

25. $g(t) = \sqrt{25 - t^2} = (25 - t^2)^{1/2}$. Differentiating $g(t)$, we have $g'(t) = \frac{1}{2}(25 - t^2)^{-1/2}(-2t) = -\dfrac{t}{\sqrt{25 - t^2}}$.

Setting $g'(t) = 0$ gives $t = 0$ as a critical number of g. The domain of g is given by solving the inequality $25 - t^2 \geq 0$ or $(5 - t)(5 + t) \geq 0$ which implies that $t \in [-5, 5]$. From the table, we conclude that $g(0) = 5$ is the absolute maximum of g and $g(-5) = 0$ and $g(5) = 0$ is the absolute minimum value of g.

t	-5	0	5
$g(t)$	0	5	0

27. $h(t) = t^3 - 6t^2$, so $h'(t) = 3t^2 - 12t = 3t(t - 4) = 0$ if $t = 0$ or $t = 4$, critical numbers of h. But only $t = 4$ lies in $(2, 5)$. From the table, we see that h has an absolute minimum at $(4, -32)$ and an absolute maximum at $(2, -16)$.

t	2	4	5
$h(t)$	-16	-32	-25

29. $f(x) = x - \dfrac{1}{x}$ on $[1, 3]$, so $f'(x) = 1 + \dfrac{1}{x^2}$. Because $f'(x)$ is never zero, f has no critical number. Calculating $f(x)$ at the endpoints, we see that $f(1) = 0$ is the absolute minimum value and $f(3) = \frac{8}{3}$ is the absolute maximum value.

© 2011 Cengage Learning. All Rights Reserved. May not be scanned, copied or duplicated, or posted to a publicly accessible website, in whole or in part.

31. $f(s) = s\sqrt{1 - s^2}$ on $[-1, 1]$. The function f is continuous on $[-1, 1]$ and differentiable on $(-1, 1)$. Next,

$$f'(s) = (1 - s^2)^{1/2} + s\left(\tfrac{1}{2}\right)(1 - s^2)^{-1/2}(-2s) = \frac{1 - 2s^2}{\sqrt{1 - s^2}}.$$ Setting $f'(s) = 0$,

we find that $s = \pm\frac{\sqrt{2}}{2}$ are critical numbers of f. From the table, we

see that $f\left(-\frac{\sqrt{2}}{2}\right) = -\frac{1}{2}$ is the absolute minimum value and

$f\left(\frac{\sqrt{2}}{2}\right) = \frac{1}{2}$ is the absolute maximum value of f.

x	-1	$-\frac{\sqrt{2}}{2}$	$\frac{\sqrt{2}}{2}$	1
$f(x)$	0	$-\frac{1}{2}$	$\frac{1}{2}$	0

33. We want to maximize $P(x) = -x^2 + 8x + 20$. Now, $P'(x) = -2x + 8 = 0$ if $x = 4$, a critical number of P. Because $P''(x) = -2 < 0$, the graph of P is concave downward. Therefore, the critical number $x = 4$ yields an absolute maximum. So, to maximize profit, the company should spend \$4000 per month on advertising.

35. a. $R'(t) = \dfrac{d}{dt}\left(0.03056t^3 - 0.45357t^2 + 4.81111t + 31.7\right) = 0.09168t^2 - 0.90714t + 4.8111$. Observe that R' is

continuous everywhere. Setting $R'(t) = 0$ gives $t = \dfrac{0.90714 \pm \sqrt{0.90714^2 - 4(0.09168)(4.81111)}}{2(0.09168)}$.

But the radicand is $-1.76 < 0$, and so there is no solution. We see that R has no extremum. Because $R'(0) = 4.81111 > 0$, we see that R is always increasing. In particular, R is increasing on $(0, 6)$.

b. The result of part (a) shows that the revenue is always increasing over the period in question.

37. a. $N'(t) = 2.25t^2 - 3t + 8.25$, so $N''(t) = 4.5t - 3 = 0$ implies $t \approx 0.67$. From the sign diagram of N'', we see that the graph of N is concave downward on $(0, 0.67)$ and concave upward on $(0.67, 4)$.

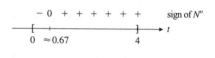

b. From the sign diagram, we see that an inflection point occurs when $t = 0.67$. Because $N(0.67) \approx 138.1$, the inflection point is approximately $(0.67, 138.1)$. The rate of increase of shipments is slowest at $t = 0.67$; that is a little after the middle of 2001.

39. $S(x) = -0.002x^3 + 0.6x^2 + x + 500$, so $S'(x) = -0.006x^2 + 1.2x + 1$ and $S''(x) = -0.012x + 1.2$. $x = 100$ is a candidate for an inflection point of S. The sign diagram for S'' shows that $(100, 4600)$ is an inflection point of S.

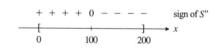

41. a. $I(t) = \dfrac{50t^2 + 600}{t^2 + 10}$, so $I'(t) = \dfrac{(t^2 + 10)(100t) - (50t^2 + 600)(2t)}{(t^2 + 10)^2} = -\dfrac{200t}{(t^2 + 10)^2} < 0$ on $(0, 10)$, and so I

is decreasing on $(0, 10)$.

b. $I''(t) = -200\left[\dfrac{(t^2 + 10)^2(1) - t(2)(t^2 + 10)(2t)}{(t^2 + 10)^4}\right] = \dfrac{-200(t^2 + 10)[(t^2 + 10) - 4t^2]}{(t^2 + 10)^4} = -\dfrac{200(10 - 3t^2)}{(t^2 + 10)^3}.$

The sign diagram of I'' for $t > 0$ shows that I is

concave downward on $\left(0, \sqrt{\tfrac{10}{3}}\right)$ and concave upward

on $\left(\sqrt{\tfrac{10}{3}}, \infty\right)$.

© 2011 Cengage Learning. All Rights Reserved. May not be scanned, copied or duplicated, or posted to a publicly accessible website, in whole or in part.

c.

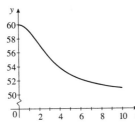

d. The rate of decline in the environmental quality of the wildlife was increasing for the first 1.8 years. After that time the rate of decline decreased.

43. $P(x) = -0.04x^2 + 240x - 10,000$, so $P'(x) = -0.08x + 240 = 0$ if $x = 3000$. The graph of P is a parabola that opens downward and so $x = 3000$ gives rise to the absolute maximum of P. Thus, to maximize profits, the company should produce 3000 cameras per month.

45. $N(t) = -2t^3 + 12t^2 + 2t$. We wish to find the inflection point of the function N. $N'(t) = -6t^2 + 24t + 2$ and $N''(t) = -12t + 24 = -12(t - 2) = 0$ if $t = 2$. Furthermore, $N''(t) > 0$ when $t < 2$ and $N''(t) < 0$ when $t > 2$. Therefore, $t = 2$ gives an inflection point of N. The average worker is performing at peak efficiency at 10 a.m.

47. $R'(x) = k\dfrac{d}{dx}x(M - x) = k[(M - x) + x(-1)] = k(M - 2x)$. Setting $R'(x) = 0$ gives $M - 2x = 0$, or $x = \frac{1}{2}M$, a critical number of R. Because $R''(x) = -2k < 0$, we see that $x = \frac{1}{2}M$ gives a maximum; that is, R is greatest when half the population is infected.

49. Suppose the radius is r and the height is h. Then the capacity is $\pi r^2 h$, and this must be equal to 32π ft^3; that is, $\pi r^2 h = 32\pi$. Let the cost per square foot for the sides be \$$c$. Then the cost of construction is

$$C = 2\pi rhc + 2(\pi r^2)(2c) = 2\pi crh + 4\pi cr^2. \text{ But } h = \frac{32\pi}{\pi r^2} = \frac{32}{r^2}, \text{ so}$$

$$C = f(r) = \frac{64c\pi}{r} + 4\pi cr^2, \text{ giving } C' = f'(r) = -\frac{64\pi c}{r^2} + 8\pi cr = \frac{-64\pi c + 8\pi cr^3}{r^2} = \frac{8\pi c(-8 + r^3)}{r^2}.$$

Setting $f'(r) = 0$ gives $r^3 = 8$, or $r = 2$. Next, $f''(r) = \frac{128\pi c}{r^3} + 8\pi c$, and so $f''(2) > 0$. Therefore, $r = 2$ minimizes f. The required dimensions are $r = 2$ and $h = \frac{32}{4} = 8$. That is, the radius is 2 ft and the height is 8 ft.

51. Let x denote the number of cases in each order. Then the average number of cases of beer in storage during the year is $\frac{1}{2}x$. The storage cost in dollars is $2\left(\frac{1}{2}x\right) = x$. Next, we see that the number of orders required is $\dfrac{800,000}{x}$, and so the ordering cost is $\dfrac{500(800,000)}{x} = \dfrac{400,000,000}{x}$. Thus, the total cost incurred by the company per year is given by $C(x) = x + \dfrac{400,000,000}{x}$. We want to minimize C in the interval $(0, \infty)$, so we calculate $C'(x) = 1 - \dfrac{400,000,000}{x^2}$. Setting $C'(x) = 0$ gives $x^2 = 400,000,000$, or $x = 20,000$ (we reject $x = -20,000$).

Next, $C''(x) = \dfrac{800,000,000}{x^3} > 0$ for all x, so C is concave upward. Thus, $x = 20,000$ gives rise to the absolute minimum of C. The company should order 20,000 cases of beer per order.

© 2011 Cengage Learning. All Rights Reserved. May not be scanned, copied or duplicated, or posted to a publicly accessible website, in whole or in part.

53. $f(x) = x^2 + ax + b$, so $f'(x) = 2x + a$. We require that $f'(2) = 0$, so $(2)(2) + a = 0$, and $a = -4$. Next, $f(2) = 7$ implies that $f(2) = 2^2 + (-4)(2) + b = 7$, so $b = 11$. Thus, $f(x) = x^2 - 4x + 11$. Because the graph of f is a parabola that opens upward, $(2, 7)$ is a relative minimum.

55. Because $(a, f(a))$ is an inflection point of f, $f''(a) = 0$. This shows that a is a critical number of f'. Next, f changes concavity at $(a, f(a))$. If the concavity changes from concave downward to concave upward [that is, $f''(x) < 0$ for $x < a$ and $f''(x) > 0$ for $x > a$], then f' has a relative minimum at a. On the other hand, if the concavity changes from concave upward to concave downward, [$f''(x) > 0$ for $x < a$ and $f''(x) < 0$ for $x > a$], then f' has a relative maximum at a. In either case, f' has a relative extremum at a.

CHAPTER 4   **Before Moving On...**   page 327

1. $f'(x) = \dfrac{(1-x)(2x) - x^2(-1)}{(1-x)^2} = \dfrac{2x - 2x^2 + x^2}{(1-x)^2} = \dfrac{x(2-x)}{(1-x)^2}$; f' is not defined at 1 and has zeros at 0 and 2.

The sign diagram of f shows that f is decreasing on $(-\infty, 0) \cup (2, \infty)$ and increasing on $(0, 1) \cup (1, 2)$.

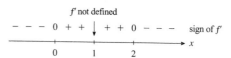

2. $f'(x) = 4x - 4x^{-2/3} = 4x^{-2/3}(x^{5/3} - 1) = \dfrac{4(x^{5/3} - 1)}{x^{2/3}}$. f' is discontinuous at $x = 0$ and has a zero where $x^{5/3} = 1$ or $x = 1$. Therefore, f has critical numbers at 0 and 1. From the sign diagram for f', we see that $x = 1$ gives a relative minimum. Because $f(1) = 2 - 12 = -10$, the relative minimum is $(1, -10)$. There is no relative maximum.

3. $f'(x) = x^2 - \frac{1}{2}x - \frac{1}{2}$ and $f''(x) = 2x - \frac{1}{2} = 0$ gives $x = \frac{1}{4}$. The sign diagram of f'' shows that f is concave downward on $\left(-\infty, \frac{1}{4}\right)$ and concave upward on $\left(\frac{1}{4}, \infty\right)$.

Because $f\left(\frac{1}{4}\right) = \frac{1}{3}\left(\frac{1}{4}\right)^3 - \frac{1}{4}\left(\frac{1}{4}\right)^2 - \frac{1}{2}\left(\frac{1}{4}\right) + 1 = \frac{83}{96}$, the inflection point is $\left(\frac{1}{4}, \frac{83}{96}\right)$.

4. $f(x) = 2x^3 - 9x^2 + 12x - 1$.

1. The domain of f is $(-\infty, \infty)$.

2. Setting $y = f(x) = 0$ gives -1 as the y-intercept of f.

3. $\lim\limits_{x \to -\infty} f(x) = -\infty$ and $\lim\limits_{x \to \infty} f(x) = \infty$.

4. There is no asymptote.

5. $f'(x) = 6x^2 - 18x + 12 = 6(x^2 - 3x + 2) = 6(x-2)(x-1)$.

The sign diagram of f' shows that f is increasing on $(-\infty, 1) \cup (2, \infty)$ and decreasing on $(1, 2)$.

6. We see that $(1, 4)$ is a relative maximum and $(2, 3)$ is a relative minimum.

© 2011 Cengage Learning. All Rights Reserved. May not be scanned, copied or duplicated, or posted to a publicly accessible website, in whole or in part.

7. $f''(x) = 12x - 18 = 6(2x - 3)$. The sign diagram of f'' shows that f is concave downward on $\left(-\infty, \frac{3}{2}\right)$ and concave upward on $\left(\frac{3}{2}, \infty\right)$.

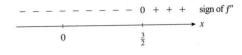

8. $f\left(\frac{3}{2}\right) = 2\left(\frac{3}{2}\right)^3 - 9\left(\frac{3}{2}\right)^2 + 12\left(\frac{3}{2}\right) - 1 = \frac{7}{2}$, so $\left(\frac{3}{2}, \frac{7}{2}\right)$ is an inflection point of f.

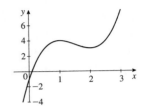

5. f is continuous on a closed interval $[-2, 3]$.

$f'(x) = 6x^2 + 6x = 6x(x+1)$, so the critical numbers of f are -1 and 0. From the table, we see that the absolute maximum value of f is 80 and the absolute minimum value is -5.

x	-2	-1	0	3
y	-5	0	-1	80

6. The amount of material used (the surface area) is $A = \pi r^2 + 2\pi rh$. But $V = \pi r^2 h = 1$,

and so $h = \dfrac{1}{\pi r^2}$. Therefore, $A = \pi r^2 + 2\pi rh \left(\dfrac{1}{\pi r^2}\right) = \pi r^2 + \dfrac{2}{r}$, so

$A' = 2\pi r - \dfrac{2}{r^2} = 0$ implies $2\pi r = \dfrac{2}{r^2}$, $r^3 = \dfrac{2}{r^2}$, $r^3 = \dfrac{1}{\pi}$, and $r = \dfrac{1}{\sqrt[3]{\pi}}$. Because

$A'' = 2\pi + \dfrac{4}{r^3} > 0$ for $r > 0$, we see that $r = \dfrac{1}{\sqrt[3]{\pi}}$ does give an absolute minimum.

Also, $h = \dfrac{1}{\pi r^2} = \dfrac{1}{\pi} \cdot \pi^{2/3} = \dfrac{1}{\pi^{1/3}} = \dfrac{1}{\sqrt[3]{\pi}}$. Therefore, the radius and height should each be $\dfrac{1}{\sqrt[3]{\pi}}$ ft.

© 2011 Cengage Learning. All Rights Reserved. May not be scanned, copied or duplicated, or posted to a publicly accessible website, in whole or in part.

5 EXPONENTIAL AND LOGARITHMIC FUNCTIONS

5.1 Problem-Solving Tips

1. Remember the order of operations when working with exponents. Note that $-5^2 \neq 25$, but rather $-5^2 = -(5)^2 = -25$. On the other hand, $(-5)^2 = 25$.

2. $b^{-x} = \dfrac{1}{b^x} = \left(\dfrac{1}{b}\right)^x$. If $b > 1$, then $0 < \dfrac{1}{b} < 1$, so the graph of b^{-x} for $b > 1$ is similar to the graph of $y = \left(\dfrac{1}{2}\right)^x$.
   (See Figure 3 in the text.)

5.1 Concept Questions   page 334

1. $f(x) = b^x$ with $b > 0$ and $b \neq 1$.

5.1 Exponential Functions   page 334

1. **a.** $4^{-3} \times 4^5 = 4^{-3+5} = 4^2 = 16$.

   b. $3^{-3} \times 3^6 = 3^{6-3} = 3^3 = 27$.

3. **a.** $9(9)^{-1/2} = \dfrac{9}{9^{1/2}} = \dfrac{9}{3} = 3$.

   b. $5(5)^{-1/2} = 5^{1/2} = \sqrt{5}$.

5. **a.** $\dfrac{(-3)^4 (-3)^5}{(-3)^8} = (-3)^{4+5-8} = (-3)^1 = -3$.

   b. $\dfrac{(2^{-4})(2^6)}{2^{-1}} = 2^{-4+6+1} = 2^3 = 8$.

7. **a.** $\dfrac{5^{3.3} \cdot 5^{-1.6}}{5^{-0.3}} = \dfrac{5^{3.3-1.6}}{5^{-0.3}} = 5^{1.7+(0.3)} = 5^2 = 25$.

   b. $\dfrac{4^{2.7} \cdot 4^{-1.3}}{4^{-0.4}} = 4^{2.7-1.3+0.4} = 4^{1.8} \approx 12.126$.

9. **a.** $\left(64x^9\right)^{1/3} = 64^{1/3}\left(x^{9/3}\right) = 4x^3$.

   b. $\left(25x^3y^4\right)^{1/2} = \left(25^{1/2}\right)\left(x^{3/2}\right)\left(y^{4/2}\right) = 5x^{3/2}y^2$
   $= 5xy^2\sqrt{x}$.

11. **a.** $\dfrac{6a^{-5}}{3a^{-3}} = 2a^{-5+3} = 2a^{-2} = \dfrac{2}{a^2}$.

    b. $\dfrac{4b^{-4}}{12b^{-6}} = \tfrac{1}{3}b^{-4+6} = \tfrac{1}{3}b^2$.

13. **a.** $\left(2x^3y^2\right)^3 = 2^3 \times x^{3(3)} \times y^{2(3)} = 8x^9y^6$.

    b. $\left(4x^2y^2z^3\right)^2 = 4^2 \cdot x^{2(2)} \cdot y^{2(2)} \cdot z^{3(2)} = 16x^4y^4z^6$.

15. **a.** $\dfrac{5^0}{\left(2^{-3}x^{-3}y^2\right)^2} = \dfrac{1}{2^{-3(2)}x^{-3(2)}y^{2(2)}} = \dfrac{2^6x^6}{y^4} = \dfrac{64x^6}{y^4}$.

    b. $\dfrac{(x+y)(x-y)}{(x-y)^0} = (x+y)(x-y)$.

17. $6^{2x} = 6^4$ if and only if $2x = 4$ or $x = 2$.

© 2011 Cengage Learning. All Rights Reserved. May not be scanned, copied or duplicated, or posted to a publicly accessible website, in whole or in part.

19. $3^{3x-4} = 3^5$ if and only if $3x - 4 = 5$, $3x = 9$, or $x = 3$.

21. $(2.1)^{x+2} = (2.1)^5$ if and only if $x + 2 = 5$, or $x = 3$.

23. $8^x = \left(\frac{1}{32}\right)^{x-2}$, $\left(2^3\right)^x = (32)^{2-x} = \left(2^5\right)^{2-x}$, so $2^{3x} = 2^{5(2-x)}$, $3x = 10 - 5x$, $8x = 10$, or $x = \frac{5}{4}$.

25. Let $y = 3^x$. Then the given equation is equivalent to $y^2 - 12y + 27 = 0$, or $(y - 9)(y - 3) = 0$, giving $y = 3$ or 9. So $3^x = 3$ or $3^x = 9$, and therefore, $x = 1$ or $x = 2$.

27. $y = 2^x$, $y = 3^x$, and $y = 4^x$.

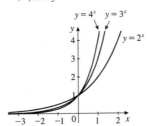

29. $y = 2^{-x}$, $y = 3^{-x}$, and $y = 4^{-x}$.

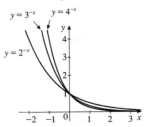

31. $y = 4^{0.5x}$, $y = 4^x$, and $y = 4^{2x}$.

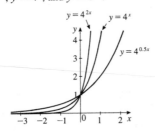

33. $y = e^{0.5x}$, $y = e^x$, $y = e^{1.5x}$.

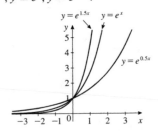

35. $y = 0.5e^{-x}$, $y = e^{-x}$, and $y = 2e^{-x}$.

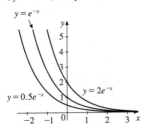

37. Because $f(0) = A = 100$ and $f(1) = 120$, we have $100e^k = 120$, and so $e^k = \frac{12}{10} = \frac{6}{5}$. Therefore, $f(x) = 100e^{kx} = 100\left(e^k\right)^x = 100\left(\frac{6}{5}\right)^x$.

© 2011 Cengage Learning. All Rights Reserved. May not be scanned, copied or duplicated, or posted to a publicly accessible website, in whole or in part.

39. $f(0) = 20$ implies that $\dfrac{1000}{1+B} = 20$, so $1000 = 20 + 20B$, or $B = \dfrac{980}{20} = 49$. Therefore,

$f(t) = \dfrac{1000}{1+49e^{-kt}}$. Next, $f(2) = 30$, so $\dfrac{1000}{1+49e^{-2t}} = 30$. We have $1 + 49e^{-2k} = \dfrac{1000}{30} = \dfrac{100}{3}$,

$49e^{-2k} = \dfrac{100}{3} - 1 = \dfrac{97}{3}$, $e^{-2k} = \dfrac{97}{147}$, and finally $e^{-k} = \left(\dfrac{97}{147}\right)^{1/2}$. Therefore, $f(t) = \dfrac{1000}{1+49\left(\frac{97}{147}\right)^{t/2}}$, so

$f(5) = \dfrac{1000}{1+49\left(\frac{97}{147}\right)^{5/2}} \approx 54.6$.

41. a. $R(t) = 26.3e^{-0.016t}$. In 1982 the rate was $R(0) = 26.3\%$, in 1986 it was $R(4) = 24.7\%$, in 1994 it was $R(12) = 21.7\%$, and in 2000 it was $R(18) = 19.7\%$.

b.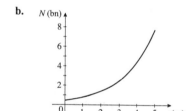

43. a.

Year	0	1	2	3	4	5
Number (billions)	0.45	0.80	1.41	2.49	4.39	7.76

b.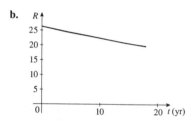

45. $N(t) = \dfrac{35.5}{1+6.89e^{-0.8674t}}$, so $N(6) = \dfrac{35.5}{1+6.89e^{-0.8674(6)}} \approx 34.2056$, or 34.21 million.

47. a. The initial concentration is given by
$N(0) = 0.08 + 0.12\left(1 - e^{-0.02\cdot 0}\right) = 0.08$, or 0.08 g/cm³.

b. The concentration after 20 seconds is given by
$N(20) = 0.08 + 0.12\left(1 - e^{-0.02\cdot 20}\right) = 0.11956$, or 0.1196 g/cm³.

c. The concentration in the long run is given by
$\lim\limits_{t\to\infty} x(t) = \lim\limits_{t\to\infty}\left[0.08 + 0.12\left(1 - e^{-0.02t}\right)\right] = 0.2$, or 0.2 g/cm³.

d.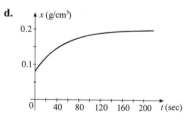

49. False. $\left(x^2 + 1\right)^3 = x^6 + 3x^4 + 3x^2 + 1$.

51. True. $f(x) = e^x$ is an increasing function and so if $x < y$, then $f(x) < f(y)$, or $e^x < e^y$.

© 2011 Cengage Learning. All Rights Reserved. May not be scanned, copied or duplicated, or posted to a publicly accessible website, in whole or in part.

5.1   Using Technology   page 337

1.

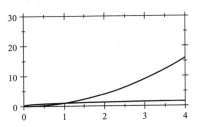

3.

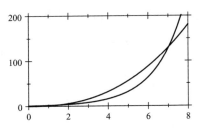

5.

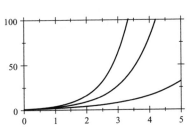

7.

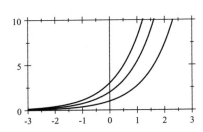

9.

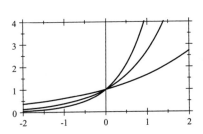

11. a.

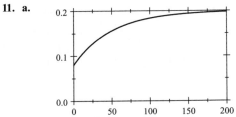

b. 0.08 g/cm³.   **c.** 0.12 g/cm³.

d. 0.2 g/cm³

13. a.

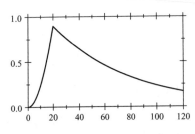

b. 20 seconds.   **c.** 35.1 seconds.

© 2011 Cengage Learning. All Rights Reserved. May not be scanned, copied or duplicated, or posted to a publicly accessible website, in whole or in part.

5.2 Problem-Solving Tips

1. Property 1 of logarithms says that $\log_b mn = \log_b m + \log_b n$. However, $\log_b (m + n) \neq \log_b m + \log_b n$ and
$$\log_b \frac{m}{n} \neq \frac{\log_b m}{\log_b n}.$$

2. When you work with logarithms be sure to distinguish between the following two operations:
$$\frac{\log 6}{\log 2} = \log 6 \div \log 2 \approx 2.585 \text{ and } \log \frac{6}{2} = \log 6 - \log 2 \approx 0.477. \text{ Property 2 of logarithms says that}$$
$$\log_b \frac{m}{n} = \log_b m - \log_b n.$$

3. The domain of the logarithmic function is $(0, \infty)$, so the logarithms of 0 and negative numbers are not defined.

5.2 Concept Questions   page 343

1. **a.** $y = \log_b x$ if and only if $x = b^y$.
   b. $f(x) = \log_b x, b > 0, b \neq 1$. Its domain is $(0, \infty)$.

3. **a.** $e^{\ln x} = x$.
   b. $\ln e^x = x$.

5.2 Logarithmic Functions   page 343

1. $\log_2 64 = 6$. 3. $\log_3 \frac{1}{9} = -2$. 5. $\log_{1/3} \frac{1}{3} = 1$. 7. $\log_{32} 8 = \frac{3}{5}$.

9. $\log_{10} 0.001 = -3$.

11. $\log 12 = \log 4 \times 3 = \log 4 + \log 3 = 0.6021 + 0.4771 = 1.0792$.

13. $\log 16 = \log 4^2 = 2 \log 4 = 2 (0.6021) = 1.2042$.

15. $\log 48 = \log (3 \cdot 4^2) = \log 3 + 2 \log 4 = 0.4771 + 2 (0.6021) = 1.6813$.

17. $2 \ln a + 3 \ln b = \ln a^2 b^3$.

19. $\ln 3 + \frac{1}{2} \ln x + \ln y - \frac{1}{3} \ln z = \ln \frac{3 \sqrt{xy}}{\sqrt[3]{z}}$.

21. $\log x (x + 1)^4 = \log x + \log (x + 1)^4 = \log x + 4 \log (x + 1)$.

23. $\log \frac{\sqrt{x + 1}}{x^2 + 1} = \log (x + 1)^{1/2} - \log (x^2 + 1) = \frac{1}{2} \log (x + 1) - \log (x^2 + 1)$.

25. $\ln x e^{-x^2} = \ln x - x^2$.

© 2011 Cengage Learning. All Rights Reserved. May not be scanned, copied or duplicated, or posted to a publicly accessible website, in whole or in part.

27. $\ln\left(\dfrac{x^{1/2}}{x^2\sqrt{1+x^2}}\right) = \ln x^{1/2} - \ln x^2 - \ln\left(1+x^2\right)^{1/2} = \frac{1}{2}\ln x - 2\ln x - \frac{1}{2}\ln\left(1+x^2\right) = -\frac{3}{2}\ln x - \frac{1}{2}\ln\left(1+x^2\right).$

29. $y = \log_3 x.$ **31.** $y = \ln 2x.$ **33.** $y = 2^x$ and $y = \log_2 x.$

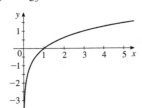   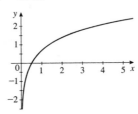   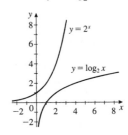

35. $e^{0.4t} = 8$, so $0.4t \ln e = \ln 8$ and thus $0.4t = \ln 8$ because $\ln e = 1$. Therefore, $t = \dfrac{\ln 8}{0.4} \approx 5.1986.$

37. $5e^{-2t} = 6$, so $e^{-2t} = \frac{6}{5} = 1.2$. Taking the logarithm, we have $-2t \ln e = \ln 1.2$, so $t = -\dfrac{\ln 1.2}{2} \approx -0.0912.$

39. $2e^{-0.2t} - 4 = 6$, so $2e^{-0.2t} = 10$. Taking the logarithm on both sides of this last equation, we have $\ln e^{-0.2t} = \ln 5$, $-0.2t \ln e = \ln 5$, $-0.2t = \ln 5$, and and $t = -\frac{\ln 5}{0.2} \approx -8.0472.$

41. $\dfrac{50}{1+4e^{0.2t}} = 20$, so $1 + 4e^{0.2t} = \dfrac{50}{20} = 2.5$, $4e^{0.2t} = 1.5$, $e^{0.2t} = \dfrac{1.5}{4} = 0.375$, $\ln e^{0.2t} = \ln 0.375$, and

$0.2t = \ln 0.375$. Thus, $t = \dfrac{\ln 0.375}{0.2} \approx -4.9041.$

43. Taking logarithms of both sides, we obtain $\ln A = \ln Be^{-t/2}$, $\ln A = \ln B + \ln e^{-t/2}$, and $\ln A - \ln B = -\dfrac{t}{2}\ln e$, so

$\ln\dfrac{A}{B} = -\dfrac{t}{2}$ and $t = -2\ln\dfrac{A}{B} = 2\ln\dfrac{B}{A}.$

45. $f(1) = 2$, so $a + b(0) = 2$. Thus, $a = 2$. Therefore, $f(x) = 2 + b\ln x$. We calculate $f(2) = 4$, so $2 + b\ln 2 = 4$.

Solving for b, we obtain $b = \dfrac{2}{\ln 2} \approx 2.8854$, so $f(x) = 2 + 2.8854\ln x.$

47. $p(x) = 19.4\ln x + 18$. For a child weighing 92 lb, we find $p(92) = 19.4\ln 92 + 18 \approx 105.7$, or approximately 106 millimeters of mercury.

49. a. $30 = 10\log\dfrac{I}{I_0}$, so $3 = \log\dfrac{I}{I_0}$, and $\dfrac{I}{I_0} = 10^3 = 1000$. Thus, $I = 1000 I_0.$

b. When $D = 80$, $I = 10^8 I_0$ and when $D = 30$, $I = 10^3 I_0$. Therefore, an 80-decibel sound is $10^8/10^3 = 10^5 = 100{,}000$ times louder than a 30-decibel sound.

c. It is $10^{15}/10^8 = 10^7 = 10{,}000{,}000$ times louder.

51. We solve the equation $\dfrac{160}{1+240e^{-0.2t}} = 80$ for t, obtaining $1 + 240e^{-0.2t} = \dfrac{160}{80}$, $240e^{-0.2t} = 2 - 1 = 1$,

$e^{-0.2t} = \dfrac{1}{240}$, $-0.2t = \ln\dfrac{1}{240}$, and $t = -\dfrac{1}{0.2}\ln\dfrac{1}{240} \approx 27.40$, or approximately 27.4 years old.

© 2011 Cengage Learning. All Rights Reserved. May not be scanned, copied or duplicated, or posted to a publicly accessible website, in whole or in part.

53. We solve the equation $200\left(1 - 0.956e^{-0.18t}\right) = 140$ for t, obtaining $1 - 0.956e^{-0.18t} = \frac{140}{200} = 0.7$,

$-0.956e^{-0.18t} = 0.7 - 1 = -0.3$, $e^{-0.18t} = \frac{0.3}{0.956}$, $-0.18t = \ln\left(\frac{0.3}{0.956}\right)$, and finally $t = -\dfrac{\ln\left(\frac{0.3}{0.956}\right)}{0.18} \approx 6.43875$.

Thus, it is approximately 6.4 years old.

55. a. We solve the equation $0.08 + 0.12e^{-0.02t} = 0.18$, obtaining $0.12e^{-0.02t} = 0.1$, $e^{-0.02t} = \frac{0.1}{0.12} = \frac{1}{1.2}$,

$\ln e^{-0.02t} = \ln\frac{1}{1.2} = \ln 1 - \ln 1.2 = -\ln 1.2$, $-0.02t = -\ln 1.2$, and $t = \frac{\ln 1.2}{0.02} \approx 9.116$, or approximately

9.1 seconds.

b. We solve the equation $0.08 + 0.12e^{-0.02t} = 0.16$, obtaining $0.12e^{-0.02t} = 0.08$, $e^{-0.02t} = \frac{0.08}{0.12} = \frac{2}{3}$,

$-0.02t = \ln\frac{2}{3}$, and $t = -\frac{1}{0.02}\ln\frac{2}{3} \approx 20.2733$, or approximately 20.3 seconds.

57. False. Take $x = e$. Then $(\ln e)^3 = 1^3 = 1 \neq 3\ln e = 3$.

59. True. $g(x) = \ln x$ is continuous and greater than zero on $(1, \infty)$. Therefore, $f(x) = \dfrac{1}{\ln x}$ is continuous on $(1, \infty)$.

61. a. Taking logarithms of both sides gives $\ln 2^x = \ln e^{kx}$, so $x\ln 2 = kx(\ln e) = kx$. Thus, $x(\ln 2 - k) = 0$ for all x, and this implies that $k = \ln 2$.

b. Proceeding as in part (a), we find that $k = \ln b$.

63. Let $\log_b m = p$. Then $m = b^p$. Therefore, $m^n = (b^p)^n = b^{np}$, and so

$\log_b m^n = \log_b b^{np} = np\log_b b = np$ (since $\log_b b = 1$) $= n\log_b m$.

5.3   Problem-Solving Tips

In applied problems involving interest rates, it is important to choose the correct interest formula to solve the problem. If the problems asks for the *future value* of an investment, then use the compound interest formula. If the problem asks for the *amount of money that needs to be invested now* to accumulate a certain sum in the future, then use the present value formula for compound interest. If the interest in the applied problem is compounded continuously then use the corresponding formulas for continuous compound interest.

5.3   Concept Questions   page 356

1. a. When simple interest is computed, the interest earned is based on the original principal. When compound interest is computed, the interest earned is periodically added to the principal and thereafter earns interest at the same rate.

b. The simple interest formula is $A = P(1 + rt)$ and the compound interest formula is $A = P\left(1 + \dfrac{r}{m}\right)^{mt}$.

3. $P = A\left(1 + \dfrac{r}{m}\right)^{-mt}$.

© 2011 Cengage Learning. All Rights Reserved. May not be scanned, copied or duplicated, or posted to a publicly accessible website, in whole or in part.

5.3   Compound Interest   page 356

1. $A = 2500 \left(1 + \dfrac{0.07}{2}\right)^{20} = 4974.47$, or \$4974.47.

3. $A = 150{,}000 \left(1 + \dfrac{0.1}{12}\right)^{48} = 223{,}403.11$, or \$223,403.11

5. a. Using the formula $r_{\text{eff}} = \left(1 + \dfrac{r}{m}\right)^m - 1$ with $r = 0.10$ and $m = 2$, we have $r_{\text{eff}} = \left(1 + \dfrac{0.10}{2}\right)^2 - 1 = 0.1025$,

 or 10.25%/yr.

 b. Using the formula $r_{\text{eff}} = \left(1 + \dfrac{r}{m}\right)^m - 1$ with $r = 0.09$ and $m = 4$, we have $r_{\text{eff}} = \left(1 + \dfrac{0.09}{4}\right)^4 - 1 = 0.09308$,

 or 9.308%/yr.

7. a. The present value is given by $P = 40{,}000 \left(1 + \dfrac{0.08}{2}\right)^{-8} = 29{,}227.61$, or \$29,227.61.

 b. The present value is given by $P = 40{,}000 \left(1 + \dfrac{0.08}{4}\right)^{-16} = 29{,}137.83$, or \$29,137.83.

9. $A = 5000 e^{0.08(4)} \approx 6885.64$, or \$6,885.64.

11. We use Formula (8) with $A = 10{,}000$, $m = 365$, $r = 0.07$, and $t = 2$. The required deposit is

 $P = 10{,}000 \left(1 + \dfrac{0.07}{365}\right)^{-365(2)} \approx 8693.699$, or \$8693.70.

13. We use Formula (11) with $A = 20{,}000$, $r = 0.06$, and $t = 3$. Jack should deposit

 $P = 20{,}000 e^{-(0.06)(3)} \approx 16{,}705.404$, or \$16,705.40.

15. We use Formula (6) with $A = 7500$, $P = 5000$, $m = 12$, and $t = 3$. Thus, $7500 = 5000 \left(1 + \dfrac{r}{12}\right)^{36}$,

 $\left(1 + \dfrac{r}{12}\right)^{36} = \dfrac{7500}{5000} = \dfrac{3}{2}$, $\ln\left(1 + \dfrac{r}{12}\right)^{36} = \ln 1.5$, $36\left(1 + \dfrac{r}{12}\right) = \ln 1.5$, $\left(1 + \dfrac{r}{12}\right) = \dfrac{\ln 1.5}{36} = 0.0112629$,

 $1 + \dfrac{r}{12} = e^{0.0112629} = 1.011327$, $\dfrac{r}{12} = 0.011327$, and $r = 0.13592$. The annual interest rate is 13.59%.

17. We use Formula (6) with $A = 8000$, $P = 5000$, $m = 2$, and $t = 4$. Thus, $8000 = 5000 \left(1 + \dfrac{r}{2}\right)^8$,

 $\left(1 + \dfrac{r}{2}\right)^8 = \dfrac{8000}{5000} = 1.6$, $\ln\left(1 + \dfrac{r}{2}\right)^8 = \ln 1.6$, $8\ln\left(1 + \dfrac{r}{2}\right) = \ln 1.6$, $\ln\left(1 + \dfrac{r}{2}\right) = \dfrac{\ln 1.6}{8} = 0.05875$,

 $1 + \dfrac{r}{2} = e^{0.05875} = 1.06051$; $\dfrac{r}{2} = 0.06051$, and so $r = 0.1210$. The required annual interest rate is 12.1%.

19. We use Formula (6) with $A = 4000$, $P = 2000$, $m = 1$, and $t = 5$. Thus, $4000 = 2000 (1 + r)^5$, $(1 + r)^5 = 2$,

 $5\ln(1 + r) = \ln 2$, $\ln(1 + r) = \dfrac{\ln 2}{5} = 0.138629$, $1 + r = e^{0.138629} = 1.148698$, and so $r = 0.1487$. The required

 annual interest rate is 14.87%.

21. We use Formula (6) with $A = 6500$, $P = 5000$, $m = 12$, and $r = 0.12$. Thus, $6500 = 5000 \left(1 + \dfrac{0.12}{12}\right)^{12t}$,

 $(1.01)^{12t} = \dfrac{6500}{5000} = 1.3$, $12t \ln(1.01) = \ln 1.3$, and so $t = \dfrac{\ln 1.3}{12 \ln 1.01} \approx 2.197$. It will take approximately 2.2 years.

© 2011 Cengage Learning. All Rights Reserved. May not be scanned, copied or duplicated, or posted to a publicly accessible website, in whole or in part.

23. We use Formula (6) with $A = 4000$, $P = 2000$, $m = 12$, and $r = 0.09$. Thus, $4000 = 2000 \left(1 + \frac{0.09}{12}\right)^{12t}$,

$\left(1 + \frac{0.09}{12}\right)^{12t} = 2$, $12t \ln\left(1 + \frac{0.09}{12}\right) = \ln 2$, and so $t = \dfrac{\ln 2}{12 \ln \left(1 + \frac{0.09}{12}\right)} \approx 7.73$. It will take approximately

7.7 years.

25. We use Formula (10) with $A = 6000$, $P = 5000$, and $t = 3$. Thus, $6000 = 5000e^{3r}$, $e^{3r} = \frac{6000}{5000} = 1.2$, $3r = \ln 1.2$,

and $r = \frac{1}{3} \ln 1.2 \approx 0.6077$. The annual interest rate is 6.08%.

27. We use Formula (10) with $A = 7000$, $P = 6000$, and $r = 0.075$. Thus, $7000 = 6000e^{0.075t}$, $e^{0.075t} = \frac{7000}{6000} = \frac{7}{6}$,

$0.075t \ln e = \ln \frac{7}{6}$, and so $t = \dfrac{\ln \frac{7}{6}}{0.075} \approx 2.055$. It will take 2.06 years.

29. The Estradas can expect to pay $180{,}000 \left(1 + 0.09\right)^4$, or approximately \$254,084.69.

31. The investment will be worth $A = 1.5 \left(1 + \dfrac{0.065}{2}\right)^{20} = 2.84376$, or approximately \$2.844 million.

33. The present value of the \$8000 loan due in 3 years is given by $P = 8000 \left(1 + \dfrac{0.10}{2}\right)^{-6} \approx 5969.72$, or \$5969.72.

The present value of the \$15,000 loan due in 6 years is given by $P = 15{,}000 \left(1 + \dfrac{0.10}{2}\right)^{-12} \approx 8352.56$, or

\$8352.56.

Therefore, the amount the proprietors of the inn will be required to pay at the end of 5 years is given by

$A = 14{,}322.28 \left(1 + \dfrac{0.10}{2}\right)^{10} \approx 23{,}329.48$, or \$23,329.48.

35. He can expect the minimum revenue for 2010 to be $240{,}000 \left(1.2\right)\left(1.3\right)\left(1.25\right)^3 \approx 731{,}250$, or \$731,250.

37. We want the value of a 2004 dollar in the year 2000. Denoting this value by x, we have

$(1.034)(1.028)(1.016)(1.023) x = 1$, so $x \approx 0.9051$. Thus, the purchasing power is approximately 91 cents.

39. The effective annual rate of return on his investment is found by solving the equation $(1 + r)^2 = \frac{32,100}{25,250}$. We find

$1 + r = \left(\frac{32,100}{25,250}\right)^{1/2}$, so $1 + r \approx 1.1275$, and $r \approx 0.1275$, or 12.75%.

41. $P = Ae^{-rt} = 59{,}673e^{-(0.08)5} \approx 40{,}000.008$, or approximately \$40,000.

43. **a.** If they invest the money at 10.5% compounded quarterly, they should set aside

$P = 70{,}000 \left(1 + \frac{0.105}{4}\right)^{-28} \approx 33{,}885.14$, or \$33,885.14.

b. If they invest the money at 10.5% compounded continuously, they should set aside

$P = 70{,}000e^{-(0.105)(7)} = 33{,}565.38$, or \$33,565.38.

45. **a.** If inflation over the next 15 years is 6%, then the first year of Eleni's pension will be worth

$P_6 = 40{,}000e^{-0.9} = 16{,}262.79$, or \$16,262.79.

b. If inflation over the next 15 years is 8%, then the first year of Eleni's pension will be worth

$P_8 = 40{,}000e^{-1.2} = 12{,}047.77$, or \$12,047.77.

© 2011 Cengage Learning. All Rights Reserved. May not be scanned, copied or duplicated, or posted to a publicly accessible website, in whole or in part.

c. If inflation over the next 15 years is 12%, then the first year of Eleni's pension will be worth
$$P_{12} = 40,000e^{-1.8} = 6611.96, \text{ or } \$6611.96.$$

47. $r_{\text{eff}} = \lim_{m \to \infty} \left(1 + \dfrac{r}{m}\right)^m - 1 = e^r - 1.$

49. The effective rate of interest at Bank A is given by $R = \left(1 + \dfrac{0.07}{4}\right)^4 - 1 = 0.07186$, or 7.186%. The effective rate at Bank B is given by $R = e^r - 1 = e^{0.07125} - 1 = 0.07385$, or 7.385%. We conclude that Bank B has the higher effective rate of interest.

51. The nominal rate of interest that yields an effective annual rate of interest of 10% when compounded continuously is found by solving the equation $R = e^r - 1$, so $0.10 = e^r - 1$, $1.10 = e^r$, $\ln 1.10 = r \ln e$, and $r = \ln 1.10 \approx 0.09531$, or 9.531%.

5.3   Using Technology   page 360

1. $5872.78   3. 8.95%/yr   5. $29,743.30

5.4   Problem-Solving Tips

1. The derivative of e^x is equal to e^x. By the chain rule, $\dfrac{d}{dx}\left(e^{3x}\right) = 3e^{3x}$ and $\dfrac{d}{dx}\left(e^{2x^2-1}\right) = 4xe^{2x^2-1}$. Note that the exponents in the original function and the derivative are the same.

2. Don't confuse functions of the type e^x with functions of the type x^r. The latter is a *power function* and its exponent is a *constant*; whereas the exponent in an *exponential function* such as e^x is a *variable*. A different rule is used to differentiate the two types of function. Thus, $\dfrac{d}{dx}\left(x^2e^x\right) = x^2\dfrac{d}{dx}\left(e^x\right) + e^x\dfrac{d}{dx}\left(x^2\right) = x^2e^x + e^x \cdot 2x = xe^x(x+2)$.

5.4   Concept Questions   page 366

1. a. $f'(x) = e^x$
   b. $g'(x) = e^{f(x)} \cdot f'(x)$

5.4   Differentiation of Exponential Functions   page 366

1. $f(x) = e^{3x}$, so $f'(x) = 3e^{3x}$

3. $g(t) = e^{-t}$, so $g'(t) = -e^{-t}$

5. $f(x) = e^x + x$, so $f'(x) = e^x + 1.$

7. $f(x) = x^3e^x$, so $f'(x) = x^3e^x + e^x\left(3x^2\right) = x^2e^x(x+3).$

© 2011 Cengage Learning. All Rights Reserved. May not be scanned, copied or duplicated, or posted to a publicly accessible website, in whole or in part.

9. $f(x) = \dfrac{2e^x}{x}$, so $f'(x) = \dfrac{x(2e^x) - 2e^x(1)}{x^2} = \dfrac{2e^x(x-1)}{x^2}$.

11. $f(x) = 3(e^x + e^{-x})$, so $f'(x) = 3(e^x - e^{-x})$.

13. $f(w) = \dfrac{e^w + 1}{e^w} = 1 + \dfrac{1}{e^w} = 1 + e^{-w}$, so $f'(w) = -e^{-w} = -\dfrac{1}{e^w}$.

15. $f(x) = 2e^{3x-1}$, so $f'(x) = 2e^{3x-1}(3) = 6e^{3x-1}$.

17. $h(x) = e^{-x^2}$, so $h'(x) = e^{-x^2}(-2x) = -2xe^{-x^2}$.

19. $f(x) = 3e^{-1/x}$, so $f'(x) = 3e^{-1/x} \cdot \dfrac{d}{dx}\left(-\dfrac{1}{x}\right) = 3e^{-1/x}\left(\dfrac{1}{x^2}\right) = \dfrac{3e^{-1/x}}{x^2}$.

21. $f(x) = (e^x + 1)^{25}$, so $f'(x) = 25(e^x + 1)^{24} e^x = 25e^x(e^x + 1)^{24}$.

23. $f(x) = e^{\sqrt{x}}$, so $f'(x) = e^{\sqrt{x}} \dfrac{d}{dx}(x^{1/2}) = e^{\sqrt{x}} \tfrac{1}{2} x^{-1/2} = \dfrac{e^{\sqrt{x}}}{2\sqrt{x}}$.

25. $f(x) = (x-1)e^{3x+2}$, so $f'(x) = (x-1)(3)e^{3x+2} + e^{3x+2} = e^{3x+2}(3x-3+1) = e^{3x+2}(3x-2)$.

27. $f(x) = \dfrac{e^x - 1}{e^x + 1}$, so $f'(x) = \dfrac{(e^x+1)(e^x) - (e^x-1)(e^x)}{(e^x+1)^2} = \dfrac{e^x(e^x + 1 - e^x + 1)}{(e^x+1)^2} = \dfrac{2e^x}{(e^x+1)^2}$.

29. $f(x) = e^{-4x} + 2e^{3x}$, so $f'(x) = -4e^{-4x} + 6e^{3x}$ and $f''(x) = 16e^{-4x} + 18e^{3x} = 2(8e^{-4x} + 9e^{3x})$.

31. $f(x) = 2xe^{3x}$, so $f'(x) = 2e^{3x} + 2xe^{3x}(3) = 2(3x+1)e^{3x}$ and $f''(x) = 6e^{3x} + 2(3x+1)e^{3x}(3) = 6(3x+2)e^{3x}$.

33. $y = f(x) = e^{2x-3}$, so $f'(x) = 2e^{2x-3}$. To find the slope of the tangent line to the graph of f at $x = \tfrac{3}{2}$, we compute $f'\left(\tfrac{3}{2}\right) = 2e^{3-3} = 2$. Next, using the point-slope form of the equation of a line, we find that $y - 1 = 2\left(x - \tfrac{3}{2}\right) = 2x - 3$, or $y = 2x - 2$.

35. $f(x) = e^{-x^2/2}$, so $f'(x) = e^{-x^2/2}(-x) = -xe^{-x^2/2}$.
Setting $f'(x) = 0$, gives $x = 0$ as the only critical point of f. From the sign diagram, we conclude that f is increasing on $(-\infty, 0)$ and decreasing on $(0, \infty)$.

$$+ \; + \; + \; + \; 0 \; - \; - \; - \; - \qquad \text{sign of } f'$$
$$\xrightarrow{\hspace{3cm}} x$$
$$0$$

37. $f(x) = \tfrac{1}{2}e^x - \tfrac{1}{2}e^{-x}$, so $f'(x) = \tfrac{1}{2}(e^x + e^{-x})$ and $f''(x) = \tfrac{1}{2}(e^x - e^{-x})$. Setting $f''(x) = 0$ gives $e^x = e^{-x}$ or $e^{2x} = 1$, and so $x = 0$. From the sign diagram for f'', we conclude that f is concave upward on $(0, \infty)$ and concave downward on $(-\infty, 0)$.

$$+ \; + \; + \; + \; 0 \; - \; - \; - \; - \qquad \text{sign of } f''$$
$$\xrightarrow{\hspace{3cm}} x$$
$$0$$

© 2011 Cengage Learning. All Rights Reserved. May not be scanned, copied or duplicated, or posted to a publicly accessible website, in whole or in part.

39. $f(x) = xe^{-2x}$, so $f'(x) = e^{-2x} + xe^{-2x}(-2) = (1-2x)e^{-2x}$ and

$f''(x) = -2e^{-2x} + (1-2x)e^{-2x}(-2) = 4(x-1)e^{-2x}$.

Observe that $f''(x) = 0$ if $x = 1$. The sign diagram of f''

shows that $(1, e^{-2})$ is an inflection point.

```
- - - - - 0 + + + +     sign of f''
─────────────┼──────────→ x
             1
```

41. $f(x) = e^{-x^2}$, so $f'(x) = -2xe^{-x^2}$ and $f''(x) = -2e^{-x^2} - 2xe^{-x^2}(-2x) = -2e^{-x^2}(1-2x^2) = 0$

implies $x = \pm\frac{\sqrt{2}}{2}$. The sign diagram of f'' shows that the

graph of f has inflection points at $\left(-\frac{\sqrt{2}}{2}, e^{-1/2}\right)$ and

$\left(\frac{\sqrt{2}}{2}, e^{-1/2}\right)$. The slope of the tangent line at $\left(-\frac{\sqrt{2}}{2}, e^{-1/2}\right)$

```
+ + + 0 - - - - - 0 + + +    sign of f''
────────┼─────────┼─────────→ x
     -√2/2    0    √2/2
```

is $f'\left(-\frac{\sqrt{2}}{2}\right) = \sqrt{2}e^{-1/2}$, and the tangent line has equation $y - e^{-1/2} = \sqrt{2}e^{-1/2}\left(x + \frac{\sqrt{2}}{2}\right)$, which can be

simplified to $y = e^{-1/2}\left(\sqrt{2}x + 2\right)$. The slope of the tangent line at $\left(\frac{\sqrt{2}}{2}, e^{-1/2}\right)$ is $f'\left(\frac{\sqrt{2}}{2}\right) = -\sqrt{2}e^{-1/2}$, and

this tangent line has equation $y - e^{-1/2} = -\sqrt{2}e^{-1/2}\left(x - \frac{\sqrt{2}}{2}\right)$ or $y = e^{-1/2}\left(-\sqrt{2}x + 2\right)$.

43. $f(x) = e^{-x^2}$, so $f'(x) = -2xe^{-x^2} = 0$ if $x = 0$, the only critical

point of f. From the table, we see that f has an absolute minimum

value of e^{-1} attained at $x = -1$ and $x = 1$. It has an absolute

maximum at $(0, 1)$.

x	-1	0	1
$f(x)$	e^{-1}	1	e^{-1}

45. $g(x) = (2x-1)e^{-x}$, so

$g'(x) = 2e^{-x} + (2x-1)e^{-x}(-1) = (3-2x)e^{-x} = 0$ if $x = \frac{3}{2}$.

The graph of g shows that $\left(\frac{3}{2}, 2e^{-3/2}\right)$ is an absolute maximum,

and $(0, -1)$ is an absolute minimum.

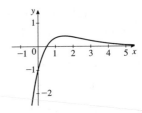

47. $f(t) = e^t - t$. We first gather the following information on f.

    1. The domain of f is $(-\infty, \infty)$.

    2. Setting $t = 0$ gives 1 as the y-intercept.

    3. $\lim_{t \to -\infty} (e^t - t) = \infty$ and $\lim_{t \to \infty} (e^t - t) = \infty$.

    4. There is no asymptote.

    5. $f'(t) = e^t - 1$ if $t = 0$, a critical point of f. From

        the sign diagram for f', we see that f is decreasing

        on $(-\infty, 0)$ and increasing on $(0, \infty)$.

```
- - - - 0 + + + +     sign of f'
───────────┼──────────→ t
           0
```

© 2011 Cengage Learning. All Rights Reserved. May not be scanned, copied or duplicated, or posted to a publicly accessible website, in whole or in part.

6. From the results of part 5, we see that $(0, 1)$ is a relative minimum of f.

7. $f''(t) = e^t > 0$ for all t, so the graph of f is concave upward on $(-\infty, \infty)$.

8. There is no inflection point.

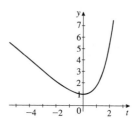

49. $f(x) = 2 - e^{-x}$. We first gather the following information on f.

1. The domain of f is $(-\infty, \infty)$.

2. Setting $x = 0$ gives 1 as the y-intercept.

3. $\lim\limits_{x \to -\infty} (2 - e^{-x}) = -\infty$ and $\lim\limits_{x \to \infty} (2 - e^{-x}) = 2$.

4. From the results of part 3, we see that $y = 2$ is a horizontal asymptote of f.

5. $f'(x) = e^{-x} > 0$ for all x in $(-\infty, \infty)$, so f is increasing on $(-\infty, \infty)$.

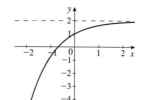

6. Because there is no critical point, f has no relative extremum.

7. $f''(x) = -e^{-x} < 0$ for all x in $(-\infty, \infty)$ and so the graph of f is concave downward on $(-\infty, \infty)$.

8. There is no inflection point.

51. $P'(t) = 20.6 (-0.009) e^{-0.009t} = -0.1854 e^{-0.009t}$, so $P'(10) = -0.1694$, $P'(20) = -0.1549$, and $P'(30) = -0.1415$. This tells us that the percentage of the total population relocating each year was decreasing at the rate of 0.17% in 1970, 0.15% in 1980, and 0.14% in 1990.

53. a. The population at the beginning of 2000 was $P(0) = 0.07$, or 70,000. The population at the beginning of 2030 will be $P(3) = 0.3537$, or approximately 353,700.

b. $P'(t) = 0.0378 e^{0.54t}$. The population was changing at the rate of $P'(0) = 0.0378$, or 37,800/decade, at the beginning of 2000. At the beginning of 2030, it was changing at the rate of $P'(3) \approx 0.191$, or increasing by approximately 191,000/decade.

55. a. The total loans outstanding in 1998 were $L(0) = 4.6$, or \$4.6 trillion. The total loans outstanding in 2004 were $L(6) = 3.6$, or \$3.6 trillion.

b. $L'(t) = -0.184 e^{-0.04t}$. The total loans outstanding were changing at the rate of $L'(0) = -0.184$, that is, they were declining at the rate of \$0.18 trillion/yr in 1998. In 2004, they were changing at $L'(16) \approx -0.145$, or declining at the rate of \$0.145 trillion/yr.

c. $L''(t) = 0.00736 e^{-0.04t}$. Because $L''(t) > 0$ on the interval $(0, 6)$, we see that L is decreasing, but at a slower rate, and this proves the assertion.

57. a. $S(t) = 20{,}000 \left(1 + e^{-0.5t}\right)$, so $S'(t) = 20{,}000 \left(-0.5 e^{-0.5t}\right) = -10{,}000 e^{-0.5t}$. Thus, $S'(1) = -10{,}000 e^{-0.5} = -6065$, or $-\$6065$/day/day; $S'(2) = -10{,}000 e^{-1} = -3679$, or $-\$3679$/day/day; $S'(3) = -10{,}000 \left(e^{-1.5}\right) = -2231$, or $-\$2231$/day/day; and $S'(4) = -10{,}000 e^{-2} = -1353$, or $-\$1353$/day/day.

© 2011 Cengage Learning. All Rights Reserved. May not be scanned, copied or duplicated, or posted to a publicly accessible website, in whole or in part.

b. $S(t) = 20,000 \left(1 + e^{-0.5t}\right) = 27,400$, so $1 + e^{-0.5t} = \frac{27,400}{20,000}$, $e^{-0.5t} = \frac{274}{200} - 1$, $-0.5t = \ln\left(\frac{274}{200} - 1\right)$, and so

$$t = \frac{\ln\left(\frac{274}{200} - 1\right)}{-0.5} \approx 2, \text{ or 2 days.}$$

59. $N(t) = 5.3e^{0.095t^2 - 0.85t}$.

a. $N'(t) = 5.3e^{0.095t^2 - 0.85t}(0.19t - 0.85)$. Because $N'(t)$ is negative for $0 \le t \le 4$, we see that $N(t)$ is decreasing over that interval.

b. To find the rate at which the number of polio cases was decreasing at the beginning of 1959, we compute $N'(0) = 5.3e^{0.095(0^2) - 0.85(0)}(0.85) \approx 5.3(-0.85) = -4.505$, or 4505 cases per year per year (t is measured in thousands). To find the rate at which the number of polio cases was decreasing at the beginning of 1962, we compute $N'(3) = 5.3e^{0.095(9) - 0.85(3)}(0.57 - 0.85) \approx (-0.28)(0.9731) \approx -0.273$, or 273 cases per year per year.

61. a. The number of deaths in 1950 was $N(0) = 130.7 + 50 = 180.7$, or approximately 181 per 100,000 people.

b. $N'(t) = (130.7)(-0.1155)(2t)e^{-0.1155t^2}$

$= -30.1917te^{-0.1155t^2}$.

The rates of change of the number of deaths per 100,000 people per decade are given in the table.

Year	1950	1960	1970	1980
Rate	0	−27	−38	−32

c. $N''(t) = -30.1917\left[e^{-0.1155t^2} + t(-0.1155)(2t)e^{-0.1155t^2}\right] = -30.1917\left(1 - 0.231t^2\right)e^{-0.1155t^2}$. Setting $N''(t) = 0$ gives $t \approx \pm 2.08$. So $t \approx 2$ gives an inflection point, and we conclude that the decline was greatest around 1970.

d. The number is given by $N(6) \approx 52.04$, or approximately 52.

63. From the results of Exercise 60, we see that $R'(x) = 100(1 - 0.0001x)e^{-0.0001x}$. Setting $R'(x) = 0$ gives $x = 10,000$, a critical point of R. From the graph of R, we see that the revenue is maximized when $x = 10,000$. So 10,000 pairs must be sold, yielding a maximum revenue of $R(10,000) = 367,879.44$, or \$367,879.

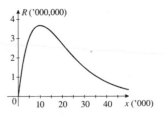

65. a. $p = 240\left(1 - \dfrac{3}{3 + e^{-0.0005x}}\right) = 240\left[1 - 3\left(3 + e^{-0.0005x}\right)^{-1}\right]$,

so $p' = 720\left(3 + e^{-0.0005x}\right)^{-2}\left(-0.0005e^{-0.0005x}\right)$. Thus,

$p'(1000) = 720\left(3 + e^{-0.0005 \cdot 1000}\right)^{-2}\left(-0.0005e^{-0.0005 \cdot 1000}\right) = -\dfrac{0.36(0.606531)}{(3 + 0.606531)^2} \approx -0.0168$, or -1.68 cents per case per case.

b. $p(1000) = 240\left(1 - \dfrac{3}{3.606531}\right) \approx 40.36$, or \$40.36/case.

© 2011 Cengage Learning. All Rights Reserved. May not be scanned, copied or duplicated, or posted to a publicly accessible website, in whole or in part.

67. a. $W = 2.4e^{1.84h}$, so if $h = 1.6$, $W = 2.4e^{1.84(1.6)} \approx 45.58$, or approximately 45.6 kg.

  b. $\Delta W \approx dW = (2.4)(1.84)e^{1.84h}\, dh$. With $h = 1.6$ and $dh = \Delta h = 1.65 - 1.6 = 0.05$, we find
  $\Delta W \approx (2.4)(1.84)e^{1.84(1.6)} \cdot (0.05) \approx 4.19$, or approximately 4.2 kg.

69. $P(t) = 80{,}000e^{\sqrt{t}/2 - 0.09t} = 80{,}000e^{(1/2)t^{1/2} - 0.09t}$, so $P'(t) = 80{,}000\left(\frac{1}{4}t^{-1/2} - 0.09\right)e^{(1/2)t^{1/2} - 0.09t}$. Setting

  $P'(t) = 0$, we have $\frac{1}{4}t^{-1/2} = 0.09$, so $t^{-1/2} = 0.36$, $\frac{1}{\sqrt{t}} = 0.36$, and $t = \left(\frac{1}{0.36}\right)^2 \approx 7.72$. Evaluating $P(t)$

  at each of its endpoints and at the point $t = 7.72$, we find $P(0) = 80{,}000$, $P(7.72) \approx 160{,}207.69$, and
  $P(8) \approx 160{,}170.71$. We conclude that P is optimized at $t = 7.72$. The optimal price is approximately \$160,208.

71. $A(t) = 0.23te^{-0.4t}$, so $A'(t) = 0.23(1 - 0.4t)e^{-0.4t}$. Setting

  $A'(t) = 0$ gives $t = \frac{1}{0.4} = \frac{5}{2}$. From the graph of A, we see that the

  proportion of alcohol is highest $2\frac{1}{2}$ hours after drinking. The level

  is given by $A\left(\frac{5}{2}\right) \approx 0.2115$, or approximately 0.21%.

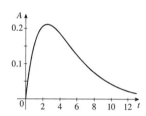

73. a. The temperature inside the house is given by $T(0) = 30 + 40e^0 = 70$, or 70°F.

  b. The reading is changing at the rate of $T'(1) = 40(-0.98)e^{-0.98t}\big|_{t=1} = -14.7$. Thus, it is dropping at the rate
  of approximately 14.7°F/min.

  c. The temperature outdoors is given by $\lim\limits_{t \to \infty} T(t) = \lim\limits_{t \to \infty}\left(30 + 40e^{-0.98t}\right) = 30 + 0 = 30$, or 30°F.

75. a. $y' = c\left(-be^{-bt} + ae^{-at}\right) = ca\left(-\frac{b}{a}e^{-bt} + e^{-at}\right) = cae^{-at}\left[-\frac{b}{a}e^{(a-b)t} + 1\right]$. Setting $y' = 0$ gives

  $-\frac{b}{a}e^{(a-b)t} + 1 = 0$, $e^{(a-b)t} = \frac{a}{b}$, $\ln e^{(a-b)t} = \ln\frac{a}{b}$, and so $t = \dfrac{\ln\frac{a}{b}}{a-b}$. Because $y(0) = 0$ and $\lim\limits_{t\to\infty} y = 0$,

  $t = \dfrac{\ln\frac{a}{b}}{a-b}$ gives the time at which the concentration is maximal.

  b. $y'' = c\left(b^2e^{-bt} - a^2e^{-at}\right) = ca^2e^{-at}\left[\frac{b^2}{a^2}e^{(a-b)t} - 1\right]$. Setting $y'' = 0$ gives $e^{(a-b)t} = \frac{a^2}{b^2}$, so $t = \dfrac{2\ln\frac{a}{b}}{a-b}$.

  From the sign diagram of y'', we see that the concentration of

  the drug is increasing most rapidly when $t = \dfrac{2\ln\frac{a}{b}}{a-b}$ seconds.

  $$+ + + + + 0\ - - - - -\quad\text{sign of } y''$$
  $$\xrightarrow{\hspace{4cm}} t$$
  $$0 \qquad\qquad \frac{2\ln\frac{a}{b}}{a-b}$$

77. We are given that $c\left(1 - e^{-at/V}\right) < m$, so $1 - e^{-at/V} < \dfrac{m}{c}$, $-e^{-at/V} < \dfrac{m}{c} - 1$, and $e^{-at/V} > 1 - \dfrac{m}{c}$. Taking

  logarithms of both sides of the inequality, we have $-\dfrac{at}{V}\ln e > \ln\dfrac{c-m}{c}$, $-\dfrac{at}{V} > \ln\dfrac{c-m}{c}$, $-t > \dfrac{V}{a}\ln\dfrac{c-m}{c}$, and

  so $t < \dfrac{V}{a}\left(-\ln\dfrac{c-m}{c}\right) = \dfrac{V}{a}\ln\dfrac{c}{c-m}$. Therefore, the liquid must not be allowed to enter the organ for longer

  than $t = \dfrac{V}{a}\ln\dfrac{c}{c-m}$ minutes.

© 2011 Cengage Learning. All Rights Reserved. May not be scanned, copied or duplicated, or posted to a publicly accessible website, in whole or in part.

79. $C'(t) = \begin{cases} 0.3 + 18e^{-t/60}\left(\frac{1}{60}\right) & \text{if } 0 < t < 20 \\ -\frac{18}{60}e^{-t/60} + \frac{12}{60}e^{-(t-20)/60} & \text{if } t > 20 \end{cases} = \begin{cases} 0.3\left(1 - e^{-t/60}\right) & \text{if } 0 < t < 20 \\ -0.3e^{-t/60} + 0.2e^{-(t-20)/60} & \text{if } t > 20 \end{cases}$

a. $C'(10) = 0.3\left(1 - e^{-10/60}\right) \approx 0.05$, or 0.05 $\text{g/cm}^3/\text{sec.}$

b. $C'(30) = -0.3e^{-30/60} + 0.2e^{-10/60} \approx -0.01$, or decreasing at the rate of 0.01 $\text{g/cm}^3/\text{sec.}$

c. On the interval $(0, 20)$, $C'(t) = 0$ implies $1 - e^{-t/60} = 0$, so $t = 0$. Therefore, C attains its absolute
maximum value at an endpoint — in this case, at $t = 20$. On the interval $[20, \infty)$, $C'(t) = 0$ implies
$-0.3e^{-t/60} = -0.2e^{-(t-20)/60}$, or $\dfrac{e^{-(t-20)/60}}{e^{-t/60}} = \dfrac{3}{2}$, or $e^{1/3} = \dfrac{3}{2}$, which is not possible. Therefore $C'(t) \neq 0$ on
$[20, \infty)$. Because $C(t) \to 0$ as $t \to \infty$, the absolute maximum of C occurs at $t = 20$. Thus, the concentration of
the drug reaches a maximum at $t = 20$ seconds.

d. The maximum concentration is $C(20) = 0.90$ g/cm^3.

81. False. $f(x) = e^{\pi}$ is a constant function and so $f'(x) = 0$.

83. True. Differentiating both sides of the equation with respect to x, we have $\dfrac{d}{dx}\left(x^2 + e^y\right) = \dfrac{d}{dx}(10)$, so
$2x + e^y\dfrac{dy}{dx} = 0$ and thus $\dfrac{dy}{dx} = -\dfrac{2x}{e^y}$.

5.4   Using Technology   page 370

1. 5.4366

3. 12.3929

5. 0.1861

7. a. The initial population of crocodiles is
$P(0) = \frac{300}{6} = 50$.

b. $\displaystyle\lim_{t\to\infty} P(t) = \lim_{t\to\infty} \dfrac{300e^{-0.024t}}{5e^{-0.024t} + 1} = \dfrac{0}{0 + 1} = 0.$

c.

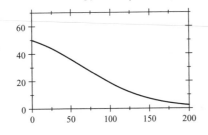

9. a.

b. We estimate $P'(1) \approx 4.272$ billion people per
half century.

11. a. Using the function evaluation capabilities of a graphing utility, we find $f(11) \approx 153.024$ and $g(11) \approx 235.181$.
This tells us that the number of violent-crime arrests will be 153,024 at the beginning of the year 2000, but if
trends like inner-city drug use and wider availability of guns continue, then the number of arrests will be 235,181.

b. Using the differentiation capability of a graphing utility, we find $f'(11) \approx -0.634$ and $g'(11) \approx 18.401$. This
tells us that the number of violent-crime arrests will be decreasing at the rate of 634 per year at the beginning of
the year 2000, but if the trends like inner-city drug use and wider availability of guns continue, then the number
of arrests will be increasing at the rate of 18,401 per year at the beginning of the year 2000.

© 2011 Cengage Learning. All Rights Reserved. May not be scanned, copied or duplicated, or posted to a publicly accessible website, in whole or in part.

13. a. $P(10) = \dfrac{74}{1 + 2.6e^{-0.166(10) + 0.04536(10)^2 - 0.0066(10)^3}} \approx 69.63\%.$

   b. $P'(10) \approx 5.09361$, or approximately $5.09\%/$decade.

5.5   Problem-Solving Tips

1. If you trying to find the derivative of a complicated function involving products, quotients, or powers, check to see if you can use logarithmic differentiation to simplify the process. Look at Exercises 41–50, and become familiar with the type of problems for which this method is especially suitable.

2. Example 7 provided us with a method for finding the derivative of the function $y = x^x$ ($x > 0$). Note that we use the Power rule $\left(\frac{d}{dx}[x^n] = nx^{n-1} \right)$ to differentiate functions of the form $y = x^n$, where the base is a variable and the exponent is a constant, the rule for differentiating exponential functions to differentiate functions of the form $y = e^x$, where the base is the constant e and the exponent is a variable, and logarithmic differentiation to differentiate functions of the form $y = x^x$ where both the base and the exponent of the function are variables. Be sure that you can distinguish between these functions and the rule to be applied in each of these cases.

5.5   Concept Questions   page 377

1. a. $f'(x) = \dfrac{1}{x}.$

b. $g'(x) = \dfrac{f'(x)}{f(x)}.$

5.5   Differentiation of Logarithmic Functions   page 377

1. $f(x) = 5\ln x$, so $f'(x) = 5\left(\dfrac{1}{x}\right) = \dfrac{5}{x}.$

3. $f(x) = \ln(x+1)$, so $f'(x) = \dfrac{1}{x+1}.$

5. $f(x) = \ln x^8$, so $f'(x) = \dfrac{8x^7}{x^8} = \dfrac{8}{x}.$

7. $f(x) = \ln x^{1/2}$, so $f'(x) = \dfrac{\frac{1}{2}x^{-1/2}}{x^{1/2}} = \dfrac{1}{2x}.$

9. $f(x) = \ln\left(\dfrac{1}{x^2}\right) = \ln x^{-2} = -2\ln x$, so $f'(x) = -\dfrac{2}{x}.$

11. $f(x) = \ln\left(4x^2 - 6x + 3\right)$, so $f'(x) = \dfrac{8x - 6}{4x^2 - 6x + 3} = \dfrac{2(4x-3)}{4x^2 - 6x + 3}.$

13. $f(x) = \ln\left(\dfrac{2x}{x+1}\right) = \ln 2x - \ln(x+1)$, so
$f'(x) = \dfrac{2}{2x} - \dfrac{1}{x+1} = \dfrac{2(x+1) - 2x}{2x(x+1)} = \dfrac{2x + 2 - 2x}{2x(x+1)} = \dfrac{2}{2x(x+1)} = \dfrac{1}{x(x+1)}.$

15. $f(x) = x^2 \ln x$, so $f'(x) = x^2\left(\dfrac{1}{x}\right) + (\ln x)(2x) = x + 2x\ln x = x(1 + 2\ln x).$

© 2011 Cengage Learning. All Rights Reserved. May not be scanned, copied or duplicated, or posted to a publicly accessible website, in whole or in part.

17. $f(x) = \dfrac{2\ln x}{x}$, so $f'(x) = \dfrac{x\left(\frac{2}{x}\right) - 2\ln x}{x^2} = \dfrac{2(1 - \ln x)}{x^2}$.

19. $f(u) = \ln(u - 2)^3$, so $f'(u) = \dfrac{3(u-2)^2}{(u-2)^3} = \dfrac{3}{u-2}$.

21. $f(x) = (\ln x)^{1/2}$, so $f'(x) = \frac{1}{2}(\ln x)^{-1/2}\left(\frac{1}{x}\right) = \dfrac{1}{2x\sqrt{\ln x}}$.

23. $f(x) = (\ln x)^3$, so $f'(x) = 3(\ln x)^2\left(\frac{1}{x}\right) = \dfrac{3(\ln x)^2}{x}$.

25. $f(x) = \ln(x^3 + 1)$, so $f'(x) = \dfrac{3x^2}{x^3 + 1}$.

27. $f(x) = e^x \ln x$, so $f'(x) = e^x \ln x + e^x\left(\frac{1}{x}\right) = \dfrac{e^x(x \ln x + 1)}{x}$.

29. $f(t) = e^{2t} \ln(t + 1)$, so $f'(t) = e^{2t}\left(\dfrac{1}{t+1}\right) + \ln(t+1) \cdot (2e^{2t}) = \dfrac{[2(t+1)\ln(t+1) + 1]e^{2t}}{t+1}$.

31. $f(x)\,\dfrac{\ln x}{x^2}$, so $f'(x) = \dfrac{x^2\left(\frac{1}{x}\right) - \ln x\,(2x)}{x^4} = \dfrac{1 - 2\ln x}{x^3}$.

33. $f'(x) = \dfrac{d}{dx}[\ln(\ln x)] = \dfrac{\frac{d}{dx}(\ln x)}{\ln x} = \dfrac{\frac{1}{x}}{\ln x} = \dfrac{1}{x \ln x}$.

35. $f(x) = \ln 2 + \ln x$, so $f'(x) = \dfrac{1}{x}$ and $f''(x) = -\dfrac{1}{x^2}$.

37. $f(x) = \ln(x^2 + 2)$, so $f'(x) = \dfrac{2x}{(x^2 + 2)}$ and $f''(x) = \dfrac{(x^2 + 2)(2) - 2x(2x)}{(x^2 + 2)^2} = \dfrac{2(2 - x^2)}{(x^2 + 2)^2}$.

39. $f'(x) = \dfrac{d}{dx}(x^2 \ln x) = \dfrac{d}{dx}(x^2)\ln x + \dfrac{d}{dx}(\ln x)x^2 = 2x \ln x + \dfrac{1}{x} \cdot x^2 = 2x \ln x + x = x(2\ln x + 1)$ and

$f''(x) = \dfrac{d}{dx}[x(2\ln x + 1)] = \dfrac{d}{dx}(x)(2\ln x + 1) + \dfrac{d}{dx}(2\ln x + 1)x = 2\ln x + 1 + \dfrac{2}{x} \cdot x = 2\ln x + 3$.

41. $y = (x + 1)^2(x + 2)^3$, so

$\ln y = \ln(x + 1)^2(x + 2)^3 = \ln(x + 1)^2 + \ln(x + 2)^3 = 2\ln(x + 1) + 3\ln(x + 2)$.

Thus, $\dfrac{y'}{y} = \dfrac{2}{x+1} + \dfrac{3}{x+2} = \dfrac{2(x+2) + 3(x+1)}{(x+1)(x+2)} = \dfrac{5x+7}{(x+1)(x+2)}$ and

$y' = \dfrac{(5x+7)(x+1)^2(x+2)^3}{(x+1)(x+2)} = (5x + 7)(x + 1)(x + 2)^2$.

© 2011 Cengage Learning. All Rights Reserved. May not be scanned, copied or duplicated, or posted to a publicly accessible website, in whole or in part.

43. $y = (x-1)^2 (x+1)^3 (x+3)^4$, so $\ln y = 2\ln(x-1) + 3\ln(x+1) + 4\ln(x+3)$. Thus,

$$\frac{y'}{y} = \frac{2}{x-1} + \frac{3}{x+1} + \frac{4}{x+3} = \frac{2(x+1)(x+3) + 3(x-1)(x+3) + 4(x-1)(x+1)}{(x-1)(x+1)(x+3)}$$

$$= \frac{2x^2 + 8x + 6 + 3x^2 + 6x - 9 + 4x^2 - 4}{(x-1)(x+1)(x+3)} = \frac{9x^2 + 14x - 7}{(x-1)(x+1)(x+3)}, \text{ and so}$$

$$y' = \frac{9x^2 + 14x - 7}{(x-1)(x+1)(x+3)} \cdot y = \frac{(9x^2 + 14x - 7)(x-1)^2(x+1)^3(x+3)^4}{(x-1)(x+1)(x+3)}$$

$$= (9x^2 + 14x - 7)(x-1)(x+1)^2(x+3)^3.$$

45. $y = \dfrac{(2x^2-1)^5}{\sqrt{x+1}}$, so $\ln y = \ln\dfrac{(2x^2-1)^5}{(x+1)^{1/2}} = 5\ln(2x^2-1) - \frac{1}{2}\ln(x+1)$. Thus,

$$\frac{y'}{y} = \frac{20x}{2x^2-1} - \frac{1}{2(x+1)} = \frac{40x(x+1) - (2x^2-1)}{2(2x^2-1)(x+1)} = \frac{38x^2 + 40x + 1}{2(2x^2-1)(x+1)}, \text{ and so}$$

$$y' = \frac{38x^2 + 40x + 1}{2(2x^2-1)(x+1)} \cdot \frac{(2x^2-1)^5}{\sqrt{x+1}} = \frac{(38x^2 + 40x + 1)(2x^2-1)^4}{2(x+1)^{3/2}}.$$

47. $y = 3^x$, so $\ln y = x\ln 3$, $\dfrac{1}{y} \cdot \dfrac{dy}{dx} = \ln 3$, and $\dfrac{dy}{dx} = y\ln 3 = 3^x \ln 3$.

49. $y = (x^2+1)^x$, so $\ln y = \ln(x^2+1)^x = x\ln(x^2+1)$,

$$\frac{y'}{y} = \ln(x^2+1) + x\left(\frac{2x}{x^2+1}\right) = \frac{(x^2+1)\ln(x^2+1) + 2x^2}{x^2+1}, \text{ and}$$

$$y' = \frac{[(x^2+1)\ln(x^2+1) + 2x^2](x^2+1)^x}{x^2+1} = (x^2+1)^{x-1}[(x^2+1)\ln(x^2+1) + 2x^2].$$

51. $\dfrac{d}{dx}(\ln y - x\ln x) = \dfrac{d}{dx}(-1)$, so $\dfrac{d}{dx}\ln y - \dfrac{d}{dx}(x\ln x) = 0$, $\dfrac{y'}{y} = \left[\ln x + x\left(\dfrac{1}{x}\right)\right] = \ln x + 1$, and

$y' = (\ln x + 1)\, y$.

53. $y = x\ln x$. The slope of the tangent line at any point is $y' = \ln x + x\left(\frac{1}{x}\right) = \ln x + 1$. In particular, the slope of the tangent line at $(1, 0)$ is $m = \ln 1 + 1 = 1$. Thus, an equation of the tangent line is $y - 0 = 1(x-1)$, or $y = x - 1$.

55. $f(x) = \ln x^2 = 2\ln x$ and so $f'(x) = 2/x$. Because $f'(x) < 0$ if $x < 0$ and $f'(x) > 0$ if $x > 0$, we see that f is decreasing on $(-\infty, 0)$ and increasing on $(0, \infty)$.

57. $f(x) = x^2 + \ln x^2$, so $f'(x) = 2x + \dfrac{2x}{x^2} = 2x + \dfrac{2}{x}$ and $f''(x) = 2 - \dfrac{2}{x^2}$. To find the intervals of concavity for f,

we first set $f''(x) = 0$, giving $2 - \dfrac{2}{x^2} = 0$, $2 = \dfrac{2}{x^2}$, $2x^2 = 2$, $x^2 = 1$, and so $x = \pm 1$.

From the sign diagram for f'', we see that f is concave upward on $(-\infty, -1) \cup (1, \infty)$ and concave downward on $(-1, 0) \cup (0, 1)$.

f'' not defined

$+ + + \ 0 \ - - \ | \ - - \ 0 \ + + + \quad$ sign of f''

$\xrightarrow{\hspace{6cm}} x$

$\qquad -1 \qquad 0 \qquad 1$

© 2011 Cengage Learning. All Rights Reserved. May not be scanned, copied or duplicated, or posted to a publicly accessible website, in whole or in part.

59. $f(x) = \ln(x^2 + 1)$, so $f'(x) = \dfrac{2x}{x^2 + 1}$ and $f''(x) = \dfrac{(x^2 + 1)(2) - (2x)(2x)}{(x^2 + 1)^2} = -\dfrac{2(x^2 - 1)}{(x^2 + 1)^2}$. Setting

$f''(x) = 0$ gives $x = \pm 1$ as candidates for inflection points of f. From the sign diagram of f'', we see that $(-1, \ln 2)$ and $(1, \ln 2)$ are inflection points of f.

$$- - - 0 + + 0 + + 0 - - - \quad \text{sign of } f''$$

$$\xrightarrow{\qquad \qquad \qquad \qquad} x$$
$$-1 \quad 0 \quad 1$$

61. $f(x) = x^2 + 2 \ln x$, so $f'(x) = 2x + \dfrac{2}{x}$ and $f''(x) = 2 - \dfrac{2}{x^2} = 0$ implies $2 - \dfrac{2}{x^2} = 0$, $x^2 = 1$, and so $x = \pm 1$. We

reject the negative root because the domain of f is $(0, \infty)$. The sign diagram of f'' shows that $(1, 1)$ is an inflection point of the graph of f. $f'(1) = 4$. So, an equation of the required tangent line is $y - 1 = 4(x - 1)$ or $y = 4x - 3$.

$$- - - - 0 + + + + + \quad \text{sign of } f''$$
$$\xleftarrow{\quad (\qquad\qquad\qquad} x$$
$$0 \quad 1$$

63. $f(x) = x - \ln x$, so $f'(x) = 1 - \dfrac{1}{x} = \dfrac{x - 1}{x} = 0$ if $x = 1$, a

critical point of f. From the table, we see that f has an absolute minimum at $(1, 1)$ and an absolute maximum at $(3, 3 - \ln 3)$.

x	$\frac{1}{2}$	1	3
$f(x)$	$\frac{1}{2} + \ln 2$	1	$3 - \ln 3$

65. $f(x) = 7.2956 \ln(0.0645012x^{0.95} + 1)$, so

$$f'(x) = 7.2956 \cdot \dfrac{\frac{d}{dx}(0.0645012x^{0.95} + 1)}{0.0645012x^{0.95} + 1} = \dfrac{7.2956(0.0645012)(0.95x^{-0.05})}{0.0645012x^{0.95} + 1} = \dfrac{0.4470462}{x^{0.05}(0.0645012x^{0.95} + 1)}.$$

Thus, $f'(100) = 0.05799$, or approximately $0.0580\%/\text{kg}$, and $f'(500) = 0.01329$, or approximately $0.0133\%/\text{kg}$.

67. a. The projected number at the beginning of 2005 is $N(1) = 34.68 + 23.88 \ln(6.35) \approx 78.82$ million.

b. $N'(t) = 23.88 \dfrac{1.05}{1.05t + 5.3} = \dfrac{25.074}{1.05t + 5.3}$. The projected rate of change is

$$N'(1) = \dfrac{25.074}{1.05 + 5.3} \approx 3.95 \text{ million/yr.}$$

69. a. $V(2) = 60{,}000\left(1 - \frac{2}{10}\right)^2 = 38{,}400$, or $\$38{,}400$.

b. $\dfrac{V'(2)}{V(2)} = \ln\left(1 - \frac{2}{10}\right) \approx -0.223$, or approximately -22.3%.

71. a. $W'(t) = \dfrac{d}{dt}(49.9 + 17.1 \ln t) = \dfrac{17.1}{t} > 0$ if $t > 0$, so $W'(t) > 0$ on $[1, 6]$ and W is increasing on $(1, 6)$.

b. $W''(t) = \dfrac{d}{dt}\left(\dfrac{17.1}{t}\right) = -\dfrac{17.1}{t^2} < 0$ on $(1, 6)$, so W is concave downward on $(1, 6)$.

© 2011 Cengage Learning. All Rights Reserved. May not be scanned, copied or duplicated, or posted to a publicly accessible website, in whole or in part.

73. a. If $0 < r < 100$, then $c = 1 - \dfrac{r}{100}$ satisfies $0 < c < 1$. It suffices to show that $A_1(n) = -\left(1 - \frac{r}{100}\right)^n$ is

increasing; that is, it suffices to show that $A_2(n) = -A_1(n) = \left(1 - \frac{r}{100}\right)^n$ is decreasing. Let $y = \left(1 - \frac{r}{100}\right)^n$.

Then $\ln y = \ln\left(1 - \frac{r}{100}\right)^n = \ln c^n = n \ln c$. Differentiating

both sides with respect to n, we find $\frac{y'}{y} = \ln c$, and so

$y' = (\ln c)\left(1 - \frac{r}{100}\right)^n$. This is negative because $\ln c < 0$ and

$\left(1 - \frac{r}{m}\right)^n > 0$ for $0 < r < 100$. Therefore, A is an increasing

function of n on $(0, \infty)$.

b.

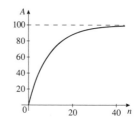

c. $\displaystyle\lim_{n\to\infty} A(n) = \lim_{n\to\infty} 100\left[1 - \left(1 - \frac{r}{100}\right)^n\right] = 100.$

75. a. $R(S_0) = k\ln\frac{S}{S_0} = k\ln 1 = 0$ because $\ln 1 = 0$.

b. $\dfrac{dR}{dS} = \dfrac{d}{dS}k\ln\dfrac{S}{S_0} = k\dfrac{d}{ds}(\ln S - \ln S_0) = k\dfrac{d}{dS}(\ln S) = \dfrac{k}{S}$, and so $\dfrac{dr}{dS}$ is inversely proportional to S with k as

the constant of proportionality. Our result says that if the stimulus is small, then a small change in S is easily felt.
But if the stimulus is larger, then a small change in S is not as discernible.

77. We differentiate $Vt = p - k\ln\left(1 - \dfrac{p}{x_0}\right)$ with respect to t, obtaining

$$V = \dfrac{dp}{dt} - k\dfrac{-\dfrac{1}{x_0}\dfrac{dp}{dt}}{1 - \dfrac{p}{x_0}} = \dfrac{dp}{dt}\left(1 + \dfrac{k}{x_0 - p}\right) = \dfrac{dp}{dt}\left(\dfrac{x_0 - p + k}{x_0 - p}\right). \text{ Thus, } \dfrac{dp}{dt} = \dfrac{V(x_0 - p)}{x_0 - p + k}.$$

79. $f(x) = 2x - \ln x$. We first gather the following information on f.

1. The domain of f is $(0, \infty)$.

2. There is no y-intercept.

3. $\displaystyle\lim_{x\to\infty}(2x - \ln x) = \infty.$

4. There is no asymptote.

5. $f'(x) = 2 - \dfrac{1}{x} = \dfrac{2x - 1}{x}$. Observe that $f'(x) = 0$ at $x = \frac{1}{2}$,

   a critical point of f. From the sign diagram of f', we conclude

   that f is decreasing on $\left(0, \frac{1}{2}\right)$ and increasing on $\left(\frac{1}{2}, \infty\right)$.

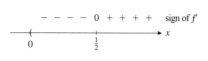

6. The results of part 5 show that $\left(\frac{1}{2}, 1 + \ln 2\right)$ is a relative

   minimum of f.

7. $f''(x) = \dfrac{1}{x^2}$ and is positive if $x > 0$, so the graph of f is

   concave upward on $(0, \infty)$.

8. The results of part 7 show that f has no inflection point.

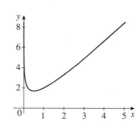

© 2011 Cengage Learning. All Rights Reserved. May not be scanned, copied or duplicated, or posted to a publicly accessible website, in whole or in part.

81. a. $f(x) = \log_b x$, so $x = b^{f(x)}$. Thus, $\dfrac{d}{dx}(x) = \dfrac{d}{dx}\left[b^{f(x)}\right]$, $1 = (\ln b)\, b^{f(x)} f'(x)$, and therefore

$$f'(x) = \frac{1}{(\ln b)\, b^{f(x)}} = \frac{1}{(\ln b)\, x}.$$

b. $f'(x) = \dfrac{d}{dx}(\log_{10} x) = \dfrac{1}{(\ln 10)\, x}.$

83. $g'(x) = \dfrac{d}{dx}\left(\dfrac{10^x}{x+1}\right) = \dfrac{(x+1)(\ln 10)\, 10^x - 10^x}{(x+1)^2} = \dfrac{[(x+1)(\ln 10) - 1]\, 10^x}{(x+1)^2}.$

85. $f(x) = 3^{x^2} + \log_2(x^2 + 1)$, so

$$f'(x) = \frac{d}{dx}\left[3^{x^2} + \log_2(x^2 + 1)\right] = (\ln 3)\, 3^{x^2}(2x) + \frac{2x}{(\ln 2)(x^2 + 1)}$$

$$= 2x\left[(\ln 3)\, 3^{x^2} + \frac{1}{(\ln 2)(x^2 + 1)}\right].$$

87. True. $f(x) = \ln a^x = x \ln a$, so $f'(x) = \frac{d}{dx}(x \cdot \ln a) = \ln a.$

89. Let $f(x) = \ln x$. Then by definition, $f'(1) = \lim\limits_{h \to 0} \dfrac{f(1+h) - f(1)}{h} = \lim\limits_{h \to 0} \dfrac{\ln(1+h) - \ln 1}{h} = \lim\limits_{h \to 0} \dfrac{\ln(h+1)}{h}.$

But $f'(x) = \dfrac{d}{dx} \ln x = \dfrac{1}{x}$, and so with $h = x$, we have $f'(1) = 1 = \lim\limits_{x \to 0} \dfrac{\ln(x+1)}{x}.$

5.6   Problem-Solving Tips

Four mathematical models were introduced in this section:

1. **Exponential growth:** $Q(t) = Q_0 e^{kt}$ describes a quantity $Q(t)$ that is initially present in the amount $Q(0) = Q_0$ and whose rate of growth at any time t is directly proportional to the amount of the quantity present at time t.

2. **Exponential decay:** $Q(t) = Q_0 e^{-kt}$ describes a quantity $Q(t)$ that is initially present in the amount $Q(0) = Q_0$ and decreases at a rate that is directly proportional to its size.

3. **Learning curves:** $Q(t) = C - Ae^{-kt}$ describes a quantity $Q(t)$, where $Q(0) = C - A$, and $Q(t)$ increases and approaches the number C as t increases without bound.

4. **Logistic growth functions:** $Q(t) = \dfrac{A}{1 + Be^{-kt}}$ describes a quantity $Q(t)$, where $Q(0) = \dfrac{A}{1 + B}$. Note that $Q(t)$ increases rapidly for small values of t but the rate of growth of $Q(t)$ decreases quickly as t increases. $Q(t)$ approaches the number A as t increases without bound.

Try to familiarize yourself with the examples and graphs for each of these models before you work through the applied problems in this section.

© 2011 Cengage Learning. All Rights Reserved. May not be scanned, copied or duplicated, or posted to a publicly accessible website, in whole or in part.

5.6   Concept Questions   page 386

1. $Q(t) = Q_0 e^{kt}$ where $k > 0$ represents exponential growth and $k < 0$ represents exponential decay. The larger the magnitude of k, the more quickly the former grows and the more quickly the latter decays.

3. $Q(t) = \dfrac{A}{1 + Be^{-kt}}$, where A, B, and k are positive constants. Q increases rapidly for small values of t but the rate of increase slows down as Q (always increasing) approaches the number A.

5.6   Exponential Functions as Mathematical Models   page 386

1. a. The growth constant is $k = 0.05$.

b. Initially, there are 400 units present.

c.

t	0	10	20	100	1000
Q	400	660	1087	59,365	2.07×10^{24}

3. a. $Q(t) = Q_0 e^{kt}$. Here $Q_0 = 100$ and so $Q(t) = 100 e^{kt}$. Because the number of cells doubles in 20 minutes, we have $Q(20) = 100 e^{20k} = 200$, $e^{20k} = 2$, $20k = \ln 2$, and so $k = \frac{1}{20} \ln 2 \approx 0.03466$. Thus, $Q(t) = 100 e^{0.03466t}$.

b. We solve the equation $100 e^{0.03466t} = 1,000,000$, obtaining $e^{0.03466t} = 10,000$, $0.03466t = \ln 10,000$, and so
$$t = \frac{\ln 10,000}{0.03466} \approx 266, \text{ or } 266 \text{ minutes.}$$

c. $Q(t) = 1000 e^{0.03466t}$.

5. a. We solve the equation $5.3 e^{0.02t} = 3(5.3)$, obtaining $e^{0.02t} = 3$, $0.02t = \ln 3$, and so $t = \dfrac{\ln 3}{0.02} \approx 54.93$. Thus, the world population will triple in approximately 54.93 years.

b. If the growth rate is 1.8%, then proceeding as before, we find $N(t) = 5.3 e^{0.018t}$. If $t = 54.93$, the population would be $N(54.93) = 5.3 e^{0.018(54.93)} \approx 14.25$, or approximately 14.25 billion.

7. $P(h) = p_0 e^{-kh}$, so $P(0) = p_0 = 15$. Thus, $P(4000) = 15 e^{-4000k} = 12.5$, $e^{-4000k} = \frac{12.5}{15}$, $-4000k = \ln\left(\frac{12.5}{15}\right)$, and so $k = 0.00004558$. Therefore, $P(12,000) = 15 e^{-0.00004558(12,000)} = 8.68$, or 8.7 lb/in^2. The rate of change of atmospheric pressure with respect to altitude is given by $P'(h) = \dfrac{d}{dh}\left(15 e^{-0.00004558h}\right) = -0.0006837 e^{-0.00004558h}$. Thus, the rate of change of atmospheric pressure with respect to altitude when the altitude is 12,000 feet is $P'(12,000) = -0.0006837 e^{-0.00004558(12,000)} \approx -0.00039566$. That is, it is declining at the rate of approximately 0.0004 lb/in^2/ft.

© 2011 Cengage Learning. All Rights Reserved. May not be scanned, copied or duplicated, or posted to a publicly accessible website, in whole or in part.

9. Suppose the amount of P-32 at time t is given by $Q(t) = Q_0 e^{-kt}$, where Q_0 is the amount present initially and k is the decay constant. Because this element has a half-life of 14.2 days, we have $\frac{1}{2}Q_0 = Q_0 e^{-14.2k}$, so $e^{-14.2k} = \frac{1}{2}$, $-14.2k = \ln\frac{1}{2}$, and $k = -\frac{\ln(1/2)}{14.2} \approx 0.0488$. Therefore, the amount of P-32 present at any time t is given by $Q(t) = 100e^{-0.0488t}$. In particular, the amount left after 7.1 days is given by $Q(7.1) = 100e^{-0.0488(7.1)} = 100e^{-0.3465} \approx 70.617$, or 70.617 grams. The rate at which the element decays is $Q'(t) = \dfrac{d}{dt}\left(100e^{-0.0488t}\right) = 100\left(-0.0488\right)e^{-0.0488t} = -4.88e^{-0.0488t}$. Therefore, $Q'(7.1) = -4.88e^{-0.0488(7.1)} \approx -3.451$; that is, it is decreasing at the rate of 3.451 g/day.

11. We solve the equation $0.2Q_0 = Q_0 e^{-0.00012t}$, obtaining $t = \dfrac{\ln 0.2}{-0.00012} \approx 13{,}412$, or approximately 13,412 years.

13.

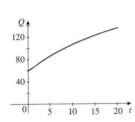

a. $Q(0) = 120\left(1 - e^0\right) + 60 = 60$, or 60 wpm.

b. $Q(10) = 120\left(1 - e^{-0.5}\right) + 60 = 107.22$, or approximately 107 wpm.

c. $Q(20) = 120\left(1 - e^{-1}\right) + 60 = 135.65$, or approximately 136 wpm.

15. a. The federal debt in 2001 was $f(1) = 5.37e^{0.078} \approx \$5.806$ trillion. In 2006, it was $f(6) = 5.37e^{0.078(6)} \approx \$8.575$ trillion.

b. $f'(t) = \dfrac{d}{dt}\left(5.37e^{0.078t}\right) = 0.41886e^{0.078t}$. In 2001, the federal debt was increasing the rate of $f'(1) = 0.41886e^{0.078} \approx \$0.45$ trillion per year. In 2006, it was increasing at the rate of $f'(6) = 0.41886e^{0.078(6)} \approx \$0.67$ trillion per year.

17.

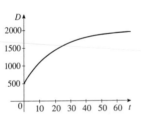

a. After 1 month, the demand is

$D(1) = 2000 - 1500e^{-0.05} \approx 573$, after 12 months it is

$D(12) = 2000 - 1500e^{-0.6} \approx 1177$, after 24 months it is

$D(24) = 2000 - 1500e^{-1.2} \approx 1548$, and after 60 months,

it is $D(60) = 2000 - 1500e^{-3} \approx 1925$.

b. $\displaystyle\lim_{t\to\infty} D(t) = \lim_{t\to\infty}\left(2000 - 1500e^{-0.05t}\right) = 2000$, and we conclude that the demand is expected to stabilize at 2000 computers per month.

c. $D'(t) = -1500e^{-0.05t}(-0.05) = 75e^{-0.05t}$. Therefore, the rate of growth after 10 months is $D'(10) = 75e^{-0.5} \approx 45.49$, or approximately 46 computers per month.

19. a. The length is given by $f(5) = 200\left(1 - 0.956e^{-0.18\cdot5}\right) \approx 122.26$, or approximately 122.3 cm.

b. $f'(t) = 200\left(-0.956\right)e^{-0.18t}(-0.18) = 34.416e^{-0.18t}$, so a 5-year-old is growing at the rate of $f'(5) = 34.416e^{-0.18(5)} \approx 13.9925$, or approximately 14 cm/yr.

c. The maximum length is given by $\displaystyle\lim_{t\to\infty} 200\left(1 - 0.956e^{-0.18t}\right) = 200$, or 200 cm.

© 2011 Cengage Learning. All Rights Reserved. May not be scanned, copied or duplicated, or posted to a publicly accessible website, in whole or in part.

21. a. The proportion of lay teachers is $f(3) = \dfrac{98}{1 + 2.77e^{-3}} \approx 86.1228$, or 86.12%.

b. $f'(t) = \dfrac{d}{dt}\left[98\left(1 + 2.77e^{-t}\right)^{-1}\right] = 98\,(-1)\left(1 + 2.77e^{-t}\right)^{-2}\left(2.77e^{-t}\right)(-1) = \dfrac{271.46e^{-t}}{\left(1 + 2.77e^{-t}\right)^2}$, so

$f'(3) = \dfrac{271.46e^{-3}}{\left(1 + 2.77e^{-3}\right)^2} \approx 10.4377$. Thus, the proportion of lay teachers was increasing at the rate of 10.44%

annually.

c. $f''(t) = 271.46\left[\dfrac{\left(1 + 2.77e^{-t}\right)^2\left(-e^{-t}\right) - e^{-t}\cdot 2\left(1 + 2.77e^{-t}\right)\left(-2.77e^{-t}\right)}{\left(1 + 2.77e^{-t}\right)^4}\right]$

$= \dfrac{271.46\left[-\left(1 + 2.77e^{-t}\right) + 5.54e^{-t}\right]}{e^t\left(1 + 2.77e^{-t}\right)^3} = \dfrac{271.46\left(2.77e^{-t} - 1\right)}{e^t\left(1 + 2.77e^{-t}\right)^3}.$

Setting $f''(t) = 0$ gives $2.77e^{-t} = 1$, so $e^{-t} = \dfrac{1}{2.77}$, $-t = \ln\frac{1}{2.77}$, and

$t \approx 1.0188$. The sign diagram of f'' shows that
$t = 1.02$ gives an inflection point of P. Thus, the
proportion of lay teachers was increasing most rapidly
in 1970.

23. The projected population of citizens aged 45–64 in 2010 is $P(20) = \dfrac{197.9}{1 + 3.274e^{-0.0361(20)}} \approx 76.3962$, or

76.4 million.

25. The first of the given conditions implies that $f(0) = 300$, that is, $300 = \dfrac{3000}{1 + Be^0} = \dfrac{3000}{1 + B}$. Thus, $1 + B = 10$,

and $B = 9$. Therefore, $f(t) = \dfrac{3000}{1 + 9e^{-kt}}$. Next, the condition $f(2) = 600$ gives the equation $600 = \dfrac{3000}{1 + 9e^{-2k}}$,

so $1 + 9e^{-2k} = 5$, $e^{-2k} = \frac{4}{9}$, and $k = -\frac{1}{2}\ln\frac{4}{9}$. Therefore, $f(t) = \dfrac{3000}{1 + 9e^{(1/2)t\cdot\ln(4/9)}} = \dfrac{3000}{1 + 9\left(\frac{4}{9}\right)^{t/2}}$. The

number of students who had heard about the policy four hours later is given by $f(4) = \dfrac{3000}{1 + 9\left(\frac{4}{9}\right)^2} = 1080$, or

1080 students. To find the rate at which the rumor was spreading at any time time, we compute

$f'(t) = \frac{d}{dt}\left[3000\left(1 + 9e^{-0.405465t}\right)^{-1}\right] = (3000)(-1)\left(1 + 9e^{-0.405465}\right)^{-2}\frac{d}{dt}\left(9e^{-0.405465t}\right)$

$= -3000\,(9)\,(-0.405465)\,e^{-0.405465t}\left(1 + 9e^{-0.405465t}\right)^{-2} = \dfrac{10947.555e^{-0.405465t}}{\left(1 + 9e^{-0.405465t}\right)^2}.$

In particular, the rate at which the rumor was spreading 4 hours after the ceremony is given by

$f'(4) = \dfrac{10947.555e^{-0.405465\cdot 4}}{\left(1 + 9e^{-0.405465\cdot 4}\right)^2} \approx 280.26$. Thus, the rumor is spreading at the rate of 280 students per hour.

27. $x(t) = \dfrac{15\left(1 - \left(\frac{2}{3}\right)^{3t}\right)}{1 - \frac{1}{4}\left(\frac{2}{3}\right)^{3t}}$, so $\lim\limits_{t\to\infty} x(t) = \lim\limits_{t\to\infty}\dfrac{15\left[1 - \left(\frac{2}{3}\right)^{3t}\right]}{1 - \frac{1}{4}\left(\frac{2}{3}\right)^{3t}} = \dfrac{15(1-0)}{1-0} = 15$, or 15 lb.

© 2011 Cengage Learning. All Rights Reserved. May not be scanned, copied or duplicated, or posted to a publicly accessible website, in whole or in part.

29. a. $C(t) = \dfrac{k}{b-a}\left(e^{-at} - e^{-bt}\right)$, so

$$C'(t) = \frac{k}{b-a}\left(-ae^{-at} + be^{-bt}\right) = \frac{kb}{b-a}\left[e^{-bt} - \left(\frac{a}{b}\right)e^{-at}\right] = \frac{kb}{b-a}e^{-bt}\left[1 - \frac{a}{b}e^{(b-a)t}\right].$$

$C'(t) = 0$ implies that $1 = \dfrac{a}{b}e^{(b-a)t}$, or $t = \dfrac{\ln\left(\frac{b}{a}\right)}{b-a}$. The sign

diagram of C' shows that this value of t gives a minimum.

$+ + + + + + 0 - - - - -$ sign of C'

$\underset{0}{\vdash}\qquad\qquad\underset{\ln\frac{b}{a}}{\overset{}{\vert}}\longrightarrow t$

$\ln\frac{b}{a} \over b-a$

b. $\displaystyle\lim_{t\to\infty} C(t) = 0.$

31. a. We solve $Q_0 e^{-kt} = \frac{1}{2}Q_0$ for t, obtaining $e^{-kt} = \frac{1}{2}$, $\ln e^{-kt} = \ln\frac{1}{2} = \ln 1 - \ln 2 = -\ln 2$, $-kt = -\ln 2$, and so $\overline{t} = \frac{\ln 2}{k}$.

b. $\overline{t} = \dfrac{\ln 2}{0.0001238} \approx 5598.927$, or approximately 5599 years.

33. a. From the results of Exercise 32, we have $Q' = kQ\left(1 - \dfrac{Q}{A}\right)$, so

$$Q'' = \frac{d}{dt}\left(kQ - \frac{k}{A}Q^2\right) = kQ' - \frac{2k}{A}QQ' = \frac{k}{A}Q'(A - 2Q). \text{ Setting } Q'' = 0 \text{ gives } Q = \frac{A}{2} \text{ since } Q' > 0 \text{ for}$$

all t. Furthermore, $Q'' > 0$ if $Q < \frac{A}{2}$ and $Q'' < 0$ if $Q > \frac{A}{2}$. So the graph of Q has an inflection point when

$Q = \frac{A}{2}$. To find the value of t, we solve the equation $\dfrac{A}{2} = \dfrac{A}{1 + Be^{-kt}}$, obtaining $1 + Be^{-kt} = 2$, $Be^{-kt} = 1$,

$e^{-kt} = \dfrac{1}{B}$, $-kt = \ln\dfrac{1}{B} = -\ln B$, and so $t = \dfrac{\ln B}{k}$.

b. The quantity Q increases most rapidly at the instant of time when it reaches one-half of the maximum quantity.

This occurs at $t = \dfrac{\ln B}{k}$.

35. We use the result of Exercise 34 with $t_1 = 14$, $t_2 = 21$, $A = 600$, $Q_1 = 76$, and $Q_2 = 167$ to obtain

$$k = \frac{1}{21 - 14}\ln\left[\frac{167(600 - 76)}{76(600 - 167)}\right] \approx 0.14.$$

5.6   Using Technology   page 391

1. a.

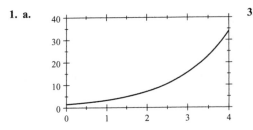

3. a.

b. $f'(3) \approx 12.146\%/\text{yr}.$

c. $f''(3) \approx 9.474\%/\text{yr}/\text{yr}.$

b. $T(0) = 666$ million; $T(8) \approx 926.8$ million.

c. $T'(8) \approx 38.3$ million/yr/yr.

© 2011 Cengage Learning. All Rights Reserved. May not be scanned, copied or duplicated, or posted to a publicly accessible website, in whole or in part.

5. a.

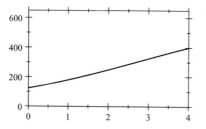

b. $P(3) \approx 325$ million.

c. $P'(3) \approx 76.84$ million per 30 years.

7. a.

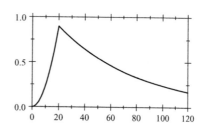

b. The initial concentration is 0.

c. $C(10) \approx 0.237$ g/cm^3.

d. $C(30) \approx 0.760$ g/cm^3.

e. $\lim\limits_{t \to \infty} C(t) = 0$.

9. a.

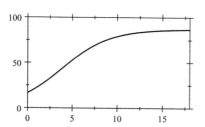

b. $f(3) \approx 36.6\%$; $f(13) \approx 84.2\%$.

c. $f'(3) \approx 7.9\%$/yr; $f'(13) \approx 1\%$/yr.

d. $f'(t)$ has a maximum at $t \approx 3.86$, near the end of 1984.

11. a. $f(t) = \dfrac{544.65}{1 + 1.65e^{-0.1846t}}$.

c. In 2006, worldwide PC shipments were increasing at the rate of $f'(1) \approx 24.52$, or approximately 24.5 million units/yr. In 2008, shipments were increasing at the rate of $f'(3) \approx 25.12$, or approximately 25.1 million units/yr.

b.

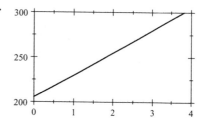

CHAPTER 5   **Concept Review**   page 393

1. power, 0, 1, exponential

3. a. $(0, \infty)$, $(-\infty, \infty)$, $(1, 0)$ **b.** $< 1, > 1$

5. accumulated amount, principal, nominal interest rate, number of conversion periods, term

7. Pe^{rt}

9. a. initially, growth   **b.** decay   **c.** time, one-half

© 2011 Cengage Learning. All Rights Reserved. May not be scanned, copied or duplicated, or posted to a publicly accessible website, in whole or in part.

1.

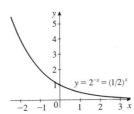

$$y = 2^{-x} = (1/2)^x$$

$\left(\dfrac{1}{2}\right)^x = \dfrac{1}{2^x} = 2^{-x}$, so the two graphs are the same.

3. $16^{-3/4} = 0.125$ is equivalent to $-\frac{3}{4} = \log_{16} 0.125$.

5. $\ln (x - 1) + \ln 4 = \ln (2x + 4) - \ln 2$, so $\ln (x - 1) - \ln (2x + 4) = -\ln 2 - \ln 4 = -(\ln 2 + \ln 4)$,

$\ln \left(\dfrac{x - 1}{2x + 4}\right) = -\ln 8 = \ln \frac{1}{8}$, $\left(\dfrac{x - 1}{2x + 4}\right) = \dfrac{1}{8}$, $8x - 8 = 2x + 4$, $6x = 12$, and so $x = 2$. Check:

LHS $= \ln (2 - 1) + \ln 4 = \ln 4$; RHS $= \ln (4 + 4) - \ln 2 = \ln 8 - \ln 2 = \ln \frac{8}{2} = \ln 4$.

7. $\ln 3.6 = \ln \frac{36}{10} = \ln 36 - \ln 10 = \ln 6^2 - \ln (2 \cdot 5) = 2 \ln 6 - \ln 2 - \ln 5 = 2 (\ln 2 + \ln 3) - \ln 2 - \ln 5$

$= 2 (x + y) - x - z = x + 2y - z.$

9. We first sketch the graph of $y = 2^{x-3}$, then reflect
this graph with respect to the line $y = x$.

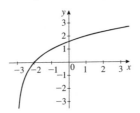

11. a. Using Formula (6) with $P = 10,000$, $r = 0.06$, $m = 365$ and $t = 2$, we have

$$A = 10,000 \left(1 + \dfrac{0.06}{365}\right)^{365(2)} = 11,274.86, \text{ or } \$11,274.86.$$

b. Using Formula (10) with $P = 10,000$, $r = 0.06$, and $t = 2$, we have $A = 10,000e^{0.06(2)} = 11,274.97$, or
$\$11,274.97$.

13. Using Formula (6) with $A = 10,000$, $P = 15,000$, $r = 0.06$, and $m = 4$, we have

$A = 10,000 \left(1 + \frac{0.06}{4}\right)^{4t} = 15,000$, or $\left(1 + \frac{0.06}{4}\right)^{4t} = 1.5$. Solving for t, we have $4t \ln (1.015) = \ln 1.5$, so

$t = \dfrac{\ln 1.5}{4 \ln 1.015} \approx 6.808$, or approximately 6.8 years.

15. $f(x) = xe^{2x}$, so $f'(x) = e^{2x} + xe^{2x}(2) = (1 + 2x)e^{2x}$.

17. $g(t) = \sqrt{t}e^{-2t}$, so $g'(t) = \frac{1}{2}t^{-1/2}e^{-2t} + \sqrt{t}e^{-2t}(-2) = \dfrac{1 - 4t}{2\sqrt{t}e^{2t}}$.

© 2011 Cengage Learning. All Rights Reserved. May not be scanned, copied or duplicated, or posted to a publicly accessible website, in whole or in part.

19. $y = \dfrac{e^{2x}}{1 + e^{-2x}}$, so $y' = \dfrac{(1 + e^{-2x}) e^{2x} (2) - e^{2x} \cdot e^{-2x} (-2)}{(1 + e^{-2x})^2} = \dfrac{2 (e^{2x} + 2)}{(1 + e^{-2x})^2}$.

21. $f(x) = xe^{-x^2}$, so $f'(x) = e^{-x^2} + xe^{-x^2} (-2x) = (1 - 2x^2) e^{-x^2}$.

23. $f(x) = x^2 e^x + e^x$, so $f'(x) = 2xe^x + x^2 e^x + e^x = (x^2 + 2x + 1) e^x = (x + 1)^2 e^x$.

25. $f(x) = \ln \left(e^{x^2} + 1\right)$, so $f'(x) = \dfrac{e^{x^2} (2x)}{e^{x^2} + 1} = \dfrac{2xe^{x^2}}{e^{x^2} + 1}$.

27. $f(x) = \dfrac{\ln x}{x + 1}$, so $f'(x) = \dfrac{(x + 1) \left(\frac{1}{x}\right) - \ln x}{(x + 1)^2} = \dfrac{1 + \frac{1}{x} - \ln x}{(x + 1)^2} = \dfrac{x - x \ln x + 1}{x (x + 1)^2}$.

29. $y = \ln \left(e^{4x} + 3\right)$, so $y' = \dfrac{e^{4x} (4)}{e^{4x} + 3} = \dfrac{4e^{4x}}{e^{4x} + 3}$.

31. $f(x) = \dfrac{\ln x}{1 + e^x}$, so

$$f'(x) = \dfrac{(1 + e^x) \dfrac{d}{dx} \ln x - \ln x \dfrac{d}{dx} (1 + e^x)}{(1 + e^x)^2} = \dfrac{(1 + e^x) \left(\frac{1}{x}\right) - (\ln x) e^x}{(1 + e^x)^2} = \dfrac{1 + e^x - xe^x \ln x}{x (1 + e^x)^2}$$

$$= \dfrac{1 + e^x (1 - x \ln x)}{x (1 + e^x)^2}.$$

33. $y = \ln (3x + 1)$, so $y' = \dfrac{3}{3x + 1}$ and $y'' = 3 \dfrac{d}{dx} (3x + 1)^{-1} = -3 (3x + 1)^{-2} (3) = -\dfrac{9}{(3x + 1)^2}$.

35. $h'(x) = g'(f(x)) f'(x)$. But $g'(x) = 1 - \dfrac{1}{x^2}$ and $f'(x) = e^x$, so $f(0) = e^0 = 1$ and $f'(0) = e^0 = 1$. Therefore, $h'(0) = g'(f(0)) f'(0) = g'(1) f'(0) = 0 \cdot 1 = 0$.

37. $y = (2x^3 + 1) (x^2 + 2)^3$, so $\ln y = \ln (2x^3 + 1) + 3 \ln (x^2 + 2)$,

$$\dfrac{y'}{y} = \dfrac{6x^2}{2x^3 + 1} + \dfrac{3 (2x)}{x^2 + 2} = \dfrac{6x^2 (x^2 + 2) + 6x (2x^3 + 1)}{(2x^3 + 1) (x^2 + 2)} = \dfrac{6x^4 + 12x^2 + 12x^4 + 6x}{(2x^3 + 1) (x^2 + 2)}$$

$$= \dfrac{18x^4 + 12x^2 + 6x}{(2x^3 + 1) (x^2 + 2)},$$

and so $y' = 6x (3x^3 + 2x + 1) (x^2 + 2)^2$.

39. $y = e^{-2x}$, so $y' = -2e^{-2x}$. This gives the slope of the tangent line to the graph of $y = e^{-2x}$ at any point (x, y). In particular, the slope of the tangent line at $(1, e^{-2})$ is $y'(1) = -2e^{-2}$. The required equation is

$$y - e^{-2} = -2e^{-2} (x - 1), \text{ or } y = \dfrac{1}{e^2} (-2x + 3).$$

© 2011 Cengage Learning. All Rights Reserved. May not be scanned, copied or duplicated, or posted to a publicly accessible website, in whole or in part.

41. $f(x) = xe^{-2x}$. We first gather the following information on f.

1. The domain of f is $(-\infty, \infty)$.

2. Setting $x = 0$ gives 0 as the y-intercept.

3. $\lim\limits_{x \to -\infty} xe^{-2x} = -\infty$ and $\lim\limits_{x \to \infty} xe^{-2x} = 0$.

4. The results of part 3 show that $y = 0$ is a horizontal asymptote.

5. $f'(x) = e^{-2x} + xe^{-2x}(-2) = (1 - 2x)e^{-2x}$. Observe that $f'(x) = 0$ at $x = \frac{1}{2}$, a critical point of f.

   The sign diagram of f' shows that f is increasing on $\left(-\infty, \frac{1}{2}\right)$ and decreasing on $\left(\frac{1}{2}, \infty\right)$.

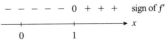

6. The results of part 5 show that $\left(\frac{1}{2}, \frac{1}{2}e^{-1}\right)$ is a relative maximum.

7. $f''(x) = -2e^{-2x} + (1 - 2x)e^{-2x}(-2) = 4(x - 1)e^{-2x} = 0$ if $x = 1$. The sign diagram of f'' shows that the graph of f is concave downward on $(-\infty, 1)$ and concave upward on $(1, \infty)$.

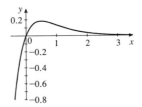

8. f has an inflection point at $\left(1, 1/e^2\right)$.

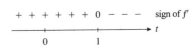

43. $f(t) = te^{-t}$, so $f'(t) = e^{-t} + t\left(-e^{-t}\right) = e^{-t}(1 - t)$. Setting $f'(t) = 0$ gives $t = 1$ as the only critical point of f. From the sign diagram of f' we see that $f(1) = e^{-1} = 1/e$ is the absolute maximum value of f.

45. We want to solve the equation $8.2 = 4.5e^{r(5)}$. We have $e^{5r} = \frac{8.2}{4.5}$, so $r = \frac{1}{5}\ln\frac{8.2}{4.5} \approx 0.120$, and so the annual rate of return is 12%.

47. $P = 119{,}346e^{-0.1(4)} \approx 80{,}000$, or \$80,000.

49. a. $Q(t) = 2000e^{kt}$. Now $Q(120) = 18{,}000$ gives $2000e^{120k} = 18{,}000$, $e^{120k} = 9$, and so $120k = \ln 9$. Thus, $k = \frac{1}{120}\ln 9 \approx 0.01831$ and $Q(t) = 2000e^{0.01831t}$.

   b. $Q(4) = 2000e^{0.01831(240)} \approx 161{,}992$, or approximately 162,000.

© 2011 Cengage Learning. All Rights Reserved. May not be scanned, copied or duplicated, or posted to a publicly accessible website, in whole or in part.

51.

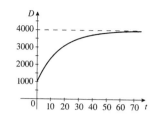

a. $D(1) = 4000 - 3000e^{-0.06} = 1175$,
$D(12) = 4000 - 3000e^{-0.72} = 2540$, and
$D(24) = 4000 - 3000e^{-1.44} = 3289$.

b. $\lim\limits_{t\to\infty} D(t) = \lim\limits_{t\to\infty} \left(4000 - 3000e^{-0.06t}\right) = 4000$.

53. $f(t) = 1.5 + 1.8te^{-1.2t}$, so $f'(t) = 1.8\frac{d}{dt}\left(te^{-1.2t}\right) = 1.8\left[e^{-1.2t} + te^{-1.2t}(-1.2)\right] = 1.8e^{-1.2t}(1 - 1.2t)$.
$f'(0) = 1.8$, $f'(1) = -0.11$, $f'(2) = -0.23$, and $f'(3) = -0.13$. Thus, measured in barrels per \$1000 of output per decade, the amount of oil used is increasing by 1.8 in 1965, decreasing by 0.11 in 1966, and so on.

55. We have $Q(10) = 90$, and so $\dfrac{3000}{1 + 499e^{-10k}} = 90$, $1 + 499e^{-10k} = \dfrac{3000}{90}$, $499e^{-10k} = \dfrac{2910}{90}$, $e^{-10k} = \dfrac{2910}{90(499)}$,

and $k = -\dfrac{1}{10}\ln\dfrac{2910}{90(499)} \approx 0.2737$. Thus, $N(t) = \dfrac{3000}{1 + 499e^{-0.2737t}}$. The number of students who have contracted

the flu by the 20th day is $N(20) = \dfrac{3000}{1 + 499e^{-0.2737(20)}} \approx 969.93$, or approximately 970 students.

57. The revenue is given by $R = px = 20e^{-0.0002x}x = 20xe^{-0.0002x}$. To find the maximum of this function, we
compute $R'(x) = 20e^{-0.0002x} - x\left(0.0002e^{-0.0002x}\right) = e^{-0.0002x}(20 - 0.0004x) = 0$

if $20 - 0.0004x = 0$, or $x = 5000$. From the sign diagram of R',
we see that $(5000, 36787.9)$ is a maximum, so 5000 pairs of socks
should be produced to yield a maximum revenue of approximately
\$36,788.

$$
\begin{array}{c}
+ \; + \; + \; 0 \; - \; - \; - \quad \text{sign of } R' \\
\vphantom{} \\
\left+\hspace{2em}+\hspace{2em}+\rightarrow x \\
0 \qquad 5000 \quad 10{,}000
\end{array}
$$

CHAPTER 5   Before Moving On...   page 396

1. $\dfrac{100}{1 + 2e^{0.3t}} = 40$, so $1 + 2e^{0.3t} = \dfrac{100}{40} = 2.5$, $2e^{0.3t} = 1.5$, $e^{0.3t} = \dfrac{1.5}{2} = 0.75$, $0.3t = \ln 0.75$, and so
$t = \dfrac{\ln 0.75}{0.3} \approx -0.959$.

2. $A = 3000\left(1 + \dfrac{0.08}{52}\right)^{4(52)} = 4130.37$, or \$4130.37.

3. $f'(x) = \dfrac{d}{dx}e^{x^{1/2}} = e^{x^{1/2}}\dfrac{d}{dx}\left(x^{1/2}\right) = e^{x^{1/2}}\left(\tfrac{1}{2}x^{-1/2}\right) = \dfrac{e^{\sqrt{x}}}{2\sqrt{x}}$.

4. $\dfrac{dy}{dx} = x\dfrac{d}{dx}\ln\left(x^2 + 1\right) + \ln\left(x^2 + 1\right)\dfrac{d}{dx}(x) = x\cdot\dfrac{2x}{x^2 + 1} + \ln\left(x^2 + 1\right) = \dfrac{2x^2}{x^2 + 1} + \ln\left(x^2 + 1\right)$. Thus,
$\dfrac{dy}{dx}\Big|_{x=1} = \dfrac{1}{1+1} + \ln 2 = 1 + \ln 2$

© 2011 Cengage Learning. All Rights Reserved. May not be scanned, copied or duplicated, or posted to a publicly accessible website, in whole or in part.

5. $y' = e^{2x} \dfrac{d}{dx} \ln 3x + \ln 3x \cdot \dfrac{d}{dx} e^{2x} = \dfrac{e^{2x}}{x} + 2e^{2x} \ln 3x$ and

$y'' = \dfrac{d}{dx} \left(x^{-1} e^{2x} \right) + 2e^{2x} \dfrac{d}{dx} \ln 3x + (\ln 3x) \dfrac{d}{dx} \left(2e^{2x} \right)$

$\quad = -x^{-2} e^{2x} + 2x^{-1} e^{2x} + 2e^{2x} \left(\dfrac{1}{x} \right) + 4e^{2x} \cdot \ln 3x$

$\quad = -\dfrac{1}{x^2} e^{2x} + \dfrac{4e^{2x}}{x} + 4 (\ln 3x) e^{2x} = e^{2x} \left(\dfrac{4x^2 \ln 3x + 4x - 1}{x^2} \right).$

6. $T(0) = 200$ gives $70 + ce^0 = 70 + C = 200$, so $C = 130$. Thus, $T(t) = 70 + 130e^{-kt}$. $T(3) = 180$ implies $70 + 130e^{-3k} = 180$, so $130e^{-3k} = 110$, $e^{-3k} = \dfrac{110}{130}$, $-3k = \ln \dfrac{11}{13}$, and $k = -\dfrac{1}{3} \ln \dfrac{11}{13} \approx 0.0557$. Therefore, $T(t) = 70 + 130e^{-0.0557t}$. So when $T(t) = 150$, we have $70 + 130e^{-0.0557t} = 150$, $130e^{-0.0557t} = 80$, $e^{-0.0557t} = \dfrac{80}{130} = \dfrac{8}{13}$, $-0.0557t = \ln \dfrac{8}{13}$, and finally $t = -\dfrac{\ln \frac{8}{13}}{0.0557} \approx 8.716$, or approximately 8.7 minutes.

© 2011 Cengage Learning. All Rights Reserved. May not be scanned, copied or duplicated, or posted to a publicly accessible website, in whole or in part.

6 INTEGRATION

6.1 Problem-Solving Tips

1. Get into the habit of using the correct notation for integration. The indefinite integral of $f(x)$ with respect to x is written $\int f(x)\,dx$. It is incorrect to write $\int f(x)$ without indicating that you are integrating with respect to x. You will appreciate how important the correct notation is if you use CAS or graphic calculator with the capability to do symbolic integration. If you don't enter this information (the variable with respect to which you are performing the integration) into your calculator or computer, the integration will not be performed.

2. If you are finding an indefinite integral, be sure to include a constant of integration in your answer. Remember that $\int f(x)\,dx$ is the family of functions given by $F(x) + C$, where $F'(x) = f(x)$.

3. It's very easy to check your answer if you are finding an indefinite integral. Just take the derivative of your answer and you should get the integrand. [You are verifying that $F'(x) = f(x)$.] If not, you know immediately that you have made an error.

6.1 Concept Questions   page 406

1. An antiderivative of a continuous function f on an interval I is a function F such that $F'(x) = f(x)$ for every x in I. For example, an antiderivative of $f(x) = x^2$ on $(-\infty, \infty)$ is the function $F(x) = \frac{1}{3}x^3$ on $(-\infty, \infty)$.

3. The indefinite integral of f is the family of functions $F(x) + C$, where F is an antiderivative of f and C is an arbitrary constant.

6.1 Antiderivatives and the Rules of Integration   page 406

1. $F(x) = \frac{1}{3}x^3 + 2x^2 - x + 2$, so $F'(x) = x^2 + 4x - 1 = f(x)$.

3. $F(x) = \left(2x^2 - 1\right)^{1/2}$, so $F'(x) = \frac{1}{2}\left(2x^2 - 1\right)^{-1/2}(4x) = 2x\left(2x^2 - 1\right)^{-1/2} = f(x)$.

173

© 2011 Cengage Learning. All Rights Reserved. May not be scanned, copied or duplicated, or posted to a publicly accessible website, in whole or in part.

5. a. $G'(x) = \dfrac{d}{dx}(2x) = 2 = f(x)$

   b. $F(x) = G(x) + C = 2x + C$

   c.

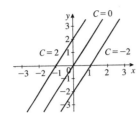

7. a. $G'(x) = \dfrac{d}{dx}\left(\frac{1}{3}x^3\right) = x^2 = f(x)$

   b. $F(x) = G(x) + C = \frac{1}{3}x^3 + C$

   c.

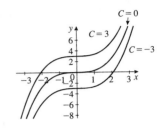

9. $\int 6\,dx = 6x + C.$

11. $\int x^3\,dx = \frac{1}{4}x^4 + C.$

13. $\int x^{-4}\,dx = -\frac{1}{3}x^{-3} + C.$

15. $\int x^{2/3}\,dx = \frac{3}{5}x^{5/3} + C.$

17. $\int x^{-5/4}\,dx = -4x^{-1/4} + C.$

19. $\displaystyle\int \frac{2}{x^2}\,dx = 2\int x^{-2}\,dx = 2\left(-x^{-1}\right) + C = -\frac{2}{x} + C.$

21. $\int \pi\sqrt{t}\,dt = \pi \int t^{1/2}\,dt = \pi\left(\frac{2}{3}t^{3/2}\right) + C$

       $= \frac{2\pi}{3}t^{3/2} + C.$

23. $\int (3 - 2x)\,dx = \int 3\,dx - 2\int x\,dx = 3x - x^2 + C.$

25. $\int \left(x^2 + x + x^{-3}\right)dx = \int x^2\,dx + \int x\,dx + \int x^{-3}\,dx = \frac{1}{3}x^3 + \frac{1}{2}x^2 - \frac{1}{2}x^{-2} + C.$

27. $\int 4e^x\,dx = 4e^x + C.$

29. $\int (1 + x + e^x)\,dx = x + \frac{1}{2}x^2 + e^x + C.$

31. $\displaystyle\int \left(4x^3 - \frac{2}{x^2} - 1\right)dx = \int \left(4x^3 - 2x^{-2} - 1\right)dx = x^4 + 2x^{-1} - x + C = x^4 + \frac{2}{x} - x + C.$

33. $\int \left(x^{5/2} + 2x^{3/2} - x\right)dx = \frac{2}{7}x^{7/2} + \frac{4}{5}x^{5/2} - \frac{1}{2}x^2 + C.$ **35.** $\int \left(x^{1/2} + 3x^{-1/2}\right)dx = \frac{2}{3}x^{3/2} + 6x^{1/2} + C.$

37. $\displaystyle\int \left(\frac{u^3 + 2u^2 - u}{3u}\right)du = \frac{1}{3}\int \left(u^2 + 2u - 1\right)du = \frac{1}{9}u^3 + \frac{1}{3}u^2 - \frac{1}{3}u + C.$

39. $\int (2t + 1)(t - 2)\,dt = \int \left(2t^2 - 3t - 2\right)dt = \frac{2}{3}t^3 - \frac{3}{2}t^2 - 2t + C.$

41. $\displaystyle\int \frac{1}{x^2}\left(x^4 - 2x^2 + 1\right)dx = \int \left(x^2 - 2 + x^{-2}\right)dx = \frac{1}{3}x^3 - 2x - x^{-1} + C = \frac{1}{3}x^3 - 2x - \frac{1}{x} + C.$

43. $\displaystyle\int \frac{ds}{(s + 1)^{-2}} = \int (s + 1)^2\,ds = \int \left(s^2 + 2s + 1\right)ds = \frac{1}{3}s^3 + s^2 + s + C.$

45. $\displaystyle\int \left(e^t + t^e\right)dt = e^t + \frac{1}{e + 1}t^{e+1} + C.$

47. $\displaystyle\int \frac{x^3 + x^2 - x + 1}{x^2}\,dx = \int \left(x + 1 - \frac{1}{x} + \frac{1}{x^2}\right)dx = \frac{1}{2}x^2 + x - \ln|x| - x^{-1} + C$

© 2011 Cengage Learning. All Rights Reserved. May not be scanned, copied or duplicated, or posted to a publicly accessible website, in whole or in part.

49. $\int \dfrac{\left(x^{1/2}-1\right)^2}{x^2}\,dx = \int \dfrac{x-2x^{1/2}+1}{x^2}\,dx = \int \left(x^{-1}-2x^{-3/2}+x^{-2}\right)dx$

$$= \ln|x| + 4x^{-1/2} - x^{-1} + C = \ln|x| + \dfrac{4}{\sqrt{x}} - \dfrac{1}{x} + C.$$

51. $\int f'(x)\,dx = \int (2x+1)\,dx = x^2 + x + C.$ The condition $f(1) = 3$ gives $f(1) = 1 + 1 + C = 3$, so $C = 1$. Therefore, $f(x) = x^2 + x + 1.$

53. $f'(x) = 3x^2 + 4x - 1$, so $f(x) = x^3 + 2x^2 - x + C.$ Using the given initial condition, we have $f(2) = 8 + 2(4) - 2 + C = 9$, so $16 - 2 + C = 9$, or $C = -5$. Therefore, $f(x) = x^3 + 2x^2 - x - 5.$

55. $f(x) = \int f'(x)\,dx = \int \left(1 + \dfrac{1}{x^2}\right)dx = \int \left(1 + x^{-2}\right)dx = x - \dfrac{1}{x} + C.$ Using the given initial condition, we have $f(1) = 1 - 1 + C = 2$, so $C = 2$. Therefore, $f(x) = x - \dfrac{1}{x} + 2.$

57. $f(x) = \int \dfrac{x+1}{x}\,dx = \int \left(1 + \dfrac{1}{x}\right)dx = x + \ln|x| + C.$ Using the initial condition, we have $f(1) = 1 + \ln 1 + C = 1 + C = 1$, so $C = 0$. Thus, $f(x) = x + \ln|x|.$

59. $f(x) = \int f'(x)\,dx = \int \frac{1}{2}x^{-1/2}\,dx = \frac{1}{2}\left(2x^{1/2}\right) + C = x^{1/2} + C$, and $f(2) = \sqrt{2} + C = \sqrt{2}$ implies $C = 0$. Thus, $f(x) = \sqrt{x}.$

61. $f'(x) = e^x + x$, so $f(x) = e^x + \frac{1}{2}x^2 + C$ and $f(0) = e^0 + \frac{1}{2}(0) + C = 1 + C.$ Thus, $3 = 1 + C$, and so $2 = C.$ Therefore, $f(x) = e^x + \frac{1}{2}x^2 + 2.$

63. The net amount on deposit in branch A is given by the area under the graph of f from $t = 0$ to $t = 180$. On the other hand, the net amount on deposit in branch B is given by the area under the graph of g over the same interval. Branch A has a larger amount on deposit because the rate at which money was deposited into branch A was always greater than the rate at which money was deposited into branch B over the period in question.

65. The position of the car is $s(t) = \int f(t)\,dt = \int 2\sqrt{t}\,dt = \int 2t^{1/2}\,dt = 2\left(\frac{2}{3}t^{3/2}\right) + C = \frac{4}{3}t^{3/2} + C.$ $s(0) = 0$ implies that $s(0) = C = 0$, so $s(t) = \frac{4}{3}t^{3/2}.$

67. $C(x) = \int C'(x)\,dx = \int \left(0.000009x^2 - 0.009x + 8\right)dx = 0.000003x^3 - 0.0045x^2 + 8x + k.$ $C(0) = k = 120$, and so $C(x) = 0.000003x^3 - 0.0045x^2 + 8x + 120.$ Thus, $C(500) = 0.000003(500)^3 - 0.0045(500)^2 + 8(500) + 120 = \$3370.$

69. $P'(x) = -0.004x + 20$, so $P(x) = -0.002x^2 + 20x + C$. Because $C = -16{,}000$, we find that $P(x) = -0.002x^2 + 20x - 16{,}000.$ The company realizes a maximum profit when $P'(x) = 0$, that is, when $x = 5000$ units. Next, $P(5000) = -0.002(5000)^2 + 20(5000) - 16{,}000 = 34{,}000.$ Thus, the maximum profit of \$34,000 is realized at a production level of 5000 units.

71. a. $f(t) = \int r(t)\,dt = \int (0.0058t + 0.159)\,dt = 0.0029t^2 + 0.159t + C.$ $f(0) = 1.6$, and so $0 + 0 + C = 1.6$, or $C = 1.6$. Therefore, $f(t) = 0.0029t^2 + 0.159t + 1.6.$

b. The national health expenditure in 2015 will be $f(13) = 0.0029\left(13^2\right) + 0.159(13) + 1.6 = 4.1571$, or approximately \$4.16 trillion.

© 2011 Cengage Learning. All Rights Reserved. May not be scanned, copied or duplicated, or posted to a publicly accessible website, in whole or in part.

73. a. The number of subscribers in year t is given by

$N(t) = \int r(t)\, dt = \int (-0.375t^2 + 2.1t + 2.45)\, dt = -0.125t^3 + 1.05t^2 + 2.45t + C$. To find C, note that

$N(0) = 1.5$. This gives $N(0) = C = 1.5$. Therefore, $N(t) = -0.125t^3 + 1.05t^2 + 2.45t + 1.5$.

b. $N(5) = -0.125\,(5^3) + 1.05\,(5^2) + 2.45\,(5) + 1.5 = 24.375$, or 24.375 million subscribers.

75. a. The approximate average credit card debt per U.S. household in year t is

$A(t) = \int D(t)\, dt = \int (-4.479t^2 + 69.8t + 279.5)\, dt = -1.493t^3 + 34.9t^2 + 279.5t + C$. Using the condition

$A(0) = 2917$, we find $A(0) = C = 2917$. Therefore, $A(t) = -1.493t^3 + 34.9t^2 + 279.5t + 2917$.

The average credit card debt per U.S. household in 2003 was

$A(13) = -1.493\,(13^3) + 34.9\,(13^2) + 279.5\,(13) + 2917 \approx 9168.479$, or approximately \$9168.

77. a. The number of gastric bypass surgeries performed in year t is

$N(t) = \int R(t)\, dt = \int (9.399t^2 - 13.4t + 14.07)\, dt = 3.133t^3 - 6.7t^2 + 14.07t + C$. Using the condition

$N(0) = 36.7$, we find $N(0) = C = 36.7$. Therefore, $N(t) = 3.133t^3 - 6.7t^2 + 14.07t + 36.7$.

b. The number of bypass surgeries performed in 2003 was

$N(3) = 3.133\,(3^3) - 6.7\,(3^2) + 14.07\,(3) + 36.7 = 103{,}201$, or approximately 103,201.

79. a. We have the initial-value problem $C'(t) = 12.288t^2 - 150.5594t + 695.23$ with $C(0) = 3142$. Integrating, we

find $C(t) = \int C'(t)\, dt = \int (12.288t^2 - 150.5594t + 695.23)\, dt = 4.096t^3 - 75.2797t^2 + 695.23t + k$.

Using the initial condition, we find $C(0) = 0 + k = 3142$, and so $k = 3142$. Therefore,

$C(t) = 4.096t^3 - 75.2797t^2 + 695.23t + 3142$.

b. The projected average out-of-pocket costs for beneficiaries is 2010 is

$C(2) = 4.096\,(2^3) - 75.2797\,(2^2) + 695.23\,(2) + 3142 = 4264.1092$, or \$4264.11.

81. The number of new subscribers at any time is $N(t) = \int (100 + 210t^{3/4})\, dt = 100t + 120t^{7/4} + C$.

The given condition implies that $N(0) = 5000$. Using this condition, we find $C = 5000$.

Therefore, $N(t) = 100t + 120t^{7/4} + 5000$. The number of subscribers 16 months from now is

$N(16) = 100\,(16) + 120\,(16)^{7/4} + 5000$, or 21,960.

83. $h(t) = \int h'(t)\, dt = \int (-3t^2 + 192t + 120)\, dt = -t^3 + 96t^2 + 120t + C = -t^3 + 96t^2 + 120t + C$.

$h(0) = C = 0$ implies $h(t) = -t^3 + 96t^2 + 120t$. The altitude 30 seconds after liftoff is

$h(30) = -30^3 + 96\,(30)^2 + 120\,(30) = 63{,}000$ ft.

85. a. $S(t) = \int R(t)\, dt = \int (3t^3 - 17.9445t^2 + 28.7222t + 26.632)\, dt$

$\qquad = 0.75t^4 - 5.9815t^3 + 14.3611t^2 + 26.632t + C$.

$S(0) = 108$, so $0 + C + 108$ and $C = 108$. Thus, $S(t) = 0.75t^4 - 5.9815t^3 + 14.3611t^2 + 26.632t + 108$.

b. The total sales of organic milk in 2004 were

$S(5) = 0.75\,(5)^5 - 5.9815\,(5)^3 + 14.3611\,(5)^2 + 26.632\,(5) + 108 = 321.25$, or \$321.25 million.

87. a. The number of health-care agencies in year t is $N(t) = \int -0.186te^{-0.02t}\, dt = 9.3e^{-0.02t} + C$. Using the

condition $N(0) = 9.3$, we find $N(0) = 9.3 + C = 9.3$, so $C = 0$. Therefore, $N(t) = 9.3e^{-0.02t}$.

b. The number of health-care agencies in 2002 is $N(14) = 9.3e^{-0.02(14)} \approx 7.03$, or 7030.

The number of health-care agencies in 2005 is $N(17) = 9.3e^{-0.02(17)} \approx 6.62$, or 6620.

© 2011 Cengage Learning. All Rights Reserved. May not be scanned, copied or duplicated, or posted to a publicly accessible website, in whole or in part.

89. $v(r) = \int v'(r)\, dr = \int -kr\, dr = -\frac{1}{2}kr^2 + C$. But $v(R) = -\frac{1}{2}kR^2 + C = 0$, so $C = \frac{1}{2}kR^2$. Therefore, $v(R) = -\frac{1}{2}kr^2 + \frac{1}{2}kR^2 = \frac{1}{2}k\left(R^2 - r^2\right)$.

91. Denote the constant deceleration by k. Then $f''(t) = -k$, so $f'(t) = v(t) = -kt + C_1$. Next, the given condition implies that $v(0) = 88$. This gives $C_1 = 88$, so $f'(t) = -kt + 88$. Now $s = f(t) = \int f'(t)\, dt = \int (-kt + 88)\, dt = -\frac{1}{2}kt^2 + 88t + C_2$, and $f(0) = 0$ gives $s = f(t) = -\frac{1}{2}kt^2 + 88t$. Because the car is brought to rest in 9 seconds, we have $v(9) = -9k + 88 = 0$, or $k = \frac{88}{9} \approx 9.78$, so the deceleration is 9.78 ft/sec^2. The distance covered is $s = f(9) = -\frac{1}{2}\left(\frac{88}{9}\right)(81) + 88\,(9) = 396$, so the stopping distance is 396 ft.

93. The time taken by runner A to cross the finish line is $t = \frac{200}{22} = \frac{100}{11}$ sec. Let a be the constant acceleration of runner B as he begins to spurt. Then $\dfrac{dv}{dt} = a$, so the velocity of runner B as he runs towards the finish line is $v = \int a\, dt = at + c$. At $t = 0$, $v = 20$ and so $v = at = 20$. Now $\frac{ds}{dt} = v = at + 20$, so $s = \int (at + 20)\, dt = \frac{1}{2}at^2 + 20t + k$, where k is the constant of integration. Next, $s(0) = 0$ gives $s = \frac{1}{2}at^2 + 20t = \left(\frac{1}{2}at + 20\right)t$. In order for runner B to cover 220 ft in $\frac{100}{11}$ sec, we must have $\left[\frac{1}{2}a\left(\frac{100}{11}\right) + 20\right]\frac{100}{11} = 220$, so $\frac{50}{11}a + 20 = \frac{220 \cdot 11}{100} = \frac{121}{5}$, $\frac{50}{11}a = \frac{121}{5} - 20 = \frac{21}{5}$, and $a = \frac{21}{5} \cdot \frac{11}{50} = 0.924$ ft/sec^2. Therefore, runner B must have an acceleration of at least 0.924 ft/sec^2.

95. a. We have the initial-value problem $R'(t) = \dfrac{8}{(t+4)^2}$ with $R(0) = 0$. Integrating, we find $R(t) = \int \dfrac{8}{(t+4)^2}\, dt = 8\int (t+4)^{-2}\, dt = -\dfrac{8}{t+4} + C$. $R(0) = 0$ implies $-\frac{8}{4} + C = 0$, so $C = 2$. Therefore, $R(t) = -\dfrac{8}{t+4} + 2 = \dfrac{-8 + 2t + 8}{t+4} = \dfrac{2t}{t+4}$.

b. After 1 hour, $R(1) = \frac{2}{5} = 0.4$, so 0.4 inches had fallen. After 2 hours, $R(2) = \frac{4}{6} = \frac{2}{3}$, so $\frac{2}{3}$ inch had fallen.

97. True. See the proof in Section 6.1 of the text.

99. True. Use the Sum Rule followed by the Constant Multiple Rule.

6.2 Problem-Solving Tips

1. Here are some tips for using the method of substitution.

    a. The idea is to replace the given integral by a simpler integral, so look for a substitution $u = g(x)$ that simplifies the integral.

    b. Check to see that $du = g'(x)\, dx$ appears in the integral.

2. Look through Problems 1−50 to familiarize yourself with the types of functions that can be integrated using the method of substitution. Even if you don't complete every problem, check to see if you can set up the given integral so that the method of substitution can be used to complete the integration.

© 2011 Cengage Learning. All Rights Reserved. May not be scanned, copied or duplicated, or posted to a publicly accessible website, in whole or in part.

6.2   Concept Questions   page 418

1. To find $I = \int f(g(x))g'(x)\,dx$ by the Method of Substitution, let $u = g(x)$, so that $du = g'(x)\,dx$. Making the substitution, we obtain $I = \int f(u)\,du$, which can be integrated with respect to u. Finally, replace u by $u = g(x)$ to evaluate the integral.

6.2   Integration by Substitution   page 418

1. Put $u = 4x + 3$, so that $du = 4\,dx$ and $dx = \frac{1}{4}\,du$. Then
$$\int 4(4x+3)^4\,dx = \int u^4\,du = \frac{1}{5}u^5 + C = \frac{1}{5}(4x+3)^5 + C.$$

3. Let $u = x^3 - 2x$, so that $du = (3x^2 - 2)\,dx$. Then
$$\int (x^3 - 2x)^2 (3x^2 - 2)\,dx = \int u^2\,du = \frac{1}{3}u^3 + C = \frac{1}{3}(x^3 - 2x)^3 + C.$$

5. Let $u = 2x^2 + 3$, so that $du = 4x\,dx$. Then
$$\int \frac{4x}{(2x^2+3)^3}\,dx = \int \frac{1}{u^3}\,du = \int u^{-3}\,du = -\frac{1}{2}u^{-2} + C = -\frac{1}{2(2x^2+3)^2} + C.$$

7. Put $u = t^3 + 2$, so that $du = 3t^2\,dt$ and $t^2\,dt = \frac{1}{3}\,du$. Then
$$\int 3t^2\sqrt{t^3+2}\,dt = \int u^{1/2}\,du = \frac{2}{3}u^{3/2} + C = \frac{2}{3}(t^3+2)^{3/2} + C.$$

9. Let $u = x^2 - 1$, so that $du = 2x\,dx$ and $x\,dx = \frac{1}{2}\,du$. Then
$$\int (x^2 - 1)^9\,x\,dx = \int \frac{1}{2}u^9\,du = \frac{1}{20}u^{10} + C = \frac{1}{20}(x^2 - 1)^{10} + C.$$

11. Let $u = 1 - x^5$, so that $du = -5x^4\,dx$ and $x^4\,dx = -\frac{1}{5}\,du$. Then
$$\int \frac{x^4}{1-x^5}\,dx = -\frac{1}{5}\int \frac{du}{u} = -\frac{1}{5}\ln|u| + C = -\frac{1}{5}\ln|1 - x^5| + C.$$

13. Let $u = x - 2$, so that $du = dx$. Then $\displaystyle\int \frac{2}{x-2}\,dx = 2\int \frac{du}{u} = 2\ln|u| + C = \ln u^2 + C = \ln(x-2)^2 + C.$

15. Let $u = 0.3x^2 - 0.4x + 2$. Then $du = (0.6x - 0.4)\,dx = 2(0.3x - 0.2)\,dx$. Thus,
$$\int \frac{0.3x - 0.2}{0.3x^2 - 0.4x + 2}\,dx = \int \frac{1}{2u}\,du = \frac{1}{2}\ln|u| + C = \frac{1}{2}\ln(0.3x^2 - 0.4x + 2) + C.$$

17. Let $u = 3x^2 - 1$, so that $du = 6x\,dx$ and $x\,dx = \frac{1}{6}\,du$. Then
$$\int \frac{x}{3x^2 - 1}\,dx = \frac{1}{6}\int \frac{du}{u} = \frac{1}{6}\ln|u| + C = \frac{1}{6}\ln|3x^2 - 1| + C.$$

19. Let $u = -2x$, so that $du = -2\,dx$ and $dx = -\frac{1}{2}\,du$. Then $\int e^{-2x}\,dx = -\frac{1}{2}\int e^u\,du = -\frac{1}{2}e^u + C = -\frac{1}{2}e^{-2x} + C.$

21. Let $u = 2 - x$, so that $du = -dx$ and $dx = -du$. Then $\int e^{2-x}\,dx = -\int e^u\,du = -e^u + C = -e^{2-x} + C.$

23. Let $u = -x^2$, so that $du = -2x\,dx$ and $x\,dx = -\frac{1}{2}\,du$. Then $\int xe^{-x^2}\,dx = \int -\frac{1}{2}e^u\,du = -\frac{1}{2}e^u + C = -\frac{1}{2}e^{-x^2} + C.$

© 2011 Cengage Learning. All Rights Reserved. May not be scanned, copied or duplicated, or posted to a publicly accessible website, in whole or in part.

25. $\int \left(e^x - e^{-x} \right) dx = \int e^x \, dx - \int e^{-x} \, dx = e^x - \int e^{-x} \, dx$. To evaluate the second integral on the right, let $u = -x$ so that $du = -dx$ and $dx = -du$. Then $\int \left(e^x - e^{-x} \right) dx = e^x + \int e^u \, du = e^x + e^u + C = e^x + e^{-x} + C$.

27. Let $u = 1 + e^x$, so that $du = e^x \, dx$. Then $\displaystyle\int \frac{e^x}{1 + e^x} \, dx = \int \frac{du}{u} = \ln |u| + C = \ln \left(1 + e^x \right) + C$.

29. Let $u = \sqrt{x} = x^{1/2}$. Then $du = \frac{1}{2} x^{-1/2} \, dx$ and $2 \, du = x^{-1/2} \, dx$, so
$$\int \frac{e^{\sqrt{x}}}{\sqrt{x}} \, dx = \int 2 e^u \, du = 2 e^u + C = 2 e^{\sqrt{x}} + C.$$

31. Let $u = e^{3x} + x^3$, so that $du = \left(3 e^{3x} + 3 x^2 \right) dx = 3 \left(e^{3x} + x^2 \right) dx$ and $\left(e^{3x} + x^2 \right) dx = \frac{1}{3} du$. Then
$$\int \frac{e^{3x} + x^2}{\left(e^{3x} + x^3 \right)^3} \, dx = \frac{1}{3} \int \frac{du}{u^3} = \frac{1}{3} \int u^{-3} \, du = -\frac{u^{-2}}{6} + C = -\frac{1}{6 \left(e^{3x} + x^3 \right)^2} + C.$$

33. Let $u = e^{2x} + 1$, so that $du = 2 e^{2x} \, dx$ and $\frac{1}{2} du = e^{2x} \, dx$. Then
$\int e^{2x} \left(e^{2x} + 1 \right)^3 dx = \int \frac{1}{2} u^3 \, du = \frac{1}{8} u^4 + C = \frac{1}{8} \left(e^{2x} + 1 \right)^4 + C$.

35. Let $u = \ln 5x$, so that $du = \frac{1}{x} \, dx$. Then $\displaystyle\int \frac{\ln 5x}{x} \, dx = \int u \, du = \frac{1}{2} u^2 + C = \frac{1}{2} \left(\ln 5x \right)^2 + C$.

37. Let $u = \ln x$, so that $du = \frac{1}{x} \, dx$. Then $\displaystyle\int \frac{1}{x \ln x} \, dx = \int \frac{du}{u} = \ln |u| + C = \ln |\ln x| + C$.

39. Let $u = \ln x$, so that $du = \frac{1}{x} \, dx$. Then $\displaystyle\int \frac{\sqrt{\ln x}}{x} \, dx = \int \sqrt{u} \, du = \frac{2}{3} u^{3/2} + C = \frac{2}{3} \left(\ln x \right)^{3/2} + C$.

41. $\displaystyle\int \left(x e^{x^2} - \frac{x}{x^2 + 2} \right) dx = \int x e^{x^2} \, dx - \int \frac{x}{x^2 + 2} \, dx$. To evaluate the first integral, let $u = x^2$, so that $du = 2x \, dx$ and $x \, dx = \frac{1}{2} du$. Then $\int x e^{x^2} \, dx = \frac{1}{2} \int e^u \, du + C_1 = \frac{1}{2} e^u + C_1 = \frac{1}{2} e^{x^2} + C_1$.
To evaluate the second integral, let $u = x^2 + 2$, so that $du = 2x \, dx$ and $x \, dx = \frac{1}{2} du$.
Then $\displaystyle\int \frac{x}{x^2 + 2} \, dx = \frac{1}{2} \int \frac{du}{u} = \frac{1}{2} \ln |u| + C_2 = \frac{1}{2} \ln \left(x^2 + 2 \right) + C_2$. Therefore,
$\displaystyle\int \left(x e^{x^2} - \frac{x}{x^2 + 2} \right) dx = \frac{1}{2} e^{x^2} - \frac{1}{2} \ln \left(x^2 + 2 \right) + C$.

43. Let $u = \sqrt{x} - 1$, so that $du = \frac{1}{2} x^{-1/2} \, dx = \frac{1}{2\sqrt{x}} \, dx$ and $dx = 2\sqrt{x} \, du$. Also, we have $\sqrt{x} = u + 1$, so that $x = (u + 1)^2 = u^2 + 2u + 1$ and $dx = 2(u + 1) \, du$. Thus,
$$\int \frac{x + 1}{\sqrt{x} - 1} \, dx = \int \frac{u^2 + 2u + 2}{u} \cdot 2(u + 1) \, du = 2 \int \frac{\left(u^3 + 3u^2 + 4u + 2 \right)}{u} \, du$$
$$= 2 \int \left(u^2 + 3u + 4 + \frac{2}{u} \right) du = 2 \left(\frac{1}{3} u^3 + \frac{3}{2} u^2 + 4u + 2 \ln |u| \right) + C$$
$$= 2 \left[\frac{1}{3} \left(\sqrt{x} - 1 \right)^3 + \frac{3}{2} \left(\sqrt{x} - 1 \right)^2 + 4 \left(\sqrt{x} - 1 \right) + 2 \ln \left| \sqrt{x} - 1 \right| \right] + C.$$

© 2011 Cengage Learning. All Rights Reserved. May not be scanned, copied or duplicated, or posted to a publicly accessible website, in whole or in part.

45. Let $u = x - 1$, so that $du = dx$. Also, $x = u + 1$, and so

$$\int x \, (x - 1)^5 \, dx = \int (u + 1) \, u^5 \, du = \int (u^6 + u^5) \, du = \tfrac{1}{7} u^7 + \tfrac{1}{6} u^6 + C = \tfrac{1}{7} (x - 1)^7 + \tfrac{1}{6} (x - 1)^6 + C$$

$$= \frac{(6x + 1) \, (x - 1)^6}{42} + C.$$

47. Let $u = 1 + \sqrt{x}$, so that $du = \tfrac{1}{2} x^{-1/2} \, dx$ and $dx = 2\sqrt{x} = 2 (u - 1) \, du$. Then

$$\int \frac{1 - \sqrt{x}}{1 + \sqrt{x}} dx = \int \left(\frac{1 - (u - 1)}{u} \right) \cdot 2 (u - 1) \, du = 2 \int \frac{(2 - u) \, (u - 1)}{u} \, du = 2 \int \frac{-u^2 + 3u - 2}{u} \, du$$

$$= 2 \int \left(-u + 3 - \frac{2}{u} \right) du = -u^2 + 6u - 4 \ln |u| + C$$

$$= - \left(1 + \sqrt{x} \right)^2 + 6 \left(1 + \sqrt{x} \right) - 4 \ln \left(1 + \sqrt{x} \right) + C$$

$$= -1 - 2\sqrt{x} - x + 6 + 6\sqrt{x} - 4 \ln \left(1 + \sqrt{x} \right) + C = -x + 4\sqrt{x} + 5 - 4 \ln \left(1 + \sqrt{x} \right) + C.$$

49. $I = \int v^2 \, (1 - v)^6 \, dv$. Let $u = 1 - v$, so $du = -dv$. Also, $1 - u = v$, and so $(1 - u)^2 = v^2$. Therefore,

$$I = \int - (1 - 2u + u^2) \, u^6 du = \int - (u^6 - 2u^7 + u^8) \, du = - \left(\tfrac{1}{7} u^7 - \tfrac{1}{4} u^8 + \tfrac{1}{9} u^9 \right) + C$$

$$= -u^7 \left(\tfrac{1}{7} - \tfrac{1}{4} u + \tfrac{1}{9} u^2 \right) + C = -\tfrac{1}{252} (1 - v)^7 \left[36 - 63 (1 - v) + 28 (1 - 2v + v^2) \right]$$

$$= -\tfrac{1}{252} (1 - v)^7 \left[36 - 63 + 63v + 28 - 56v + 28v^2 \right] = -\tfrac{1}{252} (1 - v)^7 \left(28v^2 + 7v + 1 \right) + C.$$

51. $f (x) = \int f' (x) \, dx = 5 \int (2x - 1)^4 \, dx$. Let $u = 2x - 1$, so that $du = 2 \, dx$ and $dx = \tfrac{1}{2} \, du$. Then
$f (x) = \tfrac{5}{2} \int u^4 \, du = \tfrac{1}{2} u^5 + C = \tfrac{1}{2} (2x - 1)^5 + C$. Next, $f (1) = 3$ implies $\tfrac{1}{2} + C = 3$, so $C = \tfrac{5}{2}$. Therefore,
$f (x) = \tfrac{1}{2} (2x - 1)^5 + \tfrac{5}{2}$.

53. $f (x) = \int -2xe^{-x^2 + 1} \, dx$. Let $u = -x^2 + 1$, so that $du = -2x \, dx$. Then $f (x) = \int e^u du = e^u + C = e^{-x^2 + 1} + C$.
The condition $f (1) = 0$ implies $f (1) = 1 + C = 0$, so $C = -1$. Therefore, $f (x) = e^{-x^2 + 1} - 1$.

55. The number of subscribers at time t is $N (t) = \int R (t) \, dt = \int 3.36 \, (t + 1)^{0.05} \, dt$. Let $u = t + 1$, so that $du = dt$.
Then $N = 3.36 \int u^{0.05} \, du = 3.2 u^{1.05} + C = 3.2 (t + 1)^{1.05} + C$. To find C, use the condition $N (0) = 3.2$ to
calculate $N (0) = 3.2 + C = 3.2$, so $C = 0$. Therefore, $N (t) = 3.2 (t + 1)^{1.05}$. If the projection holds true, then the
number of subscribers at the beginning of 2008 is $N (4) = 3.2 (4 + 1)^{1.05} \approx 17.341$, or 17.341 million.

57. $N' (t) = 2000 (1 + 0.2t)^{-3/2}$. Let $u = 1 + 0.2t$, so $du = 0.2 \, dt$ and $5 \, du = dt$. Then
$N (t) = (5) (2000) \int u^{-3/2} \, du = -20{,}000 u^{-1/2} + C = -20{,}000 (1 + 0.2t)^{-1/2} + C$. Next,
$N (0) = -20{,}000 (1)^{-1/2} + C = 1000$. Therefore, $C = 21{,}000$ and $N (t) = -\dfrac{20{,}000}{\sqrt{1 + 0.2t}} + 21{,}000$. In particular,

$N (5) = -\frac{20{,}000}{\sqrt{2}} + 21{,}000 \approx 6858$.

59. $p (x) = -\displaystyle\int \frac{250x}{(16 + x^2)^{3/2}} \, dx = -250 \int \frac{x}{(16 + x^2)^{3/2}} \, dx$. Let $u = 16 + x^2$, so that $du = 2x \, dx$ and $x \, dx = \tfrac{1}{2} \, du$.

Then $p (x) = -\tfrac{250}{2} \int u^{-3/2} \, du = (-125) (-2) u^{-1/2} + C = \dfrac{250}{\sqrt{16 + x^2}} + C$. $p (3) = \dfrac{250}{\sqrt{16 + 9}} + C = 50$ implies

$C = 0$, and so $p (x) = \dfrac{250}{\sqrt{16 + x^2}}$.

© 2011 Cengage Learning. All Rights Reserved. May not be scanned, copied or duplicated, or posted to a publicly accessible website, in whole or in part.

61. Let $u = 2t + 4$, so that $du = 2\,dt$. Then $r(t) = \int \dfrac{30}{\sqrt{2t+4}}\,dt = 30\int \frac{1}{2}u^{-1/2}\,du = 30u^{1/2} + C = 30\sqrt{2t+4} + C$.

$r(0) = 60 + C = 0$, so $C = -60$. Therefore, $r(t) = 30\left(\sqrt{2t+4} - 2\right)$. Then $r(16) = 30\left(\sqrt{36} - 2\right) = 120$ ft, so the polluted area is $\pi r^2 = \pi (120)^2 = 14{,}400\pi$, or $14{,}400\pi$ ft^2.

63. Let $u = 1 + 2.449e^{-0.3277t}$, so that $du = -0.802537e^{-0.3277t}\,dt$ and $e^{-0.3277t}\,dt = -1.24605\,du$. Then

$h(t) = \int \dfrac{52.8706e^{-0.3277t}}{\left(1 + 2.449e^{-0.3277t}\right)^2}\,dt = 52.8706\,(-1.24605)\int \dfrac{du}{u^2} = 65.8794u^{-1} + C = \dfrac{65.8794}{1 + 2.449e^{-0.3277t}} + C$.

$h(0) = \dfrac{65.8794}{1 + 2.449} + C = 19.4$, so $h(t) = \dfrac{65.8794}{1 + 2.449e^{-0.3277t}} + 0.3$ and hence

$h(8) = \dfrac{65.8794}{1 + 2.449e^{-0.3277(8)}} + 0.3 \approx 56.22$, or 56.22 inches.

65. $A(t) = \int A'(t)\,dt = r\int e^{-at}\,dt$. Let $u = -at$, so that $du = -a\,dt$ and $dt = -\frac{1}{a}\,du$. Then

$A(t) = r\left(-\frac{1}{a}\right)\int e^u\,du = -\frac{r}{a}e^u + C = -\frac{r}{a}e^{-at} + C$. $A(0) = 0$ implies $-\frac{r}{a} + C = 0$, so $C = \frac{r}{a}$. Therefore,

$A(t) = -\frac{r}{a}e^{-at} + \frac{r}{a} = \frac{r}{a}\left(1 - e^{-at}\right)$.

67. True. Let $I = \int xf\left(x^2\right)dx$ and put $u = x^2$. Then $du = 2x\,dx$ and $x\,dx = \frac{1}{2}\,du$, so

$I = \frac{1}{2}\int f(u)\,du = \frac{1}{2}\int f(x)\,dx$.

6.3   Problem-Solving Tips

1. In Sections 6.1 and 6.2, we found the indefinite integral of a function, that is, $\int f(x)\,dx = F(x) + C$. Note that our answer is a *family of functions* $F(x) + C$ for which $F'(x) = f(x)$. In this section, we found the definite integral of a function, $\int_a^b f(x)\,dx = \lim\limits_{n \to \infty}\left[f(x_1)\,\Delta x + f(x_2)\,\Delta x + \cdots + f(x_n)\,\Delta x\right]$. Note that the answer here is a *number*.

2. The geometric interpretation of a definite integral is as follows: If f is continuous on $[a, b]$, then $\int_a^b f(x)\,dx$ is equal to the area of the region above the x-axis between the x-axis and the graph of f over $[a, b]$ minus the area of the corresponding region below the x-axis.

6.3   Concept Questions   page 428

1. See page 425 in the text.

6.3   Area and the Definite Integral   page 428

1. $\frac{1}{3}(1.9 + 1.5 + 1.8 + 2.4 + 2.7 + 2.5) = \frac{12.8}{3} \approx 4.27$.

© 2011 Cengage Learning. All Rights Reserved. May not be scanned, copied or duplicated, or posted to a publicly accessible website, in whole or in part.

3. a.

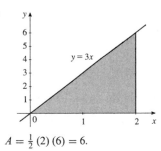

$A = \frac{1}{2}(2)(6) = 6.$

b. $\Delta x = \frac{2}{4} = \frac{1}{2}$, so $x_1 = 0$, $x_2 = \frac{1}{2}$, $x_3 = 1$, $x_4 = \frac{3}{2}$. Thus,

$$A \approx \frac{1}{2}\left[3(0) + 3\left(\frac{1}{2}\right) + 3(1) + 3\left(\frac{3}{2}\right)\right] = \frac{9}{2} = 4.5.$$

c. $\Delta x = \frac{2}{8} = \frac{1}{4}$, so $x_1 = 0, \ldots, x_8 = \frac{7}{4}$. Thus,

$$A \approx \frac{1}{4}\left[3(0) + 3\left(\frac{1}{4}\right) + 3\left(\frac{1}{2}\right) + 3\left(\frac{3}{4}\right)\right.$$
$$\left. + 3(1) + 3\left(\frac{5}{4}\right) + 3\left(\frac{3}{2}\right) + 3\left(\frac{7}{4}\right)\right]$$
$$= \frac{21}{4} = 5.25.$$

d. Yes.

5. a.

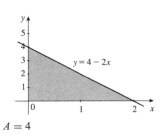

$A = 4$

b. $\Delta x = \frac{2}{5} = 0.4$, so $x_1 = 0$, $x_2 = 0.4$, $x_3 = 0.8$, $x_4 = 1.2$, $x_5 = 1.6$. Thus,

$$A \approx 0.4\,\{[4 - 2(0)] + [4 - 2(0.4)] + [4 - 2(0.8)]$$
$$+ [4 - 2(1.2)] + [4 - 2(1.6)]\} = 4.8.$$

c. $\Delta x = \frac{2}{10} = 0.2$, so $x_1 = 0$, $x_2 = 0.2$, $x_3 = 0.4, \ldots, x_{10} = 1.8$. Thus,

$$A \approx 0.2\,\{[4 - 2(0)] + [4 - 2(0.2)] + [4 - 2(0.4)] +$$
$$\cdots + [4 - 2(1.8)]\} = 4.4.$$

d. Yes.

7. a. $\Delta x = \frac{4-2}{2} = 1$, so $x_1 = 2.5$, $x_2 = 3.5$. The Riemann sum is $1\left(2.5^2 + 3.5^2\right) = 18.5$.

b. $\Delta x = \frac{4-2}{5} = 0.4$, so $x_1 = 2.2$, $x_2 = 2.6$, $x_3 = 3.0$, $x_4 = 3.4$, $x_5 = 3.8$. The Riemann sum is
$0.4\left(2.2^2 + 2.6^2 + 3.0^2 + 3.4^2 + 3.8^2\right) = 18.64$.

c. $\Delta x = \frac{4-2}{10} = 0.2$, so $x_1 = 2.1$, $x_2 = 2.3$, $x_2 = 2.5, \ldots, x_{10} = 3.9$. The Riemann sum is
$0.2\left(2.1^2 + 2.3^2 + 2.5^2 + \cdots + 3.9^2\right)] = 18.66$.

d. The area appears to be $18\frac{2}{3}$.

9. a. $\Delta x = \frac{4-2}{2} = 1$, so $x_1 = 3$, $x_2 = 4$. The Riemann sum is $(1)\left(3^2 + 4^2\right) = 25$.

b. $\Delta x = \frac{4-2}{5} = 0.4$, so $x_1 = 2.4$, $x_2 = 2.8$, $x_3 = 3.2$, $x_4 = 3.6$, $x_5 = 4$. The Riemann sum is
$0.4\left(2.4^2 + 2.8^2 + \cdots + 4^2\right) = 21.12$.

c. $\Delta x = \frac{4-2}{10} = 0.2$, so $x_1 = 2.2$, $x_2 = 2.4$, $x_3 = 2.6, \ldots, x_{10} = 4$. The Riemann sum is
$0.2\left(2.2^2 + 2.4^2 + 2.6^2 + \cdots + 4^2\right) = 19.88$.

d. The area appears to be 19.9.

11. a. $\Delta x = \frac{1}{2}$, so $x_1 = 0$, $x_2 = \frac{1}{2}$. The Riemann sum is $f(x_1)\Delta x + f(x_2)\Delta x = \left[(0)^3 + \left(\frac{1}{2}\right)^3\right]\frac{1}{2} = \frac{1}{16} = 0.0625$.

b. $\Delta x = \frac{1}{5}$, so $x_1 = 0$, $x_2 = \frac{1}{5}$, $x_3 = \frac{2}{5}$, $x_4 = \frac{3}{5}$, $x_5 = \frac{4}{5}$. The Riemann sum is

$$f(x_1)\Delta x + f(x_2)\Delta x + \cdots + f(x_5)\Delta x = \left[\left(\frac{1}{5}\right)^3 + \left(\frac{2}{5}\right)^3 + \cdots + \left(\frac{4}{5}\right)^3\right]\frac{1}{5} = \frac{100}{625} = 0.16.$$

© 2011 Cengage Learning. All Rights Reserved. May not be scanned, copied or duplicated, or posted to a publicly accessible website, in whole or in part.

c. $\Delta x = \frac{1}{10}$, so $x_1 = 0$, $x_2 = \frac{1}{10}$, $x_3 = \frac{2}{10}$, ..., $x_{10} = \frac{9}{10}$. The Riemann sum is

$$f(x_1)\,\Delta x + f(x_2)\,\Delta x + \cdots + f(x_{10})\,\Delta x = \left[\left(\tfrac{1}{10}\right)^3 + \left(\tfrac{2}{10}\right)^3 + \cdots + \left(\tfrac{9}{10}\right)^3\right]\tfrac{1}{10} = \frac{2025}{10,000} = 0.2025$$

$$\approx 0.2.$$

d. The Riemann sums seem to approach 0.2.

13. $\Delta x = \frac{2-0}{5} = \frac{2}{5}$, so $x_1 = \frac{1}{5}$, $x_2 = \frac{3}{5}$, $x_3 = \frac{5}{5}$, $x_4 = \frac{7}{5}$, $x_5 = \frac{9}{5}$. Thus,

$$A \approx \left\{\left[\left(\tfrac{1}{5}\right)^2 + 1\right] + \left[\left(\tfrac{3}{5}\right)^2 + 1\right] + \left[\left(\tfrac{5}{5}\right)^2 + 1\right] + \left[\left(\tfrac{7}{5}\right)^2 + 1\right] + \left[\left(\tfrac{9}{5}\right)^2 + 1\right]\right\}\left(\tfrac{2}{5}\right) = \frac{580}{125} = 4.64.$$

15. $\Delta x = \frac{3-1}{4} = \frac{1}{2}$, so $x_1 = \frac{3}{2}$, $x_2 = \frac{4}{2} = 2$, $x_3 = \frac{5}{2}$, $x_4 = 3$. Thus, $A \approx \left(\frac{1}{3/2} + \frac{1}{2} + \frac{1}{5/2} + \frac{1}{3}\right)\frac{1}{2} \approx 0.95$.

17. $A \approx 20\left[f(10) + f(30) + f(50) + f(70) + f(90)\right] = 20(80 + 100 + 110 + 100 + 80) = 9400\ \text{ft}^2$.

6.4 **Concept Questions**   page 438

1. See the fundamental theorem of calculus on page 430 of the text.

6.4 **The Fundamental Theorem of Calculus**   page 438

1. $A = \int_1^4 2\,dx = 2x\big|_1^4 = 2(4-1) = 6$. The region is a rectangle with area $3 \cdot 2 = 6$.

3. $A = \int_1^3 2x\,dx = x^2\big|_1^3 = 9 - 1 = 8$. The region is a parallelogram with area $\frac{1}{2}(3-1)(2+6) = 8$.

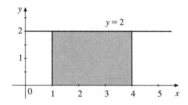

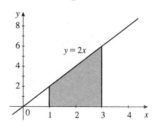

5. $A = \int_{-1}^2 (2x + 3)\,dx = (x^2 + 3x)\big|_{-1}^2 = (4+6) - (1-3) = 12$.

7. $A = \int_{-1}^2 (-x^2 + 4)\,dx = \left(-\frac{1}{3}x^3 + 4x\right)\Big|_{-1}^2 = \left(-\frac{8}{3} + 8\right) - \left(\frac{1}{3} - 4\right) = 9$.

9. $A = \int_1^2 \frac{1}{x}\,dx = \ln|x|\big|_1^2 = \ln 2 - \ln 1 = \ln 2$.

11. $A = \int_1^9 \sqrt{x}\,dx = \frac{2}{3}x^{3/2}\Big|_1^9 = \frac{2}{3}(27-1) = \frac{52}{3}$.

13. $A = \int_{-8}^{-1} (1 - x^{1/3})\,dx = \left(x - \frac{3}{4}x^{4/3}\right)\Big|_{-8}^{-1} = \left(-1 - \frac{3}{4}\right) - (-8 - 12) = \frac{73}{4}$.

© 2011 Cengage Learning. All Rights Reserved. May not be scanned, copied or duplicated, or posted to a publicly accessible website, in whole or in part.

15. $A = \int_0^2 e^x \, dx = e^x|_0^2 = e^2 - 1 \approx 6.39$.

17. $\int_2^4 3 \, dx = 3x|_2^4 = 3\,(4 - 2) = 6$.

19. $\int_1^3 (2x + 3) \, dx = (x^2 + 3x)\big|_1^3 = (9 + 9) - (1 + 3) = 14$.

21. $\int_{-1}^3 2x^2 \, dx = \frac{2}{3}x^3 \Big|_{-1}^3 = \frac{2}{3}\,(27) - \frac{2}{3}\,(-1) = \frac{56}{3}$.

23. $\int_{-2}^2 (x^2 - 1) dx = \left(\frac{1}{3}x^3 - x\right)\Big|_{-2}^2 = \left(\frac{8}{3} - 2\right) - \left(-\frac{8}{3} + 2\right) = \frac{4}{3}$.

25. $\int_1^8 4x^{1/3} \, dx = 4 \cdot \frac{3}{4}x^{4/3}\Big|_1^8 = 3\,(16 - 1) = 45$.

27. $\int_0^1 (x^3 - 2x^2 + 1) \, dx = \left(\frac{1}{4}x^4 - \frac{2}{3}x^3 + x\right)\Big|_0^1 = \frac{1}{4} - \frac{2}{3} + 1 = \frac{7}{12}$.

29. $\int_2^4 \frac{1}{x} \, dx = \ln|x| |_2^4 = \ln 4 - \ln 2 = \ln \frac{4}{2} = \ln 2$.

31. $\int_0^4 x\,(x^2 - 1) \, dx = \int_0^4 (x^3 - x) \, dx = \left(\frac{1}{4}x^4 - \frac{1}{2}x^2\right)\Big|_0^4 = 64 - 8 = 56$.

33. $\int_1^3 (t^2 - t)^2 \, dt = \int_1^3 (t^4 - 2t^3 + t^2) \, dt = \left(\frac{1}{5}t^5 - \frac{1}{2}t^4 + \frac{1}{3}t^3\right)\Big|_1^3 = \left(\frac{243}{5} - \frac{81}{2} + \frac{27}{3}\right) - \left(\frac{1}{5} - \frac{1}{2} + \frac{1}{3}\right) = \frac{512}{30} = \frac{256}{15}$.

35. $\int_{-3}^{-1} x^{-2} dx = -\frac{1}{x}\Big|_{-3}^{-1} = 1 - \frac{1}{3} = \frac{2}{3}$.

37. $\int_1^4 \left(\sqrt{x} - \frac{1}{\sqrt{x}}\right) dx = \int_1^4 (x^{1/2} - x^{-1/2}) \, dx = \left(\frac{2}{3}x^{3/2} - 2x^{1/2}\right)\Big|_1^4 = \left(\frac{16}{3} - 4\right) - \left(\frac{2}{3} - 2\right) = \frac{8}{3}$.

39. $\int_1^4 \frac{3x^3 - 2x^2 + 4}{x^2} \, dx = \int_1^4 (3x - 2 + 4x^{-2}) \, dx = \left(\frac{3}{2}x^2 - 2x - \frac{4}{x}\right)\Big|_1^4$

$\qquad = \left(24 - 8 - 1\right) - \left(\frac{3}{2} - 2 - 40\right) = \frac{39}{2}$.

41. a. $C\,(300) - C\,(0) = \int_0^{300} (0.0003x^2 - 0.12x + 20) \, dx = (0.0001x^3 - 0.06x^2 + 20x)|_0^{300}$

$\qquad = 0.0001\,(300)^3 - 0.06\,(300)^2 + 20\,(300) = 3300.$

Therefore, $C\,(300) = 3300 + C\,(0) = 3300 + 800 = \$4100$.

b. $\int_{200}^{300} C'\,(x) \, dx = (0.0001x^3 - 0.06x^2 + 20x))|_{200}^{300}$

$\qquad = [0.0001\,(300)^3 - 0.06\,(300)^2 + 20\,(300)] - [0.0001\,(200)^3 - 0.06\,(200)^2 + 20\,(200)] = \$900.$

43. a. The profit is

$\int_0^{200} (-0.0003x^2 + 0.02x + 20) \, dx + P\,(0) = (-0.0001x^3 + 0.01x^2 + 20x)|_0^{200} + P\,(0)$

$\qquad = 3600 + P\,(0) = 3600 - 800, \text{ or } \$2800.$

b. $\int_{200}^{220} P'\,(x) \, dx = P\,(220) - P\,(200) = (-0.0001x^3 + 0.01x^2 + 20x)|_{200}^{220} = 219.20, \text{ or } \$219.20.$

© 2011 Cengage Learning. All Rights Reserved. May not be scanned, copied or duplicated, or posted to a publicly accessible website, in whole or in part.

45. a. $f(t) = \int R(t)\,dt = \int 0.8256t^{-0.04}\,dt = \frac{0.8256}{0.96}t^{0.96} + C = 0.86t^{0.96} + C.$ $f(1) = 0.9$, and so $0.86 + C = 0.9$
   and $C = 0.04$. Thus, $f(t) = 0.86t^{0.96} + 0.04.$

b. In 2012, mobile phone ad spending is projected to be $f(6) = 0.86\left(6^{0.96}\right) + 0.04 \approx 4.84$, or approximately
   $4.84 billion.

47. The distance is $\int_0^{20} v(t)\,dt = \int_0^{20}\left(-t^2 + 20t + 440\right) dt = \left(-\frac{1}{3}t^3 + 10t^2 + 440t\right)\Big|_0^{20} \approx 10{,}133.3$ ft.

49. a. The percentage of these households in decade t is
   $$P(t) = \int R(t)\,dt = \int \left(0.8499t^2 - 3.872t + 5\right) dt = 0.2833t^3 - 1.936t^2 + 5t + C.$$ The condition $P(0) = 5.6$
   gives $P(0) = C = 5.6$, so $P(t) = 0.2833t^3 - 1.936t^2 + 5t + 5.6.$

b. The percentage of these households in 2010 is $P(4) = 0.2833\left(4^3\right) - 1.936\left(4^2\right) + 5(4) + 5.6 = 12.7552$, or
   approximately 12.8%.

c. The percentage of these households in 2000 was $P(3) = 0.2833\left(3^3\right) - 1.936\left(3^2\right) + 5(3) + 5.6 = 10.8251.$
   Therefore, the net increase in the percentage of these households from 1970 $(t = 0)$ to 2000 $(t = 3)$ is
   $P(3) - P(0) = 10.825 - 5.6 = 5.225$, or approximately 5.2%.

51. The number of set-top boxes shipped will be
   $\int_0^6 \left(-0.05556t^3 + 0.262t^2 + 17.46t + 63.4\right) dt = \left(-0.01389t^4 + 0.0873t^3 + 8.73t^2 + 63.4t\right)\Big|_0^6 \approx 695.53$, or
   approximately 695.5 million.

53. The increase in the senior population over the period in question is $\displaystyle\int_0^3 f(t)\,dt = \int_0^3 \frac{85}{1 + 1.859e^{-0.66t}}\,dt.$

   Multiplying the numerator and denominator of the integrand by $e^{0.66t}$ gives $\displaystyle\int_0^3 f(t)\,dt = 85\int_0^3 \frac{e^{0.66t}}{e^{0.66t} + 1.859}\,dt.$

   Now let $u = 1.859 + e^{0.66t}$, so $du = 0.66e^{0.66t}\,dt$ and $e^{0.66t}\,dt = \dfrac{du}{0.66}.$ If $t = 0$, then $u = 2.859$, and if $t = 3$, then
   $u = 9.1017.$ Substituting, we have
   $$\int_0^3 f(t)\,dt = 85\int_{2.859}^{9.1017} \frac{du}{0.66u} = \frac{85}{0.66}\ln u\,\Big|_{2.859}^{9.1017} = \frac{85}{0.66}\left(\ln 9.1017 - \ln 2.859\right) = \frac{85}{0.66}\ln\frac{9.1017}{2.859} \approx 149.135,\text{ or}$$
   approximately 149.14 million people.

55. $f(x) = x^4 - 2x^2 + 2$, so $f'(x) = 4x^3 - 4x = 4x\left(x^2 - 1\right) = 4x(x + 1)(x - 1).$ Setting $f'(x) = 0$ gives $x = -1$,
   0, and 1 as critical numbers. Now calculate $f''(x) = 12x^2 - 4 = 4\left(3x^2 - 1\right)$ and use the second derivative test:
   $f''(-1) = 8 > 0$, so $(-1, 1)$ is a relative minimum; $f''(0) = -4 < 0$, so $(0, 2)$ is a relative maximum; and
   $f''(1) = 8 > 0$, so $(1, 1)$ is a relative minimum. The graph of f is symmetric with respect to the y-axis because
   $f(-x) = (-x)^4 - 2(-x)^2 + 2 = x^4 - 2x^2 + 2 = f(x).$ Thus, the required area is the area under the graph of f
   between $x = 0$ and $x = 1$, that is, $A = \int_0^1 \left(x^4 - 2x^2 + 2\right) dx = \left(\frac{1}{5}x^5 - \frac{2}{3}x^3 + 2x\right)\Big|_0^1 = \frac{1}{5} - \frac{2}{3} + 2 = \frac{23}{15}.$

57. False. The integrand $f(x) = 1/x^3$ is discontinuous at $x = 0$.

59. False. $f(x)$ is not nonnegative on $[0, 2]$.

© 2011 Cengage Learning. All Rights Reserved. May not be scanned, copied or duplicated, or posted to a publicly accessible website, in whole or in part.

6.4   Using Technology   page 441

1. 6.1787

3. 0.7873

5. −0.5888

7. 2.7044

9. 3.9973

11. 46%, 24%

13. 333,209

15. 6,723,000

6.5   Concept Questions   page 448

1. *Approach I:* We first find the indefinite integral. Let $u = x^3 + 1$, so that $du = 3x^2dx$ and or
$x^2dx = \frac{1}{3}du$. Then $\int x^2 (x^3 + 1)^2 dx = \frac{1}{3} \int u^2 du = \frac{1}{9}u^3 + C = \frac{1}{9} (x^3 + 1)^3 + C$. Therefore,
$\int_0^1 x^2 (x^3 + 1)^2 dx = \frac{1}{9} (x^3 + 1^3) \Big|_0^1 = \frac{1}{9} (8 - 1) = \frac{7}{9}$.

Approach II: Transform the definite integral in x into an integral in u: Let $u = x^3 + 1$, so that $du = 3x^2 dx$ and
$x^2 dx = \frac{1}{3} du$. Next, find the limits of integration with respect to u. If $x = 0$, then $u = 0^3 + 1 = 1$ and if $x = 1$,
then $u = 1^3 + 1 = 2$. Therefore, $\int_0^1 x^2 (x^3 + 1)^2 dx = \frac{1}{3} \int_1^2 u^2 du = \frac{1}{9}u^3 \Big|_1^2 = \frac{1}{9} (8 - 1) = \frac{7}{9}$.

6.5   Evaluating Definite Integrals   page 448

1. Let $u = x^2 - 1$, so $du = 2x \, dx$ and $x \, dx = \frac{1}{2} du$. If $x = 0$, then $u = -1$ and if $x = 2$, then $u = 3$, so
$\int_0^2 x (x^2 - 1)^3 dx = \frac{1}{2} \int_{-1}^3 u^3 du = \frac{1}{8}u^4 \Big|_{-1}^3 = \frac{1}{8} (81) - \frac{1}{8} (1) = 10$.

3. Let $u = 5x^2 + 4$, so $du = 10x \, dx$ and $x \, dx = \frac{1}{10} du$. If $x = 0$, then $u = 4$ and if $x = 1$, then $u = 9$, so
$\int_0^1 x\sqrt{5x^2 + 4} \, dx = \frac{1}{10} \int_4^9 u^{1/2} du = \frac{1}{15}u^{3/2} \Big|_4^9 = \frac{1}{15} (27) - \frac{1}{15} (8) = \frac{19}{15}$.

5. Let $u = x^3 + 1$, so $du = 3x^2 dx$ and $x^2 dx = \frac{1}{3} du$. If $x = 0$, then $u = 1$ and if $x = 2$, then $u = 9$, so
$\int_0^2 x^2 (x^3 + 1)^{3/2} dx = \frac{1}{3} \int_1^9 u^{3/2} du = \frac{2}{15}u^{5/2} \Big|_1^9 = \frac{2}{15} (243) - \frac{2}{15} (1) = \frac{484}{15}$.

7. Let $u = 2x + 1$, so $du = 2 \, dx$ and $dx = \frac{1}{2} du$. If $x = 0$, then $u = 1$ and if $x = 1$ then $u = 3$, so
$\int_0^1 \frac{1}{\sqrt{2x + 1}} dx = \frac{1}{2} \int_1^3 \frac{1}{\sqrt{u}} du = \frac{1}{2} \int_1^3 u^{-1/2} du = u^{1/2} \Big|_1^3 = \sqrt{3} - 1$.

9. Let $u = 2x - 1$, so $du = 2 \, dx$ and $dx = \frac{1}{2} du$. If $x = 1$, then $u = 1$ and if $x = 2$, then $u = 3$, so
$\int_1^2 (2x - 1)^4 dx = \frac{1}{2} \int_1^3 u^4 du = \frac{1}{10}u^5 \Big|_1^3 = \frac{1}{10} (243 - 1) = \frac{121}{5}$.

11. Let $u = x^3 + 1$, so $du = 3x^2 dx$ and $x^2 dx = \frac{1}{3} du$. If $x = -1$, then $u = 0$ and if $x = 1$, then $u = 2$, so
$\int_{-1}^1 x^2 (x^3 + 1)^4 dx = \frac{1}{3} \int_0^2 u^4 du = \frac{1}{15}u^5 \Big|_0^2 = \frac{32}{15}$.

© 2011 Cengage Learning. All Rights Reserved. May not be scanned, copied or duplicated, or posted to a publicly accessible website, in whole or in part.

13. Let $u = x - 1$, so $du = dx$. If $x = 1$, then $u = 0$ and if $x = 5$, then $u = 4$, so

$$\int_1^5 x\sqrt{x-1}\,dx = \int_0^4 (u+1)\,u^{1/2}\,du = \int_0^4 \left(u^{3/2} + u^{1/2}\right) du = \left(\tfrac{2}{5}u^{5/2} + \tfrac{2}{3}u^{3/2}\right)\Big|_0^4 = \tfrac{2}{5}(32) + \tfrac{2}{3}(8)$$

$$= \tfrac{272}{15}.$$

15. Let $u = x^2$, so $du = 2x\,dx$ and $x\,dx = \tfrac{1}{2}\,du$. If $x = 0$, then $u = 0$ and if $x = 2$, then $u = 4$, so

$$\int_0^2 xe^{x^2}\,dx = \tfrac{1}{2}\int_0^4 e^u\,du = \tfrac{1}{2}e^u\Big|_0^4 = \tfrac{1}{2}\left(e^4 - 1\right).$$

17. $\int_0^1 \left(e^{2x} + x^2 + 1\right) dx = \left(\tfrac{1}{2}e^{2x} + \tfrac{1}{3}x^3 + x\right)\Big|_0^1 = \left(\tfrac{1}{2}e^2 + \tfrac{1}{3} + 1\right) - \tfrac{1}{2} = \tfrac{1}{2}e^2 + \tfrac{5}{6}.$

19. Put $u = x^2 + 1$, so $du = 2x\,dx$ and $x\,dx = \tfrac{1}{2}\,du$. Then $\int_{-1}^1 xe^{x^2+1}\,dx = \tfrac{1}{2}\int_2^2 e^u\,du = \tfrac{1}{2}e^u\Big|_2^2 = 0$ because the upper and lower limits are equal.

21. Let $u = x - 2$, so $du = dx$. If $x = 3$, then $u = 1$ and if $x = 6$, then $u = 4$, so

$$\int_3^6 \frac{2}{x-2}\,dx = 2\int_1^4 \frac{du}{u} = 2\ln|u|\big|_1^4 = 2\ln 4.$$

23. Let $u = x^3 + 3x^2 - 1$, so $du = (3x^2 + 6x)\,dx = 3\left(x^2 + 2x\right) dx$. If $x = 1$, then $u = 3$, and if $x = 2$, then $u = 19$, so $\int_1^2 \dfrac{x^2 + 2x}{x^3 + 3x^2 - 1}\,dx = \dfrac{1}{3}\int_3^{19} \dfrac{du}{u} = \tfrac{1}{3}\ln u\Big|_3^{19} = \tfrac{1}{3}\left(\ln 19 - \ln 3\right).$

25. $\int_1^2 \left(4e^{2u} - \dfrac{1}{u}\right) du = 2e^{2u} - \ln u\big|_1^2 = \left(2e^4 - \ln 2\right) - \left(2e^2 - 0\right) = 2e^4 - 2e^2 - \ln 2.$

27. $\int_1^2 \left(2e^{-4x} - x^{-2}\right) dx = \left(-\tfrac{1}{2}e^{-4x} + \tfrac{1}{x}\right)\Big|_1^2 = \left(-\tfrac{1}{2}e^{-8} + \tfrac{1}{2}\right) - \left(-\tfrac{1}{2}e^{-4} + 1\right) = -\tfrac{1}{2}e^{-8} + \tfrac{1}{2}e^{-4} - \tfrac{1}{2}$

$$= \tfrac{1}{2}\left(e^{-4} - e^{-8} - 1\right).$$

29. $A = \int_{-1}^2 \left(x^2 - 2x + 2\right) dx = \left(\tfrac{1}{3}x^3 - x^2 + 2x\right)\Big|_{-1}^2 = \left(\tfrac{8}{3} - 4 + 4\right) - \left(-\tfrac{1}{3} - 1 - 2\right) = 6.$

31. $A = \int_1^2 \dfrac{dx}{x^2} = \int_1^2 x^{-2}\,dx = -\dfrac{1}{x}\Big|_1^2 = \dfrac{1}{4} - (-1) = \dfrac{1}{2}.$

33. $A = \int_{-1}^2 e^{-x/2}\,dx = -2e^{-x/2}\big|_{-1}^2 = -2\left(e^{-1} - e^{1/2}\right) = 2\left(\sqrt{e} - 1/e\right).$

35. The average value is $\tfrac{1}{2}\int_0^2 (2x + 3)\,dx = \tfrac{1}{2}\left(x^2 + 3x\right)\Big|_0^2 = \tfrac{1}{2}(10) = 5.$

37. The average value is $\tfrac{1}{2}\int_1^3 \left(2x^2 - 3\right) dx = \tfrac{1}{2}\left(\tfrac{2}{3}x^3 - 3x\right)\Big|_1^3 = \tfrac{1}{2}\left(9 + \tfrac{7}{3}\right) = \tfrac{17}{3}.$

39. The average value is

$$\tfrac{1}{3}\int_{-1}^2 \left(x^2 + 2x - 3\right) dx = \tfrac{1}{3}\left(\tfrac{1}{3}x^3 + x^2 - 3x\right)\Big|_{-1}^2 = \tfrac{1}{3}\left[\left(\tfrac{8}{3} + 4 - 6\right) - \left(-\tfrac{1}{3} + 1 + 3\right)\right]$$

$$= \tfrac{1}{3}\left(\tfrac{8}{3} - 2 + \tfrac{1}{3} - 4\right) = -1.$$

© 2011 Cengage Learning. All Rights Reserved. May not be scanned, copied or duplicated, or posted to a publicly accessible website, in whole or in part.

41. The average value is $\frac{1}{4} \int_0^4 (2x + 1)^{1/2} dx = \left(\frac{1}{4} \right) \left(\frac{1}{2} \right) \left(\frac{2}{3} \right) (2x + 1)^{3/2} \Big|_0^4 = \frac{1}{12} (27 - 1) = \frac{13}{6}.$

43. The average value is $\frac{1}{2} \int_0^2 xe^{x^2} dx = \frac{1}{4} e^{x^2} \Big|_0^2 = \frac{1}{4} \left(e^4 - 1 \right).$

45. Using the substitution $u = 0.05t$, we find that the amount produced was

$$\int_0^{20} 3.5e^{0.05t} dt = \frac{3.5}{0.05} e^u \Big|_0^{20} = 70 \, (e - 1) \approx 120.3, \text{ or } 120.3 \text{ billion metric tons.}$$

47. The amount is $\int_1^2 t \left(\frac{1}{2}t^2 + 1 \right)^{1/2} dt$. Let $u = \frac{1}{2}t^2 + 1$, so $du = t \, dt$. Then

$$\int_1^2 t \left(\frac{1}{2}t^2 + 1 \right)^{1/2} dt = \int_{3/2}^3 u^{1/2} du = \frac{2}{3} u^{3/2} \Big|_{3/2}^3 = \frac{2}{3} \left[(3)^{3/2} - \left(\frac{3}{2} \right)^{3/2} \right] \approx \$2.24 \text{ million.}$$

49. The tractor depreciates by

$$\int_0^5 13{,}388.61e^{-0.22314t} \, dt = \frac{13{,}388.61}{-0.22314} e^{-0.22314t} \Big|_0^5 = -60{,}000.94 e^{-0.22314t} \Big|_0^5$$

$$= -60{,}000.94 \, (-0.672314) = 40{,}339.47, \text{ or } \$40{,}339.$$

51. The average spending per year between 2005 and 2011 is

$$A = \frac{1}{7 - 1} \int_1^7 0.86t^{0.96} \, dt = \frac{0.86}{6} \cdot \frac{1}{1.96} t^{1.96} \Big|_1^7 = \frac{0.86}{6 \, (1.96)} \left(7^{1.96} - 1 \right) \approx 3.24, \text{ or } \$3.24 \text{ billion per year.}$$

53. a. The gasoline consumption in 2017 is given by $A \, (10) = 0.014 \left(10^2 \right) + 1.93 \, (10) + 140 = 160.7$, or 160.7 billion gallons per year.

b. The average consumption per year between 2007 and 2017 is given by

$$A = \frac{1}{10-0} \int_0^{10} \left(0.014t^2 + 1.93t + 140 \right) dt = \frac{1}{10} \left(\frac{0.014}{3} t^3 + \frac{1.93}{2} t^2 + 140t \right) \Big|_0^{10} \approx 150.12, \text{ or } 150.1 \text{ billion gallons}$$
per year per year.

55. The average vacancy rate over the period under consideration is given by

$$A = \frac{1}{6-0} \int_0^6 \left(0.032t^4 - 0.26t^3 - 0.478t^2 + 5.82t + 3.8 \right) dt$$

$$= \frac{1}{6} \left(\frac{0.032}{5} t^5 - \frac{0.26}{4} t^4 - \frac{0.0478}{3} t^3 + \frac{5.82}{2} t^2 + 3.8t \right) \Big|_0^6 \approx 9.7784, \text{ or approximately } 9.8\%.$$

57. The average value is

$$A = \frac{1}{5} \int_0^5 \left(-\frac{40{,}000}{\sqrt{1 + 0.2t}} + 50{,}000 \right) dt = -8000 \int_0^5 (1 + 0.2t)^{-1/2} \, dt + 10{,}000 \int_0^5 dt. \text{ We use the substitution}$$
$u = 1 + 0.2t$ for the first integral, so $du = 0.2 \, dt$ and $dt = 5 \, du$. Thus,

$$A = -8000 \int_1^2 5u^{-1/2} \, du + \int_0^5 10{,}000 \, dt = -40{,}000 \left(2u^{1/2} \right) \Big|_1^2 + 10{,}000t \Big|_0^5$$

$$= -40{,}000 \left(2\sqrt{2} - 2 \right) + 50{,}000 \approx 16{,}863, \text{ or } 16{,}863 \text{ subscribers.}$$

59. $\int_0^5 p \, dt = \int_0^5 \left(18 - 3e^{-2t} - 6e^{-t/3} \right) dt = \frac{1}{5} \left(18t + \frac{3}{2} e^{-2t} + 18e^{-t/3} \right) \Big|_0^5$

$$= \frac{1}{5} \left[18 \, (5) + \frac{3}{2} e^{-10} + 18e^{-5/3} - \frac{3}{2} - 18 \right] = 14.78, \text{ or } \$14.78.$$

© 2011 Cengage Learning. All Rights Reserved. May not be scanned, copied or duplicated, or posted to a publicly accessible website, in whole or in part.

61. The average content of oxygen in the pond over the first 10 days is

$$A = \frac{1}{10-0}\int_0^{10} 100\left(\frac{t^2 + 10t + 100}{t^2 + 20t + 100}\right)dt = \frac{100}{10}\int_0^{10}\left[1 - \frac{10}{t+10} + \frac{100}{(t+10)^2}\right]dt$$

$$= 10\int_0^{10}\left[1 - 10(t+10)^{-1} + 100(t+10)^{-2}\right]dt.$$

Using the substitution $u = t + 10$ for the third integral, we have

$$A = 10\left[t - 10\ln(t+10) - \frac{100}{t+10}\right]\Big|_0^{10} = 10\left\{\left[10 - 10\ln 20 - \frac{100}{2p}\right] - \left[-10\ln 10 - 10\right]\right\}$$

$$= 10(10 - 10\ln 20 - 5 + 10\ln 10 + 10) \approx 80.6853, \text{ or approximately } 80.7\%.$$

63. $\int_a^a f(x)\,dx = F(x)\big|_a^a = F(a) - F(a) = 0$, where $F'(x) = f(x)$.

65. $\int_1^3 x^2\,dx = \frac{1}{3}x^3\big|_1^3 = 9 - \frac{1}{3} = \frac{26}{3} = -\int_3^1 x^2\,dx = -\frac{1}{3}x^3\big|_3^1 = -\frac{1}{3} + 9 = \frac{26}{3}$.

67. $\int_1^9 2\sqrt{x}\,dx = \frac{4}{3}x^{3/2}\big|_1^9 = \frac{4}{3}(27-1) = \frac{104}{3}$ and $2\int_1^9\sqrt{x}\,dx = 2\left(\frac{2}{3}x^{3/2}\right)\big|_1^9 = \frac{104}{3}$.

69. $\int_0^3(1+x^3)\,dx = x + \frac{1}{4}x^4\big|_0^3 = 3 + \frac{81}{4} = \frac{93}{4}$ and

$$\int_0^1(1+x^3)\,dx + \int_1^3(1+x^3)\,dx = \left(x + \tfrac{1}{4}x^4\right)\Big|_0^1 + \left(x + \tfrac{1}{4}x^4\right)\Big|_1^3 = \left(1 + \tfrac{1}{4}\right) + \left(3 + \tfrac{81}{4}\right) - \left(1 + \tfrac{1}{4}\right) = \tfrac{93}{4},$$

demonstrating Property 5.

71. $\int_3^3(1+\sqrt{x})e^{-x}\,dx = 0$ by Property 1 of the definite integral.

73. a. $\int_{-1}^2\left[2f(x) + g(x)\right]dx = 2\int_{-1}^2 f(x)\,dx + \int_{-1}^2 g(x)\,dx = 2(-2) + 3 = -1$.

b. $\int_{-1}^2\left[g(x) - f(x)\right]dx = \int_{-1}^2 g(x)\,dx - \int_{-1}^2 f(x)\,dx = 3 - (-2) = 5$.

c. $\int_{-1}^2\left[2f(x) - 3g(x)\right]dx = 2\int_{-1}^2 f(x)\,dx - 3\int_{-1}^2 g(x)\,dx = 2(-2) - 3(3) = -13$.

75. True. This follows from Property 1 of the definite integral.

77. False. Only a constant can be moved out of the integral.

79. True. This follows from Properties 3 and 4 of the definite integral.

6.5   Using Technology   page 452

1. 7.716667   **3.** 17.564865   **5.** 10,140   **7.** 60.5 mg/day

6.6   Problem-Solving Tips

Note that the formula for the area between the graphs of two continuous functions f and g, where $f(x) \geq g(x)$ on $[a, b]$, is given by $\int_a^b\left[f(x) - g(x)\right]dx$. The condition $f(x) \geq g(x)$ on $[a, b]$ tells us that we cannot interchange $f(x)$ and $g(x)$ in this formula, as that would yield a negative answer, and area cannot be negative.

© 2011 Cengage Learning. All Rights Reserved. May not be scanned, copied or duplicated, or posted to a publicly accessible website, in whole or in part.

6.6   Area between Two Curves   page 459

1. $- \int_0^6 (x^3 - 6x^2)\, dx = \left(-\frac{1}{4}x^4 + 2x^3\right)\Big|_0^6 = -\frac{1}{4}(6)^4 + 2(6)^3 = 108.$

3. $A = -\int_{-1}^0 x\sqrt{1-x^2}\, dx + \int_0^1 x\sqrt{1-x^2}\, dx = 2\int_0^1 x\left(1-x^2\right)^{1/2} dx$ by symmetry. Let $u = 1 - x^2$,

so $du = -2x\, dx$ and $x\, dx = -\frac{1}{2}\, du$. If $x = 0$, then $u = 1$ and if $x = 1$, then $u = 0$, so

$A = (2)\left(-\frac{1}{2}\right)\int_0^1 u^{1/2}\, du = -\frac{2}{3}u^{3/2}\Big|_1^0 = \frac{2}{3}.$

5. $A = -\int_0^4 \left(x - 2\sqrt{x}\right) dx = \int_0^4 \left(-x + 2x^{1/2}\right) dx = \left(-\frac{1}{2}x^2 + \frac{4}{3}x^{3/2}\right)\Big|_0^4 = 8 + \frac{32}{3} = \frac{8}{3}.$

7. The required area is given by

$$\int_{-1}^0 \left(x^2 - x^{1/3}\right) dx + \int_0^1 \left(x^{1/3} - x^2\right) dx = \left(\frac{1}{3}x^3 - \frac{3}{4}x^{4/3}\right)\Big|_{-1}^0 + \left(\frac{3}{4}x^{4/3} - \frac{1}{3}x^3\right)\Big|_0^1$$

$$= -\left(-\frac{1}{3} - \frac{3}{4}\right) + \left(\frac{3}{4} - \frac{1}{3}\right) = \frac{3}{2}.$$

9. The required area is given by $-\int_{-1}^2 -x^2\, dx = \frac{1}{3}x^3\Big|_{-1}^2 = \frac{8}{3} + \frac{1}{3} = 3.$

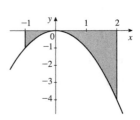

11. $y = x^2 - 5x + 4 = (x - 4)(x - 1) = 0$ if $x = 1$ or 4, the x-intercepts
of the graph of f. Thus,

$A = -\int_1^3 \left(x^2 - 5x + 4\right) dx = \left(-\frac{1}{3}x^3 + \frac{5}{2}x^2 - 4x\right)\Big|_1^3$

$= \left(-9 + \frac{45}{2} - 12\right) - \left(-\frac{1}{3} + \frac{5}{2} - 4\right) = \frac{10}{3}.$

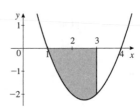

13. The required area is given by

$-\int_0^9 -\left(1 + \sqrt{x}\right) dx = \left(x + \frac{2}{3}x^{3/2}\right)\Big|_0^9 = 9 + 18 = 27.$

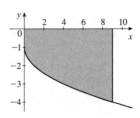

© 2011 Cengage Learning. All Rights Reserved. May not be scanned, copied or duplicated, or posted to a publicly accessible website, in whole or in part.

15. $-\int_{-2}^{4} \left(-e^{x/2}\right) dx = 2e^{x/2}\big|_{-2}^{4} = 2\left(e^2 - e^{-1}\right).$

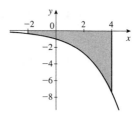

17. $A = \int_{1}^{3} \left[(x^2 + 3) - 1\right] dx = \int_{1}^{3} (x^2 + 2) \, dx = \left(\frac{1}{3}x^3 + 2x\right)\Big|_{1}^{3}$

$= (9 + 6) - \left(\frac{1}{3} + 2\right) = \frac{38}{3}.$

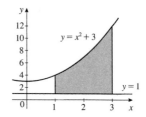

19. $A = \int_{0}^{2} \left(-x^2 + 2x + 3 + x - 3\right) dx = \int_{0}^{2} \left(-x^2 + 3x\right) dx$

$= \left(-\frac{1}{3}x^3 + \frac{3}{2}x^2\right)\Big|_{0}^{2} = -\frac{1}{3}(8) + \frac{3}{2}(4) = 6 - \frac{8}{3} = \frac{10}{3}.$

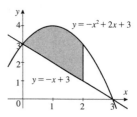

21. $A = \int_{-1}^{2} \left[(x^2 + 1) - \frac{1}{3}x^3\right] dx = \int_{-1}^{2} \left(-\frac{1}{3}x^3 + x^2 + 1\right) dx$

$= \left(-\frac{1}{12}x^4 + \frac{1}{3}x^3 + x\right)\Big|_{-1}^{2}$

$= \left(-\frac{4}{3} + \frac{8}{3} + 2\right) - \left(-\frac{1}{12} - \frac{1}{3} - 1\right) = \frac{19}{4}.$

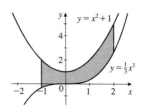

23. $A = \int_{1}^{4} \left[(2x - 1) - \frac{1}{x}\right] dx = \int_{1}^{4} \left(2x - 1 - \frac{1}{x}\right) dx$

$= \left(x^2 - x - \ln x\right)\big|_{1}^{4} = (16 - 4 - \ln 4) - (1 - 1 - \ln 1)$

$= 12 - \ln 4 \approx 10.6.$

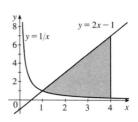

© 2011 Cengage Learning. All Rights Reserved. May not be scanned, copied or duplicated, or posted to a publicly accessible website, in whole or in part.

25. $A = \int_1^2 \left(e^x - \dfrac{1}{x}\right) dx = (e^x - \ln x)|_1^2 = (e^2 - \ln 2) - e = e^2 - e - \ln 2.$

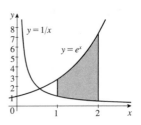

27. $A = -\int_{-1}^0 x\, dx + \int_0^2 x\, dx = -\tfrac{1}{2}x^2\Big|_{-1}^0 + \tfrac{1}{2}x^2\Big|_0^2 = \tfrac{1}{2} + 2 = \tfrac{5}{2}.$

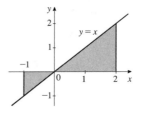

29. The x-intercepts are found by solving

$x^2 - 4x + 3 = (x - 3)(x - 1) = 0$, giving $x = 1$ or 3. Thus,

$A = -\int_{-1}^1 \left(-x^2 + 4x - 3\right) dx + \int_1^2 \left(-x^2 + 4x - 3\right) dx$

$\quad = \left(\tfrac{1}{3}x^3 - 2x^2 + 3x\right)\Big|_{-1}^1 + \left(-\tfrac{1}{3}x^3 + 2x^2 - 3x\right)\Big|_1^2$

$\quad = \left(\tfrac{1}{3} - 2 + 3\right) - \left(-\tfrac{1}{3} - 2 - 3\right)$

$\qquad\qquad + \left(-\tfrac{8}{3} + 8 - 6\right) - \left(-\tfrac{1}{3} + 2 - 3\right)$

$\quad = \tfrac{22}{3}.$

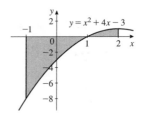

31. $A = \int_0^1 \left(x^3 - 4x^2 + 3x\right) dx - \int_1^2 \left(x^3 - 4x^2 + 3x\right) dx$

$\quad = \left(\tfrac{1}{4}x^4 - \tfrac{4}{3}x^3 + \tfrac{3}{2}x^2\right)\Big|_0^1 - \left(\tfrac{1}{4}x^4 - \tfrac{4}{3}x^3 + \tfrac{3}{2}x^2\right)\Big|_1^2 = \tfrac{3}{2}.$

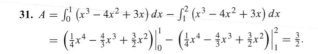

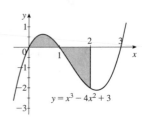

33. $A = -\int_{-1}^0 (e^x - 1)\, dx + \int_0^3 (e^x - 1) dx$

$\quad = (-e^x + x)|_{-1}^0 + (e^x - x)|_0^3$

$\quad = -1 - (-e^{-1} - 1) + (e^3 - 3) - 1 = e^3 - 4 + \tfrac{1}{e} \approx 16.5.$

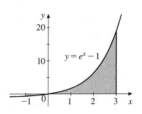

© 2011 Cengage Learning. All Rights Reserved. May not be scanned, copied or duplicated, or posted to a publicly accessible website, in whole or in part.

35. To find the points of intersection of the two curves, we solve the
equation $x^2 - 4 = x + 2$, obtaining $x^2 - x - 6 = (x - 3)(x + 2) = 0$,
so $x = -2$ or $x = 3$. Thus,

$$A = \int_{-2}^{3} \left[(x + 2) - (x^2 - 4) \right] dx = \int_{-2}^{3} \left(-x^2 + x + 6 \right) dx$$

$$= \left(-\tfrac{1}{3}x^3 + \tfrac{1}{2}x^2 + 6x \right)\Big|_{-2}^{3}$$

$$= \left(-9 + \tfrac{9}{2} + 18 \right) - \left(\tfrac{8}{3} + 2 - 12 \right) = \tfrac{125}{6}.$$

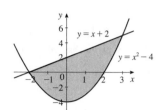

37. To find the points of intersection of the two curves, we solve the
equation $x^3 = x^2$, obtaining $x^3 - x^2 = x^2(x - 1) = 0$, so $x = 0$ or 1.

Thus, $A = -\int_0^1 \left(x^2 - x^3 \right) dx = \left(\tfrac{1}{3}x^3 - \tfrac{1}{4}x^4 \right)\Big|_0^1 = \tfrac{1}{3} - \tfrac{1}{4} = \tfrac{1}{12}.$

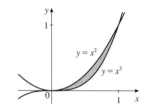

39. To find the points of intersection of the two curves, we solve the
equation $x^3 - 6x^2 + 9x = x^2 - 3x$, obtaining
$x^3 - 7x^2 + 12x = x(x - 4)(x - 3) = 0$, so $x = 0$, 3, or 4. Thus,

$$A = \int_0^3 \left[(x^3 - 6x^2 + 9x) - (x^2 + 3x) \right] dx$$

$$+ \int_3^4 \left[(x^2 - 3x) - (x^3 - 6x^2 + 9x) \right] dx$$

$$= \int_0^3 \left(x^3 - 7x^2 + 12x \right) dx - \int_3^4 \left(x^3 - 7x^2 + 12x \right) dx$$

$$= \left(\tfrac{1}{4}x^4 - \tfrac{7}{3}x^3 + 6x^2 \right)\Big|_0^3 - \left(\tfrac{1}{4}x^4 - \tfrac{7}{3}x^3 + 6x^2 \right)\Big|_3^4$$

$$= \left(\tfrac{81}{4} - 63 + 54 \right) - \left(64 - \tfrac{448}{3} + 96 \right) + \left(\tfrac{81}{4} - 63 + 54 \right) = \tfrac{71}{6}.$$

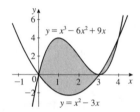

41. By symmetry, $A = 2 \int_0^3 x \left(9 - x^2 \right)^{1/2} dx$. We integrate using the
substitution $u = 9 - x^2$, so $du = -2x\, dx$. If $x = 0$, then $u = 9$ and if
$x = 3$, then $u = 0$, so

$$A = 2 \int_9^0 -\tfrac{1}{2} u^{1/2}\, du = -\int_9^0 u^{1/2}\, du = -\tfrac{2}{3} u^{3/2}\Big|_9^0 = \tfrac{2}{3}(9)^{3/2} = 18.$$

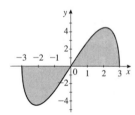

43. $S = \int_0^b \left[g(x) - f(x) \right] dx$ gives the additional revenue that the company would realize if it used a different
advertising agency.

45. The shortfall is $\int_{2010}^{2050} \left[f(t) - g(t) \right] dt.$

47. a. $\int_{T_1}^{T} \left[g(t) - f(t) \right] dt - \int_0^{T_1} \left[f(t) - g(t) \right] dt = A_2 - A_1.$
   b. The number $A_2 - A_1$ gives the distance car 2 is ahead of car 1 after t seconds.

© 2011 Cengage Learning. All Rights Reserved. May not be scanned, copied or duplicated, or posted to a publicly accessible website, in whole or in part.

49. The turbocharged model is moving at

$$A = \int_0^{10} \left[(4 + 1.2t + 0.03t^2) - (4 + 0.8t) \right] dt = \int_0^{10} \left(0.4t + 0.03t^2 \right) dt = \left(0.2t^2 + 0.1t^3 \right) \Big|_0^{10} = 20 + 10,$$

or 30 ft/sec faster than the standard model.

51. The additional number of cars is given by

$$\int_0^5 \left(5e^{0.3t} - 5 - 0.5t^{3/2} \right) dt = \left(\tfrac{5}{0.3} e^{0.3t} - 5t - 0.2t^{5/2} \right) \Big|_0^5 = \tfrac{5}{0.3} e^{1.5} - 25 - 0.2 \, (5)^{5/2} - \tfrac{5}{0.3}$$

$$= 74.695 - 25 - 0.2 \, (5)^{5/2} - \tfrac{50}{3} \approx 21.85, \text{ or } 21,850 \text{ cars.}$$

53. True. If $f(x) \geq g(x)$ on $[a, b]$, then the area of the region is
$\int_a^b \left[f(x) - g(x) \right] dx = \int_a^b |f(x) - g(x)| \, dx$. If $f(x) \leq g(x)$ on $[a, b]$, then the area of the region is
$\int_a^b \left[g(x) - f(x) \right] dx = \int_a^b \left\{ -\left[f(x) - g(x) \right] \right\} dx = \int_a^b |f(x) - g(x)| \, dx$.

55. False. Take $f(x) = x$ and $g(x) = 0$ on $[0, 1]$. Then the area bounded by the graphs of f and g on $[0, 1]$ is
$A = \int_0^1 (x - 0) \, dx = \tfrac{1}{2} x^2 \Big|_0^1 = \tfrac{1}{2}$ and so $A^2 = \tfrac{1}{4}$. However, $\int_0^1 \left[f(x) - g(x) \right]^2 dx = \int_0^1 x^2 \, dx = \tfrac{1}{3}$.

57. The area of R' is
$$A = \int_a^b \left\{ \left[f(x) + C \right] - \left[g(x) + C \right] \right\} dx = \int_a^b \left[f(x) + C - g(x) - C \right] dx = \int_a^b \left[f(x) - g(x) \right] dx.$$

6.6   Using Technology   page 464

1. a.

b. $A \approx 1074.2857$.

3. a.

b. $A \approx 0.9961$.

5. a.

b. $A \approx 5.4603$.

7. a.

b. $A \approx 25.8549$.

© 2011 Cengage Learning. All Rights Reserved. May not be scanned, copied or duplicated, or posted to a publicly accessible website, in whole or in part.

9. a.

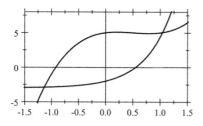

11. a.

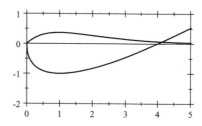

b. $A \approx 10.5144$.

b. $A \approx 3.5799$.

13. The area of the larger region is 207.43.

6.7   **Concept Questions**   page 474

1. a. See the definition on page 465 of the text.   **b.** See the definition on page 466 of the text.

3. See the definition on page 470 of the text.

6.7   **Applications of the Definite Integral to Business and Economics**   page 474

1. When $p = 4$, $-0.01x^2 - 0.1x + 6 = 4$, so $x^2 + 10x - 200 = 0$, and therefore $(x - 10)(x + 20) = 0$, giving $x = 10$ or -20. We reject the root $x = -20$ and find that the equilibrium price occurs at $x = 10$. The consumers' surplus is thus $CS = \int_0^{10} \left(-0.01x^2 - 0.1x + 6\right) dx - (4)(10) = \left(-\frac{0.01}{3}x^3 - 0.05x^2 + 6x\right)\Big|_0^{10} - 40 \approx 11.667$, or \$11,667.

3. Setting $p = 10$, we have $\sqrt{225 - 5x} = 10$, $225 - 5x = 100$, and so $x = 25$. Then $CS = \int_0^{25} \sqrt{225 - 5x}\, dx - (10)(25) = \int_0^{25} (225 - 5x)^{1/2}\, dx - 250$. To evaluate the integral, let $u = 225 - 5x$, so $du = -5\, dx$ and $dx = -\frac{1}{5}\, du$. If $x = 0$, then $u = 225$ and if $x = 25$, then $u = 100$, so $CS = -\frac{1}{5} \int_{225}^{100} u^{1/2}\, du - 250 = -\frac{2}{15}u^{3/2}\Big|_{225}^{100} - 250 = -\frac{2}{15}(1000 - 3375) - 250 = 66.667$, or \$6,667.

5. To find the equilibrium point, we solve $0.01x^2 + 0.1x + 3 = -0.01x^2 - 0.2x + 8$, finding $0.02x^2 + 0.3x - 5 = 0$, $2x^2 + 30x - 500 = (2x - 20)(x + 25) = 0$, and so $x = -25$ or 10. Thus, the equilibrium point is $(10, 5)$. Then $PS = (5)(10) - \int_0^{10} \left(0.01x^2 + 0.1x + 3\right) dx = 50 - \left(\frac{0.01}{3}x^3 + 0.05x^2 + 3x\right)\Big|_0^{10} = 50 - \frac{10}{3} - 5 - 30 = \frac{35}{3}$, or approximately \$11,667.

7. To find the market equilibrium, we solve $-0.2x^2 + 80 = 0.1x^2 + x + 40$, obtaining $0.3x^2 + x - 40 = 0$, $3x^2 + 10x - 400 = 0$, $(3x + 40)(x - 10) = 0$, and so $x = -\frac{40}{3}$ or $x = 10$. We reject the negative root. The corresponding equilibrium price is \$60, the consumers' surplus is $CS = \int_0^{10} \left(-0.2x^2 + 80\right) dx - (60)(10) = \left(-\frac{0.2}{3}x^3 + 80x\right)\Big|_0^{10} - 600 \approx 133.33$, or \$13,333, and the producers' surplus is $PS = 600 - \int_0^{10} \left(0.1x^2 + x + 40\right) dx = 600 - \left(\frac{0.1}{3}x^3 + \frac{1}{2}x^2 + 40x\right)\Big|_0^{10} \approx 116.67$, or \$11,667.

© 2011 Cengage Learning. All Rights Reserved. May not be scanned, copied or duplicated, or posted to a publicly accessible website, in whole or in part.

9. Here $P = 200,000$, $r = 0.08$, and $T = 5$, so
$$PV = \int_0^5 200,000e^{-0.08t}\, dt = -\frac{200,000}{0.08}e^{-0.08t}\Big|_0^5 = -2,500,000\left(e^{-0.4} - 1\right) \approx 824,199.85, \text{ or } \$824,200.$$

11. Here $P = 250$, $m = 12$, $T = 20$, and $r = 0.08$, so $A = \dfrac{mP}{r}\left(e^{rT} - 1\right) = \dfrac{12\,(250)}{0.08}\left(e^{1.6} - 1\right) \approx 148,238.70$, or approximately $\$148,239$.

13. Here $P = 150$, $m = 12$, $T = 15$, and $r = 0.06$, so $A = \dfrac{12\,(150)}{0.06}\left(e^{0.9} - 1\right) \approx 43,788.09$, or approximately $\$43,788$.

15. Here $P = 2000$, $m = 1$, $T = 15.75$, and $r = 0.05$, so $A = \dfrac{1\,(2000)}{0.05}\left(e^{0.7875} - 1\right) \approx 47,915.79$, or approximately $\$47,916$.

17. Here $P = 1200$, $m = 12$, $T = 15$, and $r = 0.06$, so $PV = \dfrac{12\,(1200)}{0.06}\left(1 - e^{-0.9}\right) \approx 142,423.28$, or approximately $\$142,423$.

19. We want the present value of an annuity with $P = 300$, $m = 12$, $T = 10$, and $r = 0.08$, so
$$PV = \frac{12\,(300)}{0.08}\left(1 - e^{-0.8}\right) \approx 24,780.20, \text{ or approximately } \$24,780.$$

21. a.

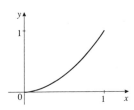

b. $f(0.4) = \frac{15}{16}(0.4)^2 + \frac{1}{16}(0.4) \approx 0.175$ and
$f(0.9) = \frac{15}{16}(0.9)^2 + \frac{1}{16}(0.9) \approx 0.816$. Thus, the lowest 40% of earners receive 17.5% of the total income and the lowest 90% of earners receive 81.6%.

23. a.

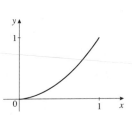

b. $f(0.3) = \frac{14}{15}(0.03)^2 + \frac{1}{15}(0.3) = 0.104$ and
$f(0.7) = \frac{14}{15}(0.7)^2 + \frac{1}{15}(0.7) \approx 0.504$.

6.7   Using Technology   page 476

1. The consumer's surplus is $\$18,000,000$ and the producer's surplus is $\$11,700,000$.

3. The consumer's surplus is $\$33,120$ and the producer's surplus is $\$2880$.

5. Investment A will generate a higher net income.

© 2011 Cengage Learning. All Rights Reserved. May not be scanned, copied or duplicated, or posted to a publicly accessible website, in whole or in part.

CHAPTER 6	Concept Review	page 478

1. a. $F'(x) = f(x)$ **b.** $F(x) + C$

3. a. unknown   **b.** function

5. a. $\int_a^b f(x)\, dx$ **b.** minus

7. a. $\dfrac{1}{b-a} \int_a^b f(x)\, dx$ **b.** area, area

9. a. $\int_0^{\bar{x}} D(x)\, dx - \bar{p}\,\bar{x}$ **b.** $\bar{p}\,\bar{x} - \int_0^{\bar{x}} S(x)\, dx$

11. $\dfrac{mP}{r} \left(e^{rT} - 1 \right)$

CHAPTER 6	Review	page 479

1. $\int \left(x^3 + 2x^2 - x \right) dx = \frac{1}{4}x^4 + \frac{2}{3}x^3 - \frac{1}{2}x^2 + C.$

3. $\int \left(x^4 - 2x^3 + \dfrac{1}{x^2} \right) dx = \dfrac{x^5}{5} - \dfrac{x^4}{2} - \dfrac{1}{x} + C.$

5. $\int x \left(2x^2 + x^{1/2} \right) dx = \int \left(2x^3 + x^{3/2} \right) dx = \frac{1}{2}x^4 + \frac{2}{5}x^{5/2} + C.$

7. $\int \left(x^2 - x + \dfrac{2}{x} + 5 \right) dx = \int x^2\, dx - \int x\, dx + 2\int \dfrac{dx}{x} + 5\int dx = \frac{1}{3}x^3 - \frac{1}{2}x^2 + 2\ln|x| + 5x + C.$

9. Let $u = 3x^2 - 2x + 1$, so $du = (6x - 2)\, dx = 2\,(3x - 1)\, dx$ or $(3x - 1)\, dx = \frac{1}{2}du.$ So
$\int (3x - 1)\left(3x^2 - 2x + 1 \right)^{1/3} dx = \frac{1}{2} \int u^{1/3} du = \frac{3}{8}u^{4/3} + C = \frac{3}{8}\left(3x^2 - 2x + 1 \right)^{4/3} + C.$

11. Let $u = x^2 - 2x + 5$, so $du = 2\,(x - 1)\, dx$ and $(x - 1)\, dx = \frac{1}{2}\, du.$ Then
$\int \dfrac{x - 1}{x^2 - 2x + 5}\, dx = \dfrac{1}{2} \int \dfrac{du}{u} = \frac{1}{2}\ln|u| + C = \frac{1}{2}\ln\left(x^2 - 2x + 5 \right) + C.$

13. Put $u = x^2 + x + 1$, so $du = (2x + 1)\, dx = 2\left(x + \frac{1}{2} \right) dx$ and $\left(x + \frac{1}{2} \right) dx = \frac{1}{2}du.$ Then
$\int \left(x + \frac{1}{2} \right) e^{x^2 + x + 1}\, dx = \frac{1}{2} \int e^u\, du = \frac{1}{2}e^u + C = \frac{1}{2}e^{x^2 + x + 1} + C.$

15. Let $u = \ln x$, so $du = \dfrac{1}{x}\, dx.$ Then $\int \dfrac{(\ln x)^5}{x}\, dx = \int u^5\, du = \frac{1}{6}u^6 + C = \frac{1}{6}(\ln x)^6 + C.$

17. Let $u = x^2 + 1$, so $x^2 = u - 1$, $du = 2x\, dx$, $x\, dx = \frac{1}{2}\, du.$ Then
$\int x^3 \left(x^2 + 1 \right)^{10} dx = \frac{1}{2} \int (u - 1)\, u^{10}\, du = \frac{1}{2} \int \left(u^{11} - u^{10} \right) du = \frac{1}{2} \left(\frac{1}{12}u^{12} - \frac{1}{11}u^{11} \right) + C$
$\qquad = \frac{1}{264}u^{11}\,(11u - 12) + C = \frac{1}{264}\left(x^2 + 1 \right)^{11}\left(11x^2 - 1 \right) + C.$

© 2011 Cengage Learning. All Rights Reserved. May not be scanned, copied or duplicated, or posted to a publicly accessible website, in whole or in part.

19. Put $u = x - 2$, so $du = dx$. Then $x = u + 2$ and

$$\int \frac{x}{\sqrt{x-2}}\,dx = \int \frac{u+2}{\sqrt{u}}\,du = \int \left(u^{1/2} + 2u^{-1/2}\right) du = \int u^{1/2}\,du + 2\int u^{-1/2}\,du = \tfrac{2}{3}u^{3/2} + 4u^{1/2} + C$$

$$= \tfrac{2}{3}u^{1/2}\,(u+6) + C = \tfrac{2}{3}\sqrt{x-2}\,(x-2+6) + C = \tfrac{2}{3}\,(x+4)\sqrt{x-2} + C.$$

21. $\int_0^1 \left(2x^3 - 3x^2 + 1\right) dx = \left(\tfrac{1}{2}x^4 - x^3 + x\right)\Big|_0^1 = \tfrac{1}{2} - 1 + 1 = \tfrac{1}{2}.$

23. $\int_1^4 \left(x^{1/2} + x^{-3/2}\right) dx = \left(\tfrac{2}{3}x^{3/2} - 2x^{-1/2}\right)\Big|_1^4 = \left(\tfrac{2}{3}x^{3/2} - \tfrac{2}{\sqrt{x}}\right)\Big|_1^4 = \left(\tfrac{16}{3} - 1\right) - \left(\tfrac{2}{3} - 2\right) = \tfrac{17}{3}.$

25. Put $u = x^3 - 3x^2 + 1$, so $du = \left(3x^2 - 6x\right) dx = 3\left(x^2 - 2x\right) dx$ and $\left(x^2 - 2x\right) dx = \tfrac{1}{3}\,du$. If $x = -1, u = -3$ and

if $x = 0, u = 1$, so $\int_{-1}^0 12\left(x^2 - 2x\right)\left(x^3 - 3x^2 + 1\right)^3 dx = (12)\left(\tfrac{1}{3}\right)\int_{-3}^1 u^3\,du = 4\left(\tfrac{1}{4}\right)u^4\Big|_{-3}^1 = 1 - 81 = -80.$

27. Let $u = x^2 + 1$, so $du = 2x\,dx$ and $x\,dx = \tfrac{1}{2}du$. If $x = 0$, then $u = 1$, and if $x = 2$, then $u = 5$, so

$$\int_0^2 \frac{x}{x^2 + 1}\,dx = \frac{1}{2}\int_1^5 \frac{du}{u} = \tfrac{1}{2}\ln u\Big|_1^5 = \tfrac{1}{2}\ln 5.$$

29. Let $u = 1 + 2x^2$, so $du = 4x\,dx$ and $x\,dx = \tfrac{1}{4}du$. If $x = 0$, then $u = 1$ and if $x = 2$, then $u = 9$, so

$$\int_0^2 \frac{4x}{\sqrt{1+2x^2}}\,dx = \int_1^9 \frac{du}{u^{1/2}} = 2u^{1/2}\Big|_1^9 = 2\,(3 - 1) = 4.$$

31. Let $u = 1 + e^{-x}$, so $du = -e^{-x}\,dx$ and $e^{-x}\,dx = -du$. Then

$$\int_{-1}^0 \frac{e^{-x}}{\left(1 + e^{-x}\right)^2}\,dx = -\int_{1+e}^2 \frac{du}{u^2} = \frac{1}{u}\Big|_{1+e}^2 = \frac{1}{2} - \frac{1}{1+e} = \frac{e-1}{2\,(1+e)}.$$

33. $f(x) = \int f'(x)\,dx = \int \left(3x^2 - 4x + 1\right) dx = 3\int x^2\,dx - 4\int x\,dx + \int dx = x^3 - 2x^2 + x + C.$ The given condition implies that $f(1) = 1$, so $1 - 2 + 1 + C = 1$, and thus $C = 1$. Therefore, the required function is $f(x) = x^3 - 2x^2 + x + 1.$

35. $f(x) = \int f'(x)\,dx = \int \left(1 - e^{-x}\right) dx = x + e^{-x} + C.$ Now $f(0) = 2$ implies $0 + 1 + C = 2$, so $C = 1$ and the required function is $f(x) = x + e^{-x} + 1.$

37. $\Delta x = \frac{2-1}{5} = \frac{1}{5}$, so $x_1 = \frac{6}{5}$, $x_2 = \frac{7}{5}$, $x_3 = \frac{8}{5}$, $x_4 = \frac{9}{5}$, $x_5 = \frac{10}{5}$. The Riemann sum is

$$f(x_1)\,\Delta x + \cdots + f(x_5)\,\Delta x = \left\{\left[-2\left(\tfrac{6}{5}\right)^2 + 1\right] + \left[-2\left(\tfrac{7}{5}\right)^2 + 1\right] + \cdots + \left[-2\left(\tfrac{10}{5}\right)^2 + 1\right]\right\}\left(\tfrac{1}{5}\right)$$

$$= \tfrac{1}{5}\,(-1.88 - 2.92 - 4.12 - 5.48 - 7) = -4.28.$$

39. a. $R(x) = \int R'(x)\,dx = \int \left(-0.03x + 60\right) dx = -0.015x^2 + 60x + C.$ $R(0) = 0$ implies that $C = 0$, so $R(x) = -0.015x^2 + 60x.$

b. From $R(x) = px$, we have $-0.015x^2 + 60x = px$, and so $p = -0.015x + 60.$

41. a. We have the initial-value problem $T'(t) = 0.15t^2 - 3.6t + 14.4$ with $T(0) = 24$. Integrating, we find $T(t) = \int T'(t)\,dt = \int \left(0.15t^2 - 3.6t + 14.4\right) dt = 0.05t^3 - 1.8t^2 + 14.4t + C.$ Using the initial condition, we find $T(0) = 24 = 0 + C$, so $C = 24$. Therefore, $T(t) = 0.05t^3 - 1.8t^2 + 14.4t + 24.$

b. The temperature at 10 a.m. was $T(4) = 0.05\,(4)^3 - 1.8\,(4)^2 + 14.4\,(4) + 24 = 56$, or $56°F.$

© 2011 Cengage Learning. All Rights Reserved. May not be scanned, copied or duplicated, or posted to a publicly accessible website, in whole or in part.

43. $C(t) = \int C'(t)\, dt = \int (0.003t^2 + 0.06t + 0.1)\, dt = 0.001t^3 + 0.03t^2 + 0.1t + k.$ But $C(0) = 2,$ so
$C(0) = k = 2.$ Therefore, $C(t) = 0.001t^3 + 0.03t^2 + 0.1t + 2.$ The pollution five years from now will be
$C(5) = 0.001(5)^3 + 0.03(5)^2 + 0.1(5) + 2 = 3.375,$ or 3.375 parts per million.

45. Using the substitution $u = 1 + 0.4t,$ we find that
$N(t) = \int 3000(1 + 0.4t)^{-1/2}\, dt = \frac{3000}{0.4} \cdot 2(1 + 0.4t)^{1/2} + C = 15{,}000\sqrt{1 + 0.4t} + C.$ $N(0) = 100{,}000$ implies
$15{,}000 + C = 100{,}000,$ so $C = 85{,}000.$ Therefore, $N(t) = 15{,}000\sqrt{1 + 0.4t} + 85{,}000.$ The number using the
subway six months from now will be $N(6) = 15{,}000\sqrt{1 + 2.4} + 85{,}000 \approx 112{,}659.$

47. Let $u = 5 - x,$ so $du = -x\, dx.$ Then
$$p(x) = \int \frac{240}{(5 - x)^2}\, dx = 240 \int (5 - x)^{-2}\, dx = 240 \int (-u^{-2})\, du = 240u^{-1} + C = \frac{240}{5 - x} + C.$$ Next, the
condition $p(2) = 50$ gives $\frac{240}{3} + C = 80 + C = 50,$ so $C = -30.$ Therefore, $p(x) = \dfrac{240}{5 - x} - 30.$

49. The number will be
$$\int_0^{10} (0.00933t^3 + 0.019t^2 - 0.10833t + 1.3467)\, dt = (0.0023325t^4 + 0.0063333t^3 - 0.054165t^2 + 1.3467t)\Big|_0^{10} = 37.7,$$
or approximately 37.7 million Americans.

51. $A = \int_{-1}^{2} (3x^2 + 2x + 1)\, dx = (x^3 + x^2 + x)\big|_{-1}^{2} = (2^3 + 2^2 + 2) - [(-1)^3 + 1 - 1] = 14 - (-1) = 15.$

53. $A = \displaystyle\int_1^3 \frac{1}{x^2}\, dx = \int_1^3 x^{-2}\, dx = -\frac{1}{x}\Big|_1^3 = -\frac{1}{3} + 1 = \frac{2}{3}.$

55. $A = \int_a^b [f(x) - g(x)]\, dx = \int_0^2 (e^x - x)\, dx = \left(e^x - \frac{1}{2}x^2\right)\Big|_0^2$
$= (e^2 - 2) - (1 - 0) = e^2 - 3.$

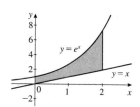

57. $A = \int_0^1 (x^3 - 3x^2 + 2x)\, dx - \int_1^2 (x^3 - 3x^2 + 2x)\, dx$
$= \left(\frac{1}{4}x^4 - x^3 + x^2\right)\Big|_0^1 - \left(\frac{1}{4}x^4 - x^3 + x^2\right)\Big|_1^2$
$= \frac{1}{4} - 1 + 1 - \left[(4 - 8 + 4) - \left(\frac{1}{4} - 1 + 1\right)\right]$
$= \frac{1}{4} + \frac{1}{4} = \frac{1}{2}.$

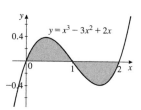

59. $A = \dfrac{1}{3}\displaystyle\int_0^3 \frac{x}{\sqrt{x^2 + 16}}\, dx = \frac{1}{3} \cdot \frac{1}{2} \cdot 2(x^2 + 16)^{1/2}\Big|_0^3 = \frac{1}{3}(x^2 + 16)^{1/2}\Big|_0^3 = \frac{1}{3}(5 - 4) = \frac{1}{3}.$

61. $\overline{A} = \frac{1}{5}\int_0^5 \left(\frac{1}{12}t^2 + 2t + 44\right)\, dt = \frac{1}{5}\left(\frac{1}{36}t^3 + t^2 + 44t\right)\Big|_0^5 = \frac{1}{5}\left(\frac{125}{36} + 25 + 220\right) = \dfrac{125 + 900 + 7920}{5(36)}$
$\approx 49.69,$ or 49.7 ft/sec.

© 2011 Cengage Learning. All Rights Reserved. May not be scanned, copied or duplicated, or posted to a publicly accessible website, in whole or in part.

63. Setting $p = 8$, we have $-0.01x^2 - 0.2x + 23 = 8$, $-0.01x^2 - 0.2x + 15 = 0$, and so
$x^2 + 20x - 1500 = (x - 30)(x + 50) = 0$, giving $x = -50$ or 30. Thus,

$$CS = \int_0^{30} \left(-0.01x^2 - 0.2x + 23\right) dx - 8(30) = \left(-\frac{0.01}{3}x^3 - 0.1x^2 + 23x\right)\Big|_0^{30} - 240$$

$$= -\frac{0.01}{3}(30)^3 - 0.1(900) + 23(30) - 240 = 270, \text{ or } \$270{,}000.$$

65. Use Equation (17) with $P = 4000$, $r = 0.08$, $T = 20$, and $m = 1$ to get $A = \dfrac{1 \cdot 4000}{0.08}\left(e^{1.6} - 1\right) \approx 197{,}651.62$.

That is, Chi-Tai will have approximately \$197,652 in his account after 20 years.

67. Here $P = 80{,}000$, $m = 1$, $T = 10$, and $r = 0.1$, so $PV = \dfrac{1 \cdot 80{,}000}{0.1}\left(1 - e^{-1}\right) \approx 505{,}696$, or approximately

\$505,696.

69. The average population will be $\frac{1}{5}\int 80{,}000 e^{0.05t}\,dt = \frac{80{,}000}{5}\left(\frac{1}{0.05}\right)e^{0.05t}\Big|_0^5 = 320{,}000\left(e^{0.25} - 1\right) \approx 90{,}888$.

CHAPTER 6   Before Moving On...   page 482

1. $\displaystyle \int \left(2x^3 + \sqrt{x} + \frac{2}{x} - \frac{2}{\sqrt{x}}\right) dx = 2\int x^3\,dx + \int x^{1/2}\,dx + 2\int \frac{1}{x}\,dx - 2\int x^{-1/2}\,dx$

$$= \tfrac{1}{2}x^4 + \tfrac{2}{3}x^{3/2} + 2\ln|x| - 4x^{1/2} + C.$$

2. $f(x) = \int f'(x)\,dx = \int (e^x + x)\,dx = e^x + \frac{1}{2}x^2 + C$. $f(0) = 2$ implies $f(0) = e^0 + 0 + C = 2$, so $C = 1$.
Therefore, $f(x) = e^x + \frac{1}{2}x^2 + 1$.

3. Let $u = x^2 + 1$, so $du = 2x\,dx$ or $x\,dx = \frac{1}{2}du$. Then

$$\int \frac{x}{\sqrt{x^2 + 1}}\,dx = \frac{1}{2}\int \frac{du}{\sqrt{u}} = \frac{1}{2}\int u^{-1/2}\,du = \frac{1}{2}\left(2u^{1/2}\right) + C = \sqrt{u} + C = \sqrt{x^2 + 1} + C.$$

4. Let $u = 2 - x^2$, so $du = -2x\,dx$ and $x\,dx = -\frac{1}{2}du$. If $x = 0$, then $u = 2$ and if $x = 1$, then $u = 1$. Therefore,

$$\int_0^1 x\sqrt{2 - x^2}\,dx = -\tfrac{1}{2}\int_2^1 u^{1/2}\,du = -\tfrac{1}{2}\left(\tfrac{2}{3}u^{3/2}\right)\Big|_2^1 = -\tfrac{1}{3}u^{3/2}\Big|_2^1 = -\tfrac{1}{3}\left(1 - 2^{3/2}\right) = \tfrac{1}{3}\left(2\sqrt{2} - 1\right).$$

5. To find the points of intersection, we solve $x^2 - 1 = 1 - x$, obtaining $x^2 + x - 2 = 0$, $(x + 2)(x - 1) = 0$, and so
$x = -2$ or $x = 1$. The points of intersection are $(-2, 3)$ and $(1, 0)$. Thus, the required area is

$$A = \int_{-2}^1 \left[(1 - x) - (x^2 - 1)\right] dx = \int_{-2}^1 \left(2 - x - x^2\right) dx = \left(2x - \tfrac{1}{2}x^2 - \tfrac{1}{3}x^3\right)\Big|_{-2}^1$$

$$= \left(2 - \tfrac{1}{2} - \tfrac{1}{3}\right) - \left(4 - 2 + \tfrac{8}{3}\right) = \tfrac{9}{2}.$$

© 2011 Cengage Learning. All Rights Reserved. May not be scanned, copied or duplicated, or posted to a publicly accessible website, in whole or in part.

7

ADDITIONAL TOPICS IN INTEGRATION

Problem-Solving Tips

1. When you integrate by parts, remember to choose u and dv so that du is simpler than u and dv is easy to integrate.

2. It is helpful to follow the pattern used in the examples in this section: $\int u\,dv = uv - \int v\,du$, where

$$u = \cdots \qquad dv = \cdots$$
$$du = \cdots \qquad v = \cdots$$

7.1 Concept Questions   page 488

1. $\int u\,dv = uv - \int v\,du$

7.1 Integration by Parts   page 488

1. $I = \int xe^{2x}\,dx$. Let $u = x$ and $dv = e^{2x}\,dx$, so $du = dx$ and $v = \frac{1}{2}e^{2x}$. Then
$I = uv - \int v\,du = \frac{1}{2}xe^{2x} - \int \frac{1}{2}e^{2x}\,dx = \frac{1}{2}xe^{2x} - \frac{1}{4}e^{2x} = \frac{1}{4}e^{2x}(2x - 1) + C.$

3. $I = \int xe^{x/4}\,dx$. Let $u = x$ and $dv = e^{x/4}\,dx$, so $du = dx$ and $v = 4e^{x/4}$. Then
$\int xe^{x/4}\,dx = uv - \int v\,du = 4xe^{x/4} - 4\int e^{x/4}\,dx = 4xe^{x/4} - 16e^{x/4} + C = 4(x - 4)e^{x/4} + C.$

5. $\int (e^x - x)^2\,dx = \int (e^{2x} - 2xe^x + x^2)\,dx = \int e^{2x}\,dx - 2\int xe^x\,dx + \int x^2\,dx.$ Using the result
$\int xe^x\,dx = (x - 1)e^x + k$, from Example 1, we see that $\int (e^x - x)^2\,dx = \frac{1}{2}e^{2x} - 2(x - 1)e^x + \frac{1}{3}x^3 + C.$

7. $I = \int (x + 1)e^x\,dx$. Let $u = x + 1$ and $dv = e^x\,dx$, so $du = dx$ and $v = e^x$. Then
$I = (x + 1)e^x - \int e^x\,dx = (x + 1)e^x - e^x + C = xe^x + C.$

9. Let $u = x$ and $dv = (x + 1)^{-3/2}\,dx$, so $du = dx$ and $v = -2(x + 1)^{-1/2}$. Then
$\int x(x + 1)^{-3/2}\,dx = uv - \int v\,du = -2x(x + 1)^{-1/2} + 2\int (x + 1)^{-1/2}\,dx$
$\qquad = -2x(x + 1)^{-1/2} + 4(x + 1)^{1/2} + C = 2(x + 1)^{-1/2}[-x + 2(x + 1)] + C = \dfrac{2(x + 2)}{\sqrt{x + 1}} + C.$

11. $I = \int x(x - 5)^{1/2}\,dx$. Let $u = x$ and $dv = (x - 5)^{1/2}\,dx$, so $du = dx$ and $v = \frac{2}{3}(x - 5)^{3/2}$. Then
$I = \frac{2}{3}x(x - 5)^{3/2} - \int \frac{2}{3}(x - 5)^{3/2}\,dx = \frac{2}{3}x(x - 5)^{3/2} - \frac{2}{3} \cdot \frac{2}{5}(x - 5)^{5/2} + C$
$\qquad = \frac{2}{3}(x - 5)^{3/2}\left[x - \frac{2}{5}(x - 5)\right] + C = \frac{2}{15}(x - 5)^{3/2}(5x - 2x + 10) + C = \frac{2}{15}(x - 5)^{3/2}(3x + 10) + C.$

© 2011 Cengage Learning. All Rights Reserved. May not be scanned, copied or duplicated, or posted to a publicly accessible website, in whole or in part.

13. $I = \int x \ln 2x \, dx$. Let $u = \ln 2x$ and $dv = x \, dx$, so $du = \frac{1}{x} dx$ and $v = \frac{1}{2}x^2$. Then
$I = \frac{1}{2}x^2 \ln 2x - \int \frac{1}{2}x \, dx = \frac{1}{2}x^2 \ln 2x - \frac{1}{4}x^2 + C = \frac{1}{4}x^2 (2 \ln 2x - 1) + C$.

15. Let $u = \ln x$ and $dv = x^3 \, dx$, so $du = \frac{1}{x} dx$, and $v = \frac{1}{4}x^4$. Then
$\int x^3 \ln x \, dx = \frac{1}{4}x^4 \ln x - \frac{1}{4} \int x^3 \, dx = \frac{1}{4}x^4 \ln x - \frac{1}{16}x^4 + C = \frac{1}{16}x^4 (4 \ln x - 1) + C$.

17. Let $u = \ln x^{1/2}$ and $dv = x^{1/2} \, dx$, so $du = \frac{1}{2x} dx$ and $v = \frac{2}{3}x^{3/2}$. Then
$\int \sqrt{x} \ln \sqrt{x} \, dx = uv - \int v \, du = \frac{2}{3}x^{3/2} \ln x^{1/2} - \frac{1}{3} \int x^{1/2} \, dx = \frac{2}{3}x^{3/2} \ln x^{1/2} - \frac{2}{9}x^{3/2} + C$
$$= \frac{2}{9}x\sqrt{x} \left(3 \ln \sqrt{x} - 1\right) + C.$$

19. Let $u = \ln x$ and $dv = x^{-2} \, dx$, so $du = \frac{1}{x} dx$ and $v = -x^{-1}$. Then
$$\int \frac{\ln x}{x^2} \, dx = uv - \int v \, du = -\frac{\ln x}{x} + \int x^{-2} \, dx = -\frac{\ln x}{x} - \frac{1}{x} + C = -\frac{1}{x} (\ln x + 1) + C.$$

21. Let $u = \ln x$ and $dv = dx$, so $du = \frac{1}{x} dx$ and $v = x$. Then
$\int \ln x \, dx = uv - \int v \, du = x \ln x - \int dx = x \ln x - x + C = x (\ln x - 1) + C$.

23. Let $u = x^2$ and $dv = e^{-x} \, dx$, so $du = 2x \, dx$ and $v = -e^{-x}$. Then
$\int x^2 e^{-x} \, dx = uv - \int v \, du = -x^2 e^{-x} + 2 \int x e^{-x} \, dx$. We can integrate by parts again or, using the result of Exercise 2, we can write
$\int x^2 e^{-x} \, dx = -x^2 e^{-x} + 2 \left[-(x+1) e^{-x} \right] + C = -x^2 e^{-x} - 2 (x+1) e^{-x} + C$
$$= -\left(x^2 + 2x + 2\right) e^{-x} + C.$$

25. $I = \int x (\ln x)^2 \, dx$. Let $u = (\ln x)^2$ and $dv = x \, dx$, so $du = 2 (\ln x) \left(\frac{1}{x}\right) = \frac{2 \ln x}{x}$ and $v = \frac{1}{2}x^2$. Then
$I = \frac{1}{2}x^2 (\ln x)^2 - \int x \ln x \, dx$. Next, we evaluate $\int x \ln x \, dx$ by letting $u = \ln x$ and $dv = x \, dx$, so
$du = \frac{1}{x} dx$ and $v = \frac{1}{2}x^2$. Then $\int x \ln x \, dx = \frac{1}{2}x^2 (\ln x) - \frac{1}{2} \int x \, dx = \frac{1}{2}x^2 \ln x - \frac{1}{4}x^2 + C$. Therefore,
$\int x (\ln x)^2 \, dx = \frac{1}{2}x^2 (\ln x)^2 - \frac{1}{2}x^2 \ln x + \frac{1}{4}x^2 + C = \frac{1}{4}x^2 \left[2 (\ln x)^2 - 2 \ln x + 1 \right] + C$.

27. $\int_0^{\ln 2} x e^x \, dx = (x - 1) e^x \big|_0^{\ln 2}$ (using the results of Example 1)
$$= (\ln 2 - 1) e^{\ln 2} - \left(-e^0\right) = 2 (\ln 2 - 1) + 1 = 2 \ln 2 - 1 \quad \text{(because } e^{\ln 2} = 2).$$

29. We first integrate $I = \int \ln x \, dx$. Using parts with $u = \ln x$ and $dv = dx$, so $du = \frac{1}{x} dx$
and $v = x$, we have $I = x \ln x - \int dx = x \ln x - x + C = x (\ln x - 1) + C$. Therefore,
$\int_1^4 \ln x \, dx = x (\ln x - 1) \big|_1^4 = 4 (\ln 4 - 1) - 1 (\ln 1 - 1) = 4 \ln 4 - 3$.

31. Let $u = x$ and $dv = e^{2x} \, dx$. Then $u = dx$ and $v = \frac{1}{2}e^{2x}$, so
$\int_0^2 x e^{2x} \, dx = \frac{1}{2}x e^{2x} \Big|_0^2 - \frac{1}{2} \int_0^2 e^{2x} \, dx = e^4 - \left(\frac{1}{4}e^{2x}\right)\Big|_0^2 = e^4 - \frac{1}{4}e^4 + \frac{1}{4} = \frac{1}{4} \left(3e^4 + 1\right)$.

33. Let $u = x$ and $dv = e^{-2x} \, dx$, so that $du = dx$ and $v = -\frac{1}{2}e^{-2x}$. Then
$f(x) = \int x e^{-2x} \, dx = -\frac{1}{2}x e^{-2x} - \frac{1}{4}e^{-2x} + C$. Solving $f(0) = -\frac{1}{4} + C = 3$, we find that $C = \frac{13}{4}$, so
$y = -\frac{1}{2}x e^{-2x} - \frac{1}{4}e^{-2x} + \frac{13}{4}$.

© 2011 Cengage Learning. All Rights Reserved. May not be scanned, copied or duplicated, or posted to a publicly accessible website, in whole or in part.

35. The required area is given by $\int_1^5 \ln x \, dx$. We first find $\int \ln x \, dx$. Using parts with $u = \ln x$ and $dv = dx$, so $du = \frac{1}{x} dx$ and $v = x$, we have $\int \ln x \, dx = x \ln x - \int dx = x \ln x - x = x (\ln x - 1) + C$. Therefore, $\int_1^5 \ln x \, dx = x (\ln x - 1) |_1^5 = 5 (\ln 5 - 1) - 1 (\ln 1 - 1) = 5 \ln 5 - 4$, and the required area is $5 \ln 5 - 4$.

37. The distance covered is given by $\int_0^{10} 100te^{-0.2t} \, dt = 100 \int_0^{10} te^{-0.2t} \, dt$. We integrate by parts with $u = t$ and $dv = e^{-0.2t} \, dt$, so $du = dt$ and $v = -\frac{1}{0.2} e^{-0.2t} = -5e^{-0.2t}$. Therefore,

$$100 \int_0^{10} te^{-0.2t} dt = 100 \left(-5te^{-0.2t}\right)\big|_0^{10} + 500 \int_0^{10} e^{-0.2t} \, dt = -5000e^{-2} - \left(2500e^{-0.2t}\right)\big|_0^{10}$$

$$= -5000e^{-2} - \left(2500e^{-2} - 2500\right) = 2500 - 7500e^{-2} \approx 1485, \text{ or } 1485 \text{ feet.}$$

39. The average concentration is $C = \frac{1}{12} \int_0^{12} 3te^{-t/3} \, dt = \frac{1}{4} \int_0^{12} te^{-t/3} \, dt$.
Let $u = t$ and $dv = e^{-t/3} \, dt$, so $du = dt$ and $v = -3e^{-t/3}$. Then
$C = \frac{1}{4} \left[-3te^{-t/3} \big|_0^{12} + 3 \int_0^{12} e^{-t/3} dt \right] = \frac{1}{4} \left[-36e^{-4} - \left(9e^{-t/3}\right)\big|_0^{12} \right] = \frac{1}{4} \left(-36e^{-4} - 9e^{-4} + 9 \right) \approx 2.04 \text{ mg/ml.}$

41. $N = 2 \int te^{-0.1t} dt$. Let $u = t$ and $dv = e^{-0.1t}$, so $du = dt$ and $v = -10e^{-0.1t}$. Then
$N(t) = 2 \left(-10te^{-0.1t} + 10 \int e^{-0.1t} \, dt \right) = 2 \left(-10te^{-0.1t} - 100e^{-0.1t} \right) + C = -20e^{-0.1t} (t + 10) + 200$ because $N(0) = 0$.

43. $PV = \int_0^5 (30,000 + 800t) e^{-0.08t} \, dt = 30,000 \int_0^5 e^{-0.08t} \, dt + 800 \int_0^5 te^{-0.08t} \, dt$. Let $I = \int te^{-0.08t} \, dt$. To evaluate I by parts, let $u = t$ and $dv = e^{-0.08t} \, dt$, so $du = dt$ and $v = -\frac{1}{0.08} e^{-0.08t} = -12.5e^{-0.08t}$. Then
$I = -12.5te^{-0.08t} + 12.5 \int e^{-0.08t} dt = -12.5te^{-0.08t} - 156.25e^{-0.08t} + C$. Thus,

$$PV = \left[-\frac{30,000}{0.08} e^{-0.08t} - 800 (12.5) te^{-0.08t} - 800 (156.25) e^{-0.08t} \right]_0^5$$

$$= -375,000e^{-0.4} + 375,000 - 50,000e^{-0.4} - 125,000e^{-0.4} + 125,000 = 500,000 - 550,000e^{-0.4}$$

$$= 131,323.97, \text{ or approximately } \$131,324.$$

45. The membership will be $N(5) = N(0) + \int_0^5 9\sqrt{t+1} \ln \sqrt{t+1} \, dt = 50 + 9 \int_0^5 \sqrt{t+1} \ln \sqrt{t+1} \, dt$.
To evaluate the integral, let $u = t + 1$, so $du = dt$. If $t = 0$, then $u = 1$ and if $t = 5$, then $u = 6$, so $9 \int_0^5 \sqrt{t+1} \ln \sqrt{t+1} \, dt = 9 \int_1^6 \sqrt{u} \ln \sqrt{u} \, du$. Using the results of Exercise 17, we find
$9 \int_1^6 \sqrt{u} \ln \sqrt{u} \, du = \left[2u\sqrt{u} \left(3 \ln \sqrt{u} - 1 \right) \right]_1^6$. Therefore, $N = 50 + 51.606 \approx 101.606$, or 101,606 people.

© 2011 Cengage Learning. All Rights Reserved. May not be scanned, copied or duplicated, or posted to a publicly accessible website, in whole or in part.

47. The average concentration from $r = r_1$ to $r = r_2$ is

$$A = \frac{1}{r^2 - r_1} \int_{r_1}^{r_2} c\,(r)\, dr = \frac{1}{r^2 - r_1} \int_{r_1}^{r_2} \left[\left(\frac{c_1 - c_2}{\ln r_1 - \ln r_2} \right) (\ln r - \ln r_2) + c_2 \right] dr$$

$$= \frac{1}{r_2 - r_1} \left(\frac{c_1 - c_2}{\ln r_1 - \ln r_2} \right) \left[\int_{r_1}^{r_2} \ln r\, dr - \int_{r_1}^{r_2} \ln r_2\, dr \right] + \frac{1}{r_2 - r_1} \int_{r_1}^{r_2} c_2\, dr.$$

We integrate $\int \ln r\, dr$ by parts, letting $u = \ln r$ and $dv = dr$, so $du = \dfrac{dr}{r}$ and $v = r$. Thus,

$$\int \ln r\, dr = r \ln r - \int r \frac{dr}{r} = r \ln r - r = r\,(\ln r - 1). \text{ Then}$$

$$A = \frac{1}{r_2 - r_1} \left(\frac{c_1 - c_2}{\ln r_1 - \ln r_2} \right) \left\{ \left[[r\,(\ln r - 1)]_{r_1}^{r_2} - (\ln r_2)|_{r_1}^{r_2} \right] \right\} + c_2$$

$$= \frac{1}{r_2 - r_1} \left(\frac{c_1 - c_2}{\ln r_1 - \ln r_2} \right) [r_2\,(\ln r_2 - 1) - r_1\,(\ln r_1 - 1) - (r_2 - r_1) \ln r_2] + c_2$$

$$= \frac{1}{r_2 - r_1} \left(\frac{c_1 - c_2}{\ln r_1 - \ln r_2} \right) [r_1\,(\ln r_2 - \ln r_1) - (r_2 - r_1)] + c_2 = (c_2 - c_1) \left(\frac{r_1}{r_2 - r_1} + \frac{1}{\ln r_1 - \ln r_2} \right) + c_2.$$

49. True. This is the integration by parts formula.

7.2   Problem-Solving Tips

The integrals in the exercise set may not have exactly the same form as those in the table of integrals. Sometimes you may need to rewrite the integral, as in Example 2 on page 493 of the text, or you may need to apply a rule more than once, as in Example 5 on page 494 of the text.

7.2   Concept Questions   page 496

1. a. Formula (19) seems appropriate.

  b. Put $a = \sqrt{2}$ and $x = u$. Then using Formula (19), we have

$$\int \frac{\sqrt{2 - x^2}}{x}\, dx = \int \frac{\sqrt{\left(\sqrt{2} \right)^2 - x^2}}{x}\, dx = \sqrt{2 - x^2} - \sqrt{2} \ln \left| \frac{\sqrt{a} + \sqrt{2 - x^2}}{x} \right| + C.$$

7.2   Integration Using Tables of Integrals   page 496

1. First we note that $\displaystyle\int \frac{2x}{2 + 3x}\, dx = 2 \int \frac{x}{2 + 3x}\, dx$. Next, we use Formula (1) with $a = 2$, $b = 3$, and $u = x$. Then

$$\int \frac{2x}{2 + 3x}\, dx = \tfrac{2}{9}\,(2 + 3x - 2 \ln |2 + 3x|) + C.$$

3. $\displaystyle\int \frac{3x^2}{2 + 4x}\, dx = \frac{3}{2} \int \frac{x^2}{1 + 2x}\, dx$. Use Formula (2) with $a = 1$ and $b = 2$, obtaining

$$\int \frac{3x^2}{2 + 4x}\, dx = \tfrac{3}{32} \left[(1 + 2x)^2 - 4\,(1 + 2x) + 2 \ln |1 + 2x| \right] + C.$$

© 2011 Cengage Learning. All Rights Reserved. May not be scanned, copied or duplicated, or posted to a publicly accessible website, in whole or in part.

5. $\int x^2\sqrt{9+4x^2}\,dx = \int x^2\sqrt{4\left(\frac{9}{4}+x^2\right)}\,dx = 2\int x^2\sqrt{\left(\frac{3}{2}\right)^2+x^2}\,dx$. Using Formula (8) with $a = \frac{3}{2}$, we find that

$\int x^2\sqrt{9+4x^2}\,dx = 2\left[\left(\frac{x}{8}\right)\left(\frac{9}{4}+2x^2\right)\sqrt{\frac{9}{4}+x^2} - \frac{81}{128}\ln\left|x+\sqrt{\frac{9}{4}+x^2}\right|\right] + C.$

7. Use Formula (6) with $a = 1$, $b = 4$, and $u = x$. Then $\displaystyle\int \frac{dx}{x\sqrt{1+4x}} = \ln\left|\frac{\sqrt{1+4x}-1}{\sqrt{1+4x}+1}\right| + C.$

9. Use Formula (9) with $a = 3$ and $u = 2x$, so $du = 2\,dx$. Then

$\displaystyle\int_0^2 \frac{dx}{\sqrt{9+4x^2}} = \frac{1}{2}\int_0^4 \frac{du}{\sqrt{3^2+u^2}} = \frac{1}{2}\ln\left|u+\sqrt{9+u^2}\right|\Big|_0^4 = \frac{1}{2}\,(\ln 9 - \ln 3) = \frac{1}{2}\ln 3.$ Note that the limits of integration change from $x = 0$ and $x = 2$ to $u = 0$ and $u = 4$ respectively.

11. Using Formula (22) with $a = 3$, we see that $\displaystyle\int \frac{dx}{(9-x^2)^{3/2}} = \frac{x}{9\sqrt{9-x^2}} + C.$

13. $I = \int x^2\sqrt{x^2-4}\,dx$. Use Formula (14) with $a = 2$ and $u = x$ to obtain

$I = \frac{x}{8}\left(2x^2-4\right)\sqrt{x^2-4} - 2\ln\left|x+\sqrt{x^2-4}\right| + C.$

15. Using Formula (19) with $a = 2$ and $u = x$, we have $\displaystyle\int \frac{\sqrt{4-x^2}}{x}\,dx = \sqrt{4-x^2} - 2\ln\left|\frac{2+\sqrt{4-x^2}}{x}\right| + C.$

17. $I = \int xe^{2x}\,dx$. Use Formula (23) with $u = x$ and $a = 2$ to obtain $I = \frac{1}{4}\,(2x-1)\,e^{2x} + C.$

19. $I = \displaystyle\int \frac{dx}{(x+1)\ln(x+1)}$. Let $u = x + 1$, so $du = dx$. Then $\displaystyle\int \frac{dx}{(x+1)\ln(x+1)} = \int \frac{du}{u\ln u}$. Now use Formula (28) with $u = x$ to obtain $\displaystyle\int \frac{du}{u\ln u} = \ln|\ln u| + C$. Therefore, $\displaystyle\int \frac{dx}{(x+1)\ln(x+1)} = \ln|\ln(x+1)| + C.$

21. $I = \displaystyle\int \frac{e^{2x}}{(1+3e^x)^2}\,dx$. Put $u = e^x$, so $du = e^x\,dx$. Then use Formula (3) with $a = 1$ and $b = 3$. Thus,

$I = \displaystyle\int \frac{u}{(1+3u)^2}\,du = \frac{1}{9}\left(\frac{1}{1+3u} + \ln|1+3u|\right) + C = \frac{1}{9}\left[\frac{1}{1+3e^x} + \ln(1+3e^x)\right] + C.$

23. $\displaystyle\int \frac{3e^x}{1+e^{x/2}}\,dx = 3\int \frac{e^{x/2}}{e^{-x/2}+1}\,dx$. Let $v = e^{x/2}$, so $dv = \frac{1}{2}e^{x/2}\,dx$ and $e^{x/2}\,dx = 2\,dv$. Then

$\displaystyle\int \frac{3e^x}{1+e^{x/2}}\,dx = 6\int \frac{dv}{(1/v)+1} = 6\int \frac{v}{v+1}\,dv.$ Use Formula (1) with $a = 1$, $b = 1$, and $u = v$, obtaining

$6\displaystyle\int \frac{v}{v+1}\,dv = 6\,(1+v-\ln|1+v|) + C.$ Thus, $\displaystyle\int \frac{3e^x}{1+e^{x/2}}\,dx = 6\left[1+e^{x/2}-\ln\left(1+e^{x/2}\right)\right] + C.$ This answer may be written in the form $6\left[e^{x/2}-\ln\left(1+e^{x/2}\right)\right] + C_1$, where $C_1 = C + 6$ is an arbitrary constant.

25. $I = \displaystyle\int \frac{\ln x}{x\,(2+3\ln x)}\,dx$. Let $v = \ln x$ so that $dv = \frac{1}{x}\,dx$. Then $I = \displaystyle\int \frac{v}{2+3v}\,dv.$ Now use Formula (1)

with $a = 2$, $b = 3$, and $u = v$ to obtain $\displaystyle\int \frac{v}{2+3v}\,dv = \frac{1}{9}\,(2+3\ln x - 2\ln|2+3\ln x|) + C.$ Thus,

$\displaystyle\int \frac{\ln x}{x\,(2+3\ln x)}\,dx = \frac{1}{9}\,(2+3\ln x - 2\ln|2+3\ln x|) + C.$

© 2011 Cengage Learning. All Rights Reserved. May not be scanned, copied or duplicated, or posted to a publicly accessible website, in whole or in part.

27. Using Formula (24) with $a = 1$, $n = 2$, and $u = x$, we have

$$\int_0^1 x^2 e^x \, dx = \left(x^2 e^x\right)\Big|_0^1 - 2\int_0^1 xe^x \, dx = \left[x^2 e^x - 2\left(xe^x - e^x\right)\right]_0^1 = \left(x^2 e^x - 2xe^x + 2e^x\right)\Big|_0^1$$

$$= e - 2e + 2e - 2 = e - 2.$$

29. $I = \int x^2 \ln x \, dx$. Use Formula (27) with $n = 2$ and $u = x$ to obtain $I = \int x^2 \ln x \, dx = \dfrac{x^3}{9}\left(3\ln x - 1\right) + C$.

31. $I = \int (\ln x)^3 \, dx$. Use Formula (29) with $n = 3$ to write $I = x\,(\ln x)^3 - 3\int (\ln x)^2 \, dx$. Now use Formula (29) again with $n = 2$ to obtain $I = x\,(\ln x)^3 - 3\left[x\,(\ln x)^2 - 2\int \ln x \, dx\right]$. Using Formula (29) one more time with $n = 1$ gives $\int (\ln x)^3 \, dx = x\,(\ln x)^3 - 3x\,(\ln x)^2 + 6\,(x\ln x - x) + C = x\,(\ln x)^3 - 3x\,(\ln x)^2 + 6x\ln x - 6x + C$.

33. Letting $p = 50$ gives $50 = \dfrac{250}{\sqrt{16 + x^2}}$, from which we deduce that $\sqrt{16 + x^2} = 5$, $16 + x^2 = 25$, and so $x = 3$.
Using Formula (9) with $u = 3$, we see that

$$CS = \int_0^3 \frac{250}{\sqrt{16 + x^2}} \, dx = 50\,(3) = 250\int_0^3 \frac{1}{\sqrt{16 + x^2}} \, dx - 150 = 250\ln\left|x + \sqrt{16 + x^2}\right|\Big|_0^3 - 150$$

$$= 250\,(\ln 8 - \ln 4) - 150 \approx 23.286795, \text{ or approximately } \$2329.$$

35. The number of visitors admitted to the amusement park by noon is found by evaluating the integral

$$\int_0^3 \frac{60}{\left(2 + t^2\right)^{3/2}} \, dt = 60\int_0^3 \frac{dt}{\left(2 + t\right)^{3/2}}. \text{ Using Formula (12) with } a = \sqrt{2} \text{ and } u = t, \text{ we find}$$

$$60\int_0^3 \frac{dt}{\left(2 + t^2\right)^{3/2}} = 60\left(\frac{t}{2\sqrt{2 + t^3}}\right)\Big|_0^3 = 60\left(\frac{3}{2\sqrt{11} - 0}\right) = \frac{90}{\sqrt{11}} \approx 27.136, \text{ or approximately } 27{,}136.$$

37. To find the average number of fruit flies over the first 10 days, use Formula (25) with $a = 0.02$ and $b = 24$. Thus,

$$\frac{1}{10}\int \frac{1000}{1 + 24e^{-0.02t}} \, dt = 100\int_0^{10} \frac{1}{1 + 24e^{-0.02t}} \, dt = 100\left[t + \tfrac{1}{0.02}\ln\left(1 + 24e^{-0.02t}\right)\right]_0^{10}$$

$$= 100\,(10 + 50\ln 20.6495 - 50\ln 25) \approx 44.09, \text{ or approximately } 44 \text{ fruit flies.}$$

Over the first 20 days, the average is

$$\frac{1}{20}\int_0^{20} \frac{1000}{1 + 24e^{-0.02t}} \, dt = 50\int_0^{20} \frac{1}{1 + 24e^{-0.02t}} \, dt = 50\left[t + \tfrac{1}{0.02}\ln\left(1 + 24e^{-0.02t}\right)\right]_0^{20}$$

$$= 50\,(20 + 50\ln 17.088 - 50\ln 25) \approx 48.71, \text{ or approximately } 49 \text{ fruit flies.}$$

39. Use Formula (25) with $a = -0.2$ and $b = 1.5$ to obtain

$$\frac{1}{5}\int_0^5 \frac{100{,}000}{2\left(1 + 1.5e^{-0.2t}\right)} \, dt = 10{,}000\int_0^5 \frac{1}{1 + 1.5e^{-0.2t}} \, dt = 10{,}000\left[t + 5\ln\left(1 + 1.5e^{-0.2t}\right)\right]_0^5$$

$$= 10{,}000\,(5 + 5\ln 1.551819162 - 5\ln 2.5) \approx 26{,}157, \text{ or about } 26{,}157 \text{ people.}$$

41. $I = e^{(0.15)(5)}\int_0^5 20{,}000te^{-0.15t} \, dt = 20{,}000e^{0.75}\int_0^5 te^{-0.15t} \, dt = 20{,}000\left[\dfrac{1}{(0.15)^2}\,(-0.15t - 1)\,e^{-0.15t}\right]_0^5$, using
Formula (23) with $a = -0.15$. Thus, $I \approx 326.222$, or approximately $\$326{,}222$.

© 2011 Cengage Learning. All Rights Reserved. May not be scanned, copied or duplicated, or posted to a publicly accessible website, in whole or in part.

7.3 Concept Questions page 508

1. In the trapezoidal rule, each region beneath (or above) the graph of f is approximated by the area of a trapezoid whose base consists of two consecutive points in the partition. Therefore, n can be odd or even. In Simpson's rule, the area of each subregion is approximated by part of a parabola passing through those points. Therefore, there are two subintervals involved in the approximations, and so n must be even.

3. If we use the trapezoidal rule and f is a linear function, then $f''(x) = 0$, so $M = 0$, and consequently the maximum error is 0. If we use Simpson's rule, then $f^{(4)}(x) = 0$, so $M = 0$, and once again the maximum error is 0.

7.3 Numerical Integration page 508

1. $\Delta x = \frac{b-a}{n} = \frac{2-0}{6} = \frac{1}{3}$, so $x_0 = 0, x_1 = \frac{1}{3}, x_2 = \frac{2}{3}, x_3 = 1, x_4 = \frac{4}{3}, x_5 = \frac{5}{3}, x_6 = 2$.
Trapezoidal rule:

$$\int_0^2 x^2\,dx \approx \frac{1}{6}\left[0 + 2\left(\tfrac{1}{3}\right)^2 + 2\left(\tfrac{2}{3}\right)^2 + 2(1)^2 + 2\left(\tfrac{4}{3}\right)^2 + 2\left(\tfrac{5}{3}\right)^2 + 2^2\right]$$

$$\approx \frac{1}{6}(0.22222 + 0.88889 + 2 + 3.55556 + 5.55556 + 4) \approx 2.7037.$$

Simpson's rule:

$$\int x^2\,dx \approx \frac{1}{9}\left[0 + 4\left(\tfrac{1}{3}\right)^2 + 2\left(\tfrac{2}{3}\right)^2 + 4(1)^2 + 2\left(\tfrac{4}{3}\right)^2 + 4\left(\tfrac{5}{3}\right)^2 + 2^2\right]$$

$$\approx \frac{1}{9}(0.44444 + 0.88889 + 4 + 3.55556 + 11.11111 + 4) \approx 2.6667.$$

Exact value: $\int_0^2 x^2\,dx = \frac{1}{3}x^3\Big|_0^2 = \frac{8}{3}$.

3. $\Delta x = \frac{b-a}{n} = \frac{1-0}{4} = \frac{1}{4}$, so $x_0 = 0, x_1 = \frac{1}{4}, x_2 = \frac{1}{2}, x_3 = \frac{3}{4}, x_4 = 1$.
Trapezoidal rule:

$$\int_0^1 x^3\,dx \approx \frac{1}{8}\left[0 + 2\left(\tfrac{1}{4}\right)^3 + 2\left(\tfrac{1}{2}\right)^3 + 2\left(\tfrac{3}{4}\right)^3 + 1^3\right] \approx \frac{1}{8}(0 + 0.3125 + 0.25 + 0.8) \approx 0.265625.$$

Simpson's rule:

$$\int_0^1 x^3\,dx \approx \frac{1}{12}\left[0 + 4\left(\tfrac{1}{4}\right)^3 + 2\left(\tfrac{1}{2}\right)^3 + 4\left(\tfrac{3}{4}\right)^3 + 1\right] \approx \frac{1}{12}(0 + 0.625 + 0.25 + 1.6875 + 1) \approx 0.25.$$

Exact value: $\int_0^1 x^3\,dx = \frac{1}{4}x^4\Big|_0^1 = \frac{1}{4} - 0 = \frac{1}{4}$.

5. Here $a = 1, b = 2$, and $n = 4$, so $\Delta x = \frac{2-1}{4} = \frac{1}{4} = 0.25$ and $x_0 = 1, x_1 = 1.25, x_2 = 1.5, x_3 = 1.75, x_4 = 2$.
Trapezoidal rule: $\int_1^2 \frac{1}{x}\,dx \approx \frac{0.25}{2}\left[1 + 2\left(\tfrac{1}{1.25}\right) + 2\left(\tfrac{1}{1.5}\right) + 2\left(\tfrac{1}{1.75}\right) + \tfrac{1}{2}\right] \approx 0.697.$

Simpson's rule: $\int_1^2 \frac{1}{x}\,dx \approx \frac{0.25}{3}\left[1 + 4\left(\tfrac{1}{1.25}\right) + 2\left(\tfrac{1}{1.5}\right) + 4\left(\tfrac{1}{1.75}\right) + \tfrac{1}{2}\right] \approx 0.6933.$

Exact value: $\int_1^2 \frac{1}{x}\,dx = \ln x|_1^2 = \ln 2 - \ln 1 = \ln 2 \approx 0.6931.$

© 2011 Cengage Learning. All Rights Reserved. May not be scanned, copied or duplicated, or posted to a publicly accessible website, in whole or in part.

7. $\Delta x = \frac{1}{4}, x_0 = 1, x_1 = \frac{5}{4}, x_2 = \frac{3}{2}, x_3 = \frac{7}{4}, x_4 = 2.$

Trapezoidal rule: $\int_1^2 \frac{1}{x^2}\, dx \approx \frac{1}{8}\left[1 + 2\left(\frac{4}{5}\right)^2 + 2\left(\frac{2}{3}\right)^2 + 2\left(\frac{4}{7}\right)^2 + \left(\frac{1}{2}\right)^2\right] \approx 0.5090.$

Simpson's rule: $\int_1^2 \frac{1}{x^2}\, dx \approx \frac{1}{12}\left[1 + 4\left(\frac{4}{5}\right)^2 + 2\left(\frac{2}{3}\right)^2 + 4\left(\frac{4}{7}\right)^2 + \left(\frac{1}{2}\right)^2\right] \approx 0.5004.$

Exact value: $\int_1^2 \frac{1}{x^2}\, dx = -\frac{1}{x}\Big|_1^2 = -\frac{1}{2} + 1 = \frac{1}{2}.$

9. $\Delta x = \frac{b-a}{n} = \frac{4-0}{8} = \frac{1}{2},$ so $x_0 = 0, x_1 = \frac{1}{2}, x_2 = \frac{2}{2}, x_3 = \frac{3}{2}, \ldots, x_8 = \frac{8}{2}.$

Trapezoidal rule: $\int_0^4 \sqrt{x}\, dx \approx \frac{1/2}{2}\left(0 + 2\sqrt{0.5} + 2\sqrt{1} + 2\sqrt{1.5} + \cdots + 2\sqrt{3.5} + \sqrt{4}\right) \approx 5.26504.$

Simpson's rule: $\int_0^4 \sqrt{x}\, dx \approx \frac{1/2}{3}\left(0 + 4\sqrt{0.5} + 2\sqrt{1} + 4\sqrt{1.5} + \cdots + 4\sqrt{3.5} + \sqrt{4}\right) \approx 5.30463.$

Exact value: $\int_0^4 \sqrt{x}\, dx \approx \frac{2}{3}x^{3/2}\Big|_0^4 = \frac{2}{3}(8) = \frac{16}{3}.$

11. $\Delta x = \frac{1-0}{6} = \frac{1}{6},$ so $x_0 = 0, x_1 = \frac{1}{6}, x_2 = \frac{2}{6}, \ldots, x_6 = \frac{6}{6}.$

Trapezoidal rule: $\int_0^1 e^{-x}\, dx \approx \frac{1/6}{2}\left(1 + 2e^{-1/6} + 2e^{-2/6} + \cdots + 2e^{-5/6} + e^{-1}\right) \approx 0.633583.$

Simpson's rule: $\int_0^1 e^{-x}\, dx \approx \frac{1/6}{3}\left(1 + 4e^{-1/6} + 2e^{-2/6} + \cdots + 4e^{-5/6} + e^{-1}\right) \approx 0.632123.$

Exact value: $\int_0^1 e^{-x}\, dx = -e^{-x}\Big|_0^1 = -e^{-1} + 1 \approx 0.632121.$

13. $\Delta x = \frac{1}{4},$ so $x_0 = 0, x_1 = \frac{5}{4}, x_2 = \frac{3}{2}, x_3 = \frac{7}{4}, x_4 = 2.$

Trapezoidal rule: $\int_1^2 \ln x\, dx \approx \frac{1}{8}\left(\ln 1 + 2\ln\frac{5}{4} + 2\ln\frac{3}{2} + 2\ln\frac{7}{4} + \ln 2\right) \approx 0.38370.$

Simpson's rule: $\int_1^2 \ln x\, dx \approx \frac{1}{12}\left(\ln 1 + 4\ln\frac{5}{4} + 2\ln\frac{3}{2} + 4\ln\frac{7}{4} + \ln 2\right) \approx 0.38626.$

Exact value: $\int_1^2 \ln x\, dx \approx x\,(\ln x - 1)|_1^2 = 2\,(\ln 2 - 1) + 1 = 2\ln 2 - 1 \approx 0.3863.$

15. $\Delta x = \frac{1-0}{4} = \frac{1}{4},$ so $x_0 = 0, x_1 = \frac{1}{4}, x_2 = \frac{2}{4}, x_3 = \frac{3}{4}, x_4 = \frac{4}{4}.$

Trapezoidal rule: $\int_0^1 \sqrt{1+x^3}\, dx \approx \frac{1/4}{2}\left[\sqrt{1} + 2\sqrt{1 + \left(\frac{1}{4}\right)^3} + \cdots + 2\sqrt{1 + \left(\frac{3}{4}\right)^3} + \sqrt{2}\right] \approx 1.1170.$

Simpson's rule:

$$\int_0^1 \sqrt{1+x^3}\, dx \approx \frac{1/4}{3}\left[\sqrt{1} + 4\sqrt{1 + \left(\frac{1}{4}\right)^3} + 2\sqrt{1 + \left(\frac{2}{4}\right)^3} + \cdots + 4\sqrt{1 + \left(\frac{3}{4}\right)^3} + \sqrt{2}\right] \approx 1.1114.$$

17. $\Delta x = \frac{2-0}{4} = \frac{1}{2},$ so $x_0 = 0, x_1 = \frac{1}{2}, x_2 = \frac{2}{2}, x_3 = \frac{3}{2}, x_4 = \frac{4}{2}.$

Trapezoidal rule:

$$\int_0^2 \frac{1}{\sqrt{x^3+1}}\, dx = \frac{1/2}{2}\left[1 + \frac{2}{\sqrt{\left(\frac{1}{2}\right)^3 + 1}} + \frac{2}{\sqrt{(1)^3 + 1}} + \frac{2}{\sqrt{\left(\frac{3}{2}\right)^3 + 1}} + \frac{1}{\sqrt{(2)^3 + 1}}\right] \approx 1.3973.$$

Simpson's rule: $$\int_0^2 \frac{1}{\sqrt{x^3+1}}\, dx = \frac{1/2}{3}\left[1 + \frac{4}{\sqrt{\left(\frac{1}{2}\right)^3 + 1}} + \frac{2}{\sqrt{(1)^3 + 1}} + \frac{4}{\sqrt{\left(\frac{3}{2}\right)^3 + 1}} + \frac{1}{\sqrt{(2)^3 + 1}}\right] \approx 1.4052.$$

© 2011 Cengage Learning. All Rights Reserved. May not be scanned, copied or duplicated, or posted to a publicly accessible website, in whole or in part.

19. $\Delta x = \frac{2}{4} = \frac{1}{2}$, so $x_0 = 0$, $x_1 = \frac{1}{2}$, $x_2 = 1$, $x_3 = \frac{3}{2}$, $x_4 = 2$.

Trapezoidal rule: $\int_0^2 e^{-x^2}\, dx = \frac{1}{4}\left[e^{-0} + 2e^{-(1/2)^2} + 2e^{-1} + 2e^{-(3/2)^2} + e^{-4}\right] \approx 0.8806$.

Simpson's rule: $\int_0^2 e^{-x^2}\, dx = \frac{1}{6}\left[e^{-0} + 4e^{-(1/2)^2} + 2e^{-1} + 4e^{-(3/2)^2} + e^{-4}\right] \approx 0.8818$.

21. $\Delta x = \frac{2-1}{4} = \frac{1}{4}$, so $x_0 = 1$, $x_1 = \frac{5}{4}$, $x_2 = \frac{6}{4}$, $x_3 = \frac{7}{4}$, $x_4 = \frac{8}{4}$.

Trapezoidal rule: $\int_1^2 x^{-1/2}e^x\, dx = \frac{1/4}{2}\left[e + \frac{2e^{5/4}}{\sqrt{5/4}} + \cdots + \frac{2e^{7/4}}{\sqrt{7/4}} + \frac{e^2}{\sqrt{2}}\right] \approx 3.7757$.

Simpson's rule: $\int_1^2 x^{-1/2}e^x\, dx = \frac{1/4}{3}\left[e + \frac{4e^{5/4}}{\sqrt{5/4}} + \cdots + \frac{4e^{7/4}}{\sqrt{7/4}} + \frac{e^2}{\sqrt{2}}\right] \approx 3.7625$.

23. a. Here $a = -1$, $b = 2$, $n = 10$, and $f(x) = x^5$. Thus, $f'(x) = 5x^4$ and $f''(x) = 20x^3$. Because $f'''(x) = 60x^2 > 0$ on $(-1, 0) \cup (0, 2)$, we see that $f''(x)$ is increasing on $(-1, 0) \cup (0, 2)$. Thus, we can take $M = f''(2) = 20\left(2^3\right) = 160$. Using Formula (7), we see that the maximum error incurred is

$$\frac{M(b-a)^3}{12n^2} = \frac{160\,[2-(-1)]^3}{12\,(100)} = 3.6.$$

b. We compute $f'''(x) = 60x^2$ and $f^{(4)}(x) = 120x$. $f^{(4)}$ is clearly increasing on $(-1, 2)$, so we can take $M = f^{(4)}(2) = 240$. Therefore, using Formula (8), we see that an error bound is

$$\frac{M(b-a)^3}{180n^4} = \frac{240\,(3)^5}{180\,(10^4)} \approx 0.0324.$$

25. a. Here $a = 1$, $b = 3$, $n = 10$, and $f(x) = \dfrac{1}{x}$. We find $f'(x) = -\dfrac{1}{x^2}$ and $f''(x) = \dfrac{2}{x^3}$. Because $f'''(x) = -\dfrac{6}{x^4} < 0$ on $(1, 3)$, we see that f'' is decreasing there. We may take $M = f''(1) = 2$. Using Formula (7), we find the error bound $\dfrac{M(b-a)^3}{12n^2} = \dfrac{2\,(3-1)^3}{12\,(100)} \approx 0.013$.

b. $f'''(x) = -\dfrac{6}{x^4}$ and $f^{(4)}(x) = \dfrac{24}{x^5}$. $f^{(4)}(x)$ is decreasing on $(1, 3)$, so we can take $M = f^{(4)}(1) = 24$. Using Formula (8), we find the error bound $\dfrac{24\,(3-1)^5}{180\,(10^4)} \approx 0.00043$.

27. a. Here $a = 0$, $b = 2$, $n = 8$, and $f(x) = (1+x)^{-1/2}$. We find $f'(x) = -\frac{1}{2}(1+x)^{-3/2}$ and $f''(x) = \frac{3}{4}(1+x)^{-5/2}$. Because f'' is positive and decreasing on $(0, 2)$, we see that $\left|f''(x)\right| \leq \frac{3}{4}$. So the maximum error is $\dfrac{\frac{3}{4}(2-0)^3}{12\,(8)^2} = 0.0078125$.

b. $f''' = -\frac{15x}{8}(1+x)^{-7/2}$ and $f^{(4)}(x) = \frac{105}{16}(1+x)^{-9/2}$. Because $f^{(4)}$ is positive and decreasing on $(0, 2)$, we find $\left|f^{(4)}(x)\right| \leq \frac{105}{16}$. Therefore, the maximum error is $\dfrac{\frac{105}{16}(2-0)^5}{180\,(8)^4} = 0.000285$.

29. The distance covered is given by

$$d = \int_0^2 V(t)\, dt = \frac{1/4}{2}\left[V(0) + 2V\left(\frac{1}{4}\right) + \cdots + 2V\left(\frac{7}{4}\right) + V(2)\right]$$

$$= \frac{1}{8}\,[19.5 + 2\,(24.3) + 2\,(34.2) + 2\,(40.5) + 2\,(38.4) + 2\,(26.2) + 2\,(18) + 2\,(16) + 8] \approx 52.84, \text{ or } 52.84 \text{ miles}.$$

© 2011 Cengage Learning. All Rights Reserved. May not be scanned, copied or duplicated, or posted to a publicly accessible website, in whole or in part.

31. $\frac{1}{13} \int_0^{13} f(t)\,dt = \left(\frac{1}{13}\right)\left(\frac{1}{2}\right)\{13.2 + 2\,[14.8 + 16.6 + 17.2 + 18.7 + 19.3 + 22.6 + 24.2 + 25$

$$+ 24.6 + 25.6 + 26.4 + 26.6] + 26.6\} \approx 21.65, \text{ or } 21.65 \text{ mpg.}$$

33. The average daily consumption of oil is $A = \dfrac{1}{b-1}\int_a^b f(t)\,dt$, where $f(t)$ has the values shown in the table, with

$t = 0$ in 1980. Using Simpson's Rule with $n = 10$ and $\Delta t = 2$, we have

$$A = \tfrac{1}{20-0}\int_0^{20} f(t)\,dt \approx \tfrac{1}{20}\cdot\tfrac{2}{3}\left[f(0) + 4f(2) + 2f(4) + 4f(6) + \cdots + 4f(18) + f(20)\right]$$

$$= \tfrac{1}{30}\,[17.1 + 4\,(15.3) + 2\,(15.7) + 4\,(16.3) + 2\,(17.3)$$

$$+ 4\,(17) + 2\,(17) + 4\,(17.7) + 2\,(18.3) + 4\,(18.9) + 19.7]$$

$$= 17.14, \text{ or } 17.14 \text{ million barrels.}$$

35. The required rate of flow is

$R = $ (rate of flow) (area of cross-section of the river) $= (4)$ (area of cross-section) $= 4\int_0^{78} y(x)\,dx$.

Approximating the integral using the trapezoidal rule, we have

$$R \approx (4)\left(\tfrac{6}{2}\right)[0.8 + 2\,(2.6) + 2\,(5.8) + 2\,(6.2) + 2\,(8.2) + 2\,(10.1) + 2\,(10.8)$$

$$+ 2\,(9.8) + 2\,(7.6) + 2\,(6.4) + 2\,(5.2) + 2\,(3.9) + 2\,(2.4) + 1.4] = 1922.4, \text{ or } 1922.4 \text{ ft}^3/\text{sec.}$$

37. We solve the equation $8 = \sqrt{0.01x^2 + 0.11x + 38}$, finding $64 = 0.01x^2 + 0.11x + 38$, $0.01x^2 + 0.11x - 26 = 0$,

$x^2 + 11x - 2600 = 0$, and so $x = \dfrac{-11 \pm \sqrt{121 + 10{,}400}}{2} \approx 45.786$ (we choose the positive root). Therefore,

$PS = (8)\,(45.786) - \int_0^{45.786} \sqrt{0.01x^2 + 0.11x + 38}\,dx$. We have $\Delta x = \dfrac{45.786}{8} = 5.72$, so $x_0 = 0$, $x_1 = 5.72$,

$x_2 = 11.44, \ldots, x_8 = 45.79$.

a. $PS = 366.288 - \tfrac{5.72}{2}\left[\sqrt{38} + 2\sqrt{0.01\,(5.72)^2 + 0.11\,(5.72) + 38} + \cdots\right.$

$$\left. + \sqrt{0.01\,(45.79)^2 + 0.11\,(45.79) + 38}\right] \approx 51{,}558, \text{ or } \$51{,}558.$$

b. $PS = 366.288 - \tfrac{5.72}{2}\left[\sqrt{38} + 4\sqrt{0.01\,(5.72)^2 + 0.11\,(5.72) + 38} + \cdots\right.$

$$\left. + \sqrt{0.01\,(45.79)^2 + 0.11\,(45.79) + 38}\right] \approx 51{,}708, \text{ or } \$51{,}708.$$

39. The average petroleum reserves from 1981 through 1990 were $A = \dfrac{1}{9-0}\int_0^9 S(t)\,dt = \dfrac{1}{9}\int_0^9 \dfrac{613.7t^2 + 1449.1}{t^2 + 6.3}\,dt$.

Using the trapezoidal rule with $a = 0$, $b = 9$, and $n = 9$, so that $\Delta t = \tfrac{9-0}{9} = 1$, we have $t_0 = 0$, $t = 1, \ldots, t_9 = 9$.

Thus,

$$A = \tfrac{1}{9}\int_0^9 S(t)\,dt = \left(\tfrac{1}{9}\right)\left(\tfrac{1}{2}\right)\left[S(0) + 2S(1) + 2f(x) + 2f(x) + \cdots + S(9)\right]$$

$$\approx \tfrac{1}{18}\,[130.02 + 2\,(282.58) + 2\,(379.02) + 2\,(455.71) + 2\,(505.30)$$

$$+ 2\,(536.47) + 2\,(556.56) + 2\,(569.99) + 2\,(579.32) + 586.01] \approx 474.77,$$

or approximately 474.77 million barrels.

41. $\Delta x = \dfrac{40{,}000 - 30{,}000}{10} = 1000$, so $x_0 = 30{,}000$, $x_1 = 31{,}000$, $x_2, \ldots,$

$x_{10} = 40{,}000$. Now we approximate $P = \dfrac{100}{2000\sqrt{2\pi}}\int_{30{,}000}^{40{,}000} e^{-0.5[(x-40{,}000)/2000]^2}\,dx$ by

$$P = \dfrac{100(1000)}{2000\sqrt{2\pi}}\left[e^{-0.5[(30{,}000-40{,}000)/2000]^2} + 4e^{-0.5[(31{,}000-40{,}000)/2000]^2} + \cdots + 1\right] \approx 0.50, \text{ or } 50\%.$$

© 2011 Cengage Learning. All Rights Reserved. May not be scanned, copied or duplicated, or posted to a publicly accessible website, in whole or in part.

43. $R = \dfrac{60D}{\int_0^T C(t)\,dt} = \dfrac{480}{\int_0^{24} C(t)\,dt}$. Now

$\int_0^{24} C(t)\,dt \approx \frac{24}{12} \cdot \frac{1}{3}\,[0 + 4\,(0) + 2\,(2.8) + 4\,(6.1) + 2\,(9.7) + 4\,(7.6) + 2\,(4.8)$
$$+ 4\,(3.7) + 2\,(1.9) + 4\,(0.8) + 2\,(0.3) + 4\,(0.1) + 0] \approx 74.8$$

and $R = \frac{480}{74.8} \approx 6.42$, or 6.42 liters/min.

45. False. The number n can be odd or even.

47. True.

49. Taking the limit and recalling the definition of a Riemann sum, we find

$\displaystyle\lim_{\Delta t \to 0} \dfrac{c\,(t_1)\,R\,\Delta t + c\,(t_2)\,R\,\Delta t + \cdots + c\,(t_n)\,R\,\Delta t}{60} = D, \ \dfrac{R}{60}\lim_{\Delta t \to 0}\,[c\,(t_1)\,\Delta t + c\,(t_2)\,\Delta t + \cdots + c\,(t_n)\,\Delta t] = D,$

$\dfrac{R}{60}\displaystyle\int_0^T c\,(t)\,dt = D$, and so $R = \dfrac{60D}{\int_0^T c\,(t)\,dt}$.

7.4 Problem-Solving Tips

1. The improper integral on the left-hand side of the equation $\int_{-\infty}^{\infty} f(x)\,dx = \int_{-\infty}^{c} f(x)\,dx + \int_{c}^{\infty} f(x)\,dx$ is convergent only if both integrals on the right-hand side of the equation converge.

2. It is often convenient to choose $c = 0$ in formula in Tip 1.

7.4 Concept Questions   page 518

1. a. $\int_a^{\infty} f(x)\,dx = \displaystyle\lim_{b \to \infty} \int_a^b f(x)\,dx$

  b. $\int_{-\infty}^{b} f(x)\,dx = \displaystyle\lim_{a \to -\infty} \int_a^b f(x)\,dx$

  c. $\int_{-\infty}^{\infty} f(x)\,dx = \int_{-\infty}^{c} f(x)\,dx + \int_c^{\infty} f(x)\,dx$, where c is any real number.

7.4 Improper Integrals   page 518

1. The required area is given by $\displaystyle\int_3^{\infty} \frac{2}{x^2}\,dx = \lim_{b \to \infty} \int_3^b \frac{2}{x^2}\,dx = \lim_{b \to \infty} \left(-\frac{2}{x}\right)\Big|_3^b = \lim_{b \to \infty}\left(-\frac{2}{b} + \frac{2}{3}\right) = \frac{2}{3}$.

3. $A = \displaystyle\int_3^{\infty} \frac{1}{(x-2)^2}\,dx = \lim_{b \to \infty} \int_3^b (x-2)^{-2}\,dx = \lim_{b \to \infty}\left(-\frac{1}{x-2}\right)\Big|_3^b = \lim_{b \to \infty}\left(-\frac{1}{b-2} + 1\right) = 1$.

5. $A = \displaystyle\int_1^{\infty} \frac{1}{x^{3/2}}\,dx = \lim_{b \to \infty} \int_1^b x^{-3/2}\,dx = \lim_{b \to \infty}\left(-\frac{2}{\sqrt{x}}\right)\Big|_1^b = \lim_{b \to \infty}\left(-\frac{2}{\sqrt{b}} + 2\right) = 2$.

© 2011 Cengage Learning. All Rights Reserved. May not be scanned, copied or duplicated, or posted to a publicly accessible website, in whole or in part.

7. $A = \int_0^\infty \dfrac{1}{(x+1)^{5/2}}\, dx = \lim\limits_{b \to \infty} \int_1^b (x+1)^{-5/2}\, dx = \lim\limits_{b \to \infty} \left[-\tfrac{2}{3}(x+1)^{-3/2} \right]_0^b$

$\quad = \lim\limits_{b \to \infty} \left[-\dfrac{2}{3(b+1)^{3/2}} + \dfrac{2}{3} \right] = \dfrac{2}{3}.$

9. $A = \int_{-\infty}^2 e^{2x}\, dx = \lim\limits_{a \to -\infty} \int_a^2 e^{2x}\, dx = \lim\limits_{a \to -\infty} \tfrac{1}{2}e^{2x} \Big|_a^2 = \lim\limits_{a \to -\infty} \left(\tfrac{1}{2}e^4 - \tfrac{1}{2}e^{2a} \right) = \tfrac{1}{2}e^4.$

11. Using symmetry, the required area is given by $2\int_0^\infty \dfrac{x}{(1+x^2)^2}\, dx = 2 \lim\limits_{b \to \infty} \int_0^\infty \dfrac{x}{(1+x^2)^2}\, dx.$

To evaluate the indefinite integral $\int \dfrac{x}{(1+x^2)^2}\, dx$, put $u = 1 + x^2$, so $du = 2x\, dx$ and

$x\, dx = \tfrac{1}{2}\, du.$ Then $\int \dfrac{x}{(1+x^2)^2}\, dx = \dfrac{1}{2}\int \dfrac{du}{u^2} = -\dfrac{1}{2u} + C = -\dfrac{1}{2(1+x^2)} + C.$ Therefore,

$2 \lim\limits_{b \to \infty} \int_0^b \dfrac{x}{(1+x^2)}\, dx = \lim\limits_{b \to \infty} \left[-\dfrac{1}{(1+x^2)^2} \right]_0^b = \lim\limits_{b \to \infty} \left[-\dfrac{1}{(1+b^2)} + 1 \right] = 1.$

13. a. $I(b) = \int_0^b \sqrt{x}\, dx = \tfrac{2}{3}x^{3/2} \Big|_0^b = \tfrac{2}{3}b^{3/2}.$

b. $\lim\limits_{b \to \infty} I(b) = \lim\limits_{b \to \infty} \tfrac{2}{3}b^{3/2} = \infty.$

15. $\int_1^\infty \dfrac{3}{x^4}\, dx = \lim\limits_{b \to \infty} \int_1^b 3x^{-4}\, dx = \lim\limits_{b \to \infty} \left(-\dfrac{1}{x^3} \right) \Big|_1^b = \lim\limits_{b \to \infty} \left(-\dfrac{1}{b^3} + 1 \right) = 1.$

17. $A = \int_4^\infty \dfrac{2}{x^{3/2}}\, dx = \lim\limits_{b \to \infty} \int_4^b 2x^{-3/2}\, dx = \lim\limits_{b \to \infty} \left(-4x^{-1/2} \right) \big|_4^b = \lim\limits_{b \to \infty} \left(-\dfrac{4}{\sqrt{b}} + 2 \right) = 2.$

19. $\int_1^\infty \dfrac{4}{x}\, dx = \lim\limits_{b \to \infty} \int_1^b \dfrac{4}{x}\, dx = \lim\limits_{b \to \infty} 4\ln x |_1^b = \lim\limits_{b \to \infty} (4\ln b) = \infty.$

21. $\int_{-\infty}^0 (x-2)^{-3}\, dx = \lim\limits_{a \to -\infty} \int_a^0 (x-2)^{-3}\, dx = \lim\limits_{a \to -\infty} \left[-\dfrac{1}{2(x-2)^2} \right]_a^0 = -\dfrac{1}{8}.$

23. $\int_1^\infty \dfrac{1}{(2x-1)^{3/2}}\, dx = \lim\limits_{b \to \infty} \int_1^b (2x-1)^{-3/2}\, dx = \lim\limits_{b \to \infty} \left[-\dfrac{1}{(2x-1)^{1/2}} \right]_1^b = \lim\limits_{b \to \infty} \left(-\dfrac{1}{\sqrt{2b-1}} + 1 \right) = 1.$

25. $\int_0^\infty e^{-x}\, dx = \lim\limits_{b \to \infty} \int_0^b e^{-x}\, dx = \lim\limits_{b \to \infty} \left(-e^{-x} \right) |_0^b = \lim\limits_{b \to \infty} \left(-e^{-b} + 1 \right) = 1.$

27. $\int_{-\infty}^0 e^{2x}\, dx = \lim\limits_{a \to -\infty} \tfrac{1}{2}e^{2x} \Big|_a^0 = \lim\limits_{a \to -\infty} \left(\tfrac{1}{2} - \tfrac{1}{2}e^{2a} \right) = \tfrac{1}{2}.$

29. We use the substitution $u = \sqrt{x}$: $\int_1^\infty \dfrac{e^{\sqrt{x}}}{\sqrt{x}}\, dx = \lim\limits_{b \to \infty} \int_1^b \dfrac{e^{\sqrt{x}}}{\sqrt{x}}\, dx = \lim\limits_{b \to \infty} \left(-2e^{\sqrt{x}} \right) \Big|_1^b = \lim\limits_{b \to \infty} \left(2e^{\sqrt{b}} - 2e \right) = \infty,$

and so the improper integral diverges.

31. Integrating by parts, we have

$\int_{-\infty}^0 xe^x\, dx = \lim\limits_{a \to -\infty} \int_a^0 xe^x\, dx = \lim\limits_{a \to -\infty} (x-1)e^x |_a^0 = \lim\limits_{a \to -\infty} \left[-1 + (a-1)e^a \right] = -1.$

© 2011 Cengage Learning. All Rights Reserved. May not be scanned, copied or duplicated, or posted to a publicly accessible website, in whole or in part.

33. $\int_{-\infty}^{\infty} x\, dx = \lim\limits_{a \to -\infty} \frac{1}{2}x^2 \Big|_{a}^{0} + \lim\limits_{b \to \infty} \frac{1}{2}x^2 \Big|_{0}^{b}$, both of which diverge, and so the integral diverges.

35. $\int_{-\infty}^{\infty} x^3 \left(1 + x^4\right)^{-2} dx = \int_{-\infty}^{0} x^3 \left(1 + x^4\right)^{-2} dx + \int_{0}^{\infty} x^3 \left(1 + x^4\right)^{-2} dx$

$$= \lim_{a \to -\infty} \int_{a}^{0} x^3 \left(1 + x^4\right)^{-2} dx + \lim_{b \to \infty} \int_{0}^{b} x^3 \left(1 + x^4\right)^{-2} dx$$

$$= \lim_{a \to -\infty} \left[-\frac{1}{4} \left(1 + x^4\right)^{-1} \right]_{a}^{0} + \lim_{b \to \infty} \left[-\frac{1}{4} \left(1 + x^4\right)^{-1} \right]_{0}^{b}$$

$$= \lim_{a \to -\infty} \left[-\frac{1}{4} + \frac{1}{4 \left(1 + a^4\right)} \right] + \lim_{b \to \infty} \left[-\frac{1}{4 \left(1 + b^4\right)} + \frac{1}{4} \right] = -\frac{1}{4} + \frac{1}{4} = 0.$$

37. $\int_{-\infty}^{\infty} x e^{1 - x^2} dx = \lim\limits_{a \to -\infty} \int_{a}^{0} x e^{1 - x^2} dx + \lim\limits_{b \to \infty} \int_{0}^{b} x e^{1 - x^2} dx = \lim\limits_{a \to -\infty} -\frac{1}{2} e^{1 - x^2} \Big|_{a}^{0} + \lim\limits_{b \to \infty} -\frac{1}{2} e^{1 - x^2} \Big|_{0}^{b}$

$$= \lim_{a \to -\infty} \left(-\frac{1}{2}e + \frac{1}{2}e^{1 - a^2} \right) + \lim_{b \to \infty} \left(-\frac{1}{2}e^{1 - b^2} + \frac{1}{2}e \right) = 0.$$

39. $\int_{-\infty}^{\infty} \dfrac{e^{-x}}{1 + e^{-x}} dx = \lim\limits_{a \to -\infty} \left[-\ln \left(1 + e^{-x}\right) \right]_{a}^{0} + \lim\limits_{b \to \infty} \left[-\ln \left(1 + e^{-x}\right) \right]_{0}^{b} = \infty$, so the integral diverges.

41. First, we find the indefinite integral $I = \int \dfrac{dx}{x \ln^3 x}$. Let $u = \ln x$, so

$du = \dfrac{1}{x} dx$. Then $I = \int \dfrac{du}{u^3} = -\dfrac{1}{2u^2} + C = -\dfrac{1}{2 \ln^2 x} + C$, so

$\int_{e}^{\infty} \dfrac{dx}{x \ln^3 x} = \lim\limits_{b \to \infty} \int_{e}^{b} \dfrac{dx}{x \ln^3 x} = \lim\limits_{b \to \infty} \left(-\dfrac{1}{2 \ln^2 x} \right) \Big|_{e}^{b} = \lim\limits_{b \to \infty} \left[-\dfrac{1}{2 (\ln b)^2} + \dfrac{1}{2} \right] = \dfrac{1}{2}$, and so the given integral
converges.

43. We want the present value PV of a perpetuity with $m = 1$, $P = 1500$, and $r = 0.08$. We find
$PV = \dfrac{(1)(1500)}{0.08} = 18{,}750$, or \$18,750.

45. Integrating by parts, we find

$PV = \int_{0}^{\infty} (10{,}000 + 4000t)\, e^{-rt}\, dt = 10{,}000 \int_{0}^{\infty} e^{-rt}\, dt + 4000 \int_{0}^{\infty} t e^{-rt}\, dt$

$$= \lim_{b \to \infty} \left[-\frac{10{,}000}{r} e^{-rt} - \frac{4000}{r^2} (rt + 1)\, e^{-rt} \right]_{0}^{b} = \frac{10{,}000}{r} + \frac{4000}{r^2} = \frac{10{,}000r + 4000}{r^2} \text{ dollars.}$$

47. True. $\int_{a}^{\infty} f(x)\, dx = \int_{a}^{b} f(x)\, dx + \int_{b}^{\infty} f(x)\, dx$, so if $\int_{a}^{\infty} f(x)\, dx$ exists, then
$\int_{b}^{\infty} f(x)\, dx = \int_{a}^{\infty} f(x)\, dx - \int_{a}^{b} f(x)\, dx$.

49. False. Let $f(x) = \begin{cases} e^{-2x} & \text{if } x \le 0 \\ e^{-x} & \text{if } x > 0 \end{cases}$. Then $\int_{-\infty}^{\infty} f(x)\, dx = \int_{-\infty}^{0} e^{2x}\, dx + \int_{0}^{\infty} e^{-x}\, dx = \frac{1}{2} + 1 = \frac{3}{2}$. But
$2 \int_{0}^{\infty} f(x)\, dx = 2 \int_{0}^{\infty} e^{-x}\, dx = 2$.

51. a. $CV \approx \int_{0}^{\infty} R e^{-it}\, dt = \lim\limits_{b \to \infty} \int_{0}^{b} R e^{-it}\, dt = \lim\limits_{b \to \infty} \left(-\dfrac{R}{i} e^{-it} \right) \Big|_{0}^{b} = \lim\limits_{b \to \infty} \left(-\dfrac{R}{i} e^{-ib} + \dfrac{R}{i} \right) = \dfrac{R}{i}$.

b. $CV \approx \dfrac{10{,}000}{0.12} \approx 83{,}333$, or \$83,333.

© 2011 Cengage Learning. All Rights Reserved. May not be scanned, copied or duplicated, or posted to a publicly accessible website, in whole or in part.

53. $\int_{-\infty}^{b} e^{px}\, dx = \lim_{a \to -\infty} \int_{a}^{b} e^{px}\, dx = \lim_{a \to -\infty} \left(\frac{1}{p} e^{px}\right)\Big|_{a}^{b} = \lim_{a \to -\infty} \left(\frac{1}{p} e^{pb} - \frac{1}{p} e^{pa}\right) = -\frac{1}{p} e^{pa}$ if $p > 0$. If $p < 0$,

the integral diverges.

7.5   Concept Questions   page 526

1. $V = \int_{a}^{b} [f(x)]^2\, dx$

3. $V = \pi \int_{a}^{b} \left\{[f(x)]^2 - [g(x)]^2\right\} dx + \pi \int_{c}^{b} \left\{[f(x)]^2 - [g(x)]^2\right\} dx$

7.5   Volumes of Solids of Revolution   page 526

1. $V = \pi \int_{0}^{2} y^2\, dx = \pi \int_{0}^{2} \left(\frac{1}{2}x\right)^2 dx = \frac{\pi}{4} \int_{0}^{2} x^2\, dx = \frac{\pi}{4} \left(\frac{1}{3}x^3\right)\Big|_{0}^{2} = \frac{2\pi}{3}.$

3. $V = \pi \int_{-1}^{2} y^2\, dx = \pi \int_{-1}^{2} \left[(x-1)^2 + 2\right]^2 dx = \pi \int_{-1}^{2} (x^4 - 4x^3 + 10x^2 - 12x + 9)\, dx$

$= \pi \left(\frac{1}{5}x^5 - x^4 + \frac{10}{3}x^3 - 6x^2 + 9x\right)\Big|_{-1}^{2} = \frac{153\pi}{5}.$

5. $V = \pi \int_{a}^{b} [f(x)]^2\, dx = \pi \int_{0}^{1} (3x)^2\, dx = 9\pi \int_{0}^{1} x^2\, dx = 3\pi x^3\big|_{0}^{1} = 3\pi.$

7. $V = \pi \int_{a}^{b} [f(x)]^2\, dx = \pi \int_{1}^{4} (\sqrt{x})^2\, dx = \pi \int_{1}^{4} x\, dx = \frac{\pi}{2}(16 - 1) = \frac{15\pi}{2}.$

9. $V = \pi \int_{a}^{b} [f(x)]^2\, dx = \pi \int_{0}^{1} \left(\sqrt{1+x^2}\right)^2 dx = \pi \int_{0}^{1} (1 + x^2)\, dx = \pi \left(x + \frac{1}{3}x^3\right)\Big|_{0}^{1} = \pi\left(1 + \frac{1}{3}\right) = \frac{4\pi}{3}.$

11. $V = \pi \int_{a}^{b} [f(x)]^2\, dx = \pi \int_{-1}^{1} (1 - x^2)^2\, dx = \pi \int_{-1}^{1} (1 - 2x^2 + x^4)\, dx = \pi \left(x - \frac{2}{3}x^3 + \frac{1}{5}\right)x^5\Big|_{-1}^{1}$

$= \pi \left[\left(1 - \frac{2}{3} + \frac{1}{5}\right) - \left(-1 + \frac{2}{3} - \frac{1}{5}\right)\right] = \frac{16\pi}{15}.$

13. $V = \pi \int_{a}^{b} [f(x)]^2\, dx = \pi \int_{0}^{1} (e^x)^2\, dx = \pi \int_{0}^{1} e^2\, dx = \frac{\pi}{2} e^{2x}\big|_{0}^{1} = \frac{\pi}{2}(e^2 - 1).$

15. $V = \pi \int_{a}^{b} \left\{[f(x)]^2 - [g(x)]^2\right\} dx = \pi \int_{0}^{1} \left[x^2 - (x^2)^2\right] dx = \pi \int_{0}^{1} (x^2 - x^4)\, dx = \pi \left(\frac{1}{3}x^3 - \frac{1}{5}x^5\right)\Big|_{0}^{1}$

$= \pi \left(\frac{1}{3} - \frac{1}{5}\right) = \frac{2\pi}{15}.$

17. $V = \pi \int_{a}^{b} \left\{[f(x)]^2 - [g(x)]^2\right\} dx = \pi \int_{a}^{b} \left\{[f(x)]^2 - [g(x)]^2\right\} dx = \pi \int_{-1}^{1} \left[(4 - x^2)^2 - 3^2\right] dx$

$= \pi \int_{-1}^{1} (16 - 8x^2 + x^4 - 9)\, dx = \pi \int_{-1}^{1} (7 - 8x^2 + x^4)\, dx = \pi \left(7x - \frac{8}{3}x^3 + \frac{1}{5}x^5\right)\Big|_{-1}^{1}$

$= \pi \left[\left(7 - \frac{8}{3} + \frac{1}{5}\right) - \left(-7 + \frac{8}{3} - \frac{1}{5}\right)\right] = \frac{136\pi}{15}.$

19. $V = \pi \int_{a}^{b} \left\{[f(x)]^2 - [g(x)]^2\right\} dx = \pi \int_{0}^{2\sqrt{2}} \left[\left(\sqrt{16 - x^2}\right)^2 - x^2\right] dx = \pi \int_{0}^{2\sqrt{2}} (16 - x^2 - x^2)\, dx$

$= \pi \int_{0}^{2\sqrt{2}} (16 - 2x^2)\, dx = \pi \left(16x - \frac{2}{3}x^3\right)\Big|_{0}^{2\sqrt{2}} = \pi \left[32\sqrt{2} - \frac{2}{3}\left(2\sqrt{2}\right)^3\right] = \pi \left[32\sqrt{2} - \frac{32\sqrt{2}}{3}\right] = \frac{64\sqrt{2}\pi}{3}.$

© 2011 Cengage Learning. All Rights Reserved. May not be scanned, copied or duplicated, or posted to a publicly accessible website, in whole or in part.

21. $V = \pi \int_a^b \left\{ [f(x)]^2 - [g(x)]^2 \right\} dx = \pi \int_0^1 \left[(e^x)^2 - (e^{-x})^2 \right] dx = \pi \int_0^1 \left(e^{3x} - e^{-2x} \right) dx = \pi \int_0^1 \left(e^{3x} - e^{-2x} \right) dx$

$= \pi \left(\frac{1}{2} e^{2x} + \frac{1}{2} e^{-2x} \right) \Big|_0^1 = \pi \left[\left(\frac{1}{2} e^2 + \frac{1}{2} e^{-2} \right) - \left(\frac{1}{2} + \frac{1}{2} \right) \right] = \frac{\pi}{2} \left(e^2 - 2 + e^{-2} \right).$

23. The points of intersection of the curves are found by solving the

simultaneous system of equations $y = x$ and $y = \sqrt{x}$, giving $x = \sqrt{x}$,

$x^2 = x$, $x^2 - x = 0$, and $x(x-1) = 0$, so $x = 0$ or $x = 1$. Thus,

$V = \pi \int_a^b \left\{ [f(x)]^2 - [g(x)]^2 \right\} dx = \pi \int_0^1 \left[(\sqrt{x})^2 - x^2 \right] dx$

$= \pi \int_0^1 (x - x^2) \, dx = \pi \left(\frac{1}{2} x^2 - \frac{1}{3} x^3 \right) \Big|_0^1 = \pi \left(\frac{1}{2} - \frac{1}{3} \right) = \frac{\pi}{6}.$

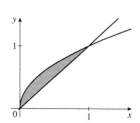

25. The points of intersection of the curves are found by solving the

simultaneous system of equations $y = x^2$ and $y = \frac{1}{2} x + 3$, so

$2x^2 - x - 6 = 0$, $(2x + 3)(x - 2) = 0$, and so $x = -\frac{3}{2}$ or $x = 2$. Thus,

$V = \pi \int_a^b \left\{ [f(x)]^2 - [g(x)]^2 \right\} dx = \pi \int_{-3/2}^2 \left[\left(\frac{1}{2} x + 3 \right)^2 - (x^2)^2 \right] dx$

$= \pi \int_{-3/2}^2 \left(\frac{1}{4} x^2 + 3x + 9 - x^4 \right) dx = \pi \left(\frac{1}{12} x^3 + \frac{3}{2} x^2 + 9x - \frac{1}{5} x^5 \right) \Big|_{-3/2}^2$

$= \left[\left(\frac{2}{3} + 6 + 18 - \frac{32}{5} \right) - \left(-\frac{27}{96} + \frac{27}{8} - \frac{27}{2} + \frac{243}{160} \right) \right] \pi = \frac{6517\pi}{240}.$

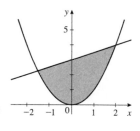

27. To find the points of intersection of the curves, we solve $y = x^2$ and

$y = 4 - x^2$, simultaneously, obtaining $x^2 = 4 - x^2$, $2x^2 = 4$, and

$x^2 = 2$. Thus, $x = \pm\sqrt{2}$. The required volume is

$V = \pi \int_a^b \left\{ [f(x)]^2 - [g(x)]^2 \right\} dx = \pi \int_{-\sqrt{2}}^{\sqrt{2}} \left[(4 - x^2)^2 - (x^2)^2 \right] dx$

$= 2\pi \int_0^{\sqrt{2}} \left[16 - 8x^2 + x^4 - x^4 \right] dx = \pi \int_{-\sqrt{2}}^{\sqrt{2}} \left[(4 - x^2)^2 - (x^2)^2 \right] dx$

$= 16\pi \left[2\sqrt{2} - \frac{1}{3} \left(2\sqrt{2} \right) \right] = \frac{64\sqrt{2}\pi}{3}.$

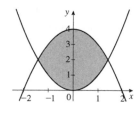

29. To find the points of intersection of $y = 2x$ and $y = \frac{1}{x}$, we solve

$2x = \frac{1}{x}$, obtaining $2x^2 = 1$, so $x = \pm\frac{1}{\sqrt{2}} = \pm\frac{\sqrt{2}}{2}$. To find the points of

intersection of $y = x$ and $y = \frac{1}{x}$, we solve $x = \frac{1}{x}$, giving $x = \pm 1$. By

symmetry,

$V = 2\pi \int_0^{\sqrt{2}/2} \left[(2x)^2 - x^2 \right] dx + 2\pi \int_{\sqrt{2}/2}^1 \left[\left(\frac{1}{x} \right)^2 - x^2 \right] dx$

$= 2\pi \int_0^{\sqrt{2}/2} \left(4x^2 - x^2 \right) dx + 2\pi \int_{\sqrt{2}/2}^1 \left(\frac{1}{x^2} - x^2 \right) dx = 2\pi \int_0^{\sqrt{2}/2} 3x^2 \, dx + 2\pi \int_{\sqrt{2}/2}^1 \left(x^{-2} - x^2 \right) dx$

$= \left[2\pi x^3 \right]_0^{\sqrt{2}/2} + \left[2\pi \left(-\frac{1}{x} - \frac{1}{3} x^3 \right) \right]_{\sqrt{2}/2}^1 = 2\pi \left(\frac{2\sqrt{2}}{8} \right) + 2\pi \left[\left(-1 - \frac{1}{3} \right) - \left(-\frac{2}{\sqrt{2}} - \frac{\sqrt{2}}{12} \right) \right] = \frac{8(\sqrt{2}-1)\pi}{3}.$

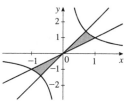

31. $V = \pi \int_a^b \left[f(x) \right]^2 dx = \pi \int_{-r}^r \left(\sqrt{r^2 - x^2} \right)^2 dx = \pi \int_{-r}^r \left(r^2 - x^2 \right) dx = 2\pi \int_0^r \left(r^2 - x^2 \right) dx$

$= 2\pi \left(r^2 x - \frac{1}{3} x^3 \right) \Big|_0^r = 2\pi \left(r^3 - \frac{1}{3} x^3 \right) = \frac{4}{3} \pi r^3.$

© 2011 Cengage Learning. All Rights Reserved. May not be scanned, copied or duplicated, or posted to a publicly accessible website, in whole or in part.

33. $V = \pi \int_{-10}^{0} x^2 \, dy = \pi \int_{-10}^{0} \left[f(y) \right]^2 dy$. Solving the given equation for x in terms of y,

we have $\dfrac{y}{10} = \left(\dfrac{x}{100} \right)^2 - 1$, so $\left(\dfrac{x}{100} \right)^2 = 1 + \dfrac{y}{10}$ and $x = 100\sqrt{1 + \dfrac{y}{10}}$. Therefore,

$V = \pi \int_{-10}^{0} 10{,}000 \left(1 + \frac{1}{10} y \right) dy = 10{,}000\pi \left(y + \frac{1}{20} y^2 \right) \Big|_{-10}^{0} = -10{,}000\pi \left(-10 + \frac{100}{20} \right) = 50{,}000\pi$ ft^3.

CHAPTER 7   **Concept Review**   page 528

1. product, $uv - \int v \, du$, u, easy to integrate

3. $\dfrac{\Delta x}{2} \left[f(x_0) + 2f(x_1) + 2f(x_2) + \cdots + 2f(x_{n-1}) + f(x_n) \right]$, $\dfrac{M(b-a)^3}{12n^2}$

5. $\lim\limits_{a \to -\infty} \int_{a}^{b} f(x) \, dx$, $\lim\limits_{b \to \infty} \int_{a}^{b} f(x) \, dx$, $\int_{-\infty}^{c} f(x) \, dx + \int_{c}^{\infty} f(x) \, dx$

CHAPTER 7   **Review**   page 529

1. Let $u = 2x$ and $dv = e^{-x} \, dx$, so $du = 2 \, dx$ and $v = -e^{-x}$. Then
$\int 2xe^{-x} \, dx = uv - \int v \, du = -2xe^{-x} + 2 \int e^{-x} \, dx = -2xe^{-x} - 2e^{-x} + C = -2(1+x)e^{-x} + C.$

3. Let $u = \ln 5x$ and $dv = dx$, so $du = \frac{1}{x} dx$ and $v = x$. Then
$\int \ln 5x \, dx = x \ln 5x \, dx - \int dx = x \ln 5x - x + C = x(\ln 5x - 1) + C.$

5. Let $u = x$ and $dv = e^{-2x} \, dx$, so $du = dx$ and $v = -\frac{1}{2} e^{-2x}$. Then
$\int_{0}^{1} xe^{-2x} \, dx = -\frac{1}{2} xe^{-2x} \Big|_{0}^{1} + \frac{1}{2} \int_{0}^{1} e^{-2x} \, dx = -\frac{1}{2} e^{-2} - \left(\frac{1}{4} e^{-2x} \right) \Big|_{0}^{1} = -\frac{1}{2} e^{-2} - \frac{1}{4} e^{-2} + \frac{1}{4} = \frac{1}{4} \left(1 - 3e^{-2} \right).$

7. $f(x) = \displaystyle\int f'(x) \, dx = \int \dfrac{\ln x}{\sqrt{x}} \, dx$. To evaluate the integral, we use parts

with $u = \ln x$, $dv = x^{-1/2} \, dx$, $du = \frac{1}{x} dx$ and $v = 2x^{1/2} \, dx$. Then

$\displaystyle\int \dfrac{\ln x}{x^{1/2}} \, dx = 2x^{1/2} \ln x - \int 2x^{-1/2} \, dx = 2x^{1/2} \ln x - 4x^{1/2} + C = 2x^{1/2} (\ln x - 2) + C = 2\sqrt{x} (\ln x - 2) + C.$

But $f(1) = -2$, giving $2\sqrt{1} (\ln 1 - 2) + C = -2$, so $C = 2$. Therefore, $f(x) = 2\sqrt{x} (\ln x - 2) + 2$.

9. Using Formula (4) with $a = 3$ and $b = 2$, we have $\displaystyle\int \dfrac{x^2}{(3 + 2x)^2} \, dx = \dfrac{1}{8} \left(3 + 2x - \dfrac{9}{3 + 2x} - 6 \ln |3 + 2x| \right) + C.$

11. Using Formula (24) with $a = 4$ and $n = 2$, we have $\int x^2 e^{4x} \, dx = \frac{1}{4} x^2 e^{4x} - \frac{1}{2} \int x e^{4x} \, dx$. Now we use

Formula (23) to obtain $\int x^2 e^{4x} \, dx = \frac{1}{4} x^2 e^{4x} - \frac{1}{2} \left[\frac{1}{16} (4x - 1) e^{4x} \right] + C = \frac{1}{32} \left(8x^2 - 4x + 1 \right) e^{4x} + C.$

13. Using Formula (17) with $a = 2$, we have $\displaystyle\int \dfrac{dx}{x^2 \sqrt{x^2 - 4}} = \dfrac{\sqrt{x^2 - 4}}{4x} + C.$

15. $\int_{0}^{\infty} e^{-2x} \, dx = \lim\limits_{b \to \infty} \int_{0}^{b} e^{-2x} \, dx = \lim\limits_{b \to \infty} \left(-\frac{1}{2} e^{-2x} \right) \Big|_{0}^{b} = \lim\limits_{b \to \infty} \left(-\frac{1}{2} e^{-2b} + \frac{1}{2} \right) = \frac{1}{2}.$

© 2011 Cengage Learning. All Rights Reserved. May not be scanned, copied or duplicated, or posted to a publicly accessible website, in whole or in part.

17. $\int_3^\infty \frac{2}{x}\,dx = \lim_{b\to\infty} \int_3^b \frac{2}{x}\,dx = \lim_{b\to\infty} 2\ln x \Big|_3^b = \lim_{b\to\infty} (2\ln b - 2\ln 3) = \infty.$

19. $\int_2^\infty \frac{dx}{(1+2x)^2} = \lim_{b\to\infty} \int_2^b (1+2x)^{-2}\,dx = \lim_{b\to\infty} \left(\frac{1}{2}\right)(-1)(1+2x)^{-1}\Big|_2^b = \lim_{b\to\infty}\left[-\frac{1}{2(1+2b)} + \frac{1}{2(5)}\right] = \frac{1}{10}.$

21. $\Delta x = \frac{b-a}{n} = \frac{3-1}{4} = \frac{1}{2}$, so $x_0 = 1$, $x_1 = \frac{3}{2}$, $x_2 = 2$, $x_3 = \frac{5}{2}$, $x_4 = 3$.

Trapezoidal rule: $\int_1^3 \frac{dx}{1+\sqrt{x}} \approx \frac{\frac{1}{2}}{2}\left[\frac{1}{2} + \frac{2}{1+\sqrt{1.5}} + \frac{2}{1+\sqrt{2}} + \frac{2}{1+\sqrt{2.5}} + \frac{1}{1+\sqrt{3}}\right] \approx 0.8421.$

Simpson's Rule $\int_1^3 \frac{dx}{1+\sqrt{x}} \approx \frac{\frac{1}{2}}{3}\left[\frac{1}{2} + \frac{4}{1+\sqrt{1.5}} + \frac{2}{1+\sqrt{2}} + \frac{4}{1+\sqrt{2.5}} + \frac{1}{1+\sqrt{3}}\right] \approx 0.8404.$

23. $\Delta x = \frac{1-(-1)}{4} = \frac{1}{2}$, so $x_0 = -1$, $x_1 = -\frac{1}{2}$, $x_2 = 0$, $x_3 = \frac{1}{2}$, $x_4 = 1$.

Trapezoidal rule: $\int_{-1}^1 \sqrt{1+x^4}\,dx \approx \frac{0.5}{2}\left[\sqrt{2} + 2\sqrt{1+(-0.5)^4} + 2 + 2\sqrt{1+(0.5)^4} + \sqrt{2}\right] \approx 2.2379.$

Simpson's rule: $\int_{-1}^1 \sqrt{1+x^4}\,dx \approx \frac{0.5}{3}\left[\sqrt{2} + 4\sqrt{1+(-0.5)^4} + 2 + 4\sqrt{1+(0.5)^4} + \sqrt{2}\right] \approx 2.1791.$

25. a. Here $a = 0$, $b = 1$, and $f(x) = \frac{1}{x+1}$. We have $f'(x) = -\frac{1}{(x+1)^2}$ and $f''(x) = \frac{2}{(x+1)^3}$.

Because f'' is positive and decreasing on $(0, 1)$, it attains its maximum value of 2 at $x = 0$, so we take $M = 2$. Using Formula (7) from Section 7.3, we see that the maximum error incurred is

$$\frac{M(b-a)^3}{12n^2} = \frac{2(1^3)}{12(8^2)} = \frac{1}{384} \approx 0.002604.$$

b. We compute $f'''(x) = -\frac{6}{(x+1)^4}$ and $f^{(4)}(x) = \frac{24}{(x+1)^5}$. Because $f^{(4)}(x)$ is positive and decreasing on

$(0, 1)$, we take $M = 24$. The maximum error is $\frac{24(1^5)}{180(8^4)} = \frac{1}{30720} \approx 0.000033.$

27. Integrate by parts with $u = t$ and $dv = e^{-0.05t}$, so $du = dt$ and $v = -20e^{-0.05t}$. Thus,

$$S(t) = -20te^{-0.05t} + \int 20e^{-0.05t}\,dt = -20te^{-0.05t} - 400e^{-0.05t} + C = -20te^{-0.05t} - 400e^{-0.05t} + C$$

$$= -20e^{-0.05t}(t+20) + C.$$

The initial condition implies that $S(0) = 0$, giving $-20(20) + C = 0$, and so $C = 400$. Therefore, $S(t) = -20e^{-0.05t}(t+20) + 400$. By the end of the first year, the number of units sold is given by $S(12) = -20e^{-0.6}(32) + 400 = 48.761$, or 48,761 cartridges.

29. Trapezoidal rule: $A \approx \frac{100}{2}(0 + 480 + 520 + 600 + 680 + 680 + 800 + 680 + 600 + 440 + 0) = 274{,}000$, or 274,000 ft^2.

Simpson's rule: $A \approx \frac{100}{3}(0 + 960 + 520 + 1200 + 680 + 1360 + 800 + 1360 + 600 + 880 + 0) = 278{,}667$, or 278,667 ft^2.

31. We want the present value of a perpetuity with $m = 1$, $P = 10{,}000$, and $r = 0.09$. We find

$$PV = \frac{(1)(10{,}000)}{0.09} \approx 111{,}111, \text{ or approximately } \$111{,}111.$$

© 2011 Cengage Learning. All Rights Reserved. May not be scanned, copied or duplicated, or posted to a publicly accessible website, in whole or in part.

33. Solving the equation $x^{1/2} = x^2$, we find that $x^{1/2} \left(x^{3/2} - 1 \right) = 0$, and so $x = 0$ or 1. Next,

$$V = \pi \int_0^1 \left(x - x^4 \right) dx = \pi \left(\tfrac{1}{2} x^2 - \tfrac{1}{5} x^5 \right) \Big|_0^1 = \pi \left(\tfrac{1}{2} - \tfrac{1}{5} \right) = \tfrac{3\pi}{10}.$$

CHAPTER 7   **Before Moving On...**   page 530

1. Let $u = \ln x$ and $dv = x^2 \, dx$, so $du = \dfrac{1}{x} \, dx$ and $v = \tfrac{1}{3} x^3$. Then

$$\int x^2 \ln x \, dx = \tfrac{1}{3} x^3 \ln x - \int \tfrac{1}{3} x^2 \, dx = \tfrac{1}{3} x^3 \ln x - \tfrac{1}{9} x^3 + C = \tfrac{1}{9} x^3 \left(3 \ln x - 1 \right) + C.$$

2. $I = \displaystyle\int \dfrac{dx}{x^2 \sqrt{8 + 2x^2}}$. Let $u = \sqrt{2} x$, so $du = \sqrt{2} \, dx$ and $dx = \tfrac{\sqrt{2}}{2} \, du$. Then

$$I = \dfrac{\sqrt{2}}{2} \int \dfrac{du}{\tfrac{1}{2} u^2 \sqrt{8 + u^2}} = \sqrt{2} \int \dfrac{du}{u^2 \sqrt{\left(2\sqrt{2} \right)^2 + u^2}}. \text{ Using Formula (11) from Section 7.2 with } a = 2\sqrt{2} \text{ and}$$

$$x = u, \; I = \sqrt{2} \int \dfrac{du}{u^2 \sqrt{\left(2\sqrt{2} \right)^2 + u^2}} = \sqrt{2} \left(-\dfrac{\sqrt{8 + u^2}}{8u} \right) + C = -\dfrac{\sqrt{8 + 2x^2}}{8x} + C.$$

3. $n = 5$, so $\Delta x = \tfrac{4-2}{5} = 0.4$ and $x_0 = 2$, $x_1 = 2.4$, $x_2 = 2.8$, $x_3 = 3.2$, $x_4 = 3.6$, $x_5 = 4$. Thus,

$$\int_2^4 \sqrt{x^2 + 1} \, dx \approx \dfrac{0.4}{2} \left[f(2) + 2f(2.4) + 2f(2.8) + 2f(3.2) + 2f(3.2) + 2f(3.6) + f(4) \right]$$

$$= 0.2 \left[2.23607 + 2(2.6) + 2(2.97321) + 2(3.35261) + 2(3.73631) + 4.12311 \right] \approx 6.3367.$$

4. $n = 6$, so $\Delta x = \tfrac{3-1}{6} = \tfrac{1}{3}$ and $x_0 = 1$, $x_1 = \tfrac{4}{3}$, $x_2 = \tfrac{5}{3}$, $x_3 = 2$, $x_4 = \tfrac{7}{3}$, $x_5 = \tfrac{8}{3}$, $x_6 = 3$.

$$\int_1^3 e^{0.2x} \, dx \approx \dfrac{1/3}{3} \left[f(1) + 4f\left(\tfrac{4}{3} \right) + 2f\left(\tfrac{5}{3} \right) + 4f(2) + 2f\left(\tfrac{7}{3} \right) + 4f\left(\tfrac{8}{3} \right) + f(3) \right]$$

$$\approx \tfrac{1}{9} \left[1.2214 + 4(1.30561) + 2(1.39561) + 4(1.49182) + 2(1.59467) + 4(1.7046) + 1.82212 \right]$$

$$\approx 3.0036.$$

5. $\displaystyle\int_1^\infty e^{-2x} \, dx = \lim_{b \to \infty} \int_1^b e^{-2x} \, dx = \lim_{b \to \infty} \left(-\tfrac{1}{2} e^{-2x} \Big|_1^b \right) = \lim_{b \to \infty} \left(-\tfrac{1}{2} e^{-2b} + \tfrac{1}{2} e^{-2} \right) = \tfrac{1}{2} e^{-2} = \dfrac{1}{2e^2}.$

6. $V = \pi \displaystyle\int_0^2 y^2 \, dx = \pi \int_0^2 \left(\sqrt{x} \right)^2 dx = \pi \int_0^2 x \, dx = \tfrac{\pi}{2} x^2 \Big|_0^2 = 2\pi.$

© 2011 Cengage Learning. All Rights Reserved. May not be scanned, copied or duplicated, or posted to a publicly accessible website, in whole or in part.

8 CALCULUS OF SEVERAL VARIABLES

8.1 Concept Questions   page 537

1. A function of two variables is a rule that assigns to each point (x, y) in a subset of the plane a unique number $f(x, y)$. For example, $f(x, y) = x^2 + 2y^2$ has the whole xy-plane as its domain.

3. **a.** The graph of $f(x, y)$ is the set $S = \{(x, y, z) \mid z = f(x, y), (x, y) \in D\}$, where D is the domain of f.
   b. The level curve of f is the projection onto the xy-plane of the trace of $f(x, y)$ in the plane $z = k$, where k is a constant in the range of f.

8.1 Functions of Several Variables   page 538

1. $f(0, 0) = 2(0) + 3(0) - 4 = -4$, $f(1, 0) = 2(1) + 3(0) - 4 = -2$, $f(0, 1) = 2(0) + 3(1) - 4 = -1$, $f(1, 2) = 2(1) + 3(2) - 4 = 4$, and $f(2, -1) = 2(2) + 3(-1) - 4 = -3$.

3. $f(1, 2) = 1^2 + 2(1)(2) - 1 + 3 = 7$, $f(2, 1) = 2^2 + 2(2)(1) - 2 + 3 = 9$, $f(-1, 2) = (-1)^2 + 2(-1)(2) - (-1) + 3 = 1$, and $f(2, -1) = 2^2 + 2(2)(-1) - 2 + 3 = 1$.

5. $g(s, t) = 3s\sqrt{t} + t\sqrt{s} + 2$, so $g(1, 2) = 3(1)\sqrt{2} + 2\sqrt{1} + 2 = 4 + 3\sqrt{2}$, $g(2, 1) = 3(2)\sqrt{1} + \sqrt{2} + 2 = 8 + \sqrt{2}$, $g(0, 4) = 0 + 0 + 2 = 2$, and $g(4, 9) = 3(4)\sqrt{9} + 9\sqrt{4} + 2 = 56$.

7. $h(1, e) = \ln e - e \ln 1 = \ln e = 1$, so $h(e, 1) = e \ln 1 - \ln e = -1$, and $h(e, e) = e \ln e - e \ln e = 0$.

9. $g(r, s, t) = re^{s/t}$, so $g(1, 1, 1) = e$, $g(1, 0, 1) = 1$, and $g(-1, -1, -1) = -e^{-1/(-1)} = -e$.

11. The domain of f is the set of all ordered pairs (x, y), where x and y are real numbers.

13. The domain is all real values of u and v except those satisfying the equation $u = v$.

15. The domain of g is the set of all ordered pairs (r, s) satisfying $rs \geq 0$, that is the set of all ordered pairs whose members have the same sign (allowing zeros).

17. The domain of h is the set of all ordered pairs (x, y) such that $x + y > 5$.

219

© 2011 Cengage Learning. All Rights Reserved. May not be scanned, copied or duplicated, or posted to a publicly accessible website, in whole or in part.

19. The graph shows level curves of
$z = f(x, y) = 2x + 3y$ for $z = -2, -1, 0, 1$, and 2.

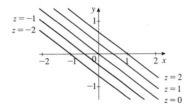

21. The graph shows level curves of
$z = f(x, y) = 2x^2 + y$ for $z = -2, -1, 0, 1$, and 2.

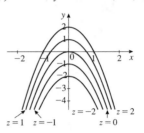

23. The graph shows level curves of
$z = f(x, y) = \sqrt{16 - x^2 - y^2}$ for $z = 0, 1, 2, 3$,
and 4.

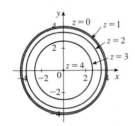

25. The level curves of f have equations
$f(x, y) = \sqrt{x^2 + y^2} = C$. An equation of the curve
containing the point $(3, 4)$ satisfies $\sqrt{3^2 + 4^2} = C$,
so $C = \sqrt{9 + 16} = 5$. Thus, an equation is
$\sqrt{x^2 + y^2} = 5$.

27. $V = f(1.5, 4) = \pi(1.5)^2(4) = 9\pi$, or 9π ft^3.

29. a. $M = \dfrac{80}{(1.8)^2} = 24.69$.

b. We must have $\dfrac{w}{(1.8)^2} < 25$; that is, $w < 25(1.8)^2 = 81$. Thus, the maximum weight is 81 kg.

31. a. $R(x, y) = xp + yq = x\left(200 - \frac{1}{5}x - \frac{1}{10}y\right) + y\left(160 - \frac{1}{10}x - \frac{1}{4}y\right) = -\frac{1}{5}x^2 - \frac{1}{4}y^2 - \frac{1}{5}xy + 200x + 160y$.

b. The domain of R is the set of all points (x, y) satisfying $200 - \frac{1}{5}x - \frac{1}{10}y \geq 0$, $160 - \frac{1}{10}x - \frac{1}{4}y \geq 0$, $x \geq 0$, and $y \geq 0$.

33. a. $R(x, y) = xp + yq = 20x - 0.005x^2 - 0.001xy + 15y - 0.001xy - 0.003y^2$
$$= -0.005x^2 - 0.003y^2 - 0.002xy + 20x + 15y.$$

b. Because p and q must both be nonnegative, the domain of R is the set of all ordered pairs (x, y) for which
$20 - 0.005x - 0.001y \geq 0$ and $15 - 0.001x - 0.003y \geq 0$.

35. a. The domain of V is the set of all ordered pairs (P, T) where P and T are positive real numbers.

b. $V = \dfrac{30.9(273)}{760} \approx 11.10$ liters.

37. The output is $f(32, 243) = 100(32^{3/5})(243)^{2/5} = 100(8)(9) = 7200$, or $7200 billion.

© 2011 Cengage Learning. All Rights Reserved. May not be scanned, copied or duplicated, or posted to a publicly accessible website, in whole or in part.

39. The number of suspicious fires is $N(100, 20) = \dfrac{100\left[1000 + 0.03\left(100^2\right)(20)\right]^{1/2}}{[5 + 0.2(20)]^2} = 103.29$, or approximately 103.

41. a. If $r = 6\%$, then $P = f(100000, 0.06, 30) = \dfrac{300{,}000\,(0.06)}{12\left[1 - \left(1 + \frac{0.06}{12}\right)^{-360}\right]} \approx 1798.65$, or \$1798.65. If $r = 8\%$,

then $P = f(300000, 0.08, 30) = \dfrac{300{,}000\,(0.08)}{12\left[1 - \left(1 + \frac{0.08}{12}\right)^{-360}\right]} \approx 2201.29$, or \$2201.29.

b. $P = f(300000, 0.08, 20) = \dfrac{300{,}000\,(0, 08)}{12\left[1 - \left(1 + \frac{0.08}{12}\right)^{-240}\right]} \approx 2509.32$, or \$2509.32.

43. $f(M, 600, 10) = \dfrac{\pi^2\,(360{,}000)\,M\,(10)}{900} \approx 39{,}478.42M$, or $\dfrac{39{,}478.42}{980} \approx 40.28$ times gravity.

45. The level curves of V have equation $\frac{kT}{P} = C$, where C is a positive constant. The level curves are the family of straight lines $T = \frac{C}{k}P$ lying in the first quadrant, because k, T, and P are positive.

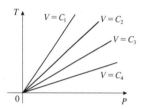

47. False. Let $h(x, y) = xy$. Then there is no pair of functions f and g such that $h(x, y) = f(x) + g(y)$.

49. False. Because $x^2 - y^2 = (x + y)(x - y)$, we see that $x^2 - y^2 = 0$ if $y = \pm x$. Therefore, the domain of f is $\{(x, y) \mid y \neq \pm x\}$.

51. False. Take $f(x, y) = \sqrt{x^2 + y^2}$, $P_1(-1, 1)$, and $P_2(1, 1)$. Then $f(x_1, y_1) = f(-1, 1) = \sqrt{(-1)^2 + 1^2} = \sqrt{2}$ and $f(x_2, y_2) = f(1, 1) = \sqrt{1^2 + 1^2} = \sqrt{2}$. So $f(x_1, y_1) = f(x_2, y_2)$, but $P(x_1, y_1) \neq P(x_2, y_2)$.

8.2	**Problem-Solving Tips**

1. The expressions f_{xy} and f_{yx} denote the second partial derivatives of the function $f(x, y)$. Note that when this notation is used, the differentiation is carried out in the order in which x and y appear (left to right).

2. The notation $\dfrac{\partial^2 f}{\partial y \partial x}$ and $\dfrac{\partial^2 f}{\partial x \partial y}$ is also used to denote the second partial derivatives of the function $f(x, y)$, but in this case the differentiation is carried out in reverse order (right to left).

© 2011 Cengage Learning. All Rights Reserved. May not be scanned, copied or duplicated, or posted to a publicly accessible website, in whole or in part.

8.2   Concept Questions   page 550

1. a. $\dfrac{\partial f}{\partial x}(a, b) = \dfrac{\partial f}{\partial x}(x, y)\Big|_{(a,b)} = \left[\lim\limits_{h \to 0} \dfrac{f(x+h, y) - f(x, y)}{h}\right]_{(a,b)}$.

b. See pages 545–546 of the text.

3. f_{xx}, f_{yy}, f_{xy}, and f_{yx}.

8.2   Partial Derivatives   page 550

1. a. $f_x(2, 1) = 4$ and $f_y(2, 1) = 4$.

b. $f_x(2, 1) = 4$ says that the slope of the tangent line to the curve of intersection of the surface $z = x^2 + 2y^2$ and the plane $y = 1$ at the point $(2, 1, 6)$ is 4. $f_y(2, 1) = 4$ says that the slope of the tangent line to the curve of intersection of the surface $z = x^2 + 2y^2$ and the plane $x = 2$ at the point $(2, 1, 6)$ is 4.

c. $f_x(2, 1) = 4$ says that the rate of change of $f(x, y)$ with respect to x with y held fixed with a value of 1 is 4 units per unit change in x. $f_y(2, 1) = 4$ says that the rate of change of $f(x, y)$ with respect to y with x held fixed with a value of 2 is 4 units per unit change in y.

3. $f(x, y) = 2x + 3y + 5$, so $f_x = 2$ and $f_y = 3$.

5. $g(x, y) = 2x^2 + 4y + 1$, so $g_x = 4x$ and $g_y = 4$.

7. $f(x, y) = \dfrac{2y}{x^2}$, so $f_x = -\dfrac{4y}{x^3}$ and $f_y = \dfrac{2}{x^2}$.

9. $g(u, v) = \dfrac{u - v}{u + v}$, so $\dfrac{\partial g}{\partial u} = \dfrac{(u + v)(1) - (u - v)(1)}{(u + v)^2} = \dfrac{2v}{(u + v)^2}$ and $\dfrac{\partial g}{\partial v} = \dfrac{(u + v)(-1) - (u - v)(1)}{(u + v)^2} = -\dfrac{2u}{(u + v)^2}$.

11. $f(s, t) = \left(s^2 - st + t^2\right)^3$, so $f_s = 3\left(s^2 - st + t^2\right)^2 (2s - t)$ and $f_t = 3\left(s^2 - st + t^2\right)^2 (2t - s)$.

13. $f(x, y) = \left(x^2 + y^2\right)^{2/3}$, so $f_x = \frac{2}{3}\left(x^2 + y^2\right)^{-1/3}(2x) = \frac{4}{3}x\left(x^2 + y^2\right)^{-1/3}$ and $f_y = \frac{4}{3}y\left(x^2 + y^2\right)^{-1/3}$.

15. $f(x, y) = e^{xy+1}$, so $f_x = ye^{xy+1}$ and $f_y = xe^{xy+1}$.

17. $f(x, y) = x \ln y + y \ln x$, so $f_x = \ln y + \dfrac{y}{x}$ and $f_y = \dfrac{x}{y} + \ln x$.

19. $g(u, v) = e^u \ln v$, so $g_u = e^u \ln v$ and $g_v = \dfrac{e^u}{v}$.

21. $f(x, y, z) = xyz + xy^2 + yz^2 + zx^2$, so $f_x = yz + y^2 + 2xz$ and $f_y = xz + 2xy + z^2$, $f_z = xy + 2yz + x^2$.

23. $h(r, s, t) = e^{rst}$, so $h_r = ste^{rst}$, $h_s = rte^{rst}$, and $h_t = rse^{rst}$.

25. $f(x, y) = x^2y + xy^2$, so $f_x(1, 2) = \left(2xy + y^2\right)\big|_{(1,2)} = 8$ and $f_y(1, 2) = \left(x^2 + 2xy\right)\big|_{(1,2)} = 5$.

© 2011 Cengage Learning. All Rights Reserved. May not be scanned, copied or duplicated, or posted to a publicly accessible website, in whole or in part.

27. $f(x, y) = x\sqrt{y} + y^2 = xy^{1/2} + y^2$, so $f_x(2, 1) = \sqrt{y}\big|_{(2,1)} = 1$ and $f_y(2, 1) = \left(\dfrac{x}{2\sqrt{y}} + 2y\right)\bigg|_{(2,1)} = 3$.

29. $f(x, y) = \dfrac{x}{y}$, so $f_x(1, 2) = \dfrac{1}{y}\bigg|_{(1,2)} = \dfrac{1}{2}$ and $f_y(1, 2) = -\dfrac{x}{y^2}\bigg|_{(1,2)} = -\dfrac{1}{4}$.

31. $f(x, y) = e^{xy}$, so $f_x(1, 1) = ye^{xy}\big|_{(1,1)} = e$ and $f_y(1, 1) = xe^{xy}\big|_{(1,1)} = e$.

33. $f(x, y, z) = x^2yz^3$, so $f_x(1, 0, 2) = 2xyz^3\big|_{(1,0,2)} = 0$, $f_y(1, 0, 2) = x^2z^3\big|_{(1,0,2)} = 8$, and
$f_z(1, 0, 2) = 3x^2yz^2\big|_{(1,0,2)} = 0$.

35. $f(x, y) = x^2y + xy^3$, so $f_x = 2xy + y^3$ and $f_y = x^2 + 3xy^2$. Therefore, $f_{xx} = 2y$, $f_{xy} = 2x + 3y^2 = f_{yx}$, and
$f_{yy} = 6xy$.

37. $f(x, y) = x^2 - 2xy + 2y^2 + x - 2y$, so $f_x = 2x - 2y + 1$ and $f_y = -2x + 4y - 2$. Therefore, $f_{xx} = 2$,
$f_{xy} = -2 = f_{yx}$, and $f_{yy} = 4$.

39. $f(x, y) = (x^2 + y^2)^{1/2}$, so $f_x = \frac{1}{2}(x^2 + y^2)^{-1/2}(2x) = x(x^2 + y^2)^{-1/2}$ and $f_y = y(x^2 + y^2)^{-1/2}$. Therefore,
$$f_{xx} = (x^2 + y^2)^{-1/2} + x\left(-\frac{1}{2}\right)(x^2 + y^2)^{-3/2}(2x) = (x^2 + y^2)^{-1/2} - x^2(x^2 + y^2)^{-3/2}$$
$$= (x^2 + y^2)^{-3/2}(x^2 + y^2 - x^2) = \frac{y^2}{(x^2 + y^2)^{3/2}},$$
$$f_{xy} = x\left(-\frac{1}{2}\right)(x^2 + y^2)^{-3/2}(2y) = -\frac{xy}{(x^2 + y^2)^{3/2}} = f_{yx}, \text{ and}$$
$$f_{yy} = (x^2 + y^2)^{-1/2} + y\left(-\frac{1}{2}\right)(x^2 + y^2)^{-3/2}(2y) = (x^2 + y^2)^{-1/2} - y^2(x^2 + y^2)^{-3/2}$$
$$= (x^2 + y^2)^{-3/2}(x^2 + y^2 - y^2) = \frac{x^2}{(x^2 + y^2)^{3/2}}.$$

41. $f(x, y) = e^{-x/y}$, so $f_x = -\dfrac{1}{y}e^{-x/y}$ and $f_y = \dfrac{x}{y^2}e^{-x/y}$. Therefore, $f_{xx} = \dfrac{1}{y^2}e^{-x/y}$,
$$f_{xy} = -\frac{x}{y^3}e^{-x/y} + \frac{1}{y^2}e^{-x/y} = \left(\frac{-x+y}{y^3}\right)e^{-x/y} = f_{yx}, \text{ and } f_{yy} = -\frac{2x}{y^3}e^{-x/y} + \frac{x^2}{y^4}e^{-x/y} = \frac{x}{y^3}\left(\frac{x}{y} - 2\right)e^{-x/y}.$$

43. a. $f(x, y) = 20x^{3/4}y^{1/4}$, so $f_x(256, 16) = 15\left(\dfrac{y}{x}\right)^{1/4}\bigg|_{(256,16)} = 15\left(\dfrac{16}{256}\right)^{1/4} = 15\left(\dfrac{1}{2}\right) = 7.5$ and

$f_y(256, 16) = 5\left(\dfrac{x}{y}\right)^{3/4}\bigg|_{(256,16)} = 5\left(\dfrac{256}{16}\right)^{3/4} = 5(80) = 40.$

b. Yes.

45. $p(x, y) = 200 - 10\left(x - \dfrac{1}{2}\right)^2 - 15(y - 1)^2$, so $\dfrac{\partial p}{\partial x}(0, 1) = -20\left(x - \dfrac{1}{2}\right)\bigg|_{(0,1)} = 10$. At the location $(0, 1)$ in the

figure, the price of land is increasing by \$10 per square foot per mile to the east. $\dfrac{\partial p}{\partial y}(0, 1) = -30(y - 1)\big|_{(0,1)} = 0$,

so at the point $(0, 1)$ in the figure, the price of land is unchanging with respect to north-south change.

© 2011 Cengage Learning. All Rights Reserved. May not be scanned, copied or duplicated, or posted to a publicly accessible website, in whole or in part.

47. $f(p, q) = 10,000 - 10p - e^{0.5q}$ and $g(p, q) = 50,000 - 4000q - 10p$. Thus, $\dfrac{\partial f}{\partial q} = -0.5e^{0.5q} < 0$ and

$\dfrac{\partial g}{\partial p} = -10 < 0$, and so the two commodities are complementary commodities.

49. $R(x, y) = -0.2x^2 - 0.25y^2 - 0.2xy + 200x + 160y$, so

$\dfrac{\partial R}{\partial x}(300, 250) = (-0.4x - 0.2y + 200)|_{(300,250)} = -0.4(300) - 0.2(250) + 200 = 30$. This says that at sales

levels of 300 finished and 250 unfinished units, revenue is increasing by $30 per week per unit increase in finished

pieces. $\dfrac{\partial R}{\partial y}(300, 250) = -0.5y - 0.2x + 160|_{(300,250)} = -0.5(250) - 0.2(300) + 160 = -25$, and this says that

at the same sales levels, revenue is decreasing by $25 per week per unit increase in unfinished pieces.

51. a. $T = f(32, 20) = 35.74 + 0.6215(32) - 35.75(20^{0.16}) + 0.4275(32)(20^{0.16}) \approx 19.99$, or approximately 20°F.

   b. $\dfrac{\partial T}{\partial s} = -35.75(0.16S^{-0.84}) + 0.4275t(0.16S^{-0.84}) = 0.16(-35.75 + 0.4275t)s^{-0.84}$, so

   $\left.\dfrac{\partial T}{\partial s}\right|_{(32,20)} = 0.16[-35.75 + 0.4275(32)]20^{-0.84} \approx -0.285$; that is, the wind chill will drop by 0.3° for each

   1 mph increase in wind speed.

53. $V = \dfrac{30.9T}{P}$, so $\dfrac{\partial V}{\partial T} = \dfrac{30.9}{P}$ and $\dfrac{\partial V}{\partial P} = -\dfrac{30.9T}{P^2}$. Therefore, $\left.\dfrac{\partial V}{\partial T}\right|_{T=300, P=800} = \dfrac{30.9}{800} = 0.039$, or 0.039 liters per

degree. $\left.\dfrac{\partial V}{\partial P}\right|_{T=300, P=800} = -\dfrac{(30.9)(300)}{800^2} \approx -0.015$, or approximately −0.015 liters per millimeter of mercury.

55. $V = \dfrac{kT}{P}$, so $\dfrac{\partial V}{\partial T} = \dfrac{k}{P}$; $T = \dfrac{VP}{k}$, so $\dfrac{\partial T}{\partial P} = \dfrac{V}{k} = \dfrac{T}{P}$; and $P = \dfrac{kT}{V}$, so $\dfrac{\partial P}{\partial V} = -\dfrac{kT}{V^2} = -kT\dfrac{P^2}{(kT)^2} = -\dfrac{P^2}{kT}$.

Therefore $\dfrac{\partial V}{\partial T} \cdot \dfrac{\partial T}{\partial P} \cdot \dfrac{\partial P}{\partial V} = \dfrac{k}{P}\left(\dfrac{T}{P}\right)\left(-\dfrac{P^2}{kT}\right) = -1$.

57. False. Let $f(x, y) = xy^{1/2}$. Then $f_x = y^{1/2}$ is defined at $(0, 0)$, but $f_y = \tfrac{1}{2}xy^{-1/2} = \dfrac{x}{2y^{1/2}}$ is not defined at $(0, 0)$.

59. True. See Section 8.2.

8.2   **Using Technology**   page 553

1. 1.3124, 0.4038.   **3.** −1.8889, 0.7778.   **5.** −0.3863, −0.8497.

© 2011 Cengage Learning. All Rights Reserved. May not be scanned, copied or duplicated, or posted to a publicly accessible website, in whole or in part.

8.3   Problem-Solving Tips

1. **To find the relative extrema of a function of several variables**, first find the critical points of $f(x, y)$ by solving the simultaneous equations $f_x = 0$ and $f_y = 0$, then use the second derivative test to classify those points.

2. **To use the second derivative test**, first evaluate the function $D(x, y) = f_{xx}f_{yy} - f_{xy}^2$ for each critical point found in Tip 1.
   - If $D(a, b) > 0$ and $f_{xx}(a, b) < 0$, then $f(x, y)$ has a relative maximum at the point (a, b).
   - If $D(a, b) > 0$ and $f_{xx}(a, b) > 0$, then $f(x, y)$ has a relative minimum at the point (a, b).
   - If $D(a, b) < 0$ then $f(x, y)$ has neither a relative maximum nor a relative minimum at the point (a, b).
   - If $D(a, b) = 0$, then the test is inconclusive.

8.3   Concept Questions   page 561

1. **a.** A function $f(x, y)$ has a relative maximum at (a, b) if $f(a, b)$ is the largest value of $f(x, y)$ for all (x, y) near (a, b).

   b. $f(a, y)$ has an absolute maximum at (a, b) if $f(a, b)$ is the largest value of $f(x, y)$ for all (x, y) in the domain of f.

3. See the procedure on page 557 of the text.

8.3   Maxima and Minima of Functions of Several Variables   page 561

1. $f(x, y) = 1 - 2x^2 - 3y^2$. To find the critical points of f, we solve the system $\begin{cases} f_x = -4x = 0 \\ f_y = -6y = 0 \end{cases}$ obtaining $(0, 0)$ as the only critical point of f. Next, $f_{xx} = -4$, $f_{xy} = 0$, and $f_{yy} = -6$. In particular, $f_{xx}(0, 0) = -4$, $f_{xy}(0, 0) = 0$, and $f_{yy}(0, 0) = -6$, giving $D(0, 0) = (-4)(-6) - 0^2 = 24 > 0$. Because $f_{xx}(0, 0) < 0$, the Second Derivative Test implies that $(0, 0)$ gives rise to a relative maximum of f. Finally, the relative maximum of f is $f(0, 0) = 1$.

3. $f(x, y) = x^2 - y^2 - 2x + 4y + 1$. To find the critical points of f, we solve the system $\begin{cases} f_x = 2x - 2 = 0 \\ f_y = -2y + 4 = 0 \end{cases}$ obtaining $x = 1$ and $y = 2$, so $(1, 2)$ is the only critical point of f. $f_{xx} = 2$, $f_{xy} = 0$, and $f_{yy} = -2$, so $D(x, y) = f_{xx}f_{yy} - f_{xy}^2 = -4$. In particular, $D(1, 2) = -4 < 0$, so $(1, 2)$ gives a saddle point of f and $f(1, 2) = 4$.

5. $f(x, y) = x^2 + 2xy + 2y^2 - 4x + 8y - 1$. To find the critical points of f, we solve the system $\begin{cases} f_x = 2x + 2y - 4 = 0 \\ f_y = 2x + 4y + 8 = 0 \end{cases}$ obtaining $(8, -6)$ as the critical point of f. Next, $f_{xx} = 2$, $f_{xy} = 2$, and $f_{yy} = 4$. In particular, $f_{xx}(8, -6) = 2$, $f_{xy}(8, -6) = 2$, and $f_{yy}(8, -6) = 4$, giving $D = 2(4) - 4 = 4 > 0$. Because $f_{xx}(8, -6) > 0$, $(8, -6)$ gives rise to a relative minimum of f. The relative minimum value of f is $f(8, -6) = -41$.

© 2011 Cengage Learning. All Rights Reserved. May not be scanned, copied or duplicated, or posted to a publicly accessible website, in whole or in part.

7. $f(x, y) = 2x^3 + y^2 - 9x^2 - 4y + 12x - 2$. To find the critical points of f, we solve the system

$$\begin{cases} f_x = 6x^2 - 18x + 12 = 0 \\ f_y = 2y - 4 = 0 \end{cases}$$

The first equation is equivalent to $x^2 - 3x + 2 = 0$, or $(x - 2)(x - 1) = 0$, giving

$x = 1$ or 2. The second equation of the system gives $y = 2$. Therefore, there are two critical points, $(1, 2)$ and $(2, 2)$. Next, we compute $f_{xx} = 12x - 18 = 6(2x - 3)$, $f_{xy} = 0$, and $f_{yy} = 2$.

At the point $(1, 2)$, $f_{xx}(1, 2) = 6(2 - 3) = -6$, $f_{xy}(1, 2) = 0$, and $f_{yy}(1, 2) = 2$, so $D = (-6)(2) - 0 = -12 < 0$ and we conclude that $(1, 2)$ gives a saddle point of f with value $f(1, 2) = -1$.

At the point $(2, 2)$, $f_{xx}(2, 2) = 6(4 - 3) = 6$, $f_{xy}(2, 2) = 0$, and $f_{yy}(2, 2) = 2$, so $D = (6)(2) - 0 = 12 > 0$. Because $f_{xx}(2, 2) > 0$, we see that $(2, 2)$ gives a relative minimum with value $f(2, 2) = -2$.

9. $f(x, y) = x^3 + y^2 - 2xy + 7x - 8y + 4$. To find the critical points of f, we solve the system

$$\begin{cases} f_x = 3x^2 - 2y + 7 = 0 \\ f_y = 2y - 2x - 8 = 0 \end{cases}$$

Adding the two equations gives $3x^2 - 2x - 1 = (3x + 1)(x - 1) = 0$. Therefore,

$x = -\frac{1}{3}$ or 1. Substituting each of these values of x into the second equation gives $y = \frac{8}{3}$ and $y = 5$, respectively. Therefore, $\left(-\frac{1}{3}, \frac{11}{3}\right)$ and $(1, 5)$ are critical points of f. Next, $f_{xx} = 6x$, $f_{xy} = -2$, and $f_{yy} = 2$, so

$D(x, y) = 12x - 4 = 4(3x - 1)$. Then $D\left(-\frac{1}{3}, \frac{11}{3}\right) = 4(-1 - 1) = -8 < 0$, and so $\left(-\frac{1}{3}, \frac{11}{3}\right)$ gives a saddle

point with value $f\left(-\frac{1}{3}, \frac{11}{3}\right) = -\frac{319}{27}$. Next, $D(1, 5) = 4(3 - 1) = 8 > 0$, and since $f_{xx}(1, 5) = 6 > 0$, we see that $(1, 5)$ gives a relative minimum with value $f(1, 5) = -13$.

11. $f(x, y) = x^3 - 3xy + y^3 - 2$. To find the critical points of f, we solve the system $\begin{cases} f_x = 3x^2 - 3y = 0 \\ f_y = -3x + 3y^2 = 0 \end{cases}$

The first equation gives $y = x^2$, and substituting this into the second equation gives $-3x + 3x^4 = 3x(x^3 - 1) = 0$. Therefore, $x = 0$ or 1. Substituting these values of x into the first equation gives $y = 0$ and $y = 1$, respectively. Therefore, $(0, 0)$ and $(1, 1)$ are critical points of f. Next, we find $f_{xx} = 6x$, $f_{xy} = -3$, and $f_{yy} = 6y$, so $D = f_{xx}f_{yy} - f_{xy}^2 = 36xy - 9$. Because $D(0, 0) = -9 < 0$, we see that $(0, 0)$ gives a saddle point of f with value $f(0, 0) = -2$. Next, $D(1, 1) = 36 - 9 = 27 > 0$, and since $f_{xx}(1, 1) = 6 > 0$, we see that $f(1, 1) = -3$ is a relative minimum value of f.

13. $f(x, y) = xy + \frac{4}{x} + \frac{2}{y}$. Solving the system of equations $\begin{cases} f_x = y - \dfrac{4}{x^2} = 0 \\ f_y = x - \dfrac{2}{y^2} = 0 \end{cases}$ we obtain $y = \dfrac{4}{x^2}$.

Therefore, $x - 2\left(\dfrac{x^4}{16}\right) = 0$ and $8x - x^4 = x(8 - x^3) = 0$, so $x = 0$ or $x = 2$. Because $x = 0$ is not in

the domain of f, $(2, 1)$ is the only critical point of f. Next, $f_{xx} = \dfrac{8}{y^3}$, $f_{xy} = 1$, and $f_{yy} = \dfrac{4}{y^3}$. Therefore,

$D(2, 1) = \left(\dfrac{32}{x^3 y^3} - 1\right)\Big|_{(2,1)} = 4 - 1 = 3 > 0$ and $f_{xx}(2, 1) = 1 > 0$, so the relative minimum value of f is f

$(2, 1) = 2 + \frac{4}{2} + \frac{2}{1} = 6$.

© 2011 Cengage Learning. All Rights Reserved. May not be scanned, copied or duplicated, or posted to a publicly accessible website, in whole or in part.

15. $f(x, y) = x^2 - e^{y^2}$. Solving the system of equations $\begin{cases} f_x = 2x = 0 \\ f_y = -2ye^{y^2} = 0 \end{cases}$ we obtain $x = 0$ and $y = 0$.

Therefore, $(0, 0)$ is the only critical point of f. Next, $f_{xx} = 2$, $f_{xy} = 0$, and $f_{yy} = -2e^{y^2} - 4y^2e^{y^2}$, so $D(0, 0) = \left[-4e^{y^2}\left(1 + 2y^2\right)\right]_{(0,0)} = -4(1) < 0$, and we conclude that $(0, 0)$ gives a saddle point with value $f(0, 0) = -1$.

17. $f(x, y) = e^{x^2+y^2}$. Solving the system $\begin{cases} f_x = 2xe^{x^2+y^2} = 0 \\ f_y = 2ye^{x^2+y^2} = 0 \end{cases}$ we see that $x = 0$ and $y = 0$

(recall that $e^{x^2+y^2} \neq 0$). Therefore, $(0, 0)$ is the only critical point of f. Next, we compute $f_{xx} = 2e^{x^2+y^2} + 2x(2x)e^{x^2+y^2} = 2\left(1 + 2x^2\right)e^{x^2+y^2}$, $f_{xy} = 2x(2y)e^{x^2+y^2} = 4xye^{x^2+y^2}$, and $f_{yy} = 2\left(1 + 2y^2\right)e^{x^2+y^2}$. In particular, at the point $(0, 0)$, $f_{xx}(0, 0) = 2$, $f_{xy}(0, 0) = 0$, and $f_{yy}(0, 0) = 2$. Therefore, $D = (2)(2) - 0 = 4 > 0$. Because $f_{xx}(0, 0) > 0$, we conclude that $(0, 0)$ gives rise to a relative minimum of f. The relative minimum value is $f(0, 0) = 1$.

19. $f(x, y) = \ln\left(1 + x^2 + y^2\right)$. We solve the system of equations $\begin{cases} f_x = \dfrac{2x}{1 + x^2 + y^2} = 0 \\ f_y = \dfrac{2y}{1 + x^2 + y^2} = 0 \end{cases}$ obtaining

$x = 0$ and $y = 0$. Therefore, $(0, 0)$ is the only critical point of f. Next,

$$f_{xx} = \frac{\left(1 + x^2 + y^2\right)2 - (2x)(2x)}{\left(1 + x^2 + y^2\right)^2} = \frac{2 + 2y^2 - 2x^2}{\left(1 + x^2 + y^2\right)^2}, \quad f_{yy} = \frac{\left(1 + x^2 + y^2\right)2 - (2y)(2y)}{\left(1 + x^2 + y^2\right)^2} = \frac{2 + 2x^2 - 2y^2}{\left(1 + x^2 + y^2\right)^2},$$

and $f_{xy} = -2x\left(1 + x^2 + y^2\right)^{-2}(2y) = -\dfrac{4xy}{\left(1 + x^2 + y^2\right)^2}$. Therefore,

$$D(x, y) = \frac{\left(2 + 2y^2 - 2x^2\right)\left(2 + 2x^2 - 2y^2\right)}{\left(1 + x^2 + y^2\right)^4} - \frac{16x^2y^2}{\left(1 + x^2 + y^2\right)^4}. \text{ Because } D(0, 0) = 4 > 0 \text{ and}$$

$f_{xx}(0, 0) = 2 > 0$, $f(0, 0) = 0$ is a relative minimum value.

21. $P(x) = -0.2x^2 - 0.25y^2 - 0.2xy + 200x + 160y - 100x - 70y - 4000$

$$= -0.2x^2 - 0.25y^2 - 0.2xy + 100x + 90y - 4000.$$

Thus, $\begin{cases} P_x = -0.4x - 0.2y + 100 = 0 \\ P_y = -0.5y - 0.2x + 90 = 0 \end{cases}$ implies that $\begin{cases} 4x + 2y = 1000 \\ 2x + 5y = 900 \end{cases}$ Solving, we find $x = 200$ and

$y = 100$. Next, $P_{xx} = -0.4$, $P_{yy} = -0.5$, $P_{xy} = -0.2$, and $D(200, 100) = (-0.4)(-0.5) - (-0.2)^2 > 0$. Because $P_{xx}(200, 100) < 0$, we conclude that $(200, 100)$ is a relative maximum of P. Thus, the company should manufacture 200 finished and 100 unfinished units per week. The maximum profit is $P(200, 100) = -0.2(200)^2 - 0.25(100)^2 - 0.2(100)(200) + 100(200) + 90(100) - 4000 = 10,500$, or $10,500.

© 2011 Cengage Learning. All Rights Reserved. May not be scanned, copied or duplicated, or posted to a publicly accessible website, in whole or in part.

23. $p(x, y) = 200 - 10\left(x - \frac{1}{2}\right)^2 - 15(y - 1)^2$. Solving the system of equations $\begin{cases} p_x = -20\left(x - \frac{1}{2}\right) = 0 \\ p_y = -30(y - 1) = 0 \end{cases}$ we

obtain $x = \frac{1}{2}$ and $y = 1$. We conclude that the only critical point of f is $\left(\frac{1}{2}, 1\right)$. Next, $p_{xx} = -20$, $p_{xy} = 0$, and

$p_{yy} = -30$, so $D\left(\frac{1}{2}, 1\right) = (-20)(-30) = 600 > 0$. Because $p_{xx} = -20 < 0$, we conclude that $f\left(\frac{1}{2}, 1\right)$ gives a

relative maximum. We conclude that the price of land is highest at $\left(\frac{1}{2}, 1\right)$.

25. We want to minimize $f(x, y) = D^2 = (x - 5)^2 + (y - 2)^2 + (x + 4)^2 + (y - 4)^2 + (x + 1)^2 + (y + 3)^2$. We

calculate $\begin{cases} f_x = 2(x - 5) + 2(x + 4) + 2(x + 1) = 6x = 0, \\ f_y = 2(y - 2) + 2(y - 4) + 2(y + 3) = 6y - 6 = 0 \end{cases}$ and conclude that $x = 0$ and $y = 1$. Also,

$f_{xx} = 6$, $f_{xy} = 0$, $f_{yy} = 6$, and $D(x, y) = (6)(6) = 36 > 0$. Because $f_{xx} > 0$, we conclude that the function is

minimized at $(0, 1)$, the desired location.

27. Refer to the figure in the text. $xy + 2xz + 2yz = 300$, so $z(2x + 2y) = 300 - xy$, The volume is given by

$$V = xyz = xy \frac{300 - xy}{2(x + y)} = \frac{300xy - x^2 y^2}{2(x + y)}. \text{We find}$$

$$\frac{\partial V}{\partial x} = \frac{1}{2} \frac{(x + y)(300y - 2xy^2) - (300xy - x^2 y^2)}{(x + y)^2} = \frac{300xy - 2x^2 y^2 + 300y^2 - 2xy^3 - 300xy + x^2 y^2}{2(x + y)^2}$$

$$= \frac{300y^2 - 2xy^3 - x^2 y^2}{2(x + y)^2} = \frac{y^2(300 - 2xy - x^2)}{2(x + y)^2}$$

and similarly $\dfrac{\partial V}{\partial y} = \dfrac{x^2(300 - 2xy - y^2)}{2(x + y)^2}$. Setting both $\dfrac{\partial V}{\partial x}$ and $\dfrac{\partial V}{\partial y}$ equal to 0 and observing that both $x > 0$ and

$y > 0$, we have the system $\begin{cases} 2yx + x^2 = 300 \\ 2yx + y^2 = 300 \end{cases}$ Subtracting, we find $y^2 - x^2 = 0$, so $(y - x)(y + x) = 0$. Thus,

$y = x$ or $y = -x$. The latter is not possible since x and y are both positive. Therefore, $y = x$. Substituting this

value into the first equation in the system gives $2x^2 + x^2 = 300$, so $x^2 = 100$ and $x = y = 10$. Substituting these

values into the expression for z gives $z = \dfrac{300 - 10^2}{2(10 + 10)} = 5$, so the dimensions are $10'' \times 10'' \times 5''$ and the volume is

500 in^3.

© 2011 Cengage Learning. All Rights Reserved. May not be scanned, copied or duplicated, or posted to a publicly accessible website, in whole or in part.

29. The heating cost is $C = 2xy + 8xz + 6yz$. But $xyz = 12,000$, so $z = \dfrac{12,000}{xy}$. Therefore,

$$C = f(x, y) = 2xy + 8x\left(\frac{12,000}{xy}\right) + 6y\left(\frac{12,000}{xy}\right) = 2xy + \frac{96,000}{y} + \frac{72,000}{x}.$$ To find the minimum

of f, we find the critical point of f by solving the system $\begin{cases} f_x = 2y - \dfrac{72,000}{x^2} = 0 \\ f_y = 2x - \dfrac{96,000}{y^2} = 0 \end{cases}$ The first equation

gives $y = \dfrac{36,000}{x^2}$, which when substituted into the second equation yields $2x - 96,000\left(\dfrac{x^2}{36,000}\right)^2 = 0$, so

$(36,000)^2 x - 48,000x^4 = 0$ and $x\left(27,000 - x^3\right) = 0$. Solving this equation, we have $x = 0$ or $x = 30$.

We reject the first root because $x = 0$ lies outside the domain of f. With $x = 30$, we find $y = 40$. Next,

$f_{xx} = \dfrac{144,000}{x^3}$ and $f_{xy} = 2$. In particular, $f_{xx}(30, 40) = 5.33$, $f_{xy} = (30, 40) = 2$, and $f_{yy}(30, 40) = 3$. Thus,

$D(30, 40) = (5.33)(3) - 4 = 11.99 > 0$, and since $f_{xx}(30, 40) > 0$, we see that $(30, 40)$ gives a relative

minimum. Physical considerations tell us that this is an absolute minimum. The minimal annual heating cost is

$$f(30, 40) = 2(30)(40) + \frac{96,000}{40} + \frac{72,000}{30} = 7200, \text{ or } \$7,200.$$

31. False. Let $f(x, y) = xy$. Then $f_x(0, 0) = 0$ and $f_y(0, 0) = 0$, but $(0, 0)$ does not give a relative extremum of
$(0, 0)$. In fact, $f_{xx} = 0$, $f_{yy} = 0$, and $f_{xy} = 1$, so $D(x, y) = f_{xx}f_{yy} - f_{xy}^2 = -1$ and $D(0, 0) = -1$, showing that
$(0, 0, 0)$ is a saddle point.

33. False. $f_x(a, b)$ and/or $f_y(a, b)$ may be undefined.

35. True. Here $f_{xx}(a, b) = -f_{yy}(a, b)$, so
$$D(a, b) = f_{xx}(a, b)f_{yy}(a, b) - f_{xy}^2(a, b) = f_{xx}(a, b)\left[-f_{xx}(a, b)\right] - 0 = -f_{xx}^2(a, b) < 0, \text{ and so } f \text{ cannot}$$
have a relative extremum at (a, b).

8.4   Problem-Solving Tips

You will find it helpful to organize the data in a least-squares problem in the form of a table like those in Examples 2
and 3 in the text.

8.4   Concept Questions   page 569

1. **a.** A scatter diagram is a graph showing the data points that describe the relationship between the two variables x
   and y.

   b. The least squares line is the straight line that best fits a set of data points when the points are scattered about a
   straight line.

© 2011 Cengage Learning. All Rights Reserved. May not be scanned, copied or duplicated, or posted to a publicly accessible website, in whole or in part.

8.4   The Method of Least Squares   page 569

1. a. We first summarize the data.

x	y	x^2	xy
1	4	1	4
2	6	4	12
3	8	9	24
4	11	16	44
Sum 10	29	30	84

b.

The normal equations are $4b + 10m = 29$ and $10b + 30m = 84$. Solving this system of equations, we obtain $m = 2.3$ and $b = 1.5$, so an equation is $y = 2.3x + 1.5$.

3. a. We first summarize the data.

x	y	x^2	xy
1	4.5	1	4.5
2	5	4	10
3	3	9	9
4	2	16	8
4	3.5	16	14
6	1	36	6
Sum 20	19	82	51.5

b.

The normal equations are $6b + 20m = 19$ and $20b + 82m = 51.5$. The solutions are $m \approx -0.7717$ and $b \approx 5.7391$, so the required equation is $y = -0.772x + 5.739$.

5. a. We first summarize the data:

x	y	x^2	xy
1	3	1	3
2	5	4	10
3	5	9	15
4	7	16	28
5	8	25	40
Sum 15	28	55	96

b.

The normal equations are $55m + 15b = 96$ and $15m + 5b = 28$. Solving, we find $m = 1.2$ and $b = 2$, so the required equation is $y = 1.2x + 2$.

© 2011 Cengage Learning. All Rights Reserved. May not be scanned, copied or duplicated, or posted to a publicly accessible website, in whole or in part.

7. a. We first summarize the data:

x	y	x^2	xy
4	0.5	16	2
4.5	0.6	20.25	2.7
5	0.8	25	4
5.5	0.9	30.25	4.95
6	1.2	36	7.2
Sum 25	4	127.5	20.85

The normal equations are $5b + 25m = 4$ and $25b + 127.5m = 20.85$. The solutions are $m = 0.34$ and $b = -0.9$, so the required equation is $y = 0.34x - 0.9$.

b.

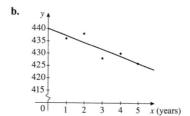

c. If $x = 6.4$, then
$y = 0.34\,(6.4) - 0.9 = 1.276$, and so 1276 completed applications can be expected.

9. a. We first summarize the data:

x	y	x^2	xy
1	436	1	436
2	438	4	876
3	428	9	1284
4	430	16	1720
5	426	25	2138
Sum 15	2158	55	6446

The normal equations are $5b + 15m = 2158$ and $15b + 55m = 6446$. Solving this system, we find $m = -2.8$ and $b = 440$. Thus, the equation of the least-squares line is $y = -2.8x + 440$.

b.

c. Two years from now, the average SAT verbal score in that area will be
$y = -2.8\,(7) + 440 = 420.4$.

11. a.

x	y	x^2	xy
1	20	1	20
2	24	4	48
3	26	9	78
4	28	16	112
5	32	25	160
Sum 15	130	55	418

The normal equations are $5b + 15m = 130$ and $15b + 55m = 418$. The solutions are $m = 2.8$ and $b = 17.6$, and so an equation of the line is $y = 2.8x + 17.6$.

b. When $x = 8$, $y = 2.8\,(8) + 17.6 = 40$. Hence, the state subsidy is expected to be $40 million for the eighth year.

© 2011 Cengage Learning. All Rights Reserved. May not be scanned, copied or duplicated, or posted to a publicly accessible website, in whole or in part.

13. a.

x	y	x^2	xy
0	126	1	0
1	144	4	144
2	171	9	343
3	191	16	573
4	216	25	864
Sum 10	848	55	1923

The normal equations are $5b + 10m = 848$ and $10b + 30m = 1923$. The solutions are $m \approx 22.7$ and $b \approx 124.2$, so the required equation is $y = 22.7x + 124.2$.

b. $y = 22.7\,(6) + 124.2 = 260.4$, or $\$260.4$ billion.

15. a.

x	y	x^2	xy
4	174	16	696
5	205	25	1025
6	228	36	1368
7	253	49	1771
8	278	64	2224
Sum 30	1138	190	7084

The normal equations are $5b + 30m = 1138$ and $30b + 190m = 7084$. The solutions are $m = 25.6$ and $b = 74$, so the required equation is $y = 25.6x + 74$.

b. $y = 25.6\,(10) + 74 = 330$, or 330 million.

17. a.

x	y	x^2	xy
0	3.7	0	0
1	4.0	1	4
2	4.4	4	8.8
3	4.8	9	14.4
4	5.2	16	20.8
5	5.8	25	29
6	6.3	36	37.8
Sum 21	34.2	91	114.8

The normal equations are $7b + 21m = 34.2$ and $21b + 91m = 114.8$. The solutions are $m \approx 0.4357$ and $b \approx 3.5786$, so the required equation is $y = 0.4357x + 3.5786$.

b. The rate of change is given by the slope of the least-squares line, that is, approximately $\$0.4357$ billion/yr.

19. a.

x	y	x^2	xy
4	1.42	16	5.68
5	1.73	25	8.65
6	1.98	36	11.88
7	2.32	49	16.24
8	2.65	64	21.2
Sum 30	10.1	190	63.65

The normal equations are $5b + 30m = 10.1$ and $30b + 190m = 63.65$. The solutions are $m \approx 0.305$ and $b \approx 0.19$, so the required equation is $y = 0.305x + 0.19$.

b. The rate of change is given by the slope of the least-squares line, that is, approximately $\$0.305$ billion/yr.

c. $f\,(10) = 0.305\,(10) + 0.19 = 3.24$, or $\$3.24$ billion

© 2011 Cengage Learning. All Rights Reserved. May not be scanned, copied or duplicated, or posted to a publicly accessible website, in whole or in part.

21. a.

x	y	x^2	xy
2	5.3	4	10.6
3	5.6	9	16.8
4	5.9	16	23.6
5	6.4	25	32
6	6.9	36	41.4
Sum 20	30.1	90	124.4

The normal equations are $5b + 20m = 30.1$ and $20b + 90m = 124.4$. The solutions are $m = 0.4$ and $b = 4.42$, so the required equation is $y = 0.4x + 4.42$.

b. The rate of change is given by the slope of the least-squares line, that is, approximately \$0.4 billion/yr.

23. a.

x	y	x^2	xy
1	87	1	87
2	87.9	4	175.8
3	90	9	270
4	94.2	16	376.8
5	97.5	25	487.5
5	102.6	36	615.6
Sum 21	559.2	91	2012.7

The normal equations are $6b + 21m = 559.2$ and $21b + 91m = 2012.7$. The solutions are $m \approx 3.17$ and $b \approx 82.1$, so the required equation is $y = 3.17x + 82.1$.

b. The FICA wage base for the year 2012 is given by $y = 3.17(10) + 82.1 = 113.8$, or \$113,800.

25. a.

x	y	x^2	xy
0	15.9	0	0
10	16.8	100	168
20	17.6	400	352
30	18.5	900	555
40	19.3	1600	772
50	20.3	2500	1015
Sum 150	108.4	5500	2862

The normal equations are $6b + 150m = 108.4$ and $150b + 5500m = 2862$. The solutions are $b = 15.90$ and $m = 0.09$, so $y = 0.09x + 15.9$.

b. The life expectancy at 65 of a male in 2040 is $y = 0.087(40) + 15.9 = 19.38$, or 19.38 years.

c. The life expectancy at 65 of a male in 2030 is $y = 0.087(30) + 15.9 = 18.51$, or 18.5 years.

27. a.

x	y	x^2	xy
1	1.4	1	1.4
2	1.6	4	3.2
3	1.8	9	5.4
4	2.1	16	8.4
5	2.3	25	11.5
6	2.5	36	15
Sum 21	11.7	91	44.9

The normal equations are $6b + 21m = 11.7$ and $21b + 91m = 44.9$. The solutions are $m = 0.23$ and $b = 1.6$, so the required equation is $y = 0.23x + 1.16$.

b. $y = 0.23(7) + 1.16 = 2.77$, or approximately 2.8 billion bushels.

29. False. See Example 1 on page 566 of the text.

31. True.

© 2011 Cengage Learning. All Rights Reserved. May not be scanned, copied or duplicated, or posted to a publicly accessible website, in whole or in part.

8.4   Using Technology   page 574

1. $y = 2.3596x + 3.8639$

3. $y = -1.1948x + 3.5525$

5. a. $y = 1.03x + 2.33$ **b.** $10.57 billion

7. a. $y = 13.321x + 72.57$ **b.** 192 million tons

9. a. $1.95x + 12.19$ **b.** $29.74 billion

8.5   Problem-Solving Tips

1. The method of Lagrange Multipliers allows us to find the critical points of the function $f(x, y)$ subject to the constraint $g(x, y) = 0$ (if extrema exist). However, it does not tell us whether those critical points lead to relative extremum. Instead we rely on the geometric or physical nature of the problem to classify these points.

2. To use the method of Lagrange Multipliers, first form the function $F(x, y, \lambda) = f(x, y) + \lambda g(x, y)$ and then

solve the system of equations $f(x) = \begin{cases} F_x = 0 \\ F_y = 0 \\ F_\lambda = 0 \end{cases}$ for all values of x, y and λ. The solutions are candidates for

extrema of f.

8.5   Concept Questions   page 583

1. A constrained relative extremum of f is an extremum of f subject to a constraint of the form $g(x, y) = 0$.

8.5   Constrained Maxima and Minima and the Method of Lagrange Multipliers   page 583

1. We form the Lagrangian function $F(x, y, \lambda) = x^2 + 3y^2 + \lambda(x + y - 1)$ and solve the system

$\begin{cases} F_x = 2x + \lambda = 0 \\ F_y = 6y + \lambda = 0 \\ F_\lambda = x + y - 1 = 0 \end{cases}$ Solving the first and the second equations for x and y in terms of λ, we obtain $x = -\frac{\lambda}{2}$

and $y = -\frac{\lambda}{6}$ which, upon substitution into the third equation, yields $-\frac{\lambda}{2} - \frac{\lambda}{6} - 1 = 0$ or $\lambda = -\frac{3}{2}$. Therefore, $x = \frac{3}{4}$ and $y = \frac{1}{4}$, which gives the point $\left(\frac{3}{4}, \frac{1}{4}\right)$ as the sole critical point of f. Therefore, $\left(\frac{3}{4}, \frac{1}{4}\right) = \frac{3}{4}$ is a minimum of f.

3. We form the Lagrangian function $F(x, y, \lambda) = 2x + 3y - x^2 - y^2 + \lambda(x + 2y - 9)$ and solve the system

$\begin{cases} F_x = 2 - 2x + \lambda = 0 \\ F_y = 3 - 2y + 2\lambda = 0 \\ F_\lambda = x + 2y - 9 = 0 \end{cases}$ Solving the first equation for λ, we obtain $\lambda = 2x - 2$. Substituting into the second

equation, we have $3 - 2y + 4x - 4 = 0$, or $4x - 2y - 1 = 0$. Adding this equation to the third equation in the system, we have $5x - 10 = 0$, or $x = 2$. Therefore, $y = \frac{7}{2}$ and $f\left(2, \frac{7}{2}\right) = -\frac{7}{4}$ is the maximum value of f.

© 2011 Cengage Learning. All Rights Reserved. May not be scanned, copied or duplicated, or posted to a publicly accessible website, in whole or in part.

5. We form the Lagrangian function $F(x, y, \lambda) = x^2 + 4y^2 + \lambda(xy - 1)$ and solve the system

$$\begin{cases} F_x = 2x + \lambda y = 0 \\ F_y = 8y + \lambda x = 0 \qquad \text{Multiplying the first and second equations by } x \text{ and } y, \text{ respectively, and subtracting} \\ F_\lambda = xy - 1 = 0 \end{cases}$$

the resulting equations, we obtain $2x^2 - 8y^2 = 0$, or $x = \pm 2y$. Substituting this into the third equation gives $2y^2 - 1 = 0$ or $y = \pm\frac{\sqrt{2}}{2}$. We conclude that $f\left(-\sqrt{2}, -\frac{\sqrt{2}}{2}\right) = f\left(\sqrt{2}, \frac{\sqrt{2}}{2}\right) = 4$ is the minimum value of f.

7. We form the Lagrangian function $F(x, y, \lambda) = x + 5y - 2xy - x^2 - 2y^2 + \lambda(2x + y - 4)$ and solve

the system $\begin{cases} F_x = 1 - 2y - 2x + 2\lambda = 0 \\ F_y = 5 - 2x - 4y + \lambda = 0 \qquad \text{Solving the last two equations for } x \text{ and } y \text{ in terms of } \lambda, \\ F_\lambda = 2x + y - 4 = 0 \end{cases}$

we obtain $y = \frac{1}{3}(1 + \lambda)$ and $x = \frac{1}{6}(11 - \lambda)$ which, upon substitution into the first equation, yields $1 - \frac{2}{3}(1 + \lambda) - \frac{1}{3}(11 - \lambda) + 2\lambda$, so $1 - \frac{2}{3} - \frac{2}{3}\lambda - \frac{11}{3} + \frac{1}{3}\lambda + 2\lambda = 0$. Hence, $\lambda = 2$, so $x = \frac{3}{2}$ and $y = 1$. The maximum of f is $f\left(\frac{3}{2}, 1\right) = \frac{3}{2} + 5 - 2\left(\frac{3}{2}\right) - \left(\frac{3}{2}\right)^2 - 2 = -\frac{3}{4}$.

9. We form the Lagrangian $F(x, y, \lambda) = xy^2 + \lambda(9x^2 + y^2 - 9)$ and solve the system

$$\begin{cases} F_x = y^2 + 18\lambda x = 0 \\ F_y = 2xy + 2\lambda y = 0 \qquad \text{The first equation gives } \lambda = -\frac{y^2}{18x}. \text{ Substituting into the second gives} \\ F_\lambda = 9x^2 + y^2 - 9 = 0 \end{cases}$$

$2xy + 2y\left(-\frac{y^2}{18x}\right) = 0$, or $18x^2y - y^3 = y(18x^2 - y^2) = 0$, giving $y = 0$ or $y = \pm 3\sqrt{2}x$. If $y = 0$, then the third equation gives $9x^2 - 9 = 0$, so $x = \pm 1$. Therefore, the points $(-1, 0)$, $(1, 0)$, $\left(-\frac{\sqrt{3}}{3}, -\sqrt{6}\right)$, $\left(-\frac{\sqrt{3}}{3}, \sqrt{6}\right)$, $\left(\frac{\sqrt{3}}{3}, -\sqrt{6}\right)$ and $\left(\frac{\sqrt{3}}{3}, \sqrt{6}\right)$ give extreme values of f subject to the given constraint. Evaluating $f(x, y)$ at each of these points, we see that $f\left(\frac{\sqrt{3}}{3}, -\sqrt{6}\right) = f\left(\frac{\sqrt{3}}{3}, \sqrt{6}\right) = 2\sqrt{3}$ is the maximum value of f.

11. We form the Lagrangian function $F(x, y, \lambda) = xy + \lambda(x^2 + y^2 - 16)$ and solve the system

$$\begin{cases} F_x = y + 2\lambda x = 0 \\ F_y = x + 2\lambda y = 0 \qquad \text{Solving the first equation for } \lambda \text{ and substituting this value into the second equation} \\ F_\lambda = x^2 + y^2 - 16 = 0 \end{cases}$$

yields $x - 2\left(\frac{y}{2x}\right)y = 0$, or $x^2 = y^2$. Substituting the last equation into the third equation in the system, yields $x^2 + x^2 - 16 = 0$, or $x^2 = 8$, that is, $x = \pm 2\sqrt{2}$. The corresponding values of y are $y = \pm 2\sqrt{2}$. Therefore the critical points of F are $\left(\pm 2\sqrt{2}, \pm 2\sqrt{2}\right)$ (all four possibilities). Evaluating f at each of these points, we find that $f\left(-2\sqrt{2}, 2\sqrt{2}\right) = f\left(2\sqrt{2}, -2\sqrt{2}\right) = -8$ are relative minima and $f\left(-2\sqrt{2}, -2\sqrt{2}\right) = f\left(2\sqrt{2}, 2\sqrt{2}\right) = 8$ are relative maxima.

© 2011 Cengage Learning. All Rights Reserved. May not be scanned, copied or duplicated, or posted to a publicly accessible website, in whole or in part.

13. We form the Lagrangian function $F(x, y, \lambda) = xy^2 + \lambda(x^2 + y^2 - 1)$ and solve the system

$$\begin{cases} F_x = y^2 + 2x\lambda = 0 \\ F_y = 2xy + 2y\lambda = 0 \\ F_\lambda = x^2 + y^2 - 1 = 0 \end{cases} \quad \text{We find that either } x = \pm\frac{\sqrt{3}}{3} \text{ and } y = \pm\frac{\sqrt{6}}{3}, \text{ or } x = \pm 1 \text{ and } y = 0. \text{ Evaluating } f \text{ at}$$

each of the critical points $\left(\pm\frac{\sqrt{3}}{3}, \pm\frac{\sqrt{6}}{3}\right)$ (all four possibilities) and $(\pm 1, 0)$, we find that $f\left(-\frac{\sqrt{3}}{3}, \pm\frac{\sqrt{6}}{3}\right) = -\frac{2\sqrt{3}}{9}$

are relative minima and $f\left(\frac{\sqrt{3}}{3}, \pm\frac{\sqrt{6}}{3}\right) = \frac{2\sqrt{3}}{9}$ are relative maxima.

15. We form the Lagrangian function $F(x, y, z, \lambda) = x^2 + y^2 + z^2 + \lambda(3x + 2y + z - 6)$ and solve the system

$$\begin{cases} F_x = 2x + 3\lambda = 0 \\ F_y = 2y + 2\lambda = 0 \\ F_z = 2x + \lambda = 0 \\ F_\lambda = 3x + 2y + z - 6 = 0 \end{cases} \quad \text{The third equation gives } \lambda = -2z. \text{ Substituting into the first two equations, we}$$

obtain $\begin{cases} 2x - 6z = 0 \\ 2y - 4z = 0 \end{cases}$ Thus, $x = 3z$ and $y = 2z$. Substituting into the third equation yields $9z + 4z + z - 6 = 0$,

or $z = \frac{3}{7}$. Therefore, $x = \frac{9}{7}$ and $y = \frac{6}{7}$, and so $f\left(\frac{9}{7}, \frac{6}{7}, \frac{3}{7}\right) = \frac{18}{7}$ is the minimum value of f.

17. We want to maximize P subject to the constraint $x + y = 200$. The Lagrangian function is
$F(x, y, \lambda) = -0.2x^2 - 0.25y^2 - 0.2xy + 100x + 90y - 4000 + \lambda(x + y - 200)$. We solve the

system $\begin{cases} F_x = -0.4x - 0.2y + 100 + \lambda = 0 \\ F_y = -0.5y - 0.2x + 90 + \lambda = 0 \\ F_\lambda = x + y - 200 = 0 \end{cases}$ Subtracting the first equation from the second yields

$0.2x - 0.3y - 10 = 0$, or $2x - 3y - 100 = 0$. Multiplying the third equation in the system by 2 and subtracting the resulting equation from the last equation, we find $-5y + 300 = 0$, so $y = 60$. Thus, $x = 140$ and the company should make 140 finished and 60 unfinished units.

19. Suppose each of the sides made of pine board is x feet long and those of steel are y feet long. Then $xy = 800$. The cost is $C = 12x + 3y$ and is to be minimized subject to the condition $xy = 800$. We form the Lagrangian function

$F(x, y, \lambda) = 12x + 3y + \lambda(xy - 800)$ and solve the system $\begin{cases} F_x = 12 + \lambda y = 0 \\ F_y = 3 + \lambda x = 0 \\ F_\lambda = xy - 800 = 0 \end{cases}$ Multiplying the first

equation by x and the second equation by y and subtracting the resulting equations, we obtain $12x - 3y = 0$, or $y = 4x$. Substituting this into the third equation of the system, we obtain $4x^2 - 800 = 0$, so $x = \pm 10\sqrt{2}$. Because x must be positive, we take $x = 10\sqrt{2}$, so $y = 40\sqrt{2}$ and the dimensions are approximately 14.14 ft by 56.56 ft.

© 2011 Cengage Learning. All Rights Reserved. May not be scanned, copied or duplicated, or posted to a publicly accessible website, in whole or in part.

21. We want to minimize the function $C(r, h)$ subject to the constraint $\pi r^2 h - 64 = 0$. We form the Lagrangian function $F(r, h, \lambda) = 8\pi r h + 6\pi r^2 - \lambda(\pi r^2 h - 64)$ and solve the system $\begin{cases} F_r = 8\pi h + 12\pi r - 2\lambda\pi r h = 0 \\ F_h = 8\pi r - \lambda r^2 = 0 \\ F_\lambda = \pi r^2 h - 64 = 0 \end{cases}$

Solving the second equation for λ yields $\lambda = 8/r$, which when substituted into the first equation yields

$8\pi h + 12\pi r - 2\pi r h \left(\dfrac{8}{r}\right) = 0$, $12\pi r = 8\pi h$, and $h = \frac{3}{2}r$. Substituting this value of h into the third equation of

the system, we find $3r^2 \left(\frac{3}{2}r\right) = 64$, $r^3 = \frac{128}{3\pi}$, so $r = \frac{4}{3}\sqrt[3]{\frac{18}{\pi}}$ and $h = 2\sqrt[3]{\frac{18}{\pi}}$.

23. Let the box have dimensions x by y by z feet. Then $xyz = 4$. We want to minimize $C = 2xz + 2yz + \frac{3}{2}(2xy) = 2xz + 2yz + 3xy$. We form the Lagrangian function

$F(x, y, z, \lambda) = 2xz + 2yz + 3xy + \lambda(xyz - 4)$ and solve the system $\begin{cases} F_x = 2z + 3y + \lambda yz = 0 \\ F_y = 2z + 3x + \lambda xz = 0 \\ F_z = 2x + 2y + \lambda xy = 0 \\ F_\lambda = xyz - 4 = 0 \end{cases}$

Multiplying the first, second, and third equations by x, y, and z respectively, we have $\begin{cases} 2xz + 3xy + \lambda xyz = 0 \\ 2yz + 3xy + \lambda xyz = 0 \\ 2xz + 2yz + \lambda xyz = 0 \end{cases}$

The first two equations imply that $2z(x - y) = 0$. Because $z \neq 0$, we see that $x = y$. The second and third equations imply that $x(3y - 2z) = 0$ or $x = \frac{3}{2}y$. Substituting these values into the fourth equation in the system,

we find $y^2 \left(\frac{3}{2}y\right) = 4$, so $y^3 = \frac{8}{3}$. Therefore, $y = \dfrac{2}{3^{1/3}} = \frac{2}{3}\sqrt[3]{9}$, $x = \frac{2}{3}\sqrt[3]{9}$, and $z = \sqrt[3]{9}$, and the dimensions (in feet) are $\frac{2}{3}\sqrt[3]{9} \times \frac{2}{3}\sqrt[3]{9} \times \sqrt[3]{9}$.

25. We want to maximize $f(x, y) = 100x^{3/4}y^{1/4}$ subject to $100x + 200y = 200,000$. We form the Lagrangian function $F(x, y, \lambda) = 100x^{3/4}y^{1/4} + \lambda(100x + 200y - 200,000)$ and solve the system

$\begin{cases} F_x = 75x^{-1/4}y^{1/4} + 100\lambda = 0 \\ F_y = 25x^{3/4}y^{-3/4} + 200\lambda = 0 \\ F_\lambda = 100x + 200y - 200,000 = 0 \end{cases}$ The first two equations imply that $150x^{-1/4}y^{1/4} - 25x^{3/4}y^{-3/4} = 0$ or,

upon multiplying by $x^{1/4}y^{3/4}$, $150y - 25x = 0$, which implies that $x = 6y$. Substituting this value of x into the third equation of the system, we have $600y + 200y - 200,000 = 0$, giving $y = 250$, and therefore $x = 1500$. So to maximize production, he should buy 1500 units of labor and 250 units of capital.

27. False. See Example 1 in Section 8.4 of the text.

29. True. We form the Lagrangian function $F(x, y, \lambda) = f(x, y) + \lambda g(x, y)$. Then $F_x = 0$, $F_y = 0$, and $F_\lambda = 0$ at (a, b) and $f_x(a, b) + \lambda g_x(a, b) = 0$, so $f_x(a, b) = -\lambda g_x(a, b)$, and $f_y(a, b) + \lambda(a, b) = 0$, so $f_y(a, b) = -\lambda g_y(a, b)$ and $g(a, b) = 0$.

© 2011 Cengage Learning. All Rights Reserved. May not be scanned, copied or duplicated, or posted to a publicly accessible website, in whole or in part.

8.6   Problem-Solving Tips

1. To evaluate the double integral $\iint\limits_{R} f(x,y)\, dA = \int_a^b \left[\int_{g_1(x)}^{g_2(x)} f(x,y)\, dy \right] dx$, first evaluate the inside integral with respect to y, treating x as a constant, then evaluate the outside integral with respect to x.

2. To evaluate the double integral $\iint\limits_{R} f(x,y)\, dA = \int_a^b \left[\int_{h_2(y)}^{h_2(y)} f(x,y)\, dx \right] dy$, first evaluate the inside integral with respect to x, treating y as a constant, then evaluate the outside integral with respect to y.

8.6   Concept Questions   page 590

1. If $z = f(x,y)$, then $dx = \Delta x$ and $dy = \Delta y$. The differential of z is $dz = \frac{\partial f}{\partial x} dx + \frac{\partial f}{\partial y} dy$.

8.6   Total Differentials   page 590

1. $f(x,y) = x^2 + 2y$, so $df = 2x\, dx + 2\, dy$.

3. $f(x,y) = 2x^2 - 3xy + 4x$, so $df = (4x - 3y + 4)\, dx - 3x\, dy$.

5. $f(x,y) = \sqrt{x^2 + y^2}$, so
$$df = \tfrac{1}{2}\left(x^2 + y^2\right)^{-1/2}(2x)\, dx + \tfrac{1}{2}\left(x^2 + y^2\right)^{-1/2}(2y)\, dy = \frac{x}{\sqrt{x^2 + y^2}}\, dx + \frac{y}{\sqrt{x^2 + y^2}}\, dy.$$

7. $f(x,y) = \dfrac{5y}{x - y}$, so
$$df = \frac{\partial}{\partial x}\left[5y(x-y)^{-1}\right] dx + \frac{\partial}{\partial x}\left(\frac{5y}{x-y}\right) dy = 5y(-1)(x-y)^{-2}\, dx + \frac{(x-y)(5) - 5y(-1)}{(x-y)^2}\, dy$$
$$= -\frac{5y}{(x-y)^2}\, dx + \frac{5x}{(x-y)^2}\, dy.$$

9. $f(x,y) = 2x^5 - ye^{-3x}$, so $df = \left(10x^4 + 3ye^{-3x}\right) dx - e^{-3x}\, dy$.

11. $f(x,y) = x^2 e^y + y \ln x$, so $df = \left(2xe^y + \frac{y}{x}\right) dx + \left(x^2 e^y + \ln x\right) dy$.

13. $f(x,y,z) = xy^2 z^3$, so $df = y^2 z^3\, dx + 2xyz^3\, dy + 3xy^2 z^2\, dz$.

15. $f(x,y,z) = \dfrac{x}{y+z}$, so
$$df = \frac{1}{y+z}\, dx + x(-1)(y+z)^{-2}\, dy + x(-1)(y+z)^{-2}\, dz = \frac{1}{y+z}\, dx - \frac{x}{(y+z)^2}\, dy - \frac{x}{(y+z)^2}\, dz.$$

17. $f(x,y,z) = xyz + xe^{yz}$, so
$$df = (yz + e^{yz})\, dx + (xz + xze^{yz})\, dy + (xy + xye^{yz})\, dz = (yz + e^{yz})\, dx + xz(1 + e^{yz})\, dy + xy(1 + e^{yz})\, dz.$$

19. $z = f(x,y) = 4x^2 - xy$, so $\Delta z \approx dz = (8x - y)\, dx - x\, dy$. If (x,y) changes from $(1,2)$ to $(1.01, 2.02)$, then $x = 1$, $y = 2$, $dx = 0.01$, and $dy = 0.02$, so $\Delta z \approx (8 - 2)(0.01) - 0.02 = 0.04$.

© 2011 Cengage Learning. All Rights Reserved. May not be scanned, copied or duplicated, or posted to a publicly accessible website, in whole or in part.

21. $z = f(x, y) = x^{2/3}y^{1/2}$, so $\Delta z \approx dz = \frac{2}{3}x^{-1/3}y^{1/2}\,dx + \frac{1}{2}x^{2/3}y^{-1/2}\,dy$. If (x, y)

changes from $(8, 9)$ to $(7.97, 9.03)$, then $x = 8$, $y = 9$, $dx = -0.03$, and $dy = 0.03$, so

$\Delta z \approx \frac{2}{3}\left(\frac{3}{2}\right)(-0.03) + \frac{1}{2}\left(\frac{4}{3}\right)(0.03) = -0.03 + 0.02 = -0.01$.

23. $f(x, y) = \dfrac{x}{x - y}$, so $f_x = \dfrac{(x - y) - x\,(1)}{(x - y)^2} = -\dfrac{y}{(x - y)^2}$ and $f_y = \dfrac{(x - y)\,(0) - x\,(-1)}{(x - y)^2} = \dfrac{x}{(x - y)^2}$. Thus,

$\Delta z \approx dz = \dfrac{-y\,dx + x\,dy}{(x - y)^2}$. Here $x = -3$, $y = -2$, $dx = -0.02$, and $dy = 0.02$, so the approximate change in

$z = f(x, y)$ is $\Delta z \approx \dfrac{-(-2)\,(-0.02) + (-3)\,(0.02)}{(-1)^2} = -0.1$.

25. $z = f(x, y) = 2xe^{-y}$, so $\Delta z \approx dz = 2e^{-y}\,dx - 2xe^{-y}\,dy$. With $x = 4$, $y = 0$, $dx = 0.03$ and $dy = 0.03$, we

have $\Delta z \approx 2e^0\,(0.03) - 2\,(4)\,(e^0)\,(0.03) = -0.18$.

27. $f(x, y) = xe^{xy} - y^2$, so $f_x = xye^{xy} + e^{xy} = e^{xy}(1 + xy)$ and $f_y = x^2e^{xy} - 2y$. Thus,

$\Delta z \approx dz = e^{xy}(1 + xy)\,dx + (x^2e^{xy} - 2y)\,dy$. With $x = -1$, $y = 0$, $dx = 0.03$, and $dy = 0.03$, we have

$\Delta z \approx e^0\,(1)\,(0.03) + (1)\,(e^0)\,(0.03) = 0.06$.

29. $f(x, y) = x\ln x + y\ln x$, so $f_x = \ln x + x\left(\frac{1}{x}\right) + \frac{y}{x} = \ln x + 1 + \frac{y}{x}$ and $f_y = \ln x$. Thus,

$\Delta z \approx dz = \left(\ln x + 1 + \frac{y}{x}\right)dx + \ln x\,dy$. With $x = 2$, $y = 3$, $dx = -0.02$, and $dy = -0.11$, we have

$\Delta z \approx (\ln 2 + 1 + 1.5)\,(-0.02) + (\ln 2)\,(-0.11) \approx -0.1401$.

31. $P(x, y) = -0.02x^2 - 15y^2 + xy + 39x + 25y - 20,000$, so

$\Delta P \approx dP = (-0.04x + y + 39)\,dx + (-30y + x + 25)\,dy$. With $x = 4000$, $y = 150$, $dx = 500$, and

$dy = -10$, we have $\Delta P \approx [-0.04\,(4000) + 150 + 39]\,(500) + [-30\,(150) + 4000 + 25]\,(-10) = 19,250$, or

$\$19,250$ per month.

33. $R(x, y) = -x^2 - 0.5y^2 + xy + 8x + 3y + 20$, so

$dR = -2x\,dx - y\,dy + y\,dx + x\,dy + 8dx + 3dy = (-2x + y + 8)\,dx + (-y + x + 3)\,dy$. If $x = 10$, $y = 15$,

$dx = 1$, and $dy = -1$, then $dR = [-2\,(10) + 15 + 8]\,(1) + (-15 + 10 + 3)\,(-1) = 3\,(1) + (-2)\,(-1) = 5$. Thus,

the revenue is expected to increase by $\$5000$ per month.

35. $R = \dfrac{x}{y}$, so $\Delta R \approx dR = \dfrac{1}{y}\,dx - \dfrac{x}{y^2}\,dy$. Therefore, when $x = 60$, $y = 4$, $dx = 2$, and $dy = -0.2$, we have

$\Delta R = \frac{1}{4}\,(2) - \frac{60}{4^2}\,(-0.2) = \frac{1}{2} + \frac{15}{4}\left(\frac{1}{5}\right) = \frac{1}{2} + \frac{3}{4} = \frac{5}{4} = 1.25$, or $\$1.25$.

37. $V = \pi r^2 h$, so $\Delta V \approx dV = 2\pi rh\,dr + \pi r^2\,dh$. If $r = 8$, $h = 20$, $dr = \pm 0.1$, and $dh = \pm 0.1$, then

$dV = 2\pi\,(8)\,(20)\,(\pm 0.1) + \pi\,(8)^2\,(\pm 0.1) = 320\pi\,(\pm 0.1) + 64\pi\,(\pm 0.1) = \pm 38.4\pi$, or approximately ± 120.6 cc.

39. Let x, y, and z denote the dimensions of the box. Then its surface area is $S = 2xy + 2xz + 2yz$, so

$dS = 2\,(y + z)\,dx + 2\,(x + z)\,dy + 2\,(x + y)\,dz$. Then, with $x = 30$, $y = 40$, $z = 60$, and $dx = dy = dz = \pm 0.2$,

we find

$|\Delta S| \approx |dS| = |2\,(40 + 60)\,(\pm 0.2) + 2\,(30 + 60)\,(\pm 0.2) + 2\,(30 + 40)\,(\pm 0.2)|$

$\leq |2\,(40 + 60)\,(\pm 0.2)| + |2\,(30 + 60)\,(\pm 0.2)| + |2\,(30 + 40)\,(\pm 0.2)| = 104$ in^2

so the maximum error is 104 in^2.

© 2011 Cengage Learning. All Rights Reserved. May not be scanned, copied or duplicated, or posted to a publicly accessible website, in whole or in part.

41. $dP = \dfrac{\partial}{\partial V}\left(\dfrac{8.314T}{V}\right)dV + \dfrac{\partial}{\partial T}\left(\dfrac{8.314T}{V}\right)dT = -\dfrac{8.314T}{V^2}\,dV + \dfrac{8.314}{V}\,dT.$ With $V = 20$, $T = 300$, $dV = 0.2$,

and $dT = -5$, we find $\Delta P \approx dP = -\dfrac{8.314\,(300)}{20^2}\,(0.2) + \dfrac{8.314}{20}\,(-5) = -3.3256$, so the pressure decreases by

approximately 3.3256 Pa.

43. Differentiating implicitly, we have $-\dfrac{1}{R^2}\,dR = -\dfrac{1}{R_1^2}\,dR_1 - \dfrac{1}{R_2^2}\,dR_2 - \dfrac{1}{R_3^2}\,dR_3$, so

$|dR| \le \left(\dfrac{R}{R_1}\right)^2|dR_1| + \left(\dfrac{R}{R_2}\right)^2|dR_2| + \left(\dfrac{R}{R_3}\right)^2|dR_3|$. If $R_1 = 100$, $R_2 = 200$, $R_3 = 300$, $|dR_1| \le 1$, $|dR_2| \le 2$,

and $|dR_3| \le 3$, then $\dfrac{1}{R} = \frac{1}{100} + \frac{1}{200} + \frac{1}{300} = \frac{6+3+2}{600} = \frac{11}{600}$, so $|dR| \le \left(\frac{6}{11}\right)^2(1) + \left(\frac{3}{11}\right)^2(2) + \left(\frac{2}{11}\right)^2(3) = 0.5455$,

or 0.55 ohms.

8.7   Concept Questions   page 597

1. An iterated integral is a single integral such as $\int_a^b f(x, y)\,dx$, where we think of y as a constant. It is evaluated as

follows: $\int_R \int f(x, y)\,dA = \int_c^d \left[\int_a^b f(x, y)\,dx\right]dy.$

3. $\int_R \int f(x, y)\,dA = \int_c^d \left[\int_{h_1(y)}^{h_2(y)} f(x, y)\,dx\right]dy.$

8.7   Double Integrals   page 597

1. $\int_1^2 \int_0^1 (y + 2x)\,dy\,dx = \int_1^2 \left(\frac{1}{2}y^2 + 2xy\right)\Big|_{y=0}^{y=1} dx = \int_1^2 \left(\frac{1}{2} + 2x\right)dx = \left(\frac{1}{2}x + x^2\right)\Big|_1^2 = 5 - \frac{3}{2} = \frac{7}{2}.$

3. $\int_{-1}^1 \int_0^1 xy^2\,dy\,dx = \int_{-1}^1 \frac{1}{3}xy^3\Big|_{y=0}^{y=1} dx = \int_{-1}^1 \frac{1}{3}x\,dx = \frac{1}{6}x^2\Big|_{-1}^1 = \frac{1}{6} - \left(\frac{1}{6}\right) = 0.$

5. $\int_{-1}^2 \int_1^{e^3} \dfrac{x}{y}\,dy\,dx = \int_{-1}^2 x \ln y\big|_{y=1}^{y=e^3} dx = \int_{-1}^2 x \ln e^3\,dx = \int_{-1}^2 3x\,dx = \frac{3}{2}x^2\Big|_{-1}^2 = \frac{3}{2}(4) - \frac{3}{2}(1) = \frac{9}{2}.$

7. $\int_{-2}^0 \int_0^1 4xe^{2x^2+y}\,dx\,dy = \int_{-2}^0 e^{2x^2+y}\big|_{x=0}^{x=1} dy = \int_{-2}^0 \left(e^{2+y} - e^y\right)dy = \left(e^{2+y} - e^y\right)\big|_{-2}^0 = \left(e^2 - 1\right) - \left(e^0 - e^{-2}\right)$

$\qquad = e^2 - 2 + e^{-2} = \left(e^2 - 1\right)\left(1 - e^{-2}\right).$

9. $\int_0^1 \int_1^e \ln y\,dy\,dx = \int_0^1 (y \ln y - y)\big|_1^e\,dx = \int_0^1 dx = 1.$

11. $\int_0^1 \int_0^x (x + 2y)\,dy\,dx = \int_0^1 (xy + y^2)\big|_{y=0}^{y=x} dx = \int_0^1 2x^2\,dx = \frac{2}{3}x^3\Big|_0^1 = \frac{2}{3}.$

13. $\int_1^3 \int_0^{x+1} (2x + 4y)\,dy\,dx = \int_1^3 (2xy + 2y^2)\big|_{y=0}^{y=x+1} dx = \int_1^3 \left[2x(x+1) + 2(x+1)^2\right]dx = \int_1^3 (4x^2 + 6x + 2)\,dx$

$\qquad = \left(\frac{4}{3}x^3 + 3x^2 + 2x\right)\Big|_1^3 = (36 + 27 + 6) - \left(\frac{4}{3} + 3 + 2\right) = \frac{188}{3}.$

© 2011 Cengage Learning. All Rights Reserved. May not be scanned, copied or duplicated, or posted to a publicly accessible website, in whole or in part.

15. $\int_0^4 \int_0^{\sqrt{y}} (x + y)\, dx\, dy = \int_0^4 \left(\frac{1}{2}x^2 + xy\right)\Big|_{x=0}^{x=\sqrt{y}}\, dy = \int_0^4 \left(\frac{1}{2}y + y^{3/2}\right) dy = \left(\frac{1}{4}y^2 + \frac{2}{5}y^{5/2}\right)\Big|_0^4 = 4 + \frac{64}{5} = \frac{84}{5}.$

17. $\int_0^2 \int_0^{\sqrt{4-y^2}} y\, dx\, dy = \int_0^2 xy\Big|_{x=0}^{x=\sqrt{4-y^2}}\, dy = \int_0^2 y\sqrt{4-y^2}\, dy = -\frac{1}{2}\left(\frac{2}{3}\right)(4-y^2)^{3/2}\Big|_0^2 = \frac{1}{3}(4^{3/2}) = \frac{8}{3}.$

19. $\int_0^1 \int_0^x 2xe^y\, dy\, dx = \int_0^1 2xe^y\big|_{y=0}^{y=x}\, dx = \int_0^1 (2xe^x - 2x)\, dx = 2(x-1)e^x - x^2\big|_0^1 = (-1) + 2 = 1.$

21. $\int_0^1 \int_x^{\sqrt{x}} ye^x\, dy\, dx = -\int_0^1 \int_x^{\sqrt{x}} ye^x\, dy\, dx = \int_0^1 \left(-\frac{1}{2}y^2 e^x\right)\Big|_{y=\sqrt{x}}^{y=x}\, dx = -\frac{1}{2}\int_0^1 (x^2 e^x - xe^x)\, dx$

$\qquad = -\frac{1}{2}\left(x^2 e^x\big|_0^1 - 2\int_0^1 xe^x\, dx - \int_0^1 xe^x\, dx\right) = -\frac{1}{2}\left(x^2 e^x\big|_0^1 - 3\int_0^1 xe^x\, dx\right)$

$\qquad = -\frac{1}{2}(x^2 e^x - 3xe^x + 3e^x)\Big|_0^1 = -\frac{1}{2}(e - 3e + 3e - 3) = \frac{1}{2}(3 - e).$

23. $\int_0^1 \int_{2x}^2 e^{y^2}\, dy\, dx = \int_0^2 \int_0^{y/2} e^{y^2}\, dx\, dy = \int_0^2 xe^{y^2}\Big|_{x=0}^{x=y/2}\, dy = \int_0^2 \frac{1}{2}ye^{y^2}\, dy = \frac{1}{4}e^{y^2}\Big|_0^2 = \frac{1}{4}(e^4 - 1).$

25. $\int_0^2 \int_{y/2}^1 ye^{x^3}\, dx\, dy = \int_0^1 \int_0^{2x} ye^{x^3}\, dy\, dx = \int_0^1 \frac{1}{2}y^2 e^{x^3}\Big|_{y=0}^{y=2x}\, dx = \int_0^1 2x^2 e^{x^3}\, dx = \frac{2}{3}e^{x^3}\Big|_0^1 = \frac{2}{3}(e - 1).$

27. False. Let $f(x, y) = \dfrac{x}{y - 2}$, $a = 0$, $b = 3$, $c = 0$, and $d = 1$. Then $\int_{R_1}\int f(x, y)\, dA$ is

defined on $R_1 = \{(x, y)\mid 0 \leq x \leq 3, 0 \leq y \leq 1\}$, but $\int_{R_2}\int f(x, y)\, dA$ is not defined on

$R_2 = \{(x, y)\mid 0 \leq x \leq 1, 0 \leq y \leq 3\}$, because f is discontinuous on R_2 where $y = 2$.

8.8   Concept Questions   page 603

1. It gives the volume of the solid region bounded above by the graph of f and below by the region R.

3. a, b. Enclose the rectangular region R representing a certain district of the city by a rectangular grid as shown in Figure 42 on page 600 of the text. The Riemann sum with general term $F(x_i, y_i)\, hk$ gives the number of people living in that part of the city corresponding to the rectangular region R_i.

  c. The integral $\int_R\int f(x, y)\, dA$ gives the actual number of people living in the district under consideration.

8.8   Applications of Double Integrals   page 604

1. $V = \int_0^4 \int_0^3 \left(4 - x + \frac{1}{2}y\right) dx\, dy = \int_0^4 \left(4x - \frac{1}{2}x^2 + \frac{1}{2}xy\right)\Big|_{x=0}^{x=3}\, dy = \int_0^4 \left(\frac{15}{2} + \frac{3}{2}y\right) dy$

$\qquad = \left(\frac{15}{2}y + \frac{3}{4}y^2\right)\Big|_0^4 = 42.$

3. $V = \int_0^2 \int_x^{4-x} 5\, dy\, dx = \int_0^2 5y\big|_{y=x}^{y=4-x}\, dx = 5\int_0^2 (4 - 2x)\, dx = 5\left(4x - x^2\right)\big|_0^2 = 20.$

5. $V = \int_0^2 \int_0^{4-x^2} 4\, dy\, dx = \int_0^2 4y\big|_{y=0}^{y=4-x^2}\, dx = \int_0^2 4(4 - x^2)\, dx = 4\int_0^2 (4 - x^2)\, dx = 4\left(4x - \frac{1}{3}x^3\right)\Big|_0^2 = \frac{64}{3}.$

7. $V = \int_0^1 \int_0^y \sqrt{1 - y^2}\, dx\, dy = \int_0^1 x\sqrt{1 - y^2}\Big|_{x=0}^{x=y}\, dy = \int_0^1 y\sqrt{1 - y^2}\, dy = \left(-\frac{1}{2}\right)\left(\frac{2}{3}\right)(1 - y^2)^{3/2}\Big|_0^1 = \frac{1}{3}.$

© 2011 Cengage Learning. All Rights Reserved. May not be scanned, copied or duplicated, or posted to a publicly accessible website, in whole or in part.

9. $V = \int_0^1 \int_0^2 (4 - 2x - y)\, dy\, dx = \int_0^1 \left(4y - 2xy - \tfrac{1}{2}y^2\right)\Big|_{y=0}^{y=2} dx = \int_0^1 (8 - 4x - 2)\, dx = \int_0^1 (6 - 4x)\, dx$

$= (6x - 2x^2)\big|_0^1 = 6 - 2 = 4.$

11. $V = \int_0^1 \int_0^2 (x^2 + y^2)\, dy\, dx = \int_0^1 \left(x^2 y + \tfrac{1}{3}y^3\right)\Big|_{y=0}^{y=2} dx = \int_0^1 \left(2x^2 + \tfrac{8}{3}\right) dx = \int_0^1 \left(\tfrac{2}{3}x^2 + \tfrac{8}{3}\right) dx$

$= \left(\tfrac{2}{3}x^3 + \tfrac{8}{3}x\right)\Big|_0^1 = \tfrac{2}{3} + \tfrac{8}{3} = \tfrac{10}{3}.$

13. $V = \int_0^2 \int_x^2 2xe^y\, dy\, dx = \int_0^2 2xe^y|_{y=x}^{y=2}\, dx = \int_0^2 (2xe^2 - 2xe^x)\, dx = \left[e^2 x^2 - 2(x-1)e^x\right]_0^2$ (by parts)

$= 4e^2 - 2e^2 - 2 = 2(e^2 - 1).$

15. $V = \int_0^1 \int_{x^2}^x 2x^2 y\, dy\, dx = \int_0^1 x^2 y^2\Big|_{y=x^2}^{y=x}\, dx = \int_0^1 (x^4 - x^6)\, dx = \tfrac{1}{5} - \tfrac{1}{7} = \tfrac{2}{35}.$

17. $A = \tfrac{1}{6} \int_0^3 \int_0^2 6x^2 y^3\, dx\, dy = \int_0^3 \tfrac{1}{3}x^3 y^3\Big|_0^2\, dy = \tfrac{8}{3}\int_0^3 y^3\, dy = \tfrac{2}{3}y^4\Big|_0^3 = 54.$

19. The area of R is $\tfrac{1}{2}(2)(1) = 1$, so the average value of f is

$\int_0^1 \int_y^{2-y} xy\, dx\, dy = \int_0^1 \tfrac{1}{2}x^2 y\Big|_{x=y}^{x=2-y}\, dy = \int_0^1 \left[\tfrac{1}{2}(2-y)^2 y - \tfrac{1}{2}y^3\right] dy = \int_0^1 (2y - 2y^2)\, dy = \left(y^2 - \tfrac{2}{3}y^3\right)\Big|_0^1 = \tfrac{1}{3}.$

21. The area of R is $\tfrac{1}{2}$, so the average value of f is

$2\int_0^1 \int_0^x xe^y\, dy\, dx = 2\int_0^1 xe^y|_{y=0}^{y=x}\, dx = 2\int_0^1 (xe^x - x)\, dx = 2\left(xe^x - e^x - \tfrac{1}{2}x^2\right)\Big|_0^1 = 2\left(e - e - \tfrac{1}{2} + 1\right) = 1.$

23. The population is

$2\int_0^5 \int_{-2}^0 \frac{10{,}000e^y}{1 + 0.5x}\, dy\, dx = 20{,}000\int_0^5 \frac{e^y}{1+0.5x}\Big|_{y=-2}^{y=0}\, dx = 20{,}000\left(1 - e^{-2}\right)\int_0^5 \frac{1}{1+0.5x}\, dx$

$= 20{,}000\left(1 - e^{-2}\right) \cdot 2\ln(1 + 0.5x)|_0^5 = 40{,}000\left(1 - e^{-2}\right)\ln 3.5 \approx 43{,}329.$

25. By symmetry, it suffices to compute the population in the first quadrant. In the first

quadrant, $f(x, y) = \dfrac{50{,}000xy}{(x^2 + 20)(y^2 + 36)}$. Therefore, the population in R is given by

$\iint_R f(x, y)\, dA = 4\int_0^{15} \left[\int_0^{20} \frac{50{,}000xy}{(x^2 + 20)(y^2 + 36)}\, dy\right] dx = 4\int_0^{15} \left[\frac{50{,}000x\left(\tfrac{1}{2}\right)\ln(y^2 + 36)}{x^2 + 20}\right]_0^{20} dx$

$= 100{,}000(\ln 436 - \ln 36)\int_0^{15} \frac{x}{x^2 + 20}\, dx = 100{,}000(\ln 436 - \ln 36)\left(\tfrac{1}{2}\right)\ln(x^2 + 20)\Big|_0^{15}$

$= 50{,}000(\ln 436 - \ln 36)(\ln 245 - \ln 20) \approx 312{,}455$, or approximately 312,455 people.

27. The average price is

$P = \tfrac{1}{2}\int_0^1 \int_0^2 \left[200 - 10\left(x - \tfrac{1}{2}\right)^2 - 15(y - 1)^2\right] dy\, dx = \tfrac{1}{2}\int_0^1 \left[200y - 10\left(x - \tfrac{1}{2}\right)^2 y - 5(y-1)^3\right]_0^2 dx$

$= \tfrac{1}{2}\int_0^1 \left[400 - 20\left(x - \tfrac{1}{2}\right)^2 - 5 - 5\right] dx = \tfrac{1}{2}\int_0^1 \left[390 - 20\left(x - \tfrac{1}{2}\right)^2\right] dx = \tfrac{1}{2}\left[390x - \tfrac{20}{3}\left(x - \tfrac{1}{2}\right)^3\right]_0^1$

$= \tfrac{1}{2}\left[390 - \tfrac{20}{3}\left(\tfrac{1}{8}\right) - \tfrac{20}{3}\left(\tfrac{1}{8}\right)\right] \approx 194.17$, or approximately \$194 per square foot.

© 2011 Cengage Learning. All Rights Reserved. May not be scanned, copied or duplicated, or posted to a publicly accessible website, in whole or in part.

29. True. The average value is $V_A = \dfrac{\int_R \int f\,(x, y)\,dA}{\int_R \int dA}$, and so $V_A \int_R \int dA = \int_R \int f\,(x, y)\,dA$. The quantity on the left-hand side is the volume of such a cylinder.

CHAPTER 8   Concept Review   page 606

1. xy, ordered pair, real number, $f\,(x, y)$

3. $z = f\,(x, y)$, f, surface

5. constant, x

7. $\le$, (a, b), $\le$, domain

9. scatter, minimizing, least-squares, normal

11. independent, $dx = \Delta x$, $dy = \Delta y$, $\frac{\partial f}{\partial x}\,dx + \frac{\partial f}{\partial y}\,dy$

13. iterated, $\int_3^5 \int_0^1 \left(2x + y^2\right) dx\,dy$

CHAPTER 8   Review   page 607

1. $f\,(x, y) = \dfrac{xy}{x^2 + y^2}$, so $f\,(0, 1) = 0$, $f\,(1, 0) = 0$, $f\,(1, 1) = \dfrac{1}{1 + 1} = \dfrac{1}{2}$, and $f\,(0, 0)$ does not exist because the point $(0, 0)$ does not lie in the domain of f.

3. $h\,(x, y, z) = xye^z + \dfrac{x}{y}$, so $h\,(1, 1, 0) = 1 + 1 = 2$, $h\,(-1, 1, 1) = -e - 1 = -(e + 1)$, and $h\,(1, -1, 1) = -e - 1 = -(e + 1)$.

5. $f\,(x, y) = \dfrac{x - y}{x + y}$, so $D = \{(x, y) \mid y \ne -x\}$.

7. $f\,(x, y, z) = \dfrac{xy\sqrt{z}}{(1 - x)(1 - y)(1 - z)}$. The domain of f is the set of all ordered triples (x, y, z) of real numbers such that $z \ge 0$, $x \ne 1$, $y \ne 1$, and $z \ne 1$.

9. $z = y - x^2$

11. $z = e^{xy}$

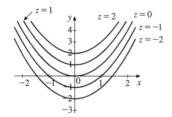

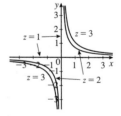

13. $f\,(x, y) = x\sqrt{y} + y\sqrt{x}$, so $f_x = \sqrt{y} + \dfrac{y}{2\sqrt{x}}$ and $f_y = \dfrac{x}{2\sqrt{y}} + \sqrt{x}$.

© 2011 Cengage Learning. All Rights Reserved. May not be scanned, copied or duplicated, or posted to a publicly accessible website, in whole or in part.

15. $f(x, y) = \dfrac{x - y}{y + 2x}$, so $f_x = \dfrac{(y + 2x) - (x - y)(2)}{(y + 2x)^2} = \dfrac{3y}{(y + 2x)^2}$ and

$f_y = \dfrac{(y + 2x)(-1) - (x - y)}{(y + 2x)^2} = \dfrac{-3x}{(y + 2x)^2}$.

17. $h(x, y) = (2xy + 3y^2)^5$, so $h_x = 10y(2xy + 3y^2)^4$ and $h_y = 10(x + 3y)(2xy + 3y^2)^4$.

19. $f(x, y) = (x^2 + y^2)e^{x^2+y^2}$, so $f_x = 2xe^{x^2+y^2} + (x^2 + y^2)(2x)e^{x^2+y^2} = 2x(x^2 + y^2 + 1)e^{x^2+y^2}$ and

$f_y = 2ye^{x^2+y^2} + (x^2 + y^2)(2y)e^{x^2+y^2} = 2y(x^2 + y^2 + 1)e^{x^2+y^2}$.

21. $f(x, y) = \ln\left(1 + \dfrac{x^2}{y^2}\right)$, so $f_x = \dfrac{2x/y^2}{1 + (x^2/y^2)} = \dfrac{2x}{x^2 + y^2}$ and $f_y = \dfrac{-2x^2/y^3}{1 + (x^2/y^2)} = -\dfrac{2x^2}{y(x^2 + y^2)}$.

23. $f(x, y) = x^4 + 2x^2y^2 - y^4$, so $f_x = 4x^3 + 4xy^2$ and $f_y = 4x^2y - 4y^3$. Therefore, $f_{xx} = 12x^2 + 4y^2$, $f_{xy} = 8xy = f_{yx}$, and $f_{yy} = 4x^2 - 12y^2$.

25. $g(x, y) = \dfrac{x}{x + y^2}$, so $g_x = \dfrac{(x + y^2) - x}{(x + y^2)^2} = \dfrac{y^2}{(x + y^2)^2}$ and

$g_y = \dfrac{-2xy}{(x + y^2)^2}$. Therefore, $g_{xx} = -2y^2(x + y^2)^{-3} = -\dfrac{2y^2}{(x + y^2)^3}$,

$g_{xy} = \dfrac{(x + y^2)2y - y^2(2)(x + y^2)2y}{(x + y^2)^4} = \dfrac{2(x + y^2)(xy + y^3 - 2y^3)}{(x + y^2)^4} = \dfrac{2y(x - y^2)}{(x + y^2)^3} = g_{yx}$, and

$g_{yy} = \dfrac{(x + y^2)^2(-2x) + 2xy(2)(x + y^2)2y}{(x + y^2)^4} = \dfrac{2x(x^2 + y^2)(-x - y^2 + 4y^2)}{(x + y^2)^4} = \dfrac{2x(3y^2 - x)}{(x + y^2)^3}$

27. $h(s, t) = \ln\left(\dfrac{s}{t}\right)$. Write $h(s, t) = \ln s - \ln t$. Then $h_s = \dfrac{1}{s}$ and $h_t = -\dfrac{1}{t}$, so $h_{ss} = -\dfrac{1}{s^2}$, $h_{st} = h_{ts} = 0$, and

$h_{tt} = \dfrac{1}{t^2}$.

29. $f(x, y) = 2x^2 + y^2 - 8x - 6y + 4$. To find the critical points of f, we solve the system

$\begin{cases} f_x = 4x - 8 = 0 \\ f_y = 2y - 6 = 0 \end{cases}$ obtaining $x = 2$ and $y = 3$. Therefore, the sole critical point of f is $(2, 3)$. Next, $f_{xx} = 4$,

$f_{xy} = 0$, and $f_{yy} = 2$, so $D(2, 3) = f_{xx}(2, 3)f_{yy}(2, 3) - f_{xy}(2, 3)^2 = 8 > 0$. Because $f_{xx}(2, 3) > 0$, we see that $f(2, 3) = -13$ is a relative minimum.

31. $f(x, y) = x^3 - 3xy + y^2$. We solve the system of equations $\begin{cases} f_x = 3x^2 - 3y = 0 \\ f_y = -3x + 2y = 0 \end{cases}$ obtaining $x^2 - y = 0$,

so $y = x^2$. Then $-3x + 2x^2 = 0$, $x(2x - 3) = 0$, and so $x = 0$ or $x = \frac{3}{2}$. The corresponding values of y are $y = 0$ and $y = \frac{9}{4}$, so the critical points are $(0, 0)$ and $\left(\frac{3}{2}, \frac{9}{4}\right)$. Next, $f_{xx} = 6x$, $f_{xy} = -3$, and $f_{yy} = 2$, and so $D(x, y) = 12x - 9 = 3(4x - 3)$. Therefore, $D(0, 0) = -9$, and so $(0, 0)$ is a saddle point and $f(0, 0) = 0$. $D\left(\frac{3}{2}, \frac{9}{4}\right) = 3(6 - 3) = 9 > 0$ and $f_{xx}\left(\frac{3}{2}, \frac{9}{4}\right) > 0$, and so $f\left(\frac{3}{2}, \frac{9}{4}\right) = \frac{27}{8} - \frac{81}{8} + \frac{81}{16} = -\frac{27}{16}$ is a relative minimum value.

© 2011 Cengage Learning. All Rights Reserved. May not be scanned, copied or duplicated, or posted to a publicly accessible website, in whole or in part.

33. $f(x, y) = f(x, y) = e^{2x^2+y^2}$. To find the critical points of f, we solve the

system $\begin{cases} f_x = 4xe^{2x^2+y^2} = 0 \\ f_y = 2ye^{2x^2+y^2} = 0 \end{cases}$ giving $(0, 0)$ as the only critical point of f. Next,

$f_{xx} = 4\left(e^{2x^2+y^2} + 4x^2e^{2x^2+y^2}\right) = 4\left(1 + 4x^2\right)e^{2x^2+y^2}$, $f_{xy} = 8xye^{2x^2+y^2} = f_{yx}$, and $f_{yy} = 2\left(1 + 2y^2\right)e^{2x^2+y^2}$,

so $D = f_{xx}(0, 0) f_{yy}(0, 0) - f_{xy}^2(0, 0) = (4)(2) - 0 = 8 > 0$. Because $f_{xx}(0, 0) > 0$, we see that $(0, 0)$ gives a relative minimum of f. The minimum value of f is $f(0, 0) = e^0 = 1$.

35. We form the Lagrangian function $F(x, y, \lambda) = -3x^2 - y^2 + 2xy + \lambda(2x + y - 4)$. Next, we solve the system

$\begin{cases} F_x = 6x + 2y + 2\lambda = 0 \\ F_y = -2y + 2x + \lambda = 0 \\ F_\lambda = 2x + y - 4 = 0 \end{cases}$ Multiplying the second equation by 2 and subtracting the resulting equation from

the first equation yields $6y - 10x = 0$ so $y = \frac{5}{3}x$. Substituting this value of y into the third equation of the system

gives $2x + \frac{5}{3}x - 4 = 0$, so $x = \frac{12}{11}$ and consequently $y = \frac{20}{11}$. Therefore, $\left(\frac{12}{11}, \frac{20}{11}\right)$ gives the maximum value

$f\left(\frac{12}{11}, \frac{20}{11}\right) = -\frac{32}{11}$ for f subject to the given constraint.

37. The Lagrangian function is $F(x, y, \lambda) = 2x - 3y + 1 + \lambda\left(2x^2 + 3y^2 - 125\right)$. Next, we solve the

system of equations $\begin{cases} F_x = 2 + 4\lambda x = 0 \\ F_y = -3 + 6\lambda y = 0 \\ F_\lambda = 2x^2 + 3y^2 - 125 = 0 \end{cases}$ Solving the first equation for x gives $x = -\frac{1}{2}\lambda$,

and the second equation gives $y = \frac{1}{2}\lambda$. Substituting these values of x and y into the third equation gives

$2\left(-\frac{1}{2\lambda}\right)^2 + 3\left(\frac{1}{2\lambda}\right)^2 - 125 = 0$, so $\dfrac{1}{2\lambda^2} + \dfrac{3}{4\lambda^2} - 125 = 0$, $2 + 3 - 500\lambda^2 = 0$, and so $\lambda = \pm\frac{1}{10}$.

Therefore, $x = \pm5$ and $y = \pm5$, and so the critical points of f are $(-5, 5)$ and $(5, -5)$. Next, we compute $f(-5, 5) = 2(-5) - 3(5) + 1 = -24$ and $f(5, -5) = 2(5) - 3(-5) + 1 = 26$. We conclude that f has a maximum value of 26 at $(5, -5)$ and a minimum value of -24 at $(-5, 5)$.

39. $f(x, y) = \left(x^2 + y^4\right)^{3/2}$, so $df = \frac{3}{2}\left(x^2 + y^4\right)^{1/2}(2x)\,dx + \frac{3}{2}\left(x^2 + y^4\right)^{1/2}\left(4y^3\right)dy$. At $(3, 2)$,

$df = 9(9 + 16)^{1/2}\,dx + 48(9 + 16)^{1/2}\,dy = 45\,dx + 240\,dy$.

41. $f(x, y) = 2x^2y^3 + 3y^2x^2 - 2xy$, so $\Delta f \approx df = \left(4xy^3 + 6y^2x - 2y\right)dx + \left(6x^2y^2 + 6yx^2 - 2x\right)dy$. If $x = 1$, $y = -1$, $dx = 0.02$, and $dy = 0.02$, we find $\Delta f \approx (-4 + 6 + 2)(0.02) + (6 - 6 - 2)(0.02) = 0.04$.

43. $\int_{-1}^{2}\int_{2}^{4}(3x - 2y)\,dx\,dy = \int_{-1}^{2}\left(\frac{3}{2}x^2 - 2xy\right)\Big|_{x=2}^{x=4}dy = \int_{-1}^{2}\left[(24 - 8y) - (6 - 4y)\right]dy = \int_{-1}^{2}(18 - 4y)\,dy$

$= \left(18y - 2y^2\right)\Big|_{-1}^{2} = (36 - 8) - (-18 - 2) = 48.$

45. $\int_{0}^{1}\int_{x^3}^{x^2}2x^2y\,dy\,dx = \int_{0}^{1}\left(x^2y^2\right)\Big|_{y=x^3}^{y=x^2}dx = \int_{0}^{1}x^2\left(x^4 - x^6\right)dx = \int_{0}^{1}\left(x^6 - x^8\right)dx = \left(\frac{1}{7}x^7 - \frac{1}{9}x^9\right)\Big|_{0}^{1} = \frac{1}{7} - \frac{1}{9}$

$= \frac{2}{63}.$

47. $\int_{0}^{2}\int_{0}^{1}\left(4x^2 + y^2\right)dy\,dx = \int_{0}^{2}\left(4x^2y + \frac{1}{3}y^3\right)\Big|_{y=0}^{y=1}dx = \int_{0}^{2}\left(4x^2 + \frac{1}{3}\right)dx = \left(\frac{4}{3}x^3 + \frac{1}{3}x\right)\Big|_{0}^{2} = \frac{32}{3} + \frac{2}{3} = \frac{34}{3}.$

© 2011 Cengage Learning. All Rights Reserved. May not be scanned, copied or duplicated, or posted to a publicly accessible website, in whole or in part.

49. The area of R is $\int_0^2 \int_{x^2}^{2x} dy \, dx = \int_0^2 y|_{y=x^2}^{y=2x} \, dx = \int_0^2 (2x - x^2) \, dx = \left(x^2 - \frac{1}{3}x^3 \right) \Big|_0^2 = \frac{4}{3}$. Thus,

$AV = \frac{1}{4/3} \int_0^2 \int_{x^2}^{2x} (xy + 1) \, dy \, dx = \frac{3}{4} \int_0^2 \left(\frac{1}{2}xy^2 + y \right) \Big|_{x^2}^{2x} \, dx = \frac{3}{4} \int_0^2 \left(-\frac{1}{2}x^5 + 2x^3 - x^2 + 2x \right) \, dx$

$= \frac{3}{4} \left(-\frac{1}{12}x^6 + \frac{1}{2}x^4 - \frac{1}{3}x^3 + x^2 \right) \Big|_0^2 = \frac{3}{4} \left(-\frac{16}{3} + 8 - \frac{8}{3} + 4 \right) = 3.$

51. a. $R(x, y) = px + qy = -0.02x^2 - 0.2xy - 0.05y^2 + 80x + 60y.$

b. The domain of R is the set of all points satisfying
$0.02x + 0.1y \le 80, 0.1x + 0.05y \le 60, x \ge 0$, and $y \ge 0$.

c. $R(100, 300) = -0.02(100)^2 - 0.2(100)(300) - 0.05(300)^2$
$$+ 80(100)^2 + 60(300)$$
$$= 15,300,$$

giving revenue of \$15,300 realized from the sale of 100
sixteen-speed and 300 ten-speed electric blenders.

53. $P(x, y) = -0.0005x^2 - 0.003y^2 - 0.002xy + 14x + 12y - 200$, so
$\Delta P \approx dP = (-0.001x - 0.002y + 14) \, dx + (-0.006y - 0.002x + 12) \, dy$. If $x = 1000, y = 1700, dx = 50$, and
$dy = -50$, then $\Delta P \approx (-1 - 3.4 + 14)(50) + (-10.2 - 2 + 12)(-50) = 490$, or \$490.

55. a. We summarize the data at right. The normal equations are
$6b + 150m = 125.5$ and $150b + 5500m = 3240$, and the solutions
are $b = 19.45$ and $m = 0.0586$. Therefore, $y = 0.059x + 19.5$.

b. In 2040, the life expectancy beyond 65 of a 65-year-old female is
$y = 0.059(40) + 19.5 = 21.86$, or 21.9 years. This is close to the
given datum of 21.8 years.

c. In 2030, the life expectancy beyond 65 of a 65-year-old female is
$y = 0.059(30) + 19.5 = 21.27$, or 21.3 years. This is close to the
given datum of 21.2 years.

x	y	x^2	xy
0	19.5	0	0
10	20	100	200
20	20.6	400	412
30	21.2	900	636
40	21.8	1600	872
50	22.4	2500	1120
Sum 150	125.5	5500	3240

57. We want to maximize the function $R(x, y) = -x^2 - 0.5y^2 + xy + 8x + 3y + 20$. To find the critical point of R, we
solve the system $\begin{cases} R_x = -2x + y + 8 = 0 \\ R_y = -y + x + 3 = 0 \end{cases}$ Adding the two equations, we obtain $-x + 11 = 0$, or $x = 11$, and
so $y = 14$. Therefore, $(11, 14)$ is a critical point of R. Next, we compute $R_{xx} = -2, R_{xy} = 1$, and $R_{yy} = -1$, so
$D(x, y) = R_{xx} R_{yy} - R_{xy}^2 = 2 - 1 = 1$. In particular, $D(11, 14) = 1 > 0$. Because $R_{xx}(11, 14) = -2 < 0$, we
see that $(11, 14)$ gives a relative maximum of R. The nature of the problem suggests that this is in fact an absolute
maximum. So the company should spend \$11,000 on advertising and employ 14 agents in order to maximize its
revenue.

© 2011 Cengage Learning. All Rights Reserved. May not be scanned, copied or duplicated, or posted to a publicly accessible website, in whole or in part.

59. We want to maximize the function Q subject to the constraint $x + y = 100$. We form the Lagrangian function $f(x, y, \lambda) = x^{3/4} y^{1/4} + \lambda(x + y - 10)$. To find the critical points of F, we solve

$$\begin{cases} F_x = \frac{3}{4} \left(\frac{y}{x}\right)^{1/4} + \lambda = 0 \\ F_y = \frac{1}{4} \left(\frac{x}{y}\right)^{3/4} + \lambda = 0 \\ F_\lambda = x + y - 100 = 0 \end{cases}$$ Solving the first equation for λ and substituting this value into the second equation

yields $\frac{1}{4} \left(\frac{x}{y}\right)^{3/4} - \frac{3}{4} \left(\frac{y}{x}\right)^{1/4} = 0$, $\left(\frac{x}{y}\right)^{3/4} = 3 \left(\frac{y}{x}\right)^{1/4}$, and so $x = 3y$. Substituting this value of x into the third equation, we have $4y = 100$, so $y = 25$ and $x = 75$. Therefore, 75 units should be spent on labor and 25 units on capital.

CHAPTER 8	Before Moving On...	page 610

1. In order for $f(x, y) = \dfrac{\sqrt{x} + \sqrt{y}}{(1 - x)(2 - y)}$ to be defined, we must have $x \geq 0$, $y \geq 0$, $x \neq 1$ and $y \neq 2$. Therefore, the domain of f is $D = \{(x, y) \mid x \geq 0, y \geq 0, x \neq 1, y \neq 2\}$.

2. $f(x, y) = x^2 y + e^{xy}$, so $f_x = 2xy + ye^{xy}$, $f_y = x^2 + xe^{xy}$, $f_{xx} = 2y + y^2 e^{xy}$, $f_{xy} = 2x + (1 + xy) e^{xy} = f_{yx}$, and $f_{yy} = x \cdot xe^{xy} = x^2 e^{xy}$.

3. $f(x, y) = 2x^3 + 2y^3 - 6xy - 5$. Solving $f_x = 6x^2 + 6y = 6(x^2 - y^2) = 0$ and $f_y = 6y^2 - 6x = 6(y^2 - x) = 0$ simultaneously gives $y = x^2$ and $x = y^2$. Therefore, $x = x^4$, $x^4 - x = x(x^3 - 1) = 0$, and so $x = 0$ or 1. The critical points of f are $(0, 0)$ and $(1, 1)$. $f_{xx} = 12x$, $f_{xy} = -6$, and $f_{yy} = 12y$, so $D(x, y) = 144x^2 + 144y^2 - 36$. In particular, $D(0, 0) = -36 < 0$, and so $(0, 0)$ does not give a relative extremum; and $D(1, 1) = 252 > 0$ and $f_{xx}(1, 1) = 12 > 0$, and so $f(1, 1)$ gives a relative minimum value of $f(1, 1) = 2(1)^3 + 2(1)^3 - 6(1)(1) - 5 = -7$.

4. We summarize the data at right. The normal equations are $5b + 11m = 36.8$ and $11b + 39m = 111.1$. Solving, we find $m = 2.036$ and $b = 2.8797$. Thus, the least-squares line has equation $y = 2.036x + 2.8797$.

x	y	x^2	xy
0	2.9	0	0
1	5.1	1	5.1
2	6.8	4	13.6
3	8.8	9	26.4
5	13.2	25	66
Sum	11	36.8	39	113.1

5. $F(x, y, \lambda) = 3x^2 + 3y^2 + 1 + \lambda(x + y - 1)$, so we solve the system $\begin{cases} F_x = 6x + \lambda = 0 \\ F_y = 6y + \lambda = 0 \\ F_\lambda = x + y - 1 = 0 \end{cases}$ We find

$\lambda = -6x - 6y$, so $y = x$. Substituting into the third equation gives $2x = 1$, so $x = \frac{1}{2}$ and $y = \frac{1}{2}$. Therefore, $\left(\frac{1}{2}, \frac{1}{2}, \frac{5}{2}\right)$ is the required minimum.

© 2011 Cengage Learning. All Rights Reserved. May not be scanned, copied or duplicated, or posted to a publicly accessible website, in whole or in part.

6. Let $z = 2x^2 - xy$.

    a. $dz = \frac{\partial f}{\partial x}\, dx + \frac{\partial f}{\partial y}\, dy = 4x\, dx - y\, dx - x\, dy = (4x - y)\, dx - x\, dy$.

    b. $dx = 0.98 - 1 = -0.02$ and $dz = 1.03 - 1 = 0.03$, so $x = 1$ and $y = 1$, we have

    $dz = [4\,(1) - 1]\,(-0.02) - 1\,(0.03) = 3\,(-0.02) - 0.03 = -0.09$.

7. $\iint\limits_{R} (1 - xy)\, dA = \int_0^1 \int_{x^2}^{x} (1 - xy)\, dy\, dx = \int_0^1 \left(y - \tfrac{1}{2}xy^2\right)\Big|_{y=x^2}^{y=x} dx = \int_0^1 \left(x - \tfrac{1}{2}x^3 - x^2 + \tfrac{1}{2}x^5\right) dx$

               $= \left(\tfrac{1}{2}x^2 - \tfrac{1}{8}x^4 - \tfrac{1}{3}x^3 + \tfrac{1}{12}x^6\right)\Big|_0^1 = \tfrac{1}{2} - \tfrac{1}{8} - \tfrac{1}{3} + \tfrac{1}{12} = \tfrac{1}{8}$.

© 2011 Cengage Learning. All Rights Reserved. May not be scanned, copied or duplicated, or posted to a publicly accessible website, in whole or in part.

9 DIFFERENTIAL EQUATIONS

9.1 Problem-Solving Tips

Familiarize yourself with the models involving differential equations that were introduced in this section:

1. **Unrestricted Growth:** $\dfrac{dQ}{dt} = kQ$. The rate of growth of Q is proportional to its current size.

2. **Restricted Growth:**

   a. $\dfrac{dQ}{dt} = k(C - Q)$. The rate of growth of Q is proportional to the difference between its current size and an upper bound C.

   b. $\dfrac{dQ}{dt} = kQ(C - Q)$. The rate of growth of Q is jointly proportional to its current size and the difference between its current size and an upper bound C.

3. **Stimulus Response:** $\dfrac{dR}{dS} = \dfrac{k}{S}$. The rate of change of a reaction R is inversely proportional to a stimulus S.

9.1 Concept Questions   page 616

1. a. A differential equation is an equation that involves an unknown function and its derivative(s).
   b. The solution of a differential equation that involves a constant c is called the general solution of the differential equation.
   c. A particular solution of a differential equation is a solution obtained by assigning a specific value to the constant c.

3. a. Unrestricted growth.   b. Restricted growth.

9.1 Differential Equations   page 617

1. Substituting $y' = 2x$ into the differential equation $xy' + y = 3x^2$ yields $x(2x) + x^2 = 3x^2$, and the given differential equation is satisfied. Therefore, $y = x^2$ is a solution of the differential equation.

3. Substituting $y = \frac{1}{2} + ce^{-x^2}$ and $y' = -2cxe^{-x^2}$ into the differential equation $y' + 2xy = x$ gives
$-2cxe^{-x^2} + 2x\left(\frac{1}{2} + ce^{-x^2}\right) = x$, and the differential equation is satisfied. Therefore, $y = \frac{1}{2} + ce^{-x^2}$ is a solution of the differential equation.

249

© 2011 Cengage Learning. All Rights Reserved. May not be scanned, copied or duplicated, or posted to a publicly accessible website, in whole or in part.

5. Substituting $y = e^{-2x}$, $y' = -2e^{-2x}$, and $y'' = 4e^{-2x}$ into the differential equation $y'' + y' - 2y = 0$ yields $4e^{-2x} - 2e^{-2x} - 2e^{-2x} = 0$, and so the differential equation is satisfied. Therefore, $y = e^{-2x}$ is a solution of the differential equation.

7. Substituting $y = C_1 e^{-2x} + C_2 x e^{-2x}$, $y' = -2C_1 e^{-2x} + C_2 e^{-2x} - 2C_2 x e^{-2x}$, and
$y'' = 4C_1 e^{-2x} - 2C_2 e^{-2x} - 2C_2 e^{-2x} + 4C_2 x e^{-2x} = 4C_1 e^{-2x} - 4C_2 e^{-2x} + 4C_2 x e^{-2x}$
into the differential equation $y'' + 4y' + 4y = 0$, we find
$4C_1 e^{-2x} - 4C_2 e^{-2x} + 4C_2 x e^{-2x} - 8C_1 e^{-2x} + 4C_2 e^{-2x} - 8C_2 x e^{-2x} + 4C_1 e^{-2x} + 4C_2 x e^{-2x} = 0$, and so the
equation is satisfied. Therefore, $y = C_1 e^{-2x} + C_2 x e^{-2x}$ is a solution of the given equation.

9. Substituting $y = \dfrac{C_1}{x} + \dfrac{C_2 \ln x}{x} = C_1 x^{-1} + C_2 x^{-1} \ln x$,
$y' = -C_1 x^{-2} + C_2 \left(-x^{-2} \ln x + x^{-2}\right) = -C_1 x^{-2} + C_2 x^{-2} (1 - \ln x)$, and
$y'' = 2C_1 x^{-3} + C_2 \left[-2x^{-3} (1 - \ln x) - x^{3}\right] = 2C_1 x^{-3} + C_2 x^{-3} (2 \ln x - 3)$ into the differential equation
$x^2 y'' + 3xy' + y = 0$ yields
$x^2 \left[2C_1 x^{-3} + C_2 x^{-3} (2 \ln x - 3)\right] + 3x \left[-C_1 x^{-2} + C_2 x^{-2} (1 - \ln x)\right] + C_1 x^{-1} + C_2 x^{-1} \ln x$
$= 2C_1 x^{-1} + C_2 x^{-1} (2 \ln x - 3) - 3C_1 x^{-1} + 3C_2 x^{-1} (1 - \ln x) + C_1 x^{-1} + C_2 x^{-1} \ln x = 0,$
and so the equation is satisfied. Therefore, $y = \dfrac{C_1}{x} + \dfrac{C_2 \ln x}{x}$ is a solution of the differential equation.

11. Substituting $y = C - Ae^{-kt}$ and $y' = kAe^{-kt}$ into the differential equation $y' = k(C - y)$, we find
$kAe^{-kt} = k\left[C - (C - Ae^{-kt})\right] = kAe^{-kt}$, and so the equation is satisfied.

13. Substituting $y = Cx^2 - 2x$ and $y' = 2Cx - 2$ into the differential equation $y' - 2\left(\dfrac{y}{x}\right) = 2$ gives

$2Cx - 2 - 2 \cdot \dfrac{1}{x} \left(Cx^2 - 2x\right) = 2Cx - 2 - 2Cx + 4 = 2$, and so the equation is satisfied. Next, $y(1) = 10$, so
$C - 2 = 10$ and $C = 12$. Therefore, a particular solution is $y = 12x^2 - 2x$.

15. Substituting $y = \dfrac{C}{x} = Cx^{-1}$ and $y' = -Cx^{-2} = -\dfrac{C}{x^2}$ into the differential equation $y' + \left(\dfrac{1}{x}\right) y = 0$ gives

$-\dfrac{C}{x^2} + \left(\dfrac{1}{x}\right)\left(\dfrac{C}{x}\right) = 0$, and so the equation is satisfied. Next, $y(1) = 1$ implies $C = 1$, and so a particular solution
is $y = 1/x$.

17. Substituting $y = \dfrac{Ce^x}{x} + \dfrac{1}{2}xe^x = Cx^{-1}e^x + \dfrac{1}{2}xe^x$ and $y' = C\left(x^{-1}e^x - x^{-2}e^x\right) + \dfrac{1}{2}(e^x + xe^x)$ into the differential

equation $y' + \left(\dfrac{1-x}{x}\right) y = e^x$ gives $C\left(x^{-1}e^x - x^{-2}e^x\right) + \dfrac{1}{2}(e^x + xe^x) + \dfrac{1-x}{x}\left(Cx^{-1}e^x + \dfrac{1}{2}xe^x\right) = e^x$, and so

the equation is satisfied. Next, $y(1) = -\dfrac{1}{2}e$, so $Ce + \dfrac{1}{2}e = -\dfrac{1}{2}e$ and $C = -1$. Therefore, a particular solution is
$y = -\dfrac{e^x}{x} + \dfrac{1}{2}xe^x$.

19. Let $Q(t)$ denote the amount of the substance present at time t. Since the substance decays at a rate directly
proportional to the amount present, we have $\dfrac{dQ}{dt} = -kQ$, where k is the (positive) constant of proportion. The side
condition is $Q(0) = Q_0$.

21. Let $A(t)$ denote the total investment at time t. Then $\dfrac{dA}{dt} = k(C - A)$, where k is the constant of proportion.

© 2011 Cengage Learning. All Rights Reserved. May not be scanned, copied or duplicated, or posted to a publicly accessible website, in whole or in part.

23. Since the rate of decrease of the concentration of the drug at time t is proportional to the concentration $C(t)$ at any time t, we have $\frac{dC}{dt} = -kC$, where k is the (positive) constant of proportion. The initial condition is $C(0) = C_0$.

25. The rate of decrease of the temperature is $\frac{dy}{dt}$. Since this is proportional to the difference between the temperature y and C, we have $\frac{dy}{dt} = -k(y - C)$, where k is a constant of proportionality. Furthermore, fact that the initial temperature is y_0 degrees translates into the condition $y(0) = y_0$.

27. Since the relative growth rate of one organ, $\dfrac{dx/dt}{x}$, is proportional to the relative growth rate of the other, $\dfrac{dy/dt}{y}$, we have $\dfrac{1}{x} \cdot \dfrac{dx}{dt} = k\dfrac{1}{y} \cdot \dfrac{dy}{dt}$, where k is a constant of proportionality.

29. True. If $y = x^2 + 2x + \dfrac{1}{x}$, then $y' = 2x + 2 - \dfrac{1}{x}$. Substituting into the differential equation $xy' + y = 3x^2 + 4x$, we verify that $x\left(2x + 2 - \dfrac{1}{x^2}\right) + x^2 + 2x + \dfrac{1}{x} = 2x^2 + 2x - \dfrac{1}{x} + x^2 + 2x + \dfrac{1}{x} = 3x^2 + 4x$.

31. False. If $y = 2 + ce^{-x^3}$, then $y' = -3cx^2 e^{-x^3}$. Substituting into the differential equation $y' + 3x^2 y = x^2$, we see that $-3cx^2 e^{-x^3} + 3x^2\left(2 + ce^{-x^3}\right) = -3cx^2 e^{-x^3} + 6x^2 + 3cx^2 e^{-x^3} = 6x^2 \neq x$.

33. False. Consider the solution to Exercise 29. In that case, if $y = 2f(x) = 2x^2 + 4x + 2x^{-1}$, then $y' = 4x + 4 - 2x^{-2}$. Substituting these values into the differential equation, we see that $x\left(4x + 4 - 2x^{-2}\right) + 2x^2 + 4x + 2x^{-1} = 6x^2 + 8x \neq 3x^2 + 4x$.

9.2   Problem-Solving Tips

A first-order separable differential equation can be expressed in the form $\dfrac{dy}{dx} = f(x)g(y)$ (1). To solve (1), follow these steps:

1. Write (1) in the form $\dfrac{dy}{g(y)} = f(x)\,dx$ (2).

2. Integrate each side of (2) with respect to the appropriate variable.

9.2   Concept Questions   page 623

1. a. The order of a differential equation is the order of the highest derivative of the unknown function appearing in the equation. The equation $y' = xe^x$ is a first-order equation, and the equation $\dfrac{d^2 y}{dt^2} + \left(\dfrac{dy}{dt}\right)^3 + ty - 8 = 0$ is a second-order equation.

  b. Separable differential equations are differential equations in which the variables can be separated.

3. An initial value problem consists of a differential equation with one or more side conditions specified at a point.

© 2011 Cengage Learning. All Rights Reserved. May not be scanned, copied or duplicated, or posted to a publicly accessible website, in whole or in part.

9.2 Separation of Variables   page 623

1. $\frac{dy}{dx} = \frac{x+1}{y^2}$, so $y^2\,dy = (x+1)\,dx$. Thus, $\int y^2\,dy = \int (x+1)\,dx$, so $\frac{1}{3}y^3 = \frac{1}{2}x^2 + x + C$.

3. $\frac{dy}{dx} = \frac{e^x}{y^2}$, so $y^2\,dy = e^x\,dx$. Thus, $\int y^2\,dy = \int e^x\,dx$, so $\frac{1}{3}y^3 = e^x + C$.

5. $\frac{dy}{dx} = 2y$, so $\frac{dy}{y} = 2\,dx$. Thus, $\int \frac{dy}{y} = \int 2\,dx$, so $\ln|y| = 2x + C_1$, $|y| = e^{2x+C_1} = e^{C_1}e^{2x} = Ce^{2x}$ (where
$C = e^{\pm C_1}$), and $y = Ce^{2x}$.

7. $\frac{dy}{dx} = xy^2$, so $\frac{dy}{y^2} = x\,dx$. Thus, $\int \frac{dy}{y^2} = \int x\,dx$, $-\frac{1}{y} = \frac{1}{2}x^2 + C$, $-\frac{1}{y} = \frac{x^2 + 2C}{2}$, and $y = -\frac{2}{x^2 + 2C}$.

9. $\frac{dy}{dx} = -2\,(3y+4)$, so $\frac{dy}{3y+4} = -2\,dx$. Thus, $\int \frac{dy}{3y+4} = \int -2\,dx$, $\frac{1}{3}\ln|3y+4| = -2x + C_1$,
$|3y+4| = C_2e^{-6x}$ (where $C_2 = e^{3C_1}$), and $y = -\frac{4}{3} + Ce^{-6x}$ (where $C = \pm\frac{1}{3}C_2$).

11. $\frac{dy}{dx} = \frac{x^2+1}{3y^2}$, so $3y^2\,dy = (x^2+1)\,dx$. Thus, $\int 3y^2\,dy = \int (x^2+1)\,dx$, and so $y^3 = \frac{1}{3}x^3 + x + C$.

13. $\frac{dy}{dx} = \sqrt{\frac{y}{x}} = \frac{y^{1/2}}{x^{1/2}}$, so $\frac{dy}{y^{1/2}} = \frac{dx}{x^{1/2}}$. Thus, $\int y^{-1/2}\,dy = \int x^{-1/2}\,dx$, $2y^{1/2} = 2x^{1/2} + C_1$, and $y^{1/2} - x^{1/2} = C$,
where $C = \frac{1}{2}C_1$.

15. $\frac{dy}{dx} = \frac{y\ln x}{x}$, so $\frac{dy}{y} = \frac{\ln x}{x}\,dx$. Thus, $\int \frac{dy}{y} = \int \frac{\ln x}{x}\,dx$, $\ln|y| = \frac{1}{2}(\ln x)^2 + C_1$, $|y| = C_2e^{(\ln x)^2/2}$ (where
$C_2 = e^{C_1}$), and $y = Ce^{(\ln x)^2/2}$ (where $C = \pm C_2$).

17. $\frac{dy}{dx} = \frac{2x}{y}$, so $\int y\,dy = \int 2x\,dx$. Thus, $\frac{1}{2}y^2 = x^2 + C_1$ and $y = \pm\sqrt{2x^2 + C}$, where $C = 2C_1$. Now $y(1) = -2$
implies that $-2 = -\sqrt{2 + C}$, so $C = 2$. Therefore, the solution is $y = -\sqrt{2x^2 + 2}$.

19. $\frac{dy}{dx} = 2 - y$, so $\int \frac{dy}{2-y} = \int dx$. Thus, $-\ln|2-y| = x + C_1$, $\ln|2-y| = -x - C_1$, $|2-y| = e^{-x-C_1} = C_2e^{-x}$
(where $C_2 = e^{-C_1}$), $2 - y = Ce^{-x}$ (where $C = \pm C_2$), and $y = 2 - Ce^{-x}$. The condition $y(0) = 3$ implies that
$3 = 2 - C$, so $C = -1$. Therefore, the solution is $y = 2 + e^{-x}$.

21. $\frac{dy}{dx} = 3xy - 2x = x\,(3y - 2)$, so $\int \frac{dy}{3y-2} = \int x\,dx$, $\frac{1}{3}\ln|3y-2| = \frac{1}{2}x^2 + C_1$, $\ln|3y-2| = \frac{3}{2}x^2 + 3C_1$,
$|3y-2| = C_2e^{3x^2/2}$ (where $C_2 = e^{3C_1}$), $3y - 2 = C_3e^{3x^2/2}$ (where $C_3 = \pm C_2$), and $y = \frac{2}{3} + Ce^{3x^2/2}$ (where
$C = \frac{1}{3}C_3$). The condition $y(0) = 1$ gives $1 = \frac{2}{3} + C$, so $C = \frac{1}{3}$ and the solution is $y = \frac{2}{3} + \frac{1}{3}e^{3x^2/2}$.

23. $\frac{dy}{dx} = \frac{xy}{x^2+1}$. Separating variables and integrating, we have $\int \frac{dy}{y} = \int \frac{x\,dx}{x^2+1}$, so $\ln|y| - \ln(x^2+1)^{1/2} = \ln C_1$,
$\ln \frac{|y|}{\sqrt{x^2+1}} = \ln C_1$, and $y = C\sqrt{x^2+1}$ (where $C = \pm C_1$). The condition $y(0) = 1$ implies that $y(0) = C = 1$,
so the solution is $y = \sqrt{x^2+1}$.

© 2011 Cengage Learning. All Rights Reserved. May not be scanned, copied or duplicated, or posted to a publicly accessible website, in whole or in part.

25. $\dfrac{dy}{dx} = xye^x$, so $\displaystyle\int \dfrac{dy}{y} = \int xe^x \, dx$. Thus, $\ln|y| = (x-1)e^x + C_1$, $|y| = C_2 e^{(x-1)e^x}$ (where $C_2 = e^{C_1}$), and

$y = Ce^{(x-1)e^x}$ (where $C = \pm C_2$). The condition $y(1) = 1$ implies that $C = 1$, so the solution is $y = e^{(x-1)e^x}$.

27. $\dfrac{dy}{dx} = 3x^2 e^{-y}$, so $e^y \, dy = 3x^2 \, dx$. Thus, $e^y = x^3 + C$, so $y = \ln(x^3 + C)$. The condition $y(0) = 1$ implies that

$1 = \ln C$, so $C = e$ and the solution is $y = \ln(x^3 + e)$.

29. Let $y = f(x)$. Then $\dfrac{dy}{dx} = \dfrac{3x^2}{2y}$, so $\int 2y \, dy = \int 3x^2 \, dx$. Thus, $y^2 = x^3 + C$, and so $y = \pm\sqrt{x^3 + C}$. The given

condition implies that $3 = \sqrt{1 + C}$, so $C = 8$ and we have $f(x) = \sqrt{x^3 + 8}$.

31. $\dfrac{dQ}{dt} = -kQ$, so $\displaystyle\int \dfrac{dQ}{Q} = \int -k \, dt$. Thus, $\ln|Q| = -kt + C_1$, $|Q| = e^{-kt + C_1} = C_2 e^{-kt}$ (where $C_2 = e^{C_1}$), and

$Q = Ce^{-kt}$ (where $C = \pm C_2$). The condition $Q(0) = Q_0$ implies that $C = Q_0$, so $Q(t) = Q_0 e^{-kt}$.

33. $\dfrac{dA}{dt} = k\left(\dfrac{C}{k} - A\right)$, so $\displaystyle\int \dfrac{dA}{\frac{C}{k} - A} = \int k \, dt$, $-\ln\left|\dfrac{C}{k} - A\right| = kt + D$ (assume that $\frac{C}{k} - A > 0$), and

$\ln\left|\dfrac{C}{k} - A\right| = -kt - D_1$. Therefore, $\dfrac{C}{k} - A = De^{-kt}$ (where $D = e^{-D_1}$) and so $A = \dfrac{C}{k} - De^{-kt}$.

35. $\dfrac{dS}{dt} = k(D - S)$, so $\displaystyle\int \dfrac{dS}{D - S} = \int k \, dt$, $-\ln(D - S) = kt + C_1$ (where $D - S > 0$ and C_1 is a constant),

$\ln(D - S) = -kt - C_1$, $D - S = Ce^{-kt}$ (where $C = e^{-C_1}$), and $S = D - Ce^{-kt}$. The condition $S(0) = S_0$ gives

$D - C = S_0$, so $C = D - S_0$. Therefore, $S(t) = D - (D - S_0)e^{-kt}$.

37. True. Rewrite the equation in the form $\dfrac{dy}{dx} = y(x-1) + 2(x-1) = (y+2)(x-1)$. It is evident that the equation

is separable.

39. True. The equation can be rewritten as $f(x)g(y)\,dx + F(x)G(y)\,dy = 0$ or $\dfrac{dy}{dx} = -\dfrac{f(x)g(y)}{F(x)G(y)} = \phi(x)\psi(y)$,

where $\phi(x) = -\dfrac{f(x)}{F(x)}$ and $\psi(y) = \dfrac{g(y)}{G(y)}$. Evidently, the equation is separable.

41. False. The equation cannot be written in the form $\dfrac{dy}{dx} = f(x)g(y)$.

9.3   Concept Questions   page 632

1. a. The rate of growth $\dfrac{dQ}{dt}$ is proportional to $Q(t)$. The rate of growth is unbounded as t approaches infinity.

   b. The model for unrestricted exponential growth with initial population Q_0 is $Q(t) = Q_0 e^{kt}$.

© 2011 Cengage Learning. All Rights Reserved. May not be scanned, copied or duplicated, or posted to a publicly accessible website, in whole or in part.

9.3   Applications of Separable Differential Equations   page 632

1. $\frac{dy}{dt} = -ky$, $y = y_0 e^{-kt}$.

3. Solving $\frac{dQ}{dt} = kQ$, we find $Q = Q_0 e^{kt}$. Here $k = 0.02$ and $Q_0 = 4.5$, so $Q(t) = 4.5e^{0.02t}$. At the beginning of 2012, the population will be $Q(32) = 4.5e^{0.02(32)} \approx 8.5$, or 8.5 billion.

5. $\frac{dL}{L} = k\,dx$. Integrating, we have $\ln L = kx + C_1$, so $L = L_0 e^{kx}$ and $L_0 = e^{C_1}$. Using the given

condition $L\left(\frac{1}{2}\right) = \frac{1}{2}L_0$, we have $\frac{1}{2}L_0 = L_0 e^{k/2}$, $e^{k/2} = \frac{1}{2}$, $\frac{1}{2}k = \ln \frac{1}{2}$, and so $k = -2\ln 2$. Therefore,

$L = L_0 e^{-(2\ln 2)x} = L_0 \cdot 2^{-2x}$. We want to find x so that $L = \frac{1}{4}L_0$, that is, $\frac{1}{4}L_0 = L_0 \cdot 2^{-2x}$, so $2^{-2x} = \frac{1}{4}$ and

$x = 1$. Therefore, $\frac{1}{2}''$ of additional material is needed.

7. $\frac{dQ}{dt} = kQ^2$, so $\int \frac{dQ}{Q^2} = \int k\,dt$ and $-\frac{1}{Q} = kt + C$. Therefore, $Q = -\frac{1}{kt + C}$. Now $Q = 50$ when

$t = 0$, so $50 = -\frac{1}{C}$ and $C = -\frac{1}{50}$. Next, $Q = -\frac{1}{kt - \frac{1}{50}}$. Since $Q = 10$ when $t = 1$, $10 = -\frac{1}{k - \frac{1}{50}}$,

$10\left(k - \frac{1}{50}\right) = -1$, $10k - \frac{1}{5} = -1$, $10k = -\frac{4}{5}$, and $k = -\frac{2}{25}$. Therefore, $Q(t) = \frac{1}{\frac{2}{25}t + \frac{1}{50}} = \frac{1}{\frac{4t+1}{50}} = \frac{50}{4t + 1}$ and

$Q(2) = \frac{50}{8 + 1} \approx 5.56$ grams.

9. Let C be the air temperature and T the temperature of the coffee at any time t. Then $\frac{dT}{dt} = k(C - T)$. The solution

of the differential equation is $T = C - Ae^{-kt}$ (see Equation (10)). With $C = 72$, we have $T = 72 - Ae^{-kt}$. Next,

$T(0) = 212$ gives $212 = 72 - A$, so $A = -140$ and $T = 72 + 140e^{-kt}$. Using the condition $T(2) = 140$, we have

$72 + 140e^{-2k} = 140$, $e^{-2k} = \frac{68}{140}$, and $k = -\frac{1}{2}\ln\left(\frac{68}{140}\right) \approx 0.3611$. Thus, $T(t) = 72 + 140e^{-0.3611t}$. When

$T = 110$, we have $72 + 140e^{-0.3611t} = 110$, $140e^{-0.3611t} = 38$, $e^{-0.3611t} = \frac{38}{140}$, $-0.3611t = \ln\left(\frac{38}{140}\right)$, and so

$t \approx 3.6$, or 3.6 minutes.

11. The differential equation is $\frac{dQ}{dt} = k(40 - Q)$. Solving as in Example 2, we find that $Q(t) = 40 - Ae^{-kt}$.

The condition $Q(0) = 0$ gives $A = 40$ and so $Q(t) = 40\left(1 - e^{-kt}\right)$. Next, $Q(2) = 10$ implies that

$40 = \left(1 - e^{-2k}\right) = 10$, so $1 - e^{-2k} = 0.25$, $e^{-2k} = 0.75$, and $k = -\frac{1}{2}\ln 0.75 \approx 0.1438$. Therefore,

$Q(t) = 40\left(1 - e^{-0.1438t}\right)$. The number of claims the average trainee can process after six weeks is

$Q(6) = 40\left(1 - e^{-0.1438(6)}\right) \approx 23$.

13. $P(32) = -\frac{0.5}{0.008} + \left(226.5 + \frac{0.5}{0.008}\right)e^{0.008(32)} \approx 310.8$, or 310.8 million.

© 2011 Cengage Learning. All Rights Reserved. May not be scanned, copied or duplicated, or posted to a publicly accessible website, in whole or in part.

15. Let $Q(t)$ denote the number of people who have heard the rumor. Then $\frac{dQ}{dt} = kQ(400 - Q)$, which has solution

$Q(t) = \dfrac{400}{1 + Ae^{-400kt}}$. The condition $Q(0) = 10$ gives $10 = \dfrac{400}{1 + A}$, so $10 + 10A = 400$ and $A = 39$. Therefore,

$Q(t) = \dfrac{400}{1 + 39e^{-400kt}}$. Next, the condition, $Q(2) = 80$ gives $\dfrac{400}{1 + 39e^{-400kt}} = 80$, $1 + 39e^{-900k} = 5$,

$39e^{-800k} = 4$, and so $k = -\frac{1}{800}\ln\left(\frac{4}{39}\right) \approx 0.0028466$. Thus, $Q(t) = \dfrac{400}{1 + 39e^{-1.1386t}}$. In particular, the number of

people who will have heard the rumor after a week is $Q(7) = \dfrac{400}{1 + 39e^{-1.1386(7)}} \approx 395$.

17. a. $\dfrac{dh}{dt} = -\dfrac{B}{A}\sqrt{2gh} = -\dfrac{B}{A}\sqrt{2g}h^{1/2}$, so $\displaystyle\int \dfrac{dh}{h^{1/2}} = -\dfrac{B}{A}\sqrt{2g}\int dt$, $2h^{1/2} = -\dfrac{B}{A}t\sqrt{2g} + C_1$, and

$h^{1/2} = -\dfrac{B}{2A}t\sqrt{2g} + C_2$, where $C_2 = \frac{1}{2}C_1$. The condition $h(0) = H$ implies that $C_2 = H^{1/2}$, so

$h^{1/2} = -\dfrac{B}{A}t\sqrt{\dfrac{g}{2}} + \sqrt{H}$ and $h(t) = \left(\sqrt{H} - \dfrac{B}{A}t\sqrt{\dfrac{g}{2}}\right)^2$.

b. We solve $h(T) = 0$, obtaining $\dfrac{B}{A}T\sqrt{\dfrac{g}{2}} = \sqrt{H}$, or $T = \dfrac{A}{B}\sqrt{\dfrac{2H}{g}}$.

c. Substituting $A = 4$, $B = \frac{1}{144}$, $H = 16$, and $g = 32$, we have $T = \dfrac{4}{\frac{1}{144}}\sqrt{\dfrac{2 \cdot 16}{32}} = 576$ s, or 9 minutes 36 seconds.

19. a. $\dfrac{dP}{dt} = cP\ln\dfrac{L}{P} \Rightarrow \displaystyle\int \dfrac{1}{P\ln\frac{L}{P}}\,dP = \int c\,dt$. Let $u = \ln\dfrac{L}{P}$. Then $du = -\dfrac{dP}{P}$, so we have $-\displaystyle\int \dfrac{du}{u} = \int c\,dt$,

$\displaystyle\int \dfrac{du}{u} = -\int c\,dt$, $\ln|u| = -ct + C_1$, $|u| = C_2 e^{-ct}$ (where $C_2 = e^{C_1}$), and $u = Ce^{-ct}$ (where $C = \pm C_2$).

Thus, $\ln\dfrac{L}{P} = Ce^{-ct}$. At $t = 0$, $\ln\dfrac{L}{P_0} = C$, so we have $\ln\dfrac{L}{P} = \ln\dfrac{L}{P_0}e^{-ct}$, $\dfrac{L}{P} = e^{\ln(L/P_0)e^{-ct}}$, and finally

$P(t) = Le^{-\ln(L/P_0)e^{-ct}}$.

b. $\displaystyle\lim_{t\to\infty} P(t) = \lim_{t\to\infty} Le^{-\ln(L/P_0)e^{-ct}} = Le^0 = L$.

c. $P'(t) = cP\ln\frac{L}{P} = cP(\ln L - \ln P)$, so

$P''(t) = c\left[P'(\ln L - \ln P) + P\left(-\dfrac{P'}{P}\right)\right] = cP'(\ln\frac{L}{P} - 1) = c^2 P\ln\frac{L}{P}(\ln\frac{L}{P} - 1)$. Setting $P''(t) = 0$ gives

$P = 0$, so $\ln\frac{L}{P} = 0$, $\frac{L}{P} = 1$, $P = L$, so $\ln\frac{L}{P} - 1 = 0$, $\ln\frac{L}{P} = 1$, $\frac{L}{P} = e$, and $P = \frac{L}{e}$.

Since $P(t) = 0$ and $P(t) = L$ are equilibrium solutions of the
differential equation, we need only consider the case where
$0 < P < L$, and from the sign diagram, we see that P' has a
maximum at $P = \frac{L}{e}$; that is, $P(t)$ is increasing most rapidly when

$P = \frac{L}{e}$.

d. Substituting the value of P found in part (c) into the expression for P found in part (a), we have

$\dfrac{L}{e} = Le^{-\ln(L/P_0)e^{-ct}}$, $e = e^{\ln(L/P_0)e^{-ct}}$, $1 = \ln\frac{L}{P_0}e^{-ct}$, $e^{ct} = \ln\frac{L}{P_0}$, $ct = \ln\ln\frac{L}{P_0}$, and finally $t = \dfrac{\ln\ln\frac{L}{P_0}}{c}$.

© 2011 Cengage Learning. All Rights Reserved. May not be scanned, copied or duplicated, or posted to a publicly accessible website, in whole or in part.

21. We have $\frac{dx}{dt} = kx\,(N - x)$, where k is the constant of proportionality. Then $\int \frac{dx}{x\,(N - x)} = \int k\,dt$,

$$\frac{1}{N} \int \left(\frac{1}{x} + \frac{1}{N - x} \right) dx = \int k\,dt, \; \frac{1}{N} \left(\ln|x| - \ln|N - x| \right) = kt + C_1, \; \frac{1}{N} \ln \left| \frac{x}{N - x} \right| = kt + C_1,$$

$\ln \left| \frac{x}{N - x} \right| = kNt + C_2$ (where $C_2 = NC_1$), $\frac{x}{N - x} = Ce^{kNt}$ (where $C = e^{C_2}$), $x = NCe^{kNt} - Cxe^{kNt}$,

$\left(1 + Ce^{kNt} \right) x = CNe^{kNt}$, and finally $x\,(t) = \frac{CNe^{kNt}}{1 + Ce^{kNt}}$. Using the condition $x\,(0) = N_0$

gives $\frac{CN}{1 + C} = N_0$, $CN = N_0 + CN_0$, $C\,(N - N_0) = N_0$, and $C = \frac{N_0}{N - N_0}$. Thus,

$$x\,(t) = \frac{CN}{C + e^{-kNt}} = \frac{\frac{N_0 N}{N - N_0}}{\frac{N_0}{N - N_0} + e^{-kNt}} = \frac{N}{1 + \left(\frac{N - N_0}{N_0} \right) e^{-kNt}}.$$

23. a. $\frac{dx}{dt} = k\,(L - x)$. Integrating, we have $\int \frac{dx}{L - x} = \int k\,dt$, $-\ln|L - x| = kt + C_1$, $\ln|L - x| = -kt - C_1$,

$|L - x| = C_2 e^{-kt}$ (where $C_2 = e^{-C_1}$), $L - x = Ce^{-kt}$ (where $C = \pm C_2$), and $x\,(t) = L - Ce^{-kt}$. The

condition $x\,(0) = x_0$ gives $L - C = x_0$, so $C = L - x_0$ and $x\,(t) = L - (L - x_0)\,e^{-kt}$.

b. We are given that $x_0 = 0.4$ and $x\,(1) = 10$. Using this

information, we find $x\,(1) = L - (L - 0.4)\,e^{-k} = 10$ and

$$e^{-k} = \frac{L - 10}{L - 0.4}, \text{ so}$$

$$x\,(t) = L - (L - x_0)\left(e^{-k} \right)^t = L - (L - 0.4) \left(\frac{L - 10}{L - 0.4} \right)^t.$$

c. Taking $L = 100$ and $x_0 = 0.4$, we
have

$$x\,(t) = 100 - (100 - 0.4) \left(\frac{100 - 10}{100 - 0.4} \right)^t$$

$$\approx 100 - 99.6\,(0.9036)^t$$

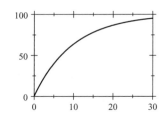

d. If $x = 40$, then we have $100 - 99.6\,(0.9036)^t = 40$, so

$(0.9036)^t = \frac{60}{99.6}$ and $t = \frac{1}{\ln 0.9036} \ln \left(\frac{60}{99.6} \right) \approx 4.9998$.

If $x = 60$, then we have

$t = \frac{1}{\ln 0.9036} \ln \left(\frac{40}{99.6} \right) \approx 8.9997$. So, on average, the

haddock caught today are between 5 and 9 years old.

25. The differential equation governing this process is $\frac{dx}{dt} = 6 - \frac{3x}{20}$. Separating variables and integrating, we

obtain $\frac{dx}{dt} = \frac{120 - 3x}{20} = \frac{3\,(40 - x)}{20}$, $\frac{dx}{40 - x} = \frac{3}{20}\,dt$, $-\ln|40 - x| = \frac{3}{20}t + C_1$, and $40 - x = Ce^{-3t/20}$

(where $C = \pm e^{-C_1}$). Therefore, $x = 40 - Ce^{-3t/20}$. The initial condition $x\,(0) = 0$ implies that $0 = 40 - C$,

so $C = 40$. Therefore, $x\,(t) = 40\left(1 - e^{-3t/20} \right)$. The amount of salt present at the end of 20 minutes is

$x\,(20) = 40\left(1 - e^{-3} \right) \approx 38$ lb. The amount of salt present in the long run is $\lim\limits_{t \to \infty} 40\left(1 - e^{-3t/20} \right) = 40$ lb.

© 2011 Cengage Learning. All Rights Reserved. May not be scanned, copied or duplicated, or posted to a publicly accessible website, in whole or in part.

9.4   Concept Questions   page 640

1. Euler's method approximates the actual solution $y = f(x)$ of the initial value problem $\frac{dy}{dx} = F(x, y)$, $y(x_0) = y_0$ for certain values of x. The values of f between two adjacent values of x are then found by linear interpolation. Geometrically, the actual solution of the differential equation is approximated by a suitable polygonal curve. See page 635 in the text.

9.4   Approximate Solutions of Differential Equations   page 640

1. **a.** Here $x_0 = 0$, $b = 1$, and $n = 4$, so $h = \frac{1}{4}$, $x_0 = 0$, $x_1 = \frac{1}{4}$, $x_2 = \frac{1}{2}$, $x_3 = \frac{3}{4}$, and $x_4 = b = 1$. Also, $F(x, y) = x + y$ and $y_0 = y(0) = 1$. Therefore, $y_0 = 1$, $y_1 = y_0 + hF(x_0, y_0) = 1 + \frac{1}{4}(0 + 1) = \frac{5}{4}$,

$y_2 = y_1 + hF(x_1, y_1) = \frac{5}{4} + \frac{1}{4}\left(\frac{1}{4} + \frac{5}{4}\right) = \frac{13}{8}$, $y_3 = y_2 + hF(x_2, y_2) = \frac{13}{8} + \frac{1}{4}\left(\frac{1}{2} + \frac{13}{8}\right) = \frac{69}{32}$, and

$y_4 = y_3 + hF(x_3, y_3) = \frac{69}{32} + \frac{1}{4}\left(\frac{3}{4} + \frac{69}{32}\right) = \frac{369}{128}$. Therefore, $y(1) \approx \frac{369}{128} \approx 2.8828$.

b. Here $n = 6$, so $h = \frac{1}{6}$, $x_0 = 0$, $x_1 = \frac{1}{6}$, $x_2 = \frac{2}{6}$, $x_3 = \frac{3}{6}$, $x_4 = \frac{4}{6}$, $x_5 = \frac{5}{6}$, and

$x_6 = 1$. Therefore, $y_0 = y(0) = 1$, $y_1 = y_0 + hF(x_0, y_0) = 1 + \frac{1}{6}(0 + 1) = \frac{7}{6}$,

$y_2 = y_1 + hF(x_1, y_1) = \frac{7}{6} + \frac{1}{6}\left(\frac{1}{6} + \frac{7}{6}\right) = \frac{50}{36} = \frac{25}{18}$, $y_3 = y_2 + hF(x_2, y_2) = \frac{25}{18} + \frac{1}{6}\left(\frac{2}{6} + \frac{25}{18}\right) = \frac{181}{108}$,

$y_4 = y_3 + hF(x_3, y_3) = \frac{181}{108} + \frac{1}{6}\left(\frac{3}{6} + \frac{181}{108}\right) = \frac{1321}{648}$, $y_5 = y_4 + hF(x_4, y_4) = \frac{1321}{648} + \frac{1}{6}\left(\frac{4}{6} + \frac{1321}{648}\right) = \frac{9679}{3888}$, and

$y_6 = y_5 + hF(x_5, y_5) = \frac{9679}{3888} + \frac{1}{6}\left(\frac{5}{6} + \frac{9679}{3888}\right) = \frac{70{,}993}{23{,}328}$. Thus, $y(1) \approx \frac{70{,}993}{23{,}328} \approx 3.043$.

3. **a.** Here $x_0 = 0$ and $b = 2$. Taking $n = 4$, we have $h = \frac{2}{4} = \frac{1}{2}$, $x_0 = 0$, $x_1 = \frac{1}{2}$, $x_2 = 1$, $x_3 = \frac{3}{2}$, and $x_4 = 2$. Also, $F(x, y) = 2x - y + 1$ and $y(0) = y_0 = 2$, so $y_0 = y(0) = 2$, $y_1 = y_0 + hF(x_0, y_0) = 2 + \frac{1}{2}(0 - 2 + 1) = \frac{3}{2}$,

$y_2 = y_1 + hF(x_1, y_1) = \frac{3}{2} + \frac{1}{2}\left(1 - \frac{3}{2} + 1\right) = \frac{7}{4}$, $y_3 = y_2 + hF(x_2, y_2) = \frac{7}{4} + \frac{1}{2}\left(2 - \frac{7}{4} + 1\right) = \frac{19}{8}$, and

$y_4 = y_3 + hF(x_3, y_3) = \frac{19}{8} + \frac{1}{2}\left(3 - \frac{19}{8} + 1\right) = \frac{51}{16}$. Therefore, $y(2) \approx \frac{51}{16} \approx 3.1875$.

b. With $n = 6$, we have $h = \frac{2}{6} = \frac{1}{3}$, so $x_0 = 0$, $x_1 = \frac{1}{3}$, $x_2 = \frac{2}{3}$, $x_3 = 1$, $x_4 = \frac{4}{3}$, $x_5 = \frac{5}{3}$,

and $x_6 = 2$. Therefore, $y_0 = y(0) = 2$, $y_1 = y_0 + hF(x_0, y_0) = 2 + \frac{1}{3}(0 - 2 + 1) = \frac{5}{3}$,

$y_2 = y_1 + hF(x_1, y_1) = \frac{5}{3} + \frac{1}{3}\left(\frac{2}{3} - \frac{5}{3} + 1\right) = \frac{5}{3}$, $y_3 = y_2 + hF(x_2, y_2) = \frac{5}{3} + \frac{1}{3}\left(\frac{4}{3} - \frac{5}{3} + 1\right) = \frac{17}{9}$,

$y_4 = y_3 + hF(x_3, y_3) = \frac{17}{9} + \frac{1}{3}\left(2 - \frac{17}{9} + 1\right) = \frac{61}{27}$, $y_5 = y_4 + hF(x_4, y_4) = \frac{61}{27} + \frac{1}{3}\left(\frac{8}{3} - \frac{61}{27} + 1\right) = \frac{221}{81}$, and

$y_6 = y_5 + hF(x_5, y_5) = \frac{221}{81} + \frac{1}{3}\left(\frac{10}{3} - \frac{221}{81} + 1\right) = \frac{793}{243}$. Therefore, $y(2) \approx \frac{793}{243} \approx 3.2634$.

5. **a.** Here $x_0 = 0$ and $b = 0.5$. Taking $n = 4$, we have $h = \frac{0.5}{4} \approx 0.125$, so $x_0 = 0$, $x_1 = 0.125$, $x_2 = 0.25$,

$x_3 = 0.375$, and $x_4 = 0.5$. Also, $F(x, y) = -2xy^2$ and $y(0) = y_0 = 1$, so $y_0 = y(0) = 1$,

$y_1 = y_0 + hF(x_0, y_0) = 1 + 0.125(0) = 1$, $y_2 = y_1 + hF(x_1, y_1) = 1 + 0.125[-2(0.125)(1)] = 0.96875$,

$y_3 = y_2 + hF(x_2, y_2) = 0.96875 + 0.125[-2(0.25)(0.96875)^2] \approx 0.910095$, and

$y_4 = y_3 + hF(x_3, y_3) \approx 0.910095 + 0.125[-2(0.375)(0.910095)^2] \approx 0.832444$. Therefore, $y(0.5) \approx 0.8324$.

© 2011 Cengage Learning. All Rights Reserved. May not be scanned, copied or duplicated, or posted to a publicly accessible website, in whole or in part.

b. With $n = 6$, we have $h = \frac{0.5}{6} = \frac{1}{12}$, so $x_0 = 0$, $x_1 = \frac{1}{12}$, $x_2 = \frac{1}{6}$, $x_3 = \frac{1}{4}$, $x_4 = \frac{1}{3}$, $x_5 = \frac{5}{12}$,

and $x_6 = b = \frac{1}{2}$. Therefore, $y_0 = y(0) = 1$, $y_1 = y_0 + hF(x_0, y_0) = 1 + 0.083333(0) = 1$,

$y_2 = y_1 + hF(x_1, y_1) = 1 + 0.083333[-2(0.083333)(1)] \approx 0.986111$,

$y_3 = y_2 + hF(x_2, y_2) = 0.986111 + 0.083333\left[-2(0.166667)(0.986111)^2\right] \approx 0.959100$,

$y_4 = y_3 + hF(x_3, y_3) = 0.959100 + 0.083333\left[-2(0.25000)(0.959100)^2\right] \approx 0.920772$,

$y_5 = y_4 + hF(x_4, y_4) = 0.920772 + 0.083333\left[-2(0.333333)(0.920772)^2\right] \approx 0.873671$, and

$y_6 = y_5 + hF(x_5, y_5) = 0.873671 + 0.083333\left[-2(0.416666)(0.873671)^2\right] \approx 0.820664$. Therefore,

$y(0.5) \approx 0.8207$.

7. a. $x_0 = 1$, $b = 1.5$, $n = 4$, and $h = 0.125$, so $x_0 = 1$, $x_1 = 1.125$, $x_2 = 1.25$, $x_3 = 1.375$, and $x_4 = 1.5$. Also,

$F(x, y) = \sqrt{x + y}$ and $y(1) = 1$, so $y_0 = y(0) = 1$, $y_1 = y_0 + hF(x_0, y_0) = 1 + 0.125\sqrt{1 + 1} \approx 1.1767767$,

$y_2 = y_1 + hF(x_1, y_1) = 1.1767767 + 0.125\sqrt{1.125 + 1.1767767} \approx 1.3664218$,

$y_3 = y_2 + hF(x_2, y_2) = 1.3664218 + 0.125\sqrt{1.25 + 1.3664218} \approx 1.5686138$, and

$y_4 = y_3 + hF(x_3, y_3) = 1.5686138 + 0.125\sqrt{1.375 + 1.5686138} \approx 1.7830758$. Therefore, $y(1.5) \approx 1.7831$.

b. With $n = 6$, we have $h = \frac{0.5}{6} = \frac{1}{12}$, so $x_0 = 1$, $x_1 = \frac{13}{12}$, $x_2 = \frac{7}{6}$, $x_3 = \frac{5}{4}$, $x_4 = \frac{4}{3}$, $x_5 = \frac{17}{12}$, and

$x_6 = \frac{3}{2}$. Thus, $y_0 = y(1) = 1$, $y_1 = y_0 + hF(x_0, y_0) = 1 + 0.0833333\sqrt{1 + 1} \approx 1.1178511$,

$y_2 = y_1 + hF(x_1, y_1) = 1.1178511 + 0.0833333\sqrt{1.0833333 + 1.1178511} \approx 1.2414876$,

$y_3 = y_2 + hF(x_2, y_2) = 1.2414876 + 0.0833333\sqrt{1.166666 + 1.2414876} \approx 1.3708061$,

$y_4 = y_3 + hF(x_3, y_3) = 1.3708061 + 0.0833333\sqrt{1.25 + 1.3708061} \approx 1.5057136$,

$y_5 = y_4 + hF(x_4, y_4) = 1.5057136 + 0.0833333\sqrt{1.333332 + 1.5057136} \approx 1.6461258$, and

$y_6 = y_5 + hF(x_5, y_5) = 1.6461258 + 0.0833333\sqrt{1.4166665 + 1.6461258} \approx 1.791966$. Therefore,

$y(1.5) \approx 1.7920$.

9. a. Here $x_0 = 0$ and $b = 1$. Taking $n = 4$, we have $h = \frac{1}{4}$, $x_0 = 0$, $x_1 = \frac{1}{4}$, $x_2 = \frac{1}{2}$, $x_3 = \frac{3}{4}$, and $x_4 = 1$.

Also, $F(x, y) = \frac{x}{y}$ and $y_0 = y(0) = 1$, so $y_0 = y(0) = 1$, $y_1 = y_0 + hF(x_0, y_0) = 1 + \frac{1}{4}(0) = 1$,

$y_2 = y_1 + hF(x_1, y_1) = 1 + \frac{1}{4}\left(\frac{1}{4}\right) = 1.0625$, $y_3 = y_2 + hF(x_2, y_2) = 1.0625 + \frac{1}{4}\left(\frac{0.5}{1.0625}\right) \approx 1.180147$, and

$y_4 = y_3 + hF(x_3, y_3) \approx 1.180147 + \frac{1}{4}\left(\frac{0.75}{1.180147}\right) \approx 1.339026$. Therefore, $y(1) \approx 1.3390$.

b. Here $n = 6$, so $h = \frac{1}{6}$, $x_0 = 0$, $x_1 = \frac{1}{6}$, $x_2 = \frac{1}{3}$, $x_3 = \frac{1}{2}$, $x_4 = \frac{2}{3}$, $x_5 = \frac{5}{6}$, and $x_6 = 1$. Thus, $y_0 = y(0) = 1$,

$y_1 = y_0 + hF(x_0, y_0) = 1 + \frac{1}{6}(0) = 1$, $y_2 = y_1 + hF(x_1, y_1) = 1 + \frac{1}{6}\left(\frac{1/6}{1}\right) \approx 1.027778$,

$y_3 = y_2 + hF(x_2, y_2) \approx 1.0277778 + \frac{1}{6}\left(\frac{1/3}{1.027778}\right) \approx 1.081832$,

$y_4 = y_3 + hF(x_3, y_3) \approx 1.081832 + \frac{1}{6}\left(\frac{1/2}{1.081832}\right) \approx 1.158862$,

$y_5 = y_4 + hF(x_4, y_4) \approx 1.158862 + \frac{1}{6}\left(\frac{2/3}{1.158862}\right) \approx 1.254742$, and

$y_6 = y_5 + hF(x_5, y_5) \approx 1.254742 + \frac{1}{6}\left(\frac{5/6}{1.254742}\right) \approx 1.365433$. Therefore, $y(1) \approx 1.3654$.

© 2011 Cengage Learning. All Rights Reserved. May not be scanned, copied or duplicated, or posted to a publicly accessible website, in whole or in part.

11. Here $x_0 = 0$ and $b = 1$. With $n = 5$, we have $h = \frac{1}{5}$, so $x_0 = 0$, $x_1 = 0.2$,

$x_2 = 0.4$, $x_3 = 0.6$, $x_4 = 0.8$, and $x_5 = 1$. Also, $F(x, y) = \frac{1}{2}xy$ and $y(0) = 1$, so

$y_0 = y(0) = 1$, $y_1 = y_0 + hF(x_0, y_0) = 1 + 0.2[0.5(0)(1)] = 1$,

$y_2 = y_1 + hF(x_1, y_1) = 1 + 0.2[0.5(0.2)(1)] = 1.02$,

$y_3 = y_2 + hF(x_2, y_2) = 1.02 + 0.2[0.5(0.4)(1.02)] = 1.0608$,

$y_4 = y_3 + hF(x_3, y_3) = 1.0608 + 0.2[0.5(0.6)(1.0608)] = 1.124448$, and

$y_5 = y_4 + hF(x_4, y_4) = 1.124448 + 0.2[0.5(0.8)(1.124448)] = 1.21440384$.

The solutions are summarized at right.

n	x_n	y_n
0	0	1
1	0.2	1
2	0.4	1.02
3	0.6	1.0608
4	0.8	1.1244
5	1	1.2144

13. Here $x_0 = 0$ and $b = 1$. With $n = 5$, we have $h = 0.2$, and so $x_0 = 0$, $x_1 = 0.2$,

$x_2 = 0.4$, $x_3 = 0.6$, $x_4 = 0.8$, and $x_5 = 1$. Also, $F(x, y) = 2x - y + 1$ and

$y(0) = 2$, so $y_0 = y(0) = 2$,

$y_1 = y_0 + hF(x_0, y_0) = 2 + 0.2[2(0) - 2 + 1] = 1.8$,

$y_2 = y_1 + hF(x_1, y_1) = 1.8 + 0.2[2(0.2) - 1.8 + 1] = 1.72$,

$y_3 = y_2 + hF(x_2, y_2) = 1.72 + 0.2[2(0.4) - 1.72 + 1] = 1.736$,

$y_4 = y_3 + hF(x_3, y_3) = 1.736 + 0.2[2(0.6) - 1.736 + 1] = 1.8288$, and

$y_5 = y_4 + hF(x_4, y_4) = 1.8288 + 0.2[2(0.8) - 1.8288 + 1] = 1.98304$. The

solutions are summarized at right.

n	x_n	y_n
0	0	2
1	0.2	1.8
2	0.4	1.72
3	0.6	1.736
4	0.8	1.8288
5	1	1.9830

15. Here $x_0 = 0$, and $b = 0.5$. With $n = 5$, we have $h = 0.1$, and so $x_0 = 0$, $x_1 = 0.1$,

$x_2 = 0.2$, $x_3 = 0.3$, $x_4 = 0.4$, and $x_5 = 0.5$. Also, $F(x, y) = x^2 + y$ and

$y(0) = 1$, so $y_0 = y(0) = 1$, $y_1 = y_0 + hF(x_0, y_0) = 1 + 0.1(0 + 1) = 1.1$,

$y_2 = y_1 + hF(x_1, y_1) = 1.1 + 0.1[(0.1)^2 + 1.1] = 1.211$,

$y_3 = y_2 + hF(x_2, y_2) = 1.211 + 0.1[(0.2)^2 + 1.211] = 1.3361$,

$y_4 = y_3 + hF(x_3, y_3) = 1.3361 + 0.1[(0.3)^2 + 1.3361] = 1.47871$, and

$y_5 = y_4 + hF(x_4, y_4) = 1.47871 + 0.1[(0.4)^2 + 1.47871] = 1.642581$. The

solutions are summarized at right.

n	x_n	y_n
0	0	1
1	0.1	1.1
2	0.2	1.211
3	0.3	1.3361
4	0.4	1.4787
5	0.5	1.6426

CHAPTER 9   Concept Review   page 641

1. a. differential equation   **b.** satisfies

3. a. highest, unknown   **b.** $f(x)g(y)$, not separable, separable   **c.** separate, variable

5. approximate, actual, polygonal

© 2011 Cengage Learning. All Rights Reserved. May not be scanned, copied or duplicated, or posted to a publicly accessible website, in whole or in part.

1. $y = C_1 e^{2x} + C_2 e^{-3x}$, so $y' = 2C_1 e^{2x} - 3C_2 e^{-3x}$ and $y'' = 4C_1 e^{2x} + 9C_2 e^{-3x}$. Substituting these values into the differential equation $y'' + y' - 6y = 0$, we have $4C_1 e^{2x} + 9C_2 e^{-3x} + 2C_2 e^{2x} - 3C_2 e^{-3x} - 6\left(C_1 e^{2x} + C_2 e^{-3x}\right) = 0$, and so the differential equation is satisfied.

3. $y = Cx^{-4/3}$, so $y' = -\frac{4}{3}Cx^{-7/3}$. Substituting these values into the given differential equation, which can be written in the form $\dfrac{dy}{dx} = -\dfrac{4xy^3}{3x^2 y^2} = -\dfrac{4y}{3x}$, we find $-\frac{4}{3}Cx^{-7/3} = -\dfrac{4\left(Cx^{-4/3}\right)}{3x}$, and we see that y is a solution of the differential equation.

5. $y = (9x + C)^{-1/3}$, so $y' = -\frac{1}{3}(9x + C)^{-4/3}(9) = -3(9x + C)^{-4/3}$. Substituting these values into the differential equation $y' = -3y^4$, we have $-3(9x + C)^{-4/3} = -3\left[(9x + C)^{-1/3}\right]^{-4} = -3(9x + C)^{-4/3}$, and we see that y is a solution. Next, using the side condition, we find that $y(0) = C^{-1/3} = \frac{1}{2}$, so $C = 8$. Therefore, the required solution is $y = (9x + 8)^{-1/3}$.

7. $\dfrac{dy}{4 - y} = 2\,dt$ implies that $-\ln|4 - y| = 2t + C_1$, so $4 - y = Ce^{-2t}$ and $y = 4 - Ce^{-2t}$, where $C = e^{-C_1}$.

9. We have $\frac{dy}{dx} = 3x^2 y^2 + y^2 = y^2\left(3x^2 + 1\right)$. Separating variables and integrating, we have $\int y^{-2}\,dy = \int \left(3x^2 + 1\right)dx$, so $-\frac{1}{y} = x^3 + x + C$. Using the side condition $y(0) = -2$, we find $\frac{1}{2} = C$, so the solution is $y = -\dfrac{1}{x^3 + x + \frac{1}{2}} = -\dfrac{2}{2x^3 + 2x + 1}$.

11. We have $\frac{dy}{dx} = -\frac{3}{2}x^2 y$. Separating variables and integrating, we have $\int y^{-1}\,dy = \int -\frac{3}{2}x^2\,dx$, so $\ln|y| = -\frac{1}{2}x^3 + C$. Using the condition $y(0) = 3$, we have $C = \ln 3$, so $\ln|y| = -\frac{1}{2}x^3 + \ln 3$ and $y = e^{-(x^3/3)+\ln 3} = 3e^{-x^3/2}$.

13. a. $x_0 = 0$, $b = 1$, and $n = 4$, so $h = 0.25$, $x_0 = 0$, $x_1 = 0.25$, $x_2 = 0.5$,
$x_3 = 0.75$, and $x_4 = 1$. Also $F(x, y) = x + y^2$ and $y(0) = 0$, so $y_0 = y(0) = 0$,
$y_1 = y_0 + hF(x_0, y_0) = 0 + 0.25\,(0) = 0$, $y_2 = y_1 + hF(x_1, y_1) = 0 + 0.25\,(0.25 + 0) = 0.0625$,
$y_3 = y_2 + hF(x_2, y_2) = 0.0625 + 0.25\left[0.5 + (0.0625)^2\right] \approx 0.1884766$, and
$y_4 = y_3 + hF(x_3, y_3) \approx 0.1884766 + 0.25\left[0.75 + (0.1884766)^2\right] \approx 0.3848575$. Therefore, $y(1) \approx 0.3849$.

b. Here $n = 6$, so $h = \frac{1}{6}$, $x_0 = 0$, $x_1 = \frac{1}{6}$, $x_2 = \frac{1}{3}$, $x_3 = \frac{1}{2}$, $x_4 = \frac{2}{3}$, $x_5 = \frac{5}{6}$, and $x_6 = 1$. Thus, $y_0 = y(0) = 0$,
$y_1 = y_0 + hF(x_0, y_0) = 0 + \frac{1}{6}(0 + 0) = 0$, $y_2 = y_1 + hF(x_1, y_1) = 1 + \frac{1}{6}\left(\frac{1}{6} + 0^2\right) \approx 0.0277778$,
$y_3 = y_2 + hF(x_2, y_2) \approx 0.0277778 + \frac{1}{6}\left[\frac{1}{3} + (0.0277778)^2\right] \approx 0.0834620$,
$y_4 = y_3 + hF(x_3, y_3) \approx 0.0834620 + \frac{1}{6}\left[\frac{1}{2} + (0.0834620)^2\right] \approx 0.1679563$,
$y_5 = y_4 + hF(x_4, y_4) \approx 0.1679563 + \frac{1}{6}\left[\frac{2}{3} + (0.1679563)^2\right] \approx 0.2837690$, and
$y_6 = y_5 + hF(x_5, y_5) \approx 0.2837690 + \frac{1}{6}\left[\frac{5}{6} + (0.2837690)^2\right] \approx 0.4360787$. Therefore, $y(1) \approx 0.4361$.

© 2011 Cengage Learning. All Rights Reserved. May not be scanned, copied or duplicated, or posted to a publicly accessible website, in whole or in part.

15. a. Here $x_0 = 0$, $b = 1$, and $n = 4$, so $h = \frac{1}{4}$, $x_0 = 0$, $x_1 = 0.25$, $x_2 = 0.5$, $x_3 = 0.75$, and $x_4 = 1$. Also,

$F(x, y) = 1 + 2xy^2$ and $y(0) = 0$, so $y_0 = y(0) = 0$, $y_1 = y_0 + hF(x_0, y_0) = 0 + 0.25(1 + 0) = 0.25$,

$y_2 = y_1 + hF(x_1, y_1) = 0.25 + 0.25[1 + 2(0.25)(0.25)^2] = 0.507813$,

$y_3 = y_2 + hF(x_2, y_2) = 0.507813 + 0.25[1 + 2(0.5)(0.507813)^2] \approx 0.822282$, and

$y_4 = y_3 + hF(x_3, y_3) = 0.822282 + 0.25[1 + 2(0.75)(0.822282)^2] \approx 1.32584$. Therefore, $y(1) \approx 1.3258$.

b. Here $n = 6$, so $h = \frac{1}{6}$, $x_0 = 0$, $x_1 = \frac{1}{6}$, $x_2 = \frac{1}{3}$, $x_3 = \frac{1}{2}$, $x_4 = \frac{2}{3}$, $x_5 = \frac{5}{6}$, and $x_6 = 1$.

Thus, $y_0 = y(0) = 0$, $y_1 = y_0 + hF(x_0, y_0) = 0 + 0.1666667(1 + 0) = 0.166667$,

$y_2 = y_1 + hF(x_1, y_1) = 1 + 0.1666667[1 + 2(0.166667)(0.166667)^2] = 0.334877$,

$y_3 = y_2 + hF(x_2, y_2) = 0.334877 + 0.1666667[1 + 2(0.333333)(0.334877)^2] \approx 0.514004$,

$y_4 = y_3 + hF(x_3, y_3) = 0.514004 + 0.1666667[1 + 2(0.50000)(0.514004)^2] \approx 0.724704$,

$y_5 = y_4 + hF(x_4, y_4) = 0.724704 + 0.1666667[1 + 2(0.666667)(0.724704)^2] \approx 1.008081$, and

$y_6 = y_5 + hF(x_5, y_5) \approx 1.008081 + \frac{1}{6}\left[1 + 2\left(\frac{5}{6}\right)(1.008081)^2\right] \approx 1.457033$. Therefore, $y(1) \approx 1.4570$.

17. Here $x_0 = 0$ and $b = 1$, so with $n = 5$, we have $h = \frac{1}{5} = 0.2$, $x_0 = 0$,

$x_1 = 0.2$, $x_2 = 0.4$, $x_3 = 0.6$, $x_4 = 0.8$, and $x_5 = 1$. Also, $F(x, y) = 2xy$ and

$y_0 = y(0) = 1$, so $y_0 = y(0) = 1$, $y_1 = y_0 + hF(x_0, y_0) = 1 + 0.2(2)(0)(1) = 1$,

$y_2 = y_1 + hF(x_1, y_1) = 1 + 0.2(2)(0.2)(1) = 1.08$, $y_3 = y_2 + hF(x_2, y_2) = 1.08 + 0.2(2)(0.4)(1.08) = 1.2528$,

$y_4 = y_3 + hF(x_3, y_3) = 1.2528 + (0.2)(2)(0.6)(1.2528) = 1.553472$, and

$y_5 = y_4 + hF(x_4, y_4) = 1.553472 + (0.2)(2)(0.8)(1.553472) \approx 2.05058$. Therefore, $y(1) \approx 2.0506$.

19. $\frac{dS}{dT} = -kS$, so $S(t) = S_0 e^{-kt} = 50,000 e^{-kt}$. The condition $S(2) = 32,000 = 50,000 e^{-2k}$ gives $e^{-2k} = \frac{32}{50} = 0.64$,

so $-2k \ln e \approx \ln 0.64$ and $k \approx 0.223144$.

a. $S = 50,000 e^{-0.223144t} = 50,000(0.8)^t$.

b. $S(5) = 50,000(0.8)^5 = 16,384$, or $16,384$.

21. $\frac{dA}{dt} = rA + P$. Separating variables and integrating, we have $\displaystyle\int \frac{dA}{rA + P} = \int dt$, so $\frac{1}{r}\ln(rA + P) = t + C_1$,

$\ln(rA + P) = rt + C_2$ ($C_2 = C_1 r$), and $rA + P = Ce^{rt}$. Thus, $A = \frac{1}{r}(Ce^{rt} - P)$. Using the condition

$A(0) = 0$, we have $0 = \frac{1}{r}(C - P)$, so $C = P$ and $A = \frac{P}{r}(e^{rt} - 1)$. The size of the fund after five years is

$A = \frac{50,000}{0.12}\left[e^{0.12(5)} - 1\right] \approx 342,549.50$, or approximately $342,549.50$.

23. According to Newton's Law of cooling, $\frac{dT}{dt} = k(350 - T)$, where T is the temperature of the roast. We also

have the conditions $T(0) = 68$ and $T(2) = 118$. Separating the variables in the differential equation and

integrating, we have $\frac{dT}{350 - T} = k\,dt$, $\ln|350 - T| = kt + C_1$, $350 - T = Ce^{kt}$, and so $T = 350 - Ce^{kt}$.

Using the condition $T(0) = 68$, we find $350 - C = 68$, so $C = 282$. Therefore, So $T = 350 - 282e^{kt}$.

Next, we use the condition $T(2) = 118$ to find $118 = 350 - 282e^{2k}$ and $e^{2k} = 0.822695035$, so

$k \approx \frac{1}{2}\ln 0.822695035 \approx -0.097584$. Thus, the desired function is $T = 350 - 282e^{-0.097584t}$. We want to find t

when $T = 150$, so we solve the equation $150 = 350 - 282e^{-0.097584t}$, obtaining $e^{-0.097584t} = 0.709219858$,

$-0.097584t = \ln 0.709219858 = -0.343589704$, and $t = 3.52096$, or approximately 3.5 hours. The temperature

of the roast is $150°F$ at approximately 7:30 p.m.

© 2011 Cengage Learning. All Rights Reserved. May not be scanned, copied or duplicated, or posted to a publicly accessible website, in whole or in part.

25. $N = \dfrac{200}{1 + 49e^{-200kt}}$. When $t = 2$, $N = 40$, and so $40 = \dfrac{200}{1 + 49e^{-400k}}$, $40 + 1960e^{-400k} = 200$,

$e^{-400k} = \frac{160}{1960}$ (0.0816327), $-400\ln k = \ln 0.0816327$, and $k \approx 0.00626$. Therefore,

$N(5) = \dfrac{200}{1 + 49e^{-200(5)(0.00626)}} \approx 183$, or 183 families.

CHAPTER 9   Before Moving On...   page 643

1. a. $y' = 4x + c$. Substituting this into the given differential equation, we have

$xy' - y = x(4x + c) - (2x^2 + cx) = 2x^2$, and the equation is satisfied. So $y = 2x^2 + cx$ is indeed a solution.

b. $y(1) = 2$ implies that $2(1) + c(1) = 2$, so $c = 0$ and the particular solution is $y = 2x^2$.

2. The equation is separable. Separating variables and integrating, we have $\displaystyle\int \frac{1+x}{x}\,dx + \int \frac{dy}{y} = 0$,

$\displaystyle\int \left(\frac{1}{x} + 1\right) dx + \int \frac{dy}{y} = 0$, $\ln|x| + x + \ln|y| = C_1$, $\ln|xy| = -x + C_1$, $|xy| = e^{-x+C_1} = C_2 e^{-x}$ (where

$C_2 = e^{C_1}$), $xy = Ce^{-x}$ (where $C = \pm C_2$), and so $y = \dfrac{C}{xe^x}$. Using the condition $y(1) = 1$ gives $1 = \dfrac{C}{e}$, so $C = e$

and the solution is $y = \dfrac{e}{xe^x} = \dfrac{1}{xe^{x-1}}$

3. Here $\frac{dP}{dt} = kP$ and $P(0) = 5000$, so $P(t) = 5000e^{kt}$. Next, $P(5) = 1000$ implies that $5000e^{5k} = 10{,}000$,

$e^{5k} = 2$, $\ln e^{5k} = 2$, $5k = \ln 2$, and so $k = \frac{\ln 2}{5} \approx 0.1386$ and $P(t) = 5000e^{0.1386t}$. To find the time it takes for the

population to reach 12,000, we solve $5000e^{0.1386t} = 12{,}000$, obtaining $e^{0.1386t} = \frac{12000}{5000} = \frac{12}{5}$, $\ln e^{0.1386t} = \ln \frac{12}{5}$,

$0.1386t = \ln \frac{12}{5}$, and $t = \frac{\ln \frac{12}{5}}{0.1386} \approx 6.32$. Thus, it will take approximately 6.3 years.

4. Here $x_0 = 0$ and $b = 0.5$. Taking $n = 5$, we find $h = \frac{0.5-0}{5} = 0.1$, $x_0 = 0$, $x_1 = 0.1$, $x_2 = 0.2$,

$x_3 = 0.3$, $x_4 = 0.4$, and $x_5 = b = 0.5$. Also, $F(x,y) = y^2 - x^2$ and $y(0) = 1$, so $y_0 = 1$,

$y_1 = y_0 + hF(x_0, y_0) = 1 + 0.1(1^2 - 0^2) = 1.1$, $y_2 = y_1 + hF(x_1, y_1) = 1.1 + 0.1\left[(1.1)^2 - (0.1)^2\right] = 1.22$,

$y_3 = y_2 + hF(x_2, y_2) = 1.22 + 0.1\left[(1.22)^2 - (0.2)^2\right] = 1.36484$,

$y_4 = y_3 + hF(x_3, y_3) = 1.36484 + 0.1\left[(1.36484)^2 - (0.3)^2\right] \approx 1.54212$, and

$y_5 = y_4 + hF(x_4, y_4) \approx 1.54212 + 0.1\left[(1.54212)^2 - (0.4)^2\right] \approx 1.76393$. Therefore, $y(0.5) \approx 1.7639$.

© 2011 Cengage Learning. All Rights Reserved. May not be scanned, copied or duplicated, or posted to a publicly accessible website, in whole or in part.

10 PROBABILITY AND CALCULUS

10.1 Problem-Solving Tips

1. Familiarize yourself with the following definitions:

   a. A *random variable* is a rule that assigns a number to each outcome of a chance experiment.

   b. A *probability distribution* of a random variable gives the distinct values of the random variable X and the probabilities associated with these values.

   c. A *histogram* is the graph of the probability distribution of a random variable.

2. **To show that a function f is a probability density function** of a random variable over a given interval, first show that the function is nonnegative over that interval, and then show that the area of the region under the graph of f over that interval is equal to 1.

10.1 Concept Questions   page 653

1. **a.** An experiment is an activity with observable results called outcomes, or sample points.

   b. The totality of all outcomes is the sample space of the experiment.

   c. A subset of the sample space is called an event of the experiment.

   d. The probability or likelihood of an event is a number between 0 and 1 and may be viewed as the proportionate number of times that the event will occur if the experiment associated with the event is repeated indefinitely and independently under similar conditions.

3. **a.** The definition of a probability density function is given on page 648 of the text.

   b. The definition of a joint probability density function is given on page 651 of the text.

10.1 Probability Distributions and Random Variables   page 654

1. $f(x) = \frac{1}{3} > 0$ on $[3, 6]$. Next, $\int_3^6 \frac{1}{3}\, dx = \frac{1}{3}x\Big|_3^6 = \frac{1}{3}(6-3) = 1$.

3. $f(x) = \frac{1}{16}x \ge 0$ on $[2, 6]$. Next $\int_2^6 \frac{1}{16}x\, dx = \frac{1}{32}x^2\Big|_2^6 = \frac{1}{32}(36-4) = 1$, and so f is a probability density function on $[2, 6]$.

5. $f(x) = \frac{2}{9}\left(3x - x^2\right) = \frac{2}{9}x(3-x)$ is nonnegative on $[0, 3]$ because the factors x and and $3 - x$ are nonnegative there. Next, we compute $\int_0^3 \frac{2}{9}\left(3x - x^2\right) dx = \frac{2}{9}\left(\frac{3}{2}x^2 - \frac{1}{3}x^3\right)\Big|_0^3 = \frac{2}{9}\left(\frac{27}{2} - 9\right) = 1$, and so f is a probability density function.

263

© 2011 Cengage Learning. All Rights Reserved. May not be scanned, copied or duplicated, or posted to a publicly accessible website, in whole or in part.

7. $f(x) = \dfrac{12-x}{72}$ is nonnegative on $[0, 12]$. Next, we see that

$$\int_0^{12} \frac{12-x}{72}\,dx = \int_0^{12} \left(\frac{1}{6} - \frac{x}{72}\right)dx = \left(\frac{1}{6}x - \frac{1}{144}x^2\right)\Big|_0^{12} = 2 - 1 = 1, \text{ and conclude that } f \text{ is a probability}$$

function on $[0, 12]$.

9. $f(x) = \dfrac{8}{7x^2}$ is nonnegative on $[1, 8]$. Next, we compute $\displaystyle\int_1^8 \frac{8}{7x^2}\,dx = -\frac{8}{7x}\Big|_1^8 = -\frac{8}{7}\left(\frac{1}{8} - 1\right) = 1$, and so f is a

probability density function on $[1, 8]$.

11. First, note that $f(x) \geq 0$ on $[0, \infty)$. Next, let $I = \int x\,(x^2+1)^{-3/2}\,dx$. Integrate I using the substitution

$\quad u = x^2 + 1$, so $du = 2x\,dx$. Then $I = \frac{1}{2}\int u^{-3/2}du = \frac{1}{2}\left(-2u^{-1/2}\right) + C = -\dfrac{1}{\sqrt{u}} + C = -\dfrac{1}{\sqrt{x^2+1}} + C.$

$\quad$ Therefore, $\displaystyle\int_0^\infty \frac{x\,dx}{(x^2+1)^{3/2}} = \lim_{b\to\infty}\left(-\frac{1}{\sqrt{x^2+1}}\right)\Big|_0^b = \lim_{b\to\infty}\left(-\frac{1}{\sqrt{b^2+1}} + 1\right) = 1$, completing the proof.

13. $\int_1^4 k\,dx = kx\big|_1^4 = 3k = 1$ implies that $k = \frac{1}{3}$.

15. $\int_0^4 k\,(4-x)\,dx = k\int_0^4 (4-x)\,dx = k\left(4x - \frac{1}{2}x^2\right)\Big|_0^4 = k\,(16-8) = 8k = 1$ implies that $k = \frac{1}{8}$.

17. $\int_0^4 kx^{1/2}\,dx = \frac{2}{3}kx^{3/2}\Big|_0^4 = \frac{16}{3}k = 1$ implies that $k = \frac{3}{16}$.

19. $\displaystyle\int_1^\infty \frac{k}{x^3}\,dx = \lim_{b\to\infty}\int_1^b kx^{-3}\,dx = \lim_{b\to\infty}\left(-\frac{k}{2x^2}\right)\Big|_1^b = \lim_{b\to\infty}\left(-\frac{k}{2b^2} + \frac{k}{2}\right) = \frac{k}{2} = 1$ implies that $k = 2$.

21. a. $P\,(2 \leq X \leq 4) = \int_2^4 \frac{1}{12}x\,dx = \frac{1}{24}x^2\Big|_2^4 = \frac{1}{24}\,(16-4) = \frac{1}{2}$.

$\quad$ **b.** $P\,(1 \leq X \leq 4) = \int_1^4 \frac{1}{12}x\,dx = \frac{1}{24}x^2\Big|_1^4 = \frac{1}{24}\,(16-1) = \frac{5}{8}$.

$\quad$ **c.** $P\,(X \geq 2) = \int_2^5 \frac{1}{12}x\,dx = \frac{1}{24}x^2\Big|_2^5 = \frac{1}{24}\,(25-4) = \frac{7}{8}$.

$\quad$ **d.** $P\,(X = 2) = \int_2^2 \frac{1}{12}x\,dx = 0.$

23. a. $P\,(-1 \leq X \leq 1) = \int_{-1}^1 \frac{3}{32}\,(4-x^2)\,dx = \frac{3}{32}\left(4x - \frac{1}{3}x^3\right)\Big|_{-1}^1 = \frac{3}{32}\left[\left(4-\frac{1}{3}\right) - \left(-4+\frac{1}{3}\right)\right] = \frac{11}{16}$.

$\quad$ **b.** $P\,(X \leq 0) = \int_{-2}^0 \frac{3}{32}\,(4-x^2)\,dx = \frac{3}{32}\left(4x - \frac{1}{3}x^3\right)\Big|_{-2}^0 = \frac{3}{32}\left[0 - \left(-8+\frac{8}{3}\right)\right] = \frac{1}{2}$.

$\quad$ **c.** $P\,(X > -1) = \int_{-1}^2 \frac{3}{32}\,(4-x^2)\,dx = \frac{3}{32}\left(4x - \frac{1}{3}x^3\right)\Big|_{-1}^2 = \frac{3}{32}\left[\left(8-\frac{8}{3}\right) - \left(-4+\frac{1}{3}\right)\right] = \frac{27}{32}$.

$\quad$ **d.** $P\,(X = 0) = \int_0^0 \frac{3}{32}\,(4-x^2)\,dx = 0.$

25. a. $P\,(X \geq 4) = \int_4^9 \frac{1}{4}x^{-1/2}\,dx = \frac{1}{2}x^{1/2}\Big|_4^9 = \frac{1}{2}\,(3-2) = \frac{1}{2}$.

$\quad$ **b.** $P\,(1 \leq X \leq 8) = \int_1^8 \frac{1}{4}x^{-1/2}\,dx = \frac{1}{2}x^{1/2}\Big|_1^8 = \frac{1}{2}\left(2\sqrt{2} - 1\right) \approx 0.9142.$

$\quad$ **c.** $P\,(X = 3) = \int_3^3 \frac{1}{4}x^{-1/2}\,dx = 0.$

$\quad$ **d.** $P\,(X \leq 4) = \int_1^4 \frac{1}{4}x^{-1/2}\,dx = \frac{1}{2}x^{1/2}\Big|_1^4 = \frac{1}{2}\,(2-1) = \frac{1}{2}$.

© 2011 Cengage Learning. All Rights Reserved. May not be scanned, copied or duplicated, or posted to a publicly accessible website, in whole or in part.

27. a. $P(0 \le X \le 4) = \int_0^4 4xe^{-2x^2}\,dx = -e^{-2x^2}\Big|_0^4 = -e^{-32} + 1 \approx 1.$

  b. $P(X \ge 1) = \int_1^\infty 4xe^{-2x^2}\,dx = \lim\limits_{b\to\infty} \int_1^b 4xe^{-2x^2}\,dx = \lim\limits_{b\to\infty}\left(-e^{-2x^2}\right)\Big|_1^b = \lim\limits_{b\to\infty}\left(-e^{-2b^2} + e^{-2}\right) = e^{-2}$

  $\approx 0.1353.$

29. a. $P\left(\tfrac{1}{2} \le X \le 1\right) = \int_{1/2}^1 x\,dx = \tfrac{1}{2}x^2\Big|_{1/2}^1 = \tfrac{1}{2}(1) - \tfrac{1}{2}\left(\tfrac{1}{4}\right) = \tfrac{1}{2} - \tfrac{1}{8} = \tfrac{3}{8}.$

  b. $P\left(\tfrac{1}{2} \le X \le \tfrac{3}{2}\right) = \int_{1/2}^1 x\,dx + \int_1^{3/2}(2-x)\,dx = \tfrac{3}{8} + \left(2x - \tfrac{1}{2}x^2\right)\Big|_1^{3/2}$ (Using the result from Exercise 9)

  $= \tfrac{3}{8} + \left[\tfrac{1}{2}x\,(4-x)\right]_1^{3/2} = \tfrac{3}{8} + \left[\tfrac{1}{2}\left(\tfrac{3}{2}\right)\left(4 - \tfrac{3}{2}\right) - \tfrac{1}{2}(3)\right] = \tfrac{3}{8} + \tfrac{3}{4}\left(\tfrac{5}{2}\right) - \tfrac{3}{2} = \tfrac{3}{4}.$

  c. $P(X \ge 1) = \int_1^2 (2-x)\,dx = \left(2x - \tfrac{1}{2}x^2\right)\Big|_1^2 = (4-2) - \left(2 - \tfrac{1}{2}\right) = \tfrac{1}{2}.$

  d. $P\left(X \le \tfrac{3}{2}\right) = \int_0^1 x\,dx + \int_1^{3/2}(2-x)\,dx = \tfrac{1}{2}x^2\Big|_0^1 + \left(2x - \tfrac{1}{2}x^2\right)\Big|_1^{3/2} = \tfrac{1}{2} + \left[2\left(\tfrac{3}{2}\right) - \tfrac{1}{2}\left(\tfrac{9}{4}\right)\right] - \left(2 - \tfrac{1}{2}\right)$

  $= \tfrac{1}{2} + \left(3 - \tfrac{9}{8}\right) - \tfrac{3}{2} = \tfrac{1}{2} + \tfrac{15}{8} - \tfrac{3}{2} = \tfrac{7}{8}.$

31. f is nonnegative on D. $\int_1^3 \int_0^2 \tfrac{1}{4}xy\,dx\,dy = \int_1^3 \tfrac{1}{4}x\Big|_0^2\,dy = \int_1^3 \tfrac{1}{2}\,dy = \tfrac{1}{2}y\Big|_1^3 = \tfrac{3}{2} - \tfrac{1}{2} = 1.$

33. f is nonnegative on D. $\tfrac{1}{3}\int_1^2 \int_0^2 xy\,dx\,dy = \tfrac{1}{3}\int_1^2 \left(\tfrac{1}{2}x^2 y\right)\Big|_{x=0}^{x=2}\,dy = \tfrac{1}{3}\int_1^2 2y\,dy = y^2\Big|_1^2 = \tfrac{1}{3}(4-1) = \tfrac{1}{3}\cdot 3 = 1.$

35. $k\int_1^2 \int_0^1 x^2 y\,dx\,dy = k\int_1^2 \tfrac{1}{3}x^3 y\Big|_{x=0}^{x=1}\,dy = k\int_1^2 \tfrac{1}{3}y\,dy = k\cdot \tfrac{1}{6}y^2\Big|_1^2 = \tfrac{1}{2}k.$ Therefore, $k = 2.$

37. $k\int_1^\infty \int_0^1 (x - x^2)e^{-2y}\,dx\,dy = k\int_1^\infty \left(\tfrac{1}{2}x^2 - \tfrac{1}{3}x^3\right)\Big|_0^1 e^{-2y}\,dy = k\left(\tfrac{1}{6}\right)\left(-\tfrac{1}{2}e^{-2y}\right)\Big|_1^\infty = -\tfrac{1}{12}ke^{-2y}\Big|_1^\infty = \tfrac{1}{12}ke^{-2}.$

  Therefore, $k = 12e^2.$

39. a. $P(0 \le X \le 1, 0 \le Y \le 1) = \int_0^1 \int_0^1 xy\,dx\,dy = \int_0^1 \left(\tfrac{1}{2}x^2 y\right)\Big|_0^1\,dy = \int_0^1 \tfrac{1}{2}y\,dy = \tfrac{1}{4}y^2\Big|_0^1 = \tfrac{1}{4}.$

  b. $P((X, Y) \mid X + 2Y \le 1) = \int_0^1 \int_0^{(1-x)/2} xy\,dy\,dx = \int_0^1 \left(\tfrac{1}{2}xy^2\right)\Big|_{y=0}^{y=(1-x)/2}\,dx = \tfrac{1}{8}\int_0^1 (x - 2x^2 + x^3)\,dx$

  $= \tfrac{1}{8}\left(\tfrac{1}{2}x^2 - \tfrac{2}{3}x^3 + \tfrac{1}{4}x^4\right)\Big|_0^1 = \tfrac{1}{8}\left(\tfrac{1}{2} - \tfrac{2}{3} + \tfrac{1}{4}\right) = \tfrac{1}{96}.$

41. a. $P(0 \le X \le 2, 0 \le Y \le 1) = \tfrac{9}{224}\int_0^1 \int_1^2 x^{1/2} y^{1/2}\,dx\,dy = \tfrac{9}{224}\int_0^1 \tfrac{2}{3}x^{3/2}y^{1/2}\Big|_1^2\,dy = \tfrac{3}{112}\int_0^1 \left(2\sqrt{2}y^{3/2} - y^{3/2}\right)dy$

  $= \tfrac{3}{112}\left(2\sqrt{2} - 1\right)\int_0^1 y^{3/2}\,dy = \tfrac{3}{112}\left(2\sqrt{2} - 1\right)\left[\tfrac{2}{3}y^{3/2}\Big|_0^1\right] = \tfrac{1}{56}\left(2\sqrt{2} - 1\right).$

  b. $P\left((X, Y) \mid 1 \le X \le 4, 0 \le Y \le \sqrt{X}\right) = \tfrac{9}{224}\int_1^4 \int_0^{\sqrt{x}} x^{1/2} y^{1/2}\,dy\,dx = \tfrac{9}{224}\int_1^4 \tfrac{2}{3}x^{1/2}y^{3/2}\Big|_{y=0}^{y=\sqrt{x}}\,dx$

  $= \tfrac{3}{112}\int_1^4 x^{1/2}\left(x^{1/2}\right)^{3/2}\,dx = \tfrac{3}{112}\int_1^4 x^{5/4}\,dx = \tfrac{3}{112}\cdot \tfrac{4}{9}x^{9/4}\Big|_1^4 = \tfrac{4}{336}\left(4^{9/4} - 1\right) \approx 0.2575.$

© 2011 Cengage Learning. All Rights Reserved. May not be scanned, copied or duplicated, or posted to a publicly accessible website, in whole or in part.

43. a. $P(X \le 100) = \int_0^{100} \frac{1}{100} e^{-x/100}\, dx = -e^{-x/100}\big|_0^{100} = -e^{-1} + 1 \approx 0.6321.$

  b. $P(X \ge 120) = \int_{120}^{\infty} \frac{1}{100} e^{-x/100}\, dx = \lim_{b\to\infty} \int_{120}^{b} \frac{1}{100} e^{-x/100}\, dx = \lim_{b\to\infty} \left(-e^{-x/100}\right)\big|_{120}^{b}$

  $= \lim_{b\to\infty} \left(-e^{-b/100} + e^{-120/100}\right) = e^{-1.2} \approx 0.3012.$

  c. $P(60 < X < 140) = \int_{60}^{140} \frac{1}{100} e^{-x/100}\, dx = -e^{-x/100}\big|_{60}^{140} = -e^{-1.4} + e^{-0.6} \approx 0.3022.$

45. a. $P(600 \le T \le 800) = \int_{600}^{800} 0.001 e^{-0.001t}\, dt = -e^{-0.001t}\big|_{600}^{800} = -e^{-0.8} + e^{-0.6} \approx 0.0995.$

  b. $P(T \ge 1200) = \int_{1200}^{\infty} 0.001 e^{-0.001t}\, dt = \lim_{b\to\infty} \int_{1200}^{b} 0.001 e^{-0.001t}\, dt = \lim_{b\to\infty} \left(-e^{-0.001t}\right)\big|_{1200}^{b}$

  $= \lim_{b\to\infty} \left(-e^{-0.001b} + e^{-1.2}\right) = e^{-1.2} \approx 0.3012.$

47. $P(T \ge 2) = \int_2^{\infty} \frac{1}{30} e^{-t/30}\, dt = \lim_{b\to\infty} \int_2^b \frac{1}{30} e^{-t/30}\, dt = \lim_{b\to\infty} \left(-e^{-t/30}\right)\big|_2^b = \lim_{b\to\infty} \left(-e^{-b/30} + e^{-1/15}\right) = e^{-1/15}$

  $\approx 0.9355.$

49. $P(1 \le X \le 2) = \int_1^2 \frac{2}{9} x(3-x)\, dx = \frac{2}{9}\left(\frac{3}{2}x^2 - \frac{1}{3}x^3\right)\Big|_1^2 = \frac{2}{9}\left[\left(6 - \frac{8}{3}\right) - \left(\frac{3}{2} - \frac{1}{3}\right)\right] = \frac{13}{27} \approx 0.4815.$

  $P(X \ge 1) = \int_1^3 \frac{2}{9} x(3-x)\, dx = \frac{2}{9}\left(\frac{3}{2}x^2 - \frac{1}{3}x^3\right)\Big|_1^3 = \frac{2}{9}\left[\left(\frac{27}{2} - 9\right) - \left(\frac{3}{2} - \frac{1}{3}\right)\right] = \frac{20}{27} \approx 0.740741.$

51. $P(T > 4) = \int_4^{\infty} 9(9+t^2)^{-3/2}\, dt = \lim_{b\to\infty} \int_4^b 9(9+t^2)^{-3/2}\, dt = \lim_{b\to\infty} \left(\frac{t}{\sqrt{9+t^2}}\right)\Big|_4^b = \lim_{b\to\infty} \left(\frac{b}{\sqrt{9+b^2}} - \frac{4}{5}\right)$

  $= 1 - \frac{4}{5} = \frac{1}{5}.$

53. $P(2 \le X \le 2.5, 1 \le Y \le 2) = \frac{9}{4000} \int_2^{2.5} \int_1^2 xy(25-x^2)^{1/2}(4-y)\, dx\, dy$

  $= \frac{9}{4000} \int_2^{2.5} \left(2y^2 - \frac{1}{3}y^3\right)\left[x(25-x^2)^{1/2}\right]_{x=1}^{x=2} dy = \frac{9}{4000} \cdot \frac{11}{3} \int_2^{2.5} x(25-x^2)^{1/2}\, dx$

  $= -\frac{11}{4000}(25-x^2)^{3/2}\Big|_2^{2.5} = -\frac{11}{4000}\left\{\left[24 - \left(\frac{5}{2}\right)^2\right]^{3/2} = (25-4)^{3/2}\right\} \approx 0.041372.$

55. False. f must be nonnegative on $[a, b]$ as well.

10.1   **Using Technology**   page 657

1.

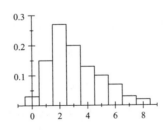

© 2011 Cengage Learning. All Rights Reserved. May not be scanned, copied or duplicated, or posted to a publicly accessible website, in whole or in part.

3. a.

x	$P(X=x)$
0	0.017
1	0.067
2	0.033
3	0.117
4	0.233
5	0.133

x	$P(X=x)$
6	0.167
7	0.1
8	0.05
9	0.067
10	0.017

b.

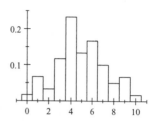

10.2 Concept Questions   page 669

1. a. The average of the n numbers $x_1, x_2, ..., x_n$ is $\bar{x} = \dfrac{x_1 + x_2 + \cdots + x_n}{n}$.

b. The expected value of a random variable X with associated probabilities $p_1, p_2, ..., p_n$ is given by
$E(X) = x_1 p_1 + x_2 p_2 + \cdots + x_n p_n$.

c. The expected value of a continuous random variable x is $E(X) = \int_a^b x f(x)\, dx$.

10.2 Expected Value and Standard Deviation   page 670

1. $\mu = \int_3^6 \frac{1}{3}x\, dx = \frac{1}{6}x^2 \Big|_3^6 = \frac{1}{6}(36-9) = \frac{9}{2}$, so Var $(X) = \int_3^6 \frac{1}{3}x^2\, dx - \frac{81}{4} = \frac{1}{9}x^3 \Big|_3^6 - \frac{81}{4} = \frac{1}{9}(216-27) - \frac{81}{4} = \frac{3}{4}$

and $\sigma = \sqrt{\frac{3}{4}} = \frac{\sqrt{3}}{2} \approx 0.8660$.

3. $\mu = \int_0^5 \frac{3}{125}x^3\, dx = \frac{3}{500}x^4 \Big|_0^5 = \frac{15}{4}$, so Var $(X) = \int_0^5 \frac{3}{125}x^4\, dx - \frac{225}{16} = \frac{3}{625}x^5 \Big|_0^5 - \frac{225}{16} = 15 - \frac{225}{16} = \frac{15}{16}$ and

$\sigma = \sqrt{\frac{15}{16}} = \frac{\sqrt{15}}{4} \approx 0.9682$.

5. $\mu = \int_1^5 \frac{3}{32}x\,(x-1)\,(5-x)\, dx = \frac{3}{32}\int_1^5 \left(-x^3 + 6x^2 - 5x\right) dx = \frac{3}{32}\left(-\frac{1}{4}x^4 + 2x^3 - \frac{5}{2}x^2\right)\Big|_1^5$

$= \frac{3}{32}\left[\left(-\frac{625}{4} + 250 - \frac{125}{2}\right) - \left(-\frac{1}{4} + 2 - \frac{5}{2}\right)\right] = 3$, so

Var $(X) = \int_1^5 \frac{3}{32}\left(-\frac{1}{4}x^4 + 6x^3 - 5x^2\right) dx - 9 = \frac{3}{32}\left(-\frac{1}{5}x^5 + \frac{3}{2}x^4 - \frac{5}{3}x^3\right)\Big|_1^5 - 9$

$= \frac{3}{32}\left[\left(-625 + \frac{1875}{2} - \frac{625}{3}\right) - \left(-\frac{1}{5} + \frac{3}{2} - \frac{5}{3}\right)\right] - 9 \approx 0.8$

and $\sigma \approx \sqrt{0.8} \approx 0.8944$.

7. $\mu = \int_1^8 \frac{8}{7x}\, dx = \frac{8}{7}\ln x \Big|_1^8 = \frac{8}{7}\ln 9 \approx 2.3765$, so

Var $(X) = \int_0^8 \frac{8}{7}dx - 5.64777 = \frac{8}{7}x \Big|_1^8 - 5.64777 = 8 - 5.647777 \approx 2.3522$ and $\sigma = \sqrt{2.3522} \approx 1.5337$.

9. $\mu = \int_1^4 \frac{3}{14}x^{3/2}dx = \frac{3}{35}x^{5/2}\Big|_1^4 = \frac{3}{35}(32-1) = \frac{93}{35}$, so

Var $(X) = \int_1^4 \frac{3}{14}x^{5/2}dx - \frac{9}{25} = \frac{3}{49}x^{7/2}\Big|_1^4 - \frac{9}{25} = \frac{3}{49}(128-1) - \frac{9}{25} \approx 0.715102$ and $\sigma = \sqrt{7.4155} \approx 0.8456$.

© 2011 Cengage Learning. All Rights Reserved. May not be scanned, copied or duplicated, or posted to a publicly accessible website, in whole or in part.

11. $\mu = \int_1^\infty \frac{3}{x^3}\,dx = \lim_{b\to\infty} \int_1^b 3x^{-3}\,dx = \lim_{b\to\infty}\left(-\frac{3}{2x^2}\right)\Big|_1^b = \lim_{b\to\infty}\left(-\frac{3}{2b^2}+\frac{3}{2}\right) = \frac{3}{2}$, so

$\text{Var}(X) = \int_1^\infty \frac{3}{x^2}\,dx - \frac{9}{4} = \lim_{b\to\infty}\int_1^b 3x^{-2}\,dx - \frac{9}{4} = \lim_{b\to\infty}\left(-\frac{3}{x}\right)\Big|_1^b - \frac{9}{4} = \lim_{b\to\infty}\left(-\frac{3}{b}+3\right)-\frac{9}{4} = \frac{3}{4}$ and

$\sigma = \sqrt{\frac{3}{4}} = \frac{\sqrt{3}}{2} \approx 0.8660$.

13. $\mu = \int_0^\infty \frac{1}{4}xe^{-x/4}\,dx = \lim_{b\to\infty}\int_0^b \frac{1}{4}xe^{-x/4}\,dx = \lim_{b\to\infty} 4\left(-\frac{1}{4}-1\right)e^{-x/4}\Big|_0^b = \lim_{b\to\infty}\left[4\left(-\frac{1}{4}b-1\right)e^{-b/4}+4\right] = 4$,

so

$\text{Var}(X) = \int_0^\infty \frac{1}{4}x^2 e^{-x/4}\,dx - 16 = \lim_{b\to\infty}\int_0^b \frac{1}{4}x^2 e^{-x/4}\,dx - 16 = \lim_{b\to\infty}\left[x^2 e^{-x/4} - 32\left(-\frac{1}{4}x-1\right)e^{-x/4}\right]_0^b - 16$

$= \lim_{b\to\infty}\left\{\left[b^2 e^{-b/4} - 32\left(-\frac{1}{4}b-1\right)e^{-b/4}\right]+32\right\} - 16 = 16$

and $\sigma = \sqrt{16} = 4$.

15. $\mu = \int_0^\infty x\cdot\frac{1}{100}e^{-x/100}\,dx = \lim_{b\to\infty}\int_0^b \frac{1}{100}xe^{-x/100}\,dx = \lim_{b\to\infty}\left[100\left(-\frac{x}{100}-1\right)e^{-x/100}\right]_0^b$

$= \lim_{b\to\infty}\left[100\left(-\frac{b}{100}-1\right)e^{-b/100}+100\right] = 100$

so a plant of this species is expected to live 100 days.

17. $\mu = \int_0^5 t\cdot\frac{2}{25}t\,dt = \frac{2}{25}\int_0^5 t^2\,dt = \frac{2}{75}t^3\Big|_0^5 = \frac{2}{75}(125) = \frac{10}{3}$, so a shopper is expected to spend 3 minutes 20 seconds in the magazine section.

19. $\mu = \int_0^3 x\cdot\frac{2}{9}x(3-x)\,dx = \frac{2}{9}\int_0^3 (3x^2-x^3)\,dx = \frac{2}{9}\left(x^3-\frac{1}{4}x^4\right)\Big|_0^3 = \frac{2}{9}\left(27-\frac{81}{4}\right) = 1.5$,

so the expected amount of snowfall is 1.5 ft.

$\text{Var}(X) = \int_0^3 x^2\cdot\frac{2}{9}x(3-x)\,dx - \left(\frac{3}{2}\right)^2 = \frac{2}{9}\int_0^3 (3x^3-x^4)\,dx - \frac{9}{4} = \frac{2}{9}\left(\frac{3}{4}x^4-\frac{1}{5}x^5\right)\Big|_0^3 - \frac{9}{4} = \frac{9}{20}$.

21. $\mu = \int_0^5 x\cdot\frac{6}{125}x(5-x)\,dx = \frac{6}{125}\int_0^5 (5x^2-x^3)\,dx = \frac{6}{125}\left(\frac{5}{3}x^3-\frac{1}{4}x^4\right)\Big|_0^5 = \frac{6}{125}\left(\frac{625}{3}-\frac{625}{4}\right) = 2.5$, so the

expected demand is 2500 lb/wk.

23. $E(X) = \int_a^b xf(x)\,dx = \int_a^b \frac{x}{b-a}\,dx = \frac{1}{2(b-a)}x^2\Big|_a^b = \frac{b^2-a^2}{2(b-a)} = \frac{(b-a)(b+a)}{2(b-a)} = \frac{b+a}{2}$ and

$\text{Var}(X) = \int_a^b x^2 f(x)\,dx - \left(\frac{b+a}{2}\right)^2 = \int_a^b \frac{x^2}{b-a}\,dx - \frac{(b+a)^2}{4} = \frac{x^3}{3(b-a)}\Big|_a^b - \frac{(b+a)^2}{4}$

$= \frac{b^3-a^3}{3(b-a)} - \frac{(b+a)^2}{4} = \frac{(b-a)(b^2+ab+a^2)}{3(b-a)} - \frac{(b+a)^2}{4} = \frac{b^2+ab+a^2}{3} - \frac{b^2+2ab+a^2}{4}$

$= \frac{(b-a)^2}{12}$.

25. Because f is a probability density function on $[1, e]$, we have

$\int_1^2 f(x)\,dx = \int_1^2 \left(ax+\frac{b}{x}\right)\,dx = \left(\frac{1}{2}ax^2+b\ln x\right)\Big|_1^2 = 1$, so $3a + (2\ln 2)b = 2$. Next, we want

$\int_1^2 xf(x)\,dx = \int_1^2 (ax^2+b)\,dx = \left(\frac{1}{3}ax^3+bx\right)\Big|_1^2 = \left(\frac{8}{3}a+2b\right)-\left(\frac{1}{3}a+b\right) = 2$, so $7a+3b = 6$. Solving these

equations simultaneously, we find $a = \dfrac{6(1-2\ln 2)}{9-14\ln 2} \approx 3.3$ and $b = \dfrac{4}{9-14\ln 2} \approx -5.7$.

© 2011 Cengage Learning. All Rights Reserved. May not be scanned, copied or duplicated, or posted to a publicly accessible website, in whole or in part.

27. a. Take $a = 0$ (6 a.m.) and $b = 60$ (7 a.m.) Then the required probability is

$P(30 \le X \le 45) = \frac{1}{60-0} \int_{30}^{45} dx = \frac{45-30}{60} = \frac{1}{4}$.

b. The expected delivery time is $E(X) = \frac{1}{2}(b + a) = \frac{1}{2}(60 + 0) = 30$; that is, 6:30 a.m.

29. $P(x \le m) = \int_{2}^{m} \frac{1}{6} dx = \frac{1}{6} x \Big|_{2}^{m} = \frac{1}{6}(m - 2) = \frac{1}{2}$. Thus, $m - 2 = 3$, and so $m = 5$.

31. $P(x \le m) = \int_{0}^{m} \frac{3}{16} x^{1/2} dx = \frac{1}{8} x^{3/2} \Big|_{0}^{m} = \frac{1}{8} m^{3/2}$. Next, we solve $\frac{1}{8} m^{3/2} = \frac{1}{2}$, obtaining $m^{3/2} = 4$ and

$m = 4^{2/3} \approx 2.5198$.

33. $P(x \le m) = \int_{1}^{m} x^{-2} dx = -\frac{1}{x} \Big|_{1}^{m} = -\frac{1}{m} + 1$. Solving $-\frac{1}{m} + 1 = \frac{1}{2}$, we obtain $-\frac{1}{m} = -\frac{1}{2}$, $-2m = -1$, and so

$m = \frac{1}{2}$.

35. True. This follows from Formula (10).

10.2 Using Technology   page 672

1. a.

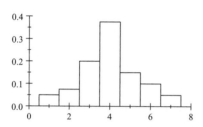

b. $\mu = 4$ and $\sigma = 1.40$.

3. a. X gives the minimum age requirement for a regular driver's license.

b.

x	15	16	17	18	19	21
$P(X = x)$	0.02	0.30	0.08	0.56	0.02	0.02

c.

d. $\mu = 17.34$ and $\sigma = 1.11$.

5. a. Let X denote the random variable that gives the weight of a carton of sugar.

b.

x	4.96	4.97	4.98	4.99	5.00	5.01	5.02	5.03	5.04	5.05	5.06
$P(X = x)$	$\frac{3}{30}$	$\frac{4}{30}$	$\frac{4}{30}$	$\frac{1}{30}$	$\frac{1}{30}$	$\frac{5}{30}$	$\frac{3}{30}$	$\frac{3}{30}$	$\frac{4}{30}$	$\frac{1}{30}$	$\frac{1}{30}$

c. $\mu = 5.00467 \approx 5.00$, $V(X) = 0.0009$, and $\sigma = \sqrt{0.0009} = 0.03$.

© 2011 Cengage Learning. All Rights Reserved. May not be scanned, copied or duplicated, or posted to a publicly accessible website, in whole or in part.

10.3   Problem-Solving Tips

1. *Normal distributions* are a special class of continuous probability distributions. A normal curve is the bell-shaped graph of a normal distribution. The standard normal curve has mean $\mu = 0$ and standard deviation $\sigma = 1$. The random variable associated with a standard normal distribution is called a *standard normal variable* and is denoted by Z. The areas under the standard normal curve to the left of the number z corresponding to the probabilities $P(Z < z)$ or $P(Z \le z)$ are given in the text in Appendix C, Table 2.

2. The area of the region under the normal curve between $x = a$ and $x = b$ is equal to the area of the region under the standard normal curve between $z = \dfrac{a - \mu}{\sigma}$ and $z = \dfrac{b - \mu}{\sigma}$. The probability of the random variable X associated with this area is $P(a < X < b) = \left(\dfrac{a - \mu}{\sigma} < Z < \dfrac{b - \mu}{\sigma} \right)$.

10.3   Concept Questions   page 681

1. **a.** The curve peaks at $x = \mu$.
   b. The curve is symmetrical with respect to the vertical line $x = \mu$.
   c. The value of the area under the normal curve is 1.

3. **a.** The standard normal curve has $\mu = 0$ and standard deviation $\sigma = 1$.
   b. The normal distribution is the distribution associated with the standard normal curve with $\mu = 0$ and standard deviation $\sigma = 1$.

10.3   Normal Distributions   page 682

1. $P(Z < 1.45) = 0.9265$.

3. $P(Z < -1.75) = 0.0401$.

5. $P(-1.32 < Z < 1.74) = P(Z < 1.74) - P(Z < -1.32) = 0.9591 - 0.0934 = 0.8657$.

7. a.

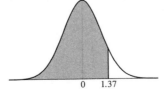

b. $P(Z < 1.37) = 0.9147$.

9. a.

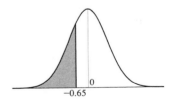

b. $P(Z < -0.65) = 0.2578$.

© 2011 Cengage Learning. All Rights Reserved. May not be scanned, copied or duplicated, or posted to a publicly accessible website, in whole or in part.

11. a.

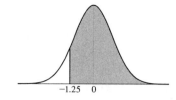

b. $P(Z > -1.25) = 1 - P(Z < -1.25)$

$$= 1 - 0.1056 = 0.8944.$$

13. a.

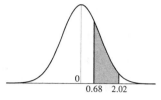

b. $P(0.68 < Z < 2.02)$

$$= P(Z < 2.02) - P(Z < 0.68)$$

$$= 0.9783 - 0.7517 = 0.2266.$$

15. a. Referring to Table (4), we see that $P(Z < z) = 0.8907$ implies that $z = 1.23$.

    b. Referring to Table (4), we see that $P(Z < z) = 0.2090$ implies that $z = -0.81$.

17. a. $P(Z > -z) = 1 - P(Z < -z) = 1 - 0.9713 = 0.0287$ implies $z = 1.9$.

    b. $P(Z < -z) = 0.9713$ implies that $z = -1.9$.

19. a. $P(X < 60) = P\left(Z < \frac{60-50}{5}\right) = P(Z < 2) = 0.9772.$

    b. $P(X > 43) = P\left(Z > \frac{43-50}{5}\right) = P(Z > -1.4) = P(Z < 1.4) = 0.9192.$

    c. $P(46 < X < 58) = P\left(\frac{46-50}{5} < Z < \frac{58-50}{5}\right) = P(-0.8 < Z < 1.6) = P(Z < 1.6) -$

    $P(Z < -0.8) = 0.9452 - 0.2119 = 0.7333.$

21. $\mu = 20$ and $\sigma = 2.6$.

    a. $P(X > 22) = P\left(Z > \frac{22-20}{2.6}\right) = P(Z > 0.7692) = P(Z < -0.7692) \approx 0.2206.$

    b. $P(X < 18) = P\left(Z < \frac{18-20}{2.6}\right) = P(Z < -0.7692) \approx 0.2206.$

    c. $P(19 < X < 21) = P\left(\frac{19-20}{2.6} < Z < \frac{21-20}{2.6}\right) = P(-0.3846 < Z < 0.3846)$

    $$= P(Z < 0.3846) - P(Z < -0.3846) = 0.6480 - 0.3520 = 0.2960.$$

23. $\mu = 750$ and $\sigma = 75$.

    a. $P(X > 900) = P\left(Z > \frac{900-750}{75}\right) = P(Z > 2) = P(Z < -2) = 0.0228.$

    b. $P(X < 600) = P\left(Z < \frac{600-750}{75}\right) = P(Z < -2) = 0.0228.$

    c. $P(750 < X < 900) = P\left(Z < \frac{750-750}{75} < Z < \frac{900-750}{75}\right) = P(0 < Z < 2) = P(Z < 2) - P(Z < 0)$

    $$= 0.9772 - 0.5000 = 0.4772.$$

    d. $P(600 < X < 800) = P\left(\frac{600-750}{75} < Z < \frac{800-750}{75}\right) = P(-2 < Z < 0.667) = P(Z < 0.667) - P(Z < -2)$

    $$= 0.7486 - 0.0228 = 0.7258.$$

© 2011 Cengage Learning. All Rights Reserved. May not be scanned, copied or duplicated, or posted to a publicly accessible website, in whole or in part.

25. $\mu = 100$ and $\sigma = 15$.

    a. $P(X > 140) = P\left(Z > \frac{140-100}{15}\right) = P(Z > 2.667) = P(Z < -2.667) = 0.0038.$

    b. $P(X > 120) = P\left(Z > \frac{120-100}{15}\right) = P(Z > 1.33) = P(Z < -1.33) = 0.0918.$

    c. $P(100 < X < 120) = P\left(\frac{100-100}{15} < Z < \frac{120-100}{15}\right) = P(0 < Z < 1.333) = P(Z < 0) - P(Z < 1.333)$

$$= 0.9082 - 0.5000 = 0.4082.$$

    d. $P(X < 90) = P\left(Z < \frac{90-100}{15}\right) = P(Z < -0.667) = 0.2514.$

27. Here $\mu = 675$ and $\sigma = 50$.

$$P(650 < X < 750) = P\left(\frac{650-675}{50} < Z < \frac{750-675}{50}\right) = P(-0.5 < Z < 1.5) = P(Z < 1.5) - P(Z < -0.5)$$

$$= 0.9332 - 0.3085 = 0.6247.$$

29. Here $\mu = 22$ and $\sigma = 4$. $P(X < 12) = P\left(Z < \frac{12-22}{4}\right) = P(Z < -2.5) = 0.0062$, or 0.62%.

31. $\mu = 70$ and $\sigma = 10$. To find the cutoff point for an A, we solve $P(Y < y) = 0.85$ for y, obtaining

$$P(Y < y) = P\left(Z < \frac{y-70}{10}\right) = 0.85 \text{ and so } \frac{y-70}{10} = 1.04 \text{ and } y = 80.4 \approx 80.$$

    For a B: $P(Y < y) = P\left(Z < \frac{y-70}{10}\right) = 0.60$, so $\frac{y-70}{10} = 0.25$ and $y \approx 73.$

    For a C: $P(Y < y) = P\left(Z < \frac{y-70}{10}\right) = 0.2$, so $\frac{y-70}{10} = -0.84$ and $y \approx 62.$

    For an D: $P(Y < y) = P\left(Z < \frac{y-70}{10}\right) = 0.05$, so $\frac{y-70}{10} = -1.64$ and $y \approx 54.$

CHAPTER 10   Concept Review   page 685

1. a. experiment

    b. outcomes, sample points

    c. sample space

    d. 0, 1

3. a. continuous

    b. nonnegative, 1, $\int_a^b f(x)\,dx$

5. a. $\dfrac{x_1 + x_2 + \cdots + x_n}{n}$

    b. central tendency, X

    c. $x_1 p_1 + x_2 p_2 + \cdots + x_n p_n$, X, probabilities

    d. $\int_a^b x f(x)\,dx$, probability density, $[a, b]$

7. $\int_a^b (x - \mu)^2 f(x)\,dx$, $\sqrt{\operatorname{Var}(X)}$

© 2011 Cengage Learning. All Rights Reserved. May not be scanned, copied or duplicated, or posted to a publicly accessible website, in whole or in part.

CHAPTER 10   Review   page 686

1. $f(x) = \frac{1}{28}(2x + 3) \ge 0$ on $[0, 4]$ and $\int_0^4 \frac{1}{28}(2x + 3)\,dx = \frac{1}{28}(x^2 + 3x)\Big|_0^4 = \frac{1}{28}(16 + 12) = 1$.

3. $f(x) = \frac{1}{4} > 0$ on $[7, 11]$ and $\int_7^{11} \frac{1}{4}\,dx = \frac{1}{4}x\Big|_7^{11} = \frac{1}{4}(11 - 7) = 1$.

5. $\int_0^9 kx^2\,dx = \frac{1}{3}kx^3\Big|_0^9 = \frac{1}{3}k(729) = 1$, so $k = \frac{1}{243}$.

7. $\int_1^3 kx^{-2}\,dx = \frac{2}{3}k = 1$, so $k = \frac{3}{2}$.

9. a. $P(X \le 4) = \int_2^4 \frac{2}{21}x\,dx = \frac{1}{21}x^2\Big|_2^4 = \frac{16}{21} - \frac{4}{21} = \frac{12}{21} = \frac{4}{7}$.

    b. $P(X = 4) = \int_4^4 \frac{2}{21}x\,dx = \frac{1}{21}x^2\Big|_4^4 = 0$.

    c. $P(3 \le X \le 4) = \int_3^4 \frac{2}{21}x\,dx = \frac{1}{21}x^2\Big|_3^4 = \frac{16}{21} - \frac{9}{21} = \frac{7}{21} = \frac{1}{3}$.

11. a. $P(1 \le X \le 3) = \frac{3}{16}\int_1^3 x^{1/2}\,dx = \frac{1}{8}x^{3/2}\Big|_1^3 = \frac{1}{8}\left(3\sqrt{3} - 1\right) \approx 0.52$.

    b. $P(X \le 3) = \frac{3}{16}\int_0^3 x^{1/2}\,dx = \frac{1}{8}x^{3/2}\Big|_0^3 = \frac{1}{8}\left(3\sqrt{3} - 0\right) \approx 0.65$.

    c. $P(X = 2) = \frac{3}{16}\int_2^2 x^{1/2}\,dx = 0$.

13. $\mu = \frac{1}{5}\int_2^7 x\,dx = \frac{1}{10}x^2\Big|_2^7 = \frac{1}{10}(49 - 4) = \frac{9}{2}$, so

$\text{Var}(X) = \frac{1}{5}\int_2^7 x^2\,dx - \left(\frac{9}{2}\right)^2 = \frac{1}{15}x^3\Big|_2^7 - (4.5)^2 = \frac{1}{15}(343 - 8) - (4.5)^2 \approx 2.083$ and $\sigma = \sqrt{2.083} \approx 1.44$.

15. $\mu = \frac{1}{4}\int_{-1}^1 x(3x^2 + 1)\,dx = \frac{1}{4}\int_{-1}^1 (3x^3 + x)\,dx = \frac{1}{4}\left(\frac{3}{4}x^4 + \frac{1}{2}x^2\right)\Big|_{-1}^1 = 0$, so

$\text{Var}(X) = \int_{-1}^1 x^2(3x^2 - 1)\,dx - 0 = \frac{1}{4}\int_{-1}^1 (3x^4 + x^2)\,dx = \frac{1}{4}\left(\frac{3}{5}x^5 + \frac{1}{3}x^3\right)\Big|_{-1}^1 = \frac{7}{15}$ and $\sigma = \sqrt{\frac{7}{15}} \approx 0.6831$.

17. $P(Z < 2.24) \approx 0.9875$.

19. $P(0.24 \le Z \le 1.28) = P(Z \le 1.28) - P(Z \le 0.24) \approx 0.8997 - 0.5948 = 0.3049$.

21. f is nonnegative in D and

$$\iint_D f(x, y)\,dA = \int_0^8 \int_0^4 \frac{1}{64}x^{1/2}y^{1/3}\,dx\,dy = \frac{1}{64}\int_0^8 \frac{2}{3}x^{3/2}y^{1/3}\Big|_0^4\,dy = \frac{1}{64} \cdot \frac{2}{3} \cdot 8 \int_0^8 y^{1/3}\,dy = \frac{1}{12} \cdot \frac{3}{4}y^{4/3}\Big|_0^8$$

$$= \frac{1}{16} \cdot 16 = 1$$

This shows that f is a joint probability density function on D.

23. a. $P(X \le 84) = P\left(Z \le \frac{84 - 80}{8}\right) = P(Z \le 0.5) = 0.6915$.

    b. $P(X \ge 70) = P\left(\frac{70 - 80}{8}\right) = P(Z \ge -1.25) = P(Z \le 1.25) = 0.8944$.

    c. $P(75 \le X \le 85) = P\left(\frac{75 - 80}{8} \le Z \le \frac{85 - 80}{8}\right) = P(-0.625 \le Z \le 0.625) = P(Z \le 0.625) - P(Z \le -0.625)$

$$\approx 0.7341 - 0.2660 = 0.4681.$$

© 2011 Cengage Learning. All Rights Reserved. May not be scanned, copied or duplicated, or posted to a publicly accessible website, in whole or in part.

25. a. $P(T > 6) = \int_6^\infty \frac{1}{4}e^{-t/4}\, dt = \lim_{b \to \infty} \int_6^b \left(-e^{-t/4}\right) dt = \lim_{b \to \infty} \left(-e^{-t/4}\right)\Big|_6^b = \lim_{b \to \infty} \left(-e^{-b/4}\right) + e^{-6/4} \approx 0.2231.$

b. $P(T < 2) = \int_0^2 \frac{1}{4}e^{-t/4}\, dt = -e^{-t/4}\big|_0^2 = -e^{-1/2} + e^0 \approx -0.60653 + 1 = 0.39345.$

c. $\mu = \int_0^\infty \frac{1}{4}te^{-t/4}dt = \lim_{b \to \infty} \int_0^b \frac{1}{4}te^{-t/4}dt = \lim_{b \to \infty} 4\left(-\frac{1}{4}t - 1\right)e^{-t/4}\Big|_0^b = 0 + 4 = 4.$

| CHAPTER 10 | Before Moving On... | page 687 |

1. $\int_0^1 k\left(x + 2x^2\right) = x\left(\frac{1}{2}x^2 + \frac{2}{3}x^3\right)\Big|_0^1 = k\left(\frac{1}{2} + \frac{2}{3}\right) = \frac{7}{6}k$, so $k = \frac{6}{7}$.

2. $\int_5^{10} 0.1e^{-0.1x}\, dx = -e^{-0.1x}\big|_5^{10} = -e^{-1} + e^{-0.5} \approx 0.2387.$

3. $\int_1^9 x \cdot \frac{1}{4}x^{-1/2}\, dx = \frac{1}{4}\int_1^9 x^{1/2}\, dx = \frac{1}{4}\left(\frac{2}{3}x^{3/2}\right)\Big|_1^9 = \frac{13}{3}$, so

$\quad \text{Var}(X) = \int_1^9 \left(x - \frac{13}{3}\right)^2 \cdot \frac{1}{4}x^{-1/2}\, dx = \frac{1}{4}\int_1^9\left(x^{3/2} - \frac{26}{3}x^{1/2} + \frac{169}{9}x^{-1/2}\right) dx = \frac{1}{4}\left(\frac{2}{5}x^{5/2} - \frac{52}{9}x^{3/2} + \frac{338}{9}x^{1/2}\right)\Big|_1^9$

$\quad \approx 5.42$

and $\sigma = \sqrt{5.42} = 2.33.$

4. a. $P(X < 50) = P\left(Z < \frac{50-40}{4}\right) = P(Z < 2.5) \approx 0.9938.$

b. $P(X > 35) = P\left(Z < \frac{35-40}{4}\right) = P(Z > -1.25) = P(Z < 1.25) \approx 0.8944.$

c. $P(30 < X < 50) = P\left(\frac{30-40}{4} < Z < \frac{50-40}{4}\right) = P(-2.5 < Z < 2.5) = P(2.5) - P(-2.5) \approx 0.9876.$

© 2011 Cengage Learning. All Rights Reserved. May not be scanned, copied or duplicated, or posted to a publicly accessible website, in whole or in part.

11 TAYLOR POLYNOMIALS AND INFINITE SERIES

11.1 Problem-Solving Tips

Be sure you are familiar with the definition of a Taylor polynomial and the error bound in a Taylor polynomial approximation.

1. The nth Taylor polynomial of f at $x = a$, where f and its first n derivatives are defined at $x = a$, is the polynomial

$$P_n(x) = f(a) + f'(a)(x - a) + \frac{f''(a)}{2!}(x - a)^2 + \cdots + \frac{f^{(n)}(a)}{n!}(x - a)^n$$

which coincides with $f(x)$, $f'(x)$, ..., and $f^{(n)}(x)$ at $x = a$; that is,

$$P_n(a) = f(a), \, P_n'(a) = f'(a), \ldots, P_n^{(n)}(a) = f^{(n)}(a)$$

2. If a function f has derivatives up to order $N + 1$ on an interval I containing the number a, then for each x in I,

$$|R_N(x)| \leq \frac{M}{(N+1)!}|x - a|^{N+1}$$

where M is any number such that $\left|f^{(N+1)}(t)\right| \leq M$ for all t lying between a and x.

11.1 Concept Questions   page 698

1. a. A Taylor polynomial of a function f at $x = a$ is a polynomial of the form

$P_n(x) = f(a) + f'(a)(x - a) + \cdots + \frac{f^{(n)}(a)}{n!}(x - a)^n$, where n is a positive integer.

b. $a_i = \frac{f^{(i)}}{i!}$.

11.1 Taylor Polynomials   page 698

1. $f(x) = e^{-x}$, $f'(x) = -e^{-x}$, $f''(x) = e^{-x}$, and $f'''(x) = -e^{-x}$, so $f(0) = 1$, $f'(0) = -1$, $f''(0) = 1$, and
$f'''(0) = -1$. Therefore, $P_1(x) = f(0) + f'(0)x = 1 - x$, $P_2(x) = f(0) = f'(0)x + \frac{f''(0)}{2!}x^2 = 1 - x + \frac{1}{2}x^2$,
and $P_3(x) = f(0) + f'(0)x + \frac{f''(0)}{2!}x^2 + \frac{f'''(0)}{3!}x^3 = 1 - x + \frac{1}{2}x^2 - \frac{1}{6}x^3$.

3. $f(x) = \frac{1}{x+1} = (x+1)^{-1}$, $f'(x) = -(x+1)^{-2}$, $f''(x) = 2(x+1)^{-3}$, and $f'''(x) = -6(x+1)^{-4}$,
so $f(0) = 1$, $f'(0) = -1$, $f''(0) = 2$, and $f'''(0) = -6$. Therefore, $P_1(x) = 1 - x$,
$P_2(x) = 1 - x + \frac{2}{2!}x^2 = 1 - x + x^2$, and $P_3(x) = 1 - x + \frac{2}{2!}x^2 - \frac{6}{3!}x^3 = 1 - x + x^2 - x^3$.

275

© 2011 Cengage Learning. All Rights Reserved. May not be scanned, copied or duplicated, or posted to a publicly accessible website, in whole or in part.

5. $f(x) = \dfrac{1}{x} = x^{-1}$, $f'(x) = -x^{-2} = -\dfrac{1}{x^2}$, $f''(x) = (-1)(-2)x^{-3} = \dfrac{2}{x^3}$, and $f'''(x) = -\dfrac{6}{x^4}$, so $f(1) = 1$,

$f'(1) = -1$, $f''(1) = 2$, and $f'''(1) = -6$. Therefore, $P_1(x) = f(1) + f'(x)(x-1) = 1 - (x-1)$,

$P_2(x) = f(1) + f'(1)(x-1) + \dfrac{f''(1)}{2!}(x-1)^2 = 1 - (x-1) + (x-1)^2$, and

$P_3(x) = f(1) + f'(1)(x-1) + \dfrac{f''(1)}{2!}(x-1)^2 + \dfrac{f'''(1)}{3!}(x-1)^3 = 1 - (x-1) + (x-1)^2 - (x-1)^3$.

7. $f(x) = (1-x)^{1/2}$, $f'(x) = -\frac{1}{2}(1-x)^{-1/2}$, $f''(x) = -\frac{1}{4}(1-x)^{-3/2}$, and $f'''(x) = -\frac{3}{8}(1-x)^{-5/2}$,

so $f(0) = 1$, $f'(0) = -\frac{1}{2}$, $f''(0) = -\frac{1}{4}$, and $f'''(0) = -\frac{3}{8}$. Therefore, $P_1 = 1 - \frac{1}{2}x$,

$P_2(x) = 1 - \frac{1}{2}x + \dfrac{-1/4}{2!}x^2 = 1 - \frac{1}{2}x - \frac{1}{8}x^2$, and $P_3 = 1 - \frac{1}{2}x - \frac{1}{8}x^2 + \dfrac{-3/8}{3!}x^3 = 1 - \frac{1}{2}x - \frac{1}{8}x^2 - \frac{1}{16}x^3$.

9. $f(x) = \ln(1-x)$, $f'(x) = -\dfrac{1}{1-x} = -(1-x)^{-1}$, $f''(x) = -(1-x)^{-2}$, and $f'''(x) = -2(1-x)^{-3}$,

so $f(0) = 0$, $f'(0) = -1$, $f''(0) = -1$, and $f'''(0) = -2$. Therefore, $P_1(x) = -x$,

$P_2(x) = -x + \dfrac{(-1)}{2!}x^2 = -x - \frac{1}{2}x^2$, and $P_3(x) = -x - \frac{1}{2}x^2 + \dfrac{(-2)}{3!}x^3 = -x - \frac{1}{2}x^2 - \frac{1}{3}x^3$.

11. $f(x) = x^4$, $f'(x) = 4x^3$, and $f''(x) = 12x^2$, so $f(2) = 16$, $f'(2) = 32$, and $f''(2) = 48$. Therefore,

$P_2(x) = 16 + 32(x-2) + \dfrac{48}{2!}(x-2)^2 = 16 + 32(x-2) + 24(x-2)^2$.

13. $f(x) = \ln x$, $f'(x) = \dfrac{1}{x}$, $f''(x) = -\dfrac{1}{x^2}$, $f'''(x) = \dfrac{2}{x^3}$, and $f^{(4)}(x) = -\dfrac{6}{x^4}$, so

$f(1) = 0$, $f'(1) = 1$, $f''(1) = -1$, $f'''(1) = 2$, and $f^{(4)}(1) = -6$. Therefore,

$P_4(x) = (x-1) + \dfrac{(-1)}{2!}(x-1)^2 + \dfrac{2}{3!}(x-1)^3 + \dfrac{(-6)}{4!}(x-1)^4 = (x-1) - \frac{1}{2}(x-1)^2 + \frac{1}{3}(x-1)^3 - \frac{1}{4}(x-1)^4$.

15. $f(x) = e^x$, so $f'(x) = f''(x) = f'''(x) = f^{(4)}(x) = e^x$ and $f(1) = f'(1) = f''(1) = f'''(x) = f^{(4)}(x) = e$.

Therefore,

$$P_4(x) = e + e(x-1) + \dfrac{e}{2!}(x-1)^2 + \dfrac{e}{3!}(x-1)^3 + \dfrac{e}{4!}(x-1)^4$$

$$= e + e(x-1) + \frac{1}{2}e(x-1)^2 + \frac{1}{6}e(x-1)^3 + \frac{1}{24}e(x-1)^4.$$

17. $f(x) = (1-x)^{1/3}$, $f'(x) = -\frac{1}{3}(1-x)^{-2/3}$, $f''(x) = -\frac{2}{9}(1-x)^{-5/3}$, and

$f'''(x) = -\frac{10}{27}(1-x)^{-8/3}$, so $f(0) = 1$, $f'(0) = -\frac{1}{3}$, $f''(0) = -\frac{2}{9}$, and $f'''(0) = -\frac{10}{27}$. Therefore,

$P_3(x) = 1 - \frac{1}{3}x - \dfrac{2/9}{2!}x^2 - \dfrac{10/27}{3!}x^3 = 1 - \frac{1}{3}x - \frac{1}{9}x^2 - \frac{5}{81}x^3$.

19. $f(x) = \dfrac{1}{2x+3} = (2x+3)^{-1}$, $f'(x) = -2(2x+3)^{-2}$, $f''(x) = 8(2x+3)^{-3}$, and

$f'''(x) = -48(2x+3)^{-4}$, so $f(0) = \frac{1}{3}$, $f'(0) = -\frac{2}{9}$, $f''(0) = \frac{8}{27}$, and $f'''(0) = -\frac{16}{27}$. Therefore,

$P_3(x) = \frac{1}{3} - \frac{2}{9}x + \dfrac{\left(\frac{8}{27}\right)}{2!}x^2 + \dfrac{\left(-\frac{16}{27}\right)}{3!}x^3 = \frac{1}{3} - \frac{2}{9}x + \frac{4}{27}x^2 - \frac{8}{81}x^3$.

© 2011 Cengage Learning. All Rights Reserved. May not be scanned, copied or duplicated, or posted to a publicly accessible website, in whole or in part.

21. $f(x) = \dfrac{1}{1+x} = (1+x)^{-1}$, $f'(x) = -(1+x)^{-2}$, $f''(x) = 2(1+x)^{-3}$, $f'''(x) = -6(1+x)^{-4}$,

and $f^{(4)}(x) = 24(1+x)^{-5}$, so $f(0) = 1$, $f'(0) = -1$, $f''(0) = 2$, $f'''(0) = -6$, and $f^{(4)}(0) = 24$.

Therefore, $P_n(x) = 1 - x + x^2 - x^3 + \cdots + (-1)^n x^n$. In particular, $P_4(x) = 1 - x + x^2 - x^3 + x^4$, and so

$P_4(0.1) = 1 - 0.1 + 0.01 - 0.001 + 0.0001 = 0.9091$. The actual value is $f(0.1) = \dfrac{1}{1.1} = 0.909090\dots.$

23. $f(x) = e^{-x/2}$, $f'(x) = -\frac{1}{2}e^{-x/2}$, $f''(x) = \frac{1}{4}e^{-x/2}$, $f'''(x) = -\frac{1}{8}e^{-x/2}$, and $f^{(4)}(x) = \frac{1}{16}e^{-x/2}$,

so $f(0) = 1$, $f'(0) = -\frac{1}{2}$, $f''(0) = \frac{1}{4}$, $f'''(0) = -\frac{1}{8}$, and $f^{(4)}(0) = \frac{1}{16}$. Therefore,

$P_4(x) = 1 - \frac{1}{2}x + \frac{1}{8}x^2 - \frac{1}{48}x^3 + \frac{1}{384}x^4$. To estimate $e^{-0.1}$, we take $x = 0.02$ to obtain

$f(0.2) = e^{-0.1} \approx P_4(0.2) = 1 - \frac{1}{2}(0.2) + \frac{1}{8}(0.2)^2 - \frac{1}{48}(0.2)^3 + \frac{1}{384}(0.2)^4 \approx 0.90484$.

25. $f(x) = x^{1/2}$, $f'(x) = \frac{1}{2}x^{-1/2}$, and $f''(x) = -\frac{1}{4}x^{-3/2}$, so $f(16) = 4$, $f'(16) = \frac{1}{8}$, and $f''(16) = -\frac{1}{256}$.

Therefore, $P_2(x) = 4 + \frac{1}{8}(x - 16) - \frac{1}{512}(x - 16)^2$, so $\sqrt{15.6} = f(15.6) \approx P(15.6) \approx 3.9496875$.

27. $f(x) = \ln(x + 1)$. We first calculate $P_3(x) = x - \frac{1}{2}x^2 + \frac{1}{3}x^3$. Next,

$\int_0^{1/2} \ln(x + 1)\, dx = \int_0^{1/2} \left(x - \frac{1}{2}x^2 + \frac{1}{3}x^3 \right) dx = \left(\frac{1}{2}x^2 - \frac{1}{6}x^3 + \frac{1}{12}x^4 \right)\Big|_0^{1/2} = \frac{1}{8} - \frac{1}{48} + \frac{1}{192} \approx 0.109$. The actual

value is $\int_0^{1/2} \ln(x + 1)\, dx = (x + 1)\ln(x + 1) - (x + 1)|_0^{1/2} = \frac{3}{2}\ln\frac{3}{2} - \frac{3}{2} + 1 = \frac{3}{2}\ln\frac{3}{2} - \frac{1}{2} \approx 0.108$.

29. $f(x) = x^{1/2}$, $f'(x) = \frac{1}{2}x^{-1/2}$, $f''(x) = -\frac{1}{4}x^{-3/2}$, $f'''(x) = \frac{3}{8}x^{-5/2}$, and $f(4) = 2$,

so $f'(4) = \frac{1}{4}$ and $f''(4) = -\frac{1}{32}$. Therefore, $P_2(x) = 2 + \frac{1}{4}(x - 4) - \frac{1}{64}(x - 4)^2$.

$\sqrt{4.06} = f(4.06) = P_2(4.06) = 2 + \frac{1}{4}(0.06) - \frac{1}{64}(0.06)^2 = 2.01494375$. To find a bound for the approximation,

observe that $f'''(x) = \dfrac{3}{8x^{5/2}}$ is decreasing on $[4, 4.06]$, so f''' attains its absolute maximum value at $x = 4$. Thus,

$|f'''(x)| \le \dfrac{3}{8(4)^{5/2}} = \dfrac{3}{256}$, and so $|R_2(4.06)| \le \dfrac{3/256}{3!}(0.06)^3 \approx 0.000000421$.

31. $f(x) = (1-x)^{-1}$, $f'(x) = (1-x)^{-2}$, $f''(x) = 2(1-x)^{-3}$, $f'''(x) = 6(1-x)^{-4}$, $f^{(4)}(x) = 24(1-x)^{-5}$,

and $f^{(5)}(x) = 120(1-x)^{-6} > 0$, so $f(0) = 1$, $f'(0) = 1$, $f''(0) = 2$, $f'''(0) = 6$, and $f^{(4)}(0) = 24$.

Therefore, $P_3(x) = 1 + x + x^2 + x^3$ and so $f(0.2) \approx P_3(0.2) = 1 + 0.2 + 0.04 + 0.008 = 1.248$. Because

$f^{(5)}(x) > 0$, $f^{(4)}(x)$ is increasing and its maximum is $M = \dfrac{24}{(1 - 0.2)^5} \approx 73.2421875$. Thus, the error bound is

$\dfrac{73.2421875}{4!}(0.2)^4 \approx 0.0048828125$. The exact value is $f(0.2) = \dfrac{1}{1 - 0.2} = 1.25$.

33. $f(x) = \ln x$, $f'(x) = \dfrac{1}{x}$, $f''(x) = -\dfrac{1}{x^2}$, $f'''(x) = \dfrac{2}{x^3}$, and $f^{(4)} = -\dfrac{6}{x^4}$, so $f(1) = 0$, $f'(1) = 1$, and

$f''(1) = -1$. Therefore, $P_2(x) = (x - 1) - \frac{1}{2}(x - 1)^2$ and so $f(1.1) = (1.1 - 1) - \frac{1}{2}(0.1)^2 = 0.095$.

Because $f^{(4)}(x) < 0$, $f'''(x)$ is decreasing and we may take $M = \dfrac{2}{x^3}\Big|_{x=1} = 2$. Thus, the error bound is

$\dfrac{2}{3!}(0.1)^3 \approx 0.00033$.

© 2011 Cengage Learning. All Rights Reserved. May not be scanned, copied or duplicated, or posted to a publicly accessible website, in whole or in part.

35. a. $f(x) = e^{-x}$, $f'(x) = -e^{-x}$, $f''(x) = e^{-x}$, $f'''(x) = -e^{-x}$, and $f^{(4)}(x) = e^{-x}$. Because
$f^{(5)}(x) = -e^{-x} < 0$, $f^{(4)}(x)$ is decreasing and the maximum value of $f^{(4)}(x)$ is attained at $x = 0$.
Thus, we may take $M = \left| f^{(4)}(0) \right| = e^0 = 1$ and so a bound on the approximation of $f(x)$ by $P_3(x)$ is
$\frac{1}{4!}(1 - 0)^4 = 0.04167$.

b. See Exercise 1 for a calculation of $P_3(x)$. We have

$$\int_0^1 e^{-x} dx \approx \int_0^1 P_3(x) dx = \int_0^1 \left(1 - x + \tfrac{1}{2}x^2 - \tfrac{1}{6}x^3\right) dx = \left(x - \tfrac{1}{2}x^2 + \tfrac{1}{6}x^3 - \tfrac{1}{24}x^4\right)\Big|_0^1 = 1 - \tfrac{1}{2} + \tfrac{1}{6} - \tfrac{1}{24}$$

$$= 0.625.$$

c. Using the result from part (a), we find that the error bound on the approximation in part (b) is
$\int_0^1 0.04167 \, dx = 0.04167x\big|_0^1 = 0.04167$.

d. Because $\int_0^1 e^{-x} dx = -e^{-x}\big|_0^1 = -e^{-1} + 1 \approx 0.632121$, the actual error is $0.632121 - 0.625 = 0.007121$.

37. The required area is $A = \int_0^{0.5} e^{-x^2/2} dx$. Now $e^u = 1 + u + \tfrac{1}{2}u^2$ is the second Taylor polynomial at $u = 0$,
so $e^{-x^2/2} \approx 1 - \tfrac{1}{2}x^2 + \tfrac{1}{2}\left(-\tfrac{1}{2}x^2\right)^2 = 1 - \tfrac{1}{2}x^2 + \tfrac{1}{8}x^4$ is the fourth Taylor polynomial at $x = 0$. Thus,
$A = \int_0^{0.5} e^{-x^2/2} dx = \int_0^{0.5} \left(1 - \tfrac{1}{2}x^2 + \tfrac{1}{8}x^4\right) dx = \left(x - \tfrac{1}{6}x^3 + \tfrac{1}{40}x^5\right)\Big|_0^{0.5} = 0.5 - \tfrac{1}{6}(0.5)^3 + \tfrac{1}{40}(0.5)^5 \approx 0.47995$.

39. The percentage of the nonfarm work force in the service industries t decades from now is given by
$P(t) = \int 6e^{1/(2t+1)} dt$. Let $u = \dfrac{1}{2t+1}$, so that $2t + 1 = \dfrac{1}{u}$ and $t = \dfrac{1}{2}\left(\dfrac{1}{u} - 1\right)$. Therefore, $dt = -\dfrac{1}{2u^2} du$, so
we have

$$P(u) = 6 \int e^u \left(\frac{du}{-2u^2}\right) = -3 \int \frac{e^u}{u^2} du = -3 \int \frac{1}{u^2}\left(1 + u + \frac{u^2}{2} + \frac{u^3}{6} + \frac{u^4}{24}\right) du$$

$$= -3 \int \left(\frac{1}{u^2} + \frac{1}{u} + \frac{1}{2} + \frac{u}{6} + \frac{u^2}{24}\right) du = -3\left(-\frac{1}{u} + \ln u + \frac{u}{2} + \frac{u^2}{12} + \frac{u^3}{72}\right) + C.$$

Therefore, $P(t) = -3\left[-(2t+1) + \ln\left(\dfrac{1}{2t+1}\right) + \dfrac{1}{2(2t+1)} + \dfrac{1}{12(2t+1)^2} + \dfrac{1}{72(2t+1)^3}\right] + C$. Using
the condition $P(0) = 30$, we find $P(0) = -3\left(-1 + \ln 1 + \tfrac{1}{2} + \tfrac{1}{12} + \tfrac{1}{72}\right) = 30$, so $C = 28.79$. Thus,
$P(t) = -3\left[-2(2t+1) + \ln\left(\dfrac{1}{2t+1}\right) + \dfrac{1}{2(2t+1)} + \dfrac{1}{12(2t+1)^2} + \dfrac{1}{72(2t+1)^3}\right] + 28.79$. Two decades
from now, the percentage of nonfarm workers will be $P(2) \approx -3\left(-5 + \ln\tfrac{1}{5} + \tfrac{1}{10} + \tfrac{1}{300} + \tfrac{1}{9000}\right) + 28.79 \approx 48.31$,
or approximately 48%.

© 2011 Cengage Learning. All Rights Reserved. May not be scanned, copied or duplicated, or posted to a publicly accessible website, in whole or in part.

41. The average enrollment between $t = 0$ and $t = 2$ is given by

$$A = \frac{1}{2} \int_0^2 \left(-\frac{20{,}000}{\sqrt{1 + 0.2t}} + 21{,}000 \right) dt = -10{,}000 \int_0^2 (1 + 0.2t)^{-1/2} \, dt + \left[\frac{1}{2} (21{,}000t) \right]_0^2$$

$$= -10{,}000 \int_0^2 (1 + 0.2t)^{-1/2} \, dt + 21{,}000.$$

To evaluate the integral, we approximate the integrand its the second Taylor polynomial at $t = 0$. We compute $f(t) = (1 + 0.2t)^{-1/2}$, $f'(t) = -0.1(1 + 0.2t)^{-3/2}$, and $f''(t) = 0.03(1 + 0.2t)^{-5/2}$, so $f(0) = 1$, $f'(0) = -0.1$, and $f''(0) = 0.03$. Thus, $P_2(t) = 1 - 0.1t + 0.015t^2$ and so

$$A \approx -10{,}000 \int_0^2 (1 - 0.1t + 0.015t^2) \, dt + 21{,}000 = -10{,}000 \left(t - 0.05t^2 + 0.005t^3 \right) \Big|_0^2 + 21{,}000$$

$$= -10{,}000 (2 - 0.2 + 0.04) + 21{,}000 = 2600.$$

Thus, the average enrollment between $t = 0$ and $t = 2$ is approximately 2600.

43. False. The fourth Taylor polynomial $P_4(x)$ coincides with $f(x)$ for all values of x in this case. To see this, note that

$$R_4(x) = \frac{f^{(5)}(c)}{5!} (x - a)^5 = 0 \text{ since } f^{(5)}(x) = 0.$$

45. True. Because $f^{(n+1)}(x) = 0$ for all values of x, it follows that $R_n(x) = \dfrac{f^{(n+1)}(c)}{(n+1)!} (x - a)^{n+1} = 0$ for all values of x.

11.2   Concept Questions   page 707

1. a. A sequence is a function whose domain is the set of positive integers. For example, $2, 4, 6, \ldots, 2n, \ldots$ is a sequence.

   b. A sequence $\{a_n\}$ is convergent if there exists a number L such that a_n can be made as close to L as we please by taking n large enough. For example, the sequence $\left\{ \frac{1}{n} \right\}$ converges to 0.

   c. A sequence is divergent if it is not convergent, as defined in part (b). For example, the sequence $1, 2, 3, 4, \ldots, n, \ldots$ is divergent.

   d. If a sequence $\{a_n\}$ is convergent, as defined in part (b), then the number L is called the *limit* of the sequence.

11.2   Infinite Sequences   page 707

1. $1, 2, 4, 8, 16$.

3. $\dfrac{1-1}{1+1}, \dfrac{2-1}{2+1}, \dfrac{3-1}{3+1}, \dfrac{4-1}{4+1}, \dfrac{5-1}{5+1}$; that is, $0, \frac{1}{3}, \frac{2}{4}, \frac{3}{5}, \frac{4}{6}$.

5. $a_1 = \dfrac{2^0}{1!} = 1$, $a_2 = \dfrac{2^1}{2!} = 1$, $a_3 = \dfrac{2^2}{3!} = \dfrac{4}{6}$, $a_4 = \dfrac{2^3}{4!} = \dfrac{8}{24}$, and $a_5 = \dfrac{2^4}{5!} = \dfrac{16}{120}$.

7. $a_1 = e$, $a_2 = \dfrac{e^2}{8}$, $a_3 = \dfrac{e^3}{27}$, $a_4 = \dfrac{e^4}{64}$, and $a_5 = \dfrac{e^5}{125}$.

© 2011 Cengage Learning. All Rights Reserved. May not be scanned, copied or duplicated, or posted to a publicly accessible website, in whole or in part.

9. $a_1 = \dfrac{3 - 1 + 1}{2 + 1} = 1$, $a_2 = \dfrac{12 - 2 + 1}{8 + 1} = \dfrac{11}{9}$, $a_3 = \dfrac{27 - 3 + 1}{18 + 1} = \dfrac{25}{19}$, $a_4 = \dfrac{48 - 4 + 1}{32 + 1} = \dfrac{15}{11}$, and

$a_5 = \dfrac{75 - 5 + 1}{50 + 1} = \dfrac{71}{51}$.

11. $a_n = 3n - 2$.   **13.** $a_n = \dfrac{1}{n^3}$.   **15.** $a_n = 2\left(\dfrac{4}{5}\right)^{n-1} = \dfrac{2^{2n-1}}{5^{n-1}}$.

17. $a_n = \left(-\dfrac{1}{2}\right)^{n-1} = \dfrac{(-1)^{n+1}}{2^{n-1}}$.   **19.** $a_n = \dfrac{n}{(n+1)(n+2)}$.   **21.** $a_n = \dfrac{e^{n-1}}{(n-1)!}$.

23. $\left\{\dfrac{2n}{n+1}\right\}$

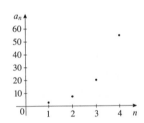

25. $\left\{\sqrt{n}\right\}$

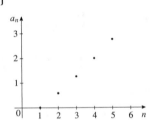

27. $\{e^n\}$

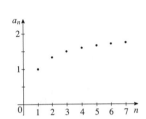

29. $\left\{n - \sqrt{n}\right\}$

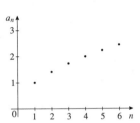

31. $\lim\limits_{n\to\infty} \dfrac{n+1}{2n} = \lim\limits_{n\to\infty} \dfrac{1 + \frac{1}{n}}{2} = \dfrac{1}{2}$.   **33.** $\lim\limits_{n\to\infty} \dfrac{(-1)^n}{\sqrt{n}} = 0$.

35. $\lim\limits_{n\to\infty} \dfrac{\sqrt{n} - 1}{\sqrt{n} + 1} = \lim\limits_{n\to\infty} \dfrac{1 - \dfrac{1}{\sqrt{n}}}{1 + \dfrac{1}{\sqrt{n}}} = 1$.

37. $\lim\limits_{n\to\infty} \dfrac{2n^4 - 1}{n^3 + 2n + 1} = \lim\limits_{n\to\infty} \dfrac{2n - \dfrac{1}{n^3}}{1 + \dfrac{2}{n^2} + \dfrac{1}{n^3}}$ does not exist.

39. $\lim\limits_{n\to\infty}\left(2 - \dfrac{1}{2^n}\right) = \lim\limits_{n\to\infty} 2 - \lim\limits_{n\to\infty} \dfrac{1}{2^n} = 2 - 0 = 2$.

41. $\lim\limits_{n\to\infty} \dfrac{2^n}{3^n} = \lim\limits_{n\to\infty}\left(\dfrac{2}{3}\right)^n = 0$.

© 2011 Cengage Learning. All Rights Reserved. May not be scanned, copied or duplicated, or posted to a publicly accessible website, in whole or in part.

43. $\lim\limits_{n\to\infty} \dfrac{n}{\sqrt{2n^2+3}} = \lim\limits_{n\to\infty} \dfrac{1}{\sqrt{2+\dfrac{3}{n^2}}} = \dfrac{1}{\sqrt{2}} = \dfrac{\sqrt{2}}{2}.$

45. a. $a_1 = 0.015$, $a_{10} = 0.14027$, $a_{100} \approx 0.77939$, $a_{1000} \approx 0.999999727$.

   b. $\lim\limits_{n\to\infty} a_n = \lim\limits_{n\to\infty} \left[1 - (0.985)^n \right] = 1 - \lim\limits_{n\to\infty} (0.985)^n = 1 - 0 = 1.$

47. a. The result follows from the compound interest formula.

   b. $a_n = 100\,(1.05)^n$.

   c. $a_{24} = 100\,(1.05)^{24} \approx 112.72$, or \$112.72. Therefore, the accumulated amount at the end of two years is \$112.72.

49. True. $\lim\limits_{n\to\infty} a_n b_n = \left(\lim\limits_{n\to\infty} a_n \right)\left(\lim\limits_{n\to\infty} b_n \right) = L \cdot 0 = 0.$

51. False. Let $\lim\limits_{n\to\infty} a_n = (-1)^n$ and $b_n = 1 + \dfrac{1}{n}$. Then $|a_n| \le 1$ for all n and $\lim\limits_{n\to\infty} b_n = \lim\limits_{n\to\infty} \left(1 + \dfrac{1}{n} \right) = 1$, but

$\lim\limits_{n\to\infty} a_n b_n = \lim\limits_{n\to\infty} (-1)^n \left(1 + \dfrac{1}{n} \right)$ does not exist because if n is very large, $a_n b_n$ is close to 1 for even n but close to -1 for odd n. Therefore, $a_n b_n$ does approach a specific number as n approaches infinity.

11.3   Problem-Solving Tips

1. Be sure you know the difference between a series and a sequence. A *sequence* is a *succession of terms*, whereas a *series* is a *sum of terms*.

2. Consider the infinite series $\sum_{n=1}^{\infty} a_n = a_1 + a_2 + a_3 + \cdots$ and its sequence of partial sums $\{S_n\}$. If $\lim\limits_{N\to\infty} S_N = S$, then the series is said to converge to S, the *sum* of the series. If $\{S_N\}$ diverges, then the series $\sum_{n=1}^{\infty} a_n$ diverges.

3. The geometric series $\sum_{n=0}^{\infty} ar^n = a + ar + ar^2 + \cdots$ converges to $\dfrac{a}{1-r}$ if $|r| < 1$ and diverges if $|r| \ge 1$.

11.3   Concept Questions   page 718

1. a. A sequence $\{a_n\}$ is an ordered array of numbers $a_1, a_2, a_3, \ldots, a_n, \ldots$, whereas a series is an expression of the form $\sum\limits_{n=1}^{\infty} a_n = a_1 + a_2 + \cdots + a_n + \cdots$.

   b. A sequence $\{a_n\}$ is convergent if a_n approaches a number L as n approaches infinity. A series $\sum_{n=1}^{\infty} a_n$ is convergent if its sequence of partial sums $\{S_N\}$ is convergent, where $S_N = \sum_{n=1}^{N} a_n$.

   c. A sequence is divergent if it is not convergent. The corresponding statement about series is also true.

   d. If a sequence $\{a_n\}$ is convergent, then the number L to which the terms of the sequence approach as n tends to infinity is called the limit. If $\{S_N\}$ converges and has limit S, then S is the sum of $\sum_{n=1}^{\infty} a_n$.

© 2011 Cengage Learning. All Rights Reserved. May not be scanned, copied or duplicated, or posted to a publicly accessible website, in whole or in part.

11.3   Infinite Series   page 718

1. $S_1 = -2$, $S_2 = -2 + 4 = 2$, $S_3 = -2 + 4 - 8 = -6$, $S_4 = -6 + 16 = 10$, and so $\lim\limits_{N \to \infty} S_N$ does not exist and the

  series $\sum\limits_{n=1}^{\infty} (-2)^n$ is divergent.

3. $S_N = \sum\limits_{n=1}^{N} \dfrac{1}{n^2 + 3n + 2} = \sum\limits_{n=1}^{N} \left(\dfrac{1}{n+1} - \dfrac{1}{n+2} \right) = \left(\dfrac{1}{2} - \dfrac{1}{3} \right) + \left(\dfrac{1}{3} - \dfrac{1}{4} \right) + \cdots + \left(\dfrac{1}{N} - \dfrac{1}{N+1} \right) +$

  $\left(\dfrac{1}{N+1} - \dfrac{1}{N+2} \right) = \dfrac{1}{2} - \dfrac{1}{N+2}$. Thus, $\lim\limits_{N \to \infty} S_N = \lim\limits_{N \to \infty} \left(\dfrac{1}{2} - \dfrac{1}{N+2} \right) = \dfrac{1}{2}$ and $\sum\limits_{n=1}^{\infty} \dfrac{1}{n^2 + 3n + 2} = \dfrac{1}{2}$.

5. $\sum\limits_{n=0}^{\infty} \left(\dfrac{1}{3} \right)^n = 1 + \dfrac{1}{3} + \left(\dfrac{1}{3} \right)^2 + \cdots = \dfrac{1}{1 - \frac{1}{3}} = \dfrac{1}{\frac{2}{3}} = \dfrac{3}{2}$.

7. This is a geometric series with $r = 1.01 > 1$ and so it diverges.

9. $\sum\limits_{n=0}^{\infty} \dfrac{(-2)^n}{3^n} = \sum\limits_{n=0}^{\infty} \left(-\dfrac{2}{3} \right)^n$ is a geometric series with $|r| = \left| -\dfrac{2}{3} \right| = \dfrac{2}{3} < 1$, and so it converges. The sum is

  $\dfrac{1}{1 - \left(-\frac{2}{3} \right)} = \dfrac{1}{1 + \frac{2}{3}} = \dfrac{3}{5}$.

11. $\sum\limits_{n=0}^{\infty} \dfrac{2^n}{3^{n+2}} = \sum\limits_{n=0}^{\infty} \dfrac{1}{9} \left(\dfrac{2}{3} \right)^n = \dfrac{1}{9} \cdot \dfrac{1}{1 - \frac{2}{3}} = \dfrac{1}{9} \cdot 3 = \dfrac{1}{3}$.

13. $\sum\limits_{n=0}^{\infty} e^{-0.2n} = \sum\limits_{n=0}^{\infty} \left(e^{-0.2} \right)^n = \sum\limits_{n=0}^{\infty} \left(\dfrac{1}{e^{0.2}} \right)^n = \dfrac{1}{1 - \frac{1}{e^{0.2}}} \approx 5.52$.

15. Here $r = -\dfrac{3}{\pi}$ and because $|r| = \left| -\dfrac{3}{\pi} \right| < 1$, the series converges. In fact,

  $\sum\limits_{n=0}^{\infty} \left(-\dfrac{3}{\pi} \right)^n = \dfrac{1}{1 - \left(-\frac{3}{\pi} \right)} = \dfrac{1}{1 + \frac{3}{\pi}} = \dfrac{1}{\frac{\pi+3}{\pi}} = \dfrac{\pi}{\pi + 3}$.

17. $8 + 4 + \dfrac{1}{2} + \dfrac{1}{4} + \dfrac{1}{8} + \cdots = 8 + 4 + \dfrac{1}{2} \left[1 + \dfrac{1}{2} + \left(\dfrac{1}{2} \right)^2 + \cdots \right] = 12 + \dfrac{1}{2} \cdot \dfrac{1}{1 - (1/2)} = 13$.

19. $3 - \dfrac{1}{3} + \dfrac{1}{9} - \dfrac{1}{27} + \cdots = 3 - \dfrac{1}{3} \left[1 - \dfrac{1}{3} + \left(-\dfrac{1}{3} \right)^2 - \cdots \right] = 3 - \dfrac{1}{3} \cdot \dfrac{1}{1 + (1/3)} = 3 - \dfrac{1}{3} \cdot \dfrac{3}{4} = 3 - \dfrac{1}{4} = \dfrac{11}{4}$.

21. $\sum\limits_{n=0}^{\infty} \dfrac{3 + 2^n}{3^n} = 3 \sum\limits_{n=0}^{\infty} \left(\dfrac{1}{3^n} \right) + \sum\limits_{n=0}^{\infty} \left(\dfrac{2}{3} \right)^n = 3 \cdot \dfrac{1}{1 - \frac{1}{3}} + \dfrac{1}{1 - \frac{2}{3}} = \dfrac{9}{2} + 3 = \dfrac{15}{2}$.

23. $\sum\limits_{n=0}^{\infty} \dfrac{3 \cdot 2^n + 4^n}{3^n} = \sum\limits_{n=0}^{\infty} \left[3 \left(\dfrac{2}{3} \right)^n + \left(\dfrac{4}{3} \right)^n \right]$ diverges because the series $\sum\limits_{n=0}^{\infty} \left(\dfrac{4}{3} \right)^n$ is geometric with constant ratio

  $\dfrac{4}{3} > 1$.

© 2011 Cengage Learning. All Rights Reserved. May not be scanned, copied or duplicated, or posted to a publicly accessible website, in whole or in part.

25. Because $\dfrac{e}{\pi} < 1$, $\displaystyle\sum_{n=1}^{\infty} \left(\dfrac{e}{\pi}\right)^n$ converges. Also, $\dfrac{\pi}{e^2} < 1$ and so $\displaystyle\sum_{n=1}^{\infty} \left(\dfrac{\pi}{e^2}\right)^n$ converges. Therefore,

$$\sum_{n=1}^{\infty}\left[\left(\dfrac{e}{\pi}\right)^n + \left(\dfrac{\pi}{e^2}\right)^n\right] = \sum_{n=1}^{\infty}\left(\dfrac{e}{\pi}\right)^n + \sum_{n=1}^{\infty}\left(\dfrac{\pi}{e^2}\right)^n = \dfrac{e}{\pi}\cdot\dfrac{1}{1-\dfrac{e}{\pi}} + \dfrac{\pi}{e^2}\cdot\dfrac{1}{1-\dfrac{\pi}{e^2}} = \dfrac{e}{\pi - e} + \dfrac{\pi}{e^2 - \pi}$$

$$= \dfrac{e^3 - \pi e + \pi^2 - \pi e}{(\pi - e)\left(e^2 - \pi\right)} = \dfrac{e^3 - 2\pi e + \pi^2}{(\pi - e)\left(e^2 - \pi\right)}.$$

27. $0.3333\ldots = 0.3 + 0.03 + 0.003 + \cdots = \dfrac{3}{10} + \dfrac{3}{10^2} + \dfrac{3}{10^3} + \cdots = \dfrac{3}{10}\displaystyle\sum_{n=0}^{\infty}\left(\dfrac{1}{10}\right)^n = \dfrac{\frac{3}{10}}{1 - \frac{1}{10}} = \dfrac{3}{10}\cdot\dfrac{10}{9} = \dfrac{1}{3}.$

29. $1.213213213\ldots = 1 + 0.213\left[1 + (0.001) + (0.001)^2 + \cdots\right] = 1 + \dfrac{213}{1000}\cdot\dfrac{1}{1 - 0.001}$

$$= 1 + \dfrac{213}{1000}\cdot\dfrac{1000}{999} = 1 + \dfrac{213}{999} = \dfrac{1212}{999} = \dfrac{404}{333}.$$

31. $\displaystyle\sum_{n=0}^{\infty}(-x)^n$ is a geometric series with $r = (-x)^n$. Therefore, it converges provided $|-x| < 1$; that is, $|x| < 1$, or

$-1 < x < 1$. In that case, $\displaystyle\sum_{n=0}^{\infty}(-x)^n = \dfrac{1}{1 - (-x)} = \dfrac{1}{1 + x}.$

33. $\displaystyle\sum_{n=0}^{\infty} 2^n (x-1)^n$ is a geometric series with $r = 2(x-1)$ and so it converges provided

$|2(x-1)| < 1$; that is, $|x - 1| < \frac{1}{2}$, $-\frac{1}{2} < x - 1 < \frac{1}{2}$, or $\frac{1}{2} < x < \frac{3}{2}$. In that case,

$$\sum_{n=1}^{\infty} 2^n (x-1)^n = 2(x-1)\cdot\dfrac{1}{1 - 2(x-1)} = 2(x-1)\cdot\dfrac{1}{3 - 2x} = \dfrac{2(x-1)}{3 - 2x}.$$

35. The additional spending generated by the proposed tax cut will be

$$(0.91)(30) + (0.91)^2(30) + (0.91)^3(30) + \cdots = (0.91)(30)\left[1 + 0.91 + (0.91)^2 + (0.91)^3 + \cdots\right]$$
$$= 27.3\left[\dfrac{1}{1 - 0.91}\right] \approx 303.33, \text{ or approximately } \$303 \text{ billion.}$$

37. $p = \frac{1}{6} + \left(\frac{1}{6}\right)\left(\frac{5}{6}\right)^2 + \frac{1}{6}\left(\frac{5}{6}\right)^4 + \cdots = \frac{1}{6}\left\{1 + \left(\frac{5}{6}\right)^2 + \left[\left(\frac{5}{6}\right)^2\right]^2 + \cdots\right\}\left(r = \left(\frac{5}{6}\right)^2\right) = \frac{1}{6}\cdot\dfrac{1}{1 - (5/6)^2}$

$= \frac{1}{6}\cdot\dfrac{1}{1 - (25/36)} = \frac{1}{6}\cdot\dfrac{36}{36 - 25} = \dfrac{6}{11}.$

39. The required upper bound is no larger than

$B = a_1 + aNa_1 + aNa_2 + aNa_3 + \cdots = a_1 + aNa_1 + aN(ra_1) + aN(r^2 a_1) + \cdots$. This is a geometric series:

$B = a_1 + aNa_1(1 + r + r^2 + \cdots) = a_1 + \dfrac{aNa_1}{1 - r} = \dfrac{a_1 - a_1(1 + aN - b) + aNa_1}{1 - (1 + aN - b)} = \dfrac{a_1 b}{b - aN}.$

41. $A = Pe^{-r} + Pe^{-2r} + Pe^{-3r} + \cdots = Pe^{-r}\left(1 + e^{-r} + e^{-2r} + \cdots\right) = \dfrac{Pe^{-r}}{1 - e^{-r}} = \dfrac{P}{e^r - 1}.$

© 2011 Cengage Learning. All Rights Reserved. May not be scanned, copied or duplicated, or posted to a publicly accessible website, in whole or in part.

43. a. The residual concentration is $R = Ce^{-kt} + Ce^{-2kt} + Ce^{-3kt} + \cdots = Ce^{-kt}\left(1 + e^{-kt} + e^{-2kt} + \cdots\right) = \dfrac{Ce^{-kt}}{1 - e^{-kt}}$.

b. $C + R \leq S$ implies that $C + \dfrac{Ce^{-kt}}{1 - e^{-kt}} \leq S$, so $\dfrac{C - Ce^{-kt} + Ce^{-kt}}{1 - e^{-kt}} \leq S$, $\dfrac{C}{1 - e^{-kt}} \leq S$, $1 - e^{-kt} \geq \dfrac{C}{S}$,

$e^{-kt} \leq 1 - \dfrac{C}{S} = \dfrac{S - C}{S}$, $-kt \leq \ln \dfrac{S - C}{S}$, $kt \geq -\ln \dfrac{S - C}{S} = \ln \dfrac{S}{S - C}$, and so $t \geq \dfrac{1}{k} \ln \dfrac{S}{S - C}$. Therefore,

the minimum time between dosages should be $\dfrac{1}{k} \ln \dfrac{S}{S - C}$ hours.

45. False. Take $a_n = -n$ and $b_n = n$. Then both $\sum_{n=0}^{\infty} a_n$ and $\sum_{n=0}^{\infty} b_n$ diverge, but $\sum_{n=0}^{\infty} (a_n + b_n) = \sum_{n=0}^{\infty} 0$ clearly converges to zero.

47. True. This is a convergent geometric series with common ratio $|r|$. If $|r| < 1$, then $\sum_{n=0}^{\infty} |r|^n = \dfrac{1}{1-|r|}$.

11.4   Concept Questions   page 728

1. a.

$f(1) = 1, f(2) = \dfrac{1}{2}, f(3) = \dfrac{1}{3}, \ldots, f(n) = \dfrac{1}{n}, \ldots$

b. $\int_1^n \frac{1}{x}\,dx$ is the area under the graph of $f(x) = \frac{1}{x}$ on the interval $[1, n]$. From the figure, we see that

$$\int_1^n \frac{1}{x}\,dx = \lim_{b \to \infty} \int_1^b \frac{1}{x}\,dx = \lim_{b \to \infty} \ln b = \infty.$$

c. $\int_1^\infty \frac{1}{x}\,dx = \lim_{b \to \infty} \int_1^b \frac{1}{x}\,dx = \lim_{b \to \infty} \ln b = \infty$. Because

$$S_{n-1} \geq \int_1^n \frac{1}{x}\,dx \text{ and } \lim_{n \to \infty} \int_1^n \frac{1}{x}\,dx = \infty, \text{ and}$$

$$\lim_{n \to \infty} S_{n-1} = \infty \text{ as well.}$$

3. a. Both $\sum b_n$ and $\sum c_n$ must be convergent. In fact, since $b_n \geq 0$ for all n, and $b_n + c_n \leq a_n$, we must have $c_n \leq a_n$ for all n and so $\sum c_n$ converges by the comparison test. Similarly, $\sum b_n$ must converge.

b. At least one of the two series $\sum b_n$ and $\sum c_n$ must be divergent. Otherwise, suppose both $\sum b_n$ and $\sum c_n$ converge. Because $a_n \leq b_n + c_n$, then $\sum a_n$ must converge, a contradiction.

11.4   Series with Positive Terms   page 729

1. Here $a_n = \dfrac{n}{n+1}$, and because $\lim_{n \to \infty} \dfrac{n}{n+1} = \lim_{n \to \infty} \dfrac{1}{1 + \frac{1}{n}} = 1 \neq 0$, the series is divergent.

3. Here $a_n = \dfrac{2n}{3n+1}$, and because $\lim_{n \to \infty} \dfrac{2n}{3n+1} = \lim_{n \to \infty} \dfrac{2}{3 + \frac{1}{n}} = \dfrac{2}{3} \neq 0$, the series is divergent.

5. Here $a_n = 2(1.5)^n$, and because $\lim_{n \to \infty} 2(1.5)^n$ does not exist, the series is divergent.

7. Here $a_n = \dfrac{1}{2 + 3^{-n}}$, and because $\lim_{n \to \infty} \dfrac{1}{2 + 3^{-n}} = \dfrac{1}{2} \neq 0$, the series diverges.

9. Here $a_n = \left(-\dfrac{\pi}{3}\right)^n = (-1)^n \left(\dfrac{\pi}{3}\right)^n$, and because $\dfrac{\pi}{3} > 1$, we see that $\lim_{n \to \infty} a_n$ does not exist, and so the series diverges.

© 2011 Cengage Learning. All Rights Reserved. May not be scanned, copied or duplicated, or posted to a publicly accessible website, in whole or in part.

11. Take $f(x) = \dfrac{1}{x+1}$ and note that f is nonnegative and decreasing for $x \geq 1$. Next,

$$\int_1^\infty f(x)\,dx = \lim_{b\to\infty} \int_1^b \frac{dx}{x+1} = \lim_{b\to\infty} \ln(x+1)|_1^b = \lim_{b\to\infty} [\ln(b+1) - \ln 2] = \infty, \text{ and so the series is divergent.}$$

13. First note that $f(x) = \dfrac{x}{2x^2+1}$ is nonnegative and decreasing for $x \geq 1$. Next,

$$\int_1^\infty f(x)\,dx = \lim_{b\to\infty} \int_1^b \frac{x\,dx}{2x^2+1} = \lim_{b\to\infty} \tfrac{1}{4}\ln(2x^2+1)\Big|_1^b = \lim_{b\to\infty} \left[\tfrac{1}{4}\ln(2b^2+1) - \tfrac{1}{4}\ln 3\right] = \infty, \text{ and so the}$$

series is divergent.

15. Let $f(x) = xe^{-x}$. Studying f', we see that f is nonnegative and decreasing for $x \geq 1$. Integrating by parts, we

have $\int_1^\infty f(x)\,dx = \lim_{b\to\infty} \int_1^b xe^{-x}\,dx = \lim_{b\to\infty} \left[-(x+1)e^{-x}\right]_1^b = \lim_{b\to\infty} \left[-(b+1)e^{-b} + 2e^{-b}\right] = 0$. (We can

verify that be^{-b} approaches 0 graphically.). Therefore, the integral converges.

17. Let $f(x) = \dfrac{x}{(x^2+1)^{3/2}}$. Then f is nonnegative and decreasing on $[1, \infty)$. Integrating, we find

$$\int_1^\infty f(x)\,dx = \lim_{b\to\infty} \int_1^b x\,(x^2+1)^{-3/2}\,dx = \lim_{b\to\infty} \left(-\frac{1}{\sqrt{x^2+1}}\right)\Big|_1^b = \lim_{b\to\infty} \left(-\frac{1}{\sqrt{b^2+1}} + \frac{1}{\sqrt{2}}\right) = \frac{1}{\sqrt{2}}, \text{ and so}$$

the series converges.

19. Let $f(x) = \dfrac{1}{x\ln^3 x}$, which is nonnegative and decreasing on $[9, \infty)$. Substituting $u = \ln x$, we have

$$\int_9^\infty f(x)\,dx = \lim_{b\to\infty} \int_9^b \frac{dx}{x(\ln x)^3} = \lim_{b\to\infty} \int_9^\infty \frac{(\ln x)^{-3}}{x}\,dx = \lim_{b\to\infty} \left(-\frac{1}{2(\ln x)^2}\right)\Big|_9^b$$

$$= \lim_{b\to\infty} \left[-\frac{1}{2(\ln b)^2} + \frac{1}{2(\ln 9)^2}\right] = \frac{1}{2(\ln 9)^2}, \text{ and so the series is convergent.}$$

21. Here $p = 3 > 1$, and so the series is convergent.

23. Here $p = 1.01 > 1$, and so the series is convergent.

25. Here $p = \pi > 1$, and so the series is convergent.

27. Let $a_n = \dfrac{1}{2n^2+1}$. Then $a_n = \dfrac{1}{2n^2+1} < \dfrac{1}{2n^2} < \dfrac{1}{n^2} = b_n$. Because $\sum b_n$ is a convergent p-series, $\sum a_n$

converges by the comparison test.

29. Let $a_n = \dfrac{1}{n-2}$. Then $a_n = \dfrac{1}{n-2} > \dfrac{1}{n} = b_n$. Because $\sum b_n$ is divergent, the comparison test implies that $\sum a_n$

diverges as well.

31. Let $a_n = \dfrac{1}{\sqrt{n^2-1}}$. Then $a_n = \dfrac{1}{\sqrt{n^2-1}} > \dfrac{1}{\sqrt{n^2}} = \dfrac{1}{n} = b_n$. Because the harmonic series $\sum b_n$ diverges, the

comparison test shows that $\sum a_n$ also diverges.

33. Let $a_n = \dfrac{2^n}{3^n+1} < \dfrac{2^n}{3^n} = \left(\dfrac{2}{3}\right)^n = b_n$. Because $\sum_{n=0}^\infty \left(\dfrac{2}{3}\right)^n$ is a convergent geometric series, $\sum_{n=0}^\infty a_n$ converges

by the comparison test.

© 2011 Cengage Learning. All Rights Reserved. May not be scanned, copied or duplicated, or posted to a publicly accessible website, in whole or in part.

35. Let $a_n = \dfrac{\ln n}{n}$. Because $\ln n > 1$ for $n > 3$, we see that $a_n = \dfrac{\ln n}{n} > \dfrac{1}{n} = b_n$. But $\sum b_n$ is the divergent harmonic

series, and the comparison test implies that $\displaystyle\sum_{n=2}^{\infty} \dfrac{\ln n}{n}$ diverges as well.

37. We use the Integral Test with $f(x) = \dfrac{1}{\sqrt{x+1}}$, which is nonnegative and decreasing on $(0, \infty)$. We find

$\displaystyle\lim_{b \to \infty} \left[2(b+1)^{1/2} - 2 \right] = \infty$, and so the series diverges.

39. Let $a_n = \dfrac{1}{n\sqrt{n^2+1}}$. Observe that if n is large, $n^2 + 1$ behaves like n^2. This suggests we compare $\sum a_n$ with $\sum b_n$,

where $b_n = \dfrac{1}{n\sqrt{n^2}} = \dfrac{1}{n^2}$. Now $\sum b_n$ is a convergent p-series with $p = 2$. Because $0 < \dfrac{1}{n\sqrt{n^2+1}} < \dfrac{1}{n^2}$, $\sum a_n$

converges by the comparison test.

41. $\displaystyle\sum_{n=1}^{\infty} \dfrac{1}{n\sqrt{n}} = \sum_{n=1}^{\infty} \dfrac{1}{n^{3/2}}$ is a convergent p-series with $p = \tfrac{3}{2}$. Next, $\displaystyle\sum_{n=1}^{\infty} \dfrac{2}{n^2}$ is a convergent p-series with $p = 2$.

Therefore, $\displaystyle\sum_{n=1}^{\infty} \left(\dfrac{1}{n\sqrt{n}} + \dfrac{2}{n^2} \right)$ is convergent.

43. We know that $\ln n > 1$ if $n > 3$, and so $a_n = \dfrac{\ln n}{\sqrt{n}} > \dfrac{1}{\sqrt{n}} = b_n$ if $n > 3$. But $\displaystyle\sum_{n=2}^{\infty} b_n = \sum_{n=2}^{\infty} \dfrac{1}{n^{1/2}}$ is a divergent

p-series with $p = \tfrac{1}{2}$. Therefore, by the comparison test, the given series is divergent.

45. We use the integral test with $f(x) = \dfrac{1}{x(\ln x)^2}$. Observe that n is nonnegative and decreasing on $(2, \infty)$.

Substituting $u = \ln x$, we have $\displaystyle\int_2^{\infty} f(x)\,dx = \int_2^{\infty} \dfrac{dx}{x(\ln x)^2} = \lim_{b \to \infty} \left(-\dfrac{1}{\ln x} \right)\Big|_2^b = \dfrac{1}{\ln 2}$, and so the given series

converges.

47. We use the comparison test. Because $a_n = \dfrac{1}{\sqrt{n}+4} > \dfrac{1}{6\sqrt{n}} = b_n$ for $n \geq 1$ and $\sum b_n$ diverges, we conclude that

$\sum a_n$ is also divergent.

49. We use the Integral Test with $f(x) = \dfrac{1}{x(\ln x)^p}$. Observe that f is nonnegative

and decreasing on $(2, \infty)$. Next, substituting $u = \ln x$, we compute

$\displaystyle\int_2^{\infty} f(x)\,dx = \lim_{b \to \infty} \int_2^b \dfrac{(\ln x)^{-p}}{x}\,dx = \lim_{b \to \infty} \left[\dfrac{(\ln x)^{-p+1}}{1-p} \right]_2^b = \lim_{b \to \infty} \left[\dfrac{(\ln b)^{1-p}}{1-p} - \dfrac{(\ln 2)^{1-p}}{1-p} \right] = \dfrac{(\ln 2)^{1-p}}{p-1}$ if

$p > 1$ and diverges if $p \leq 1$. Thus, $\displaystyle\sum_{n=2}^{\infty} \dfrac{1}{n(\ln n)^p}$ converges for $p > 1$. (See Exercise 45.)

© 2011 Cengage Learning. All Rights Reserved. May not be scanned, copied or duplicated, or posted to a publicly accessible website, in whole or in part.

51. Denoting the Nth partial sum of the series by S_N, we have

$$S_N = \sum_{n=1}^{N} \left(\frac{a}{n+1} - \frac{1}{n+2} \right) = \left(\frac{a}{2} - \frac{1}{3} \right) + \left(\frac{a}{3} - \frac{1}{4} \right) + \cdots + \left(\frac{a}{N+1} - \frac{1}{N+2} \right)$$

$$= \frac{a}{2} + \frac{a-1}{3} + \frac{a-1}{4} + \cdots + \frac{a-1}{N+1} - \frac{1}{N+2}.$$

If $a = 1$, then $S_N = \frac{1}{2} - \frac{1}{N+2}$ is the Nth partial sum of a telescoping series that converges to $\frac{1}{2}$. If $a \neq 0$, then S_N is the Nth partial sum of a series akin to the harmonic series $\sum_{n=1}^{\infty} 1/n$, and in this case, the series diverges. Therefore, the series converges only for $a = 1$.

53. $\displaystyle \int_{1}^{\infty} \frac{dx}{x^p} = \lim_{b \to \infty} \int_{1}^{b} x^{-p}\, dx = \lim_{b \to \infty} \left. \frac{x^{1-p}}{1-p} \right|_{1}^{b} = \lim_{b \to \infty} \left(\frac{b^{1-p}}{1-p} - \frac{1}{1-p} \right) = \frac{1}{p-1}$ if $p > 1$, and so the

series converges if $p > 1$. If $p < 1$, then $\displaystyle \int_{1}^{\infty} \frac{dx}{x^p} = \infty$ and so the series diverges. If $p = 1$, then we have

$\displaystyle \int_{1}^{\infty} \frac{dx}{x} = \lim_{b \to \infty} \int_{1}^{b} \frac{dx}{x} = \lim_{b \to \infty} \ln |x| \big|_{1}^{b} = \lim_{b \to \infty} (\ln b - \ln 1)$, and so the integral diverges in this case as well.

55. True. Compare it with the divergent harmonic series $\displaystyle \sum_{n=1}^{\infty} \frac{1}{n}$. In fact, if $\displaystyle \sum_{n=1}^{\infty} \frac{x}{n}$ converges for $x \neq 0$, then

$\displaystyle \sum_{n=1}^{\infty} \frac{x}{n} = x \sum_{n=1}^{\infty} \frac{1}{n}$ converges, a contradiction.

57. False. The harmonic series $\displaystyle \sum_{n=1}^{\infty} a_n$ with $a_n = \frac{1}{n}$ is divergent, but $\displaystyle \lim_{n \to \infty} a_n = 0$.

59. False. Let $a_n = \frac{1}{n^2}$ and $b_n = \frac{2}{n^2}$. Clearly $b_n \geq a_n$ for $n \geq 1$ but both $\sum a_n$ and $\sum b_n$ converge.

11.5 Concept Questions   page 738

1. a. A power series in x is an infinite series of the form $\sum_{n=0}^{\infty} a_n x^n$.

b. A power series in $x - a$ is an infinite series of the form $\sum_{n=0}^{\infty} a_n (x - a)^n$.

3. Because $x = \frac{3}{2}$ lies in the interval $(-2, 2)$, the series $\displaystyle \sum_{n=0}^{\infty} a_n x^n$ is convergent.

5. a. A Taylor series at $x = a$ is a series of the form $\displaystyle \sum_{n=0}^{\infty} \frac{f^{(n)}(a)}{n!} (x - a)^n$.

b. $f^{(n)}(c) = n!\, a_n$.

c. $f^{(5)}(1) = 5!\, a_5 = 5! \left[\dfrac{(-1)^5}{5+1} \right] = -\dfrac{5 \cdot 4 \cdot 3 \cdot 2 \cdot 1}{6} = -20.$

© 2011 Cengage Learning. All Rights Reserved. May not be scanned, copied or duplicated, or posted to a publicly accessible website, in whole or in part.

11.5   Power Series and Taylor Series   page 738

1. $R = \lim\limits_{n \to \infty} \left| \dfrac{a_n}{a_{n+1}} \right| = \lim\limits_{n \to \infty} 1 = 1$. Therefore, $R = 1$ and the interval of convergence is $(0, 2)$.

3. $R = \lim\limits_{n \to \infty} \left| \dfrac{a_n}{a_{n+1}} \right| = \lim\limits_{n \to \infty} \dfrac{n^2}{(n+1)^2} = \lim\limits_{n \to \infty} \dfrac{n^2}{n^2 + 2n + 1} = 1$, so the interval of convergence is $(-1, 1)$.

5. $R = \lim\limits_{n \to \infty} \left| \dfrac{a_n}{a_{n+1}} \right| = \lim\limits_{n \to \infty} \dfrac{\frac{1}{4^n}}{\frac{1}{4^{n+1}}} = \lim\limits_{n \to \infty} 4 = 4$, so the interval of convergence is $(-4, 4)$.

7. $R = \lim\limits_{n \to \infty} \left| \dfrac{a_n}{a_{n+1}} \right| = \lim\limits_{n \to \infty} \dfrac{1}{n!\, 2^n} (n+1)!\, 2^{n+1} = \lim\limits_{n \to \infty} 2(n+1) = \infty$, so $R = \infty$ and the interval of convergence is $(-\infty, \infty)$.

9. $R = \lim\limits_{n \to \infty} \left| \dfrac{a_n}{a_{n+1}} \right| = \lim\limits_{n \to \infty} \dfrac{\frac{n!}{2^n}}{\frac{(n+1)!}{2^{n+1}}} = \lim\limits_{n \to \infty} \dfrac{2}{n+1} = 0$, so the interval of convergence is $\{-2\}$.

11. $R = \lim\limits_{n \to \infty} \left| \dfrac{a_n}{a_{n+1}} \right| = \lim\limits_{n \to \infty} \dfrac{\frac{1}{(n+1)^2}}{\frac{1}{(n+2)^2}} = \lim\limits_{n \to \infty} \dfrac{n^2 + 4n + 4}{n^2 + 2n + 1} = 1$, so the interval of convergence is $(-4, -2)$.

13. $R = \lim\limits_{n \to \infty} \left| \dfrac{a_n}{a_{n+1}} \right| = \lim\limits_{n \to \infty} \dfrac{2n(n+2)!}{(n+1)!(2n+2)} = \infty$, so the interval of convergence is $(-\infty, \infty)$.

15. $R = \lim\limits_{n \to \infty} \left| \dfrac{a_n}{a_{n+1}} \right| = \lim\limits_{n \to \infty} \dfrac{\frac{n2^n}{n+1}}{\frac{(n+1)2^{n+1}}{n+2}} = \lim\limits_{n \to \infty} \dfrac{n(n+2)}{2(n+1)^2} = \lim\limits_{n \to \infty} \dfrac{n^2 + 2n}{2(n^2 + 2n + 1)} = \frac{1}{2}$, so the interval of convergence is $\left(-\frac{1}{2}, \frac{1}{2} \right)$.

17. $R = \lim\limits_{n \to \infty} \left| \dfrac{a_n}{a_{n+1}} \right| = \lim\limits_{n \to \infty} \dfrac{\frac{n!}{3^n}}{\frac{(n+1)!}{3^{n+1}}} = \lim\limits_{n \to \infty} \dfrac{3}{n+1} = 0$, so the interval of convergence is $\{-1\}$.

19. $R = \lim\limits_{n \to \infty} \left| \dfrac{a_n}{a_{n+1}} \right| = \lim\limits_{n \to \infty} \dfrac{n^3 (3^{n+1})}{3^n (n+1)^3} = \lim\limits_{n \to \infty} \dfrac{3}{\left(\frac{n+1}{n} \right)^3} = 3$, so the interval of convergence is $(0, 6)$.

21. $f(x) = x^{-1}$, $f'(x) = -x^{-2}$, $f''(x) = 2x^{-3}$, $f'''(x) = -3 \cdot 2x^{-4}$, $\ldots$, $f^{(n)}(x) = (-1)^n n!\, x^{-n-1}$.
Therefore, $f(1) = 1$, $f'(1) = -1$, $f''(x) = 2$, $f'''(1) = -3!$, $\ldots$, $f^{(n)}(1) = (-1)^n n!$, and so
$$f(x) = 1 - (x-1) + \frac{2}{2!}(x-1)^2 + \cdots + \frac{(-1)^n n!}{n!}(x-1)^n + \cdots = \sum_{n=0}^{\infty} (-1)^n (x-1)^n. \text{ For this series, } R = 1$$
and the interval of convergence is $(0, 2)$.

© 2011 Cengage Learning. All Rights Reserved. May not be scanned, copied or duplicated, or posted to a publicly accessible website, in whole or in part.

23. $f(x) = (x+1)^{-1}$, $f'(x) = -(x+1)^{-2}$, $f''(x) = 2(x+1)^{-3}$, $f'''(x) = -3!(x+1)^{-4}, \ldots,$,

$f^{(n)}(x) = (-1)^n n! (x+1)^{-n-1}$. Therefore, $f(2) = \dfrac{1}{3}$, $f'(2) = -\dfrac{1}{3^2}$, $f''(2) = \dfrac{2}{3^3}, \ldots, f^{(n)} = \dfrac{(-1)^n n!}{3^{n+1}}$, and so

$f(x) = \dfrac{1}{3} - \dfrac{1}{3^2}(x-2) + \dfrac{1}{3^3}(x-2)^2 + \cdots + \dfrac{(-1)^n}{3^{n+1}}(x-2)^n + \cdots = \displaystyle\sum_{n=0}^{\infty} (-1)^n \dfrac{(x-2)^n}{3^{n+1}}$. For this series,

$R = 3$, and the interval of convergence is $(-1, 5)$.

25. $f(x) = (1-x)^{-1}$, $f'(x) = (1-x)^{-2}$, $f''(x) = 2(1-x)^{-3}$,

$f'''(x) = 6(1-x)^{-4}, \ldots, f^{(n)}(x) = n!(1-x)^{-n-1}$. Therefore, $f(2) = -1$,

$f'(2) = 1$, $f''(2) = -2$, $f'''(2) = 6, \ldots, f^{(n)}(2) = (-1)^{n+1} n!$. Thus,

$f(x) = -1 + (x-2) - (x-2)^2 + \cdots + (-1)^{n+1}(x-2)^n + \cdots = \displaystyle\sum_{n=0}^{\infty} (-1)^{n+1}(x-2)^n$. For this series, $R = 1$

and the interval of convergence is $(1, 3)$.

27. $f(x) = x^{1/2}$, $f'(x) = \frac{1}{2}x^{-1/2}$, $f''(x) = -\frac{1}{4}x^{-3/2}$, $f'''(x) = \frac{3}{8}x^{-5/2}$, $f^{(4)}(x) = -\frac{3\cdot5}{16}x^{-7/2}, \ldots$. Thus,

$f(1) = 1$, $f'(1) = \dfrac{1}{2}$, $f''(1) = -\dfrac{1}{2^2}$, $f'''(1) = \dfrac{1\cdot3}{2^3}, \ldots, f^{(n)}(1) = (-1)^{n+1}\dfrac{1\cdot3\cdot5\cdots\cdots(2n-3)}{2^n}$ for $n \geq 2$.

Thus, $f(x) = 1 + \frac{1}{2}(x-1) + \displaystyle\sum_{n=2}^{\infty} (-1)^{n+1} \dfrac{1\cdot3\cdot5\cdots\cdots(2n-3)}{n!\,2^n}(x-1)^n$. For this series, $R = 1$ and the interval

of convergence is $(0, 2)$.

29. $f(x) = e^{2x}$, $f'(x) = 2e^{2x}$, $f''(x) = 2^2 e^{2x}, \ldots, f^{(n)}(x) = 2^n e^{2x}$. $f(0) = 1$, $f'(0) = 2$,

$f''(0) = 2^2, \ldots, f^{(n)}(0) = 2^n$. Therefore, $f(x) = 1 + 2x + 2^2 x^2 + \cdots + \dfrac{2^n}{n!}x^n + \cdots = \displaystyle\sum_{n=0}^{\infty} \dfrac{2^n}{n!}x^n$. For this

series, $R = \infty$ and the interval of convergence is $(-\infty, \infty)$.

31. $f(x) = (x+1)^{-1/2}$, $f'(x) = -\frac{1}{2}(x+1)^{-3/2}$, $f''(x) = \frac{3}{4}(x+1)^{-5/2}$, $f'''(x) = -\dfrac{1\cdot3\cdot5}{2^3}x^{-7/2}, \ldots,$

$f^{(n)}(x) = (-1)^n \dfrac{1\cdot3\cdot5\cdots\cdots(2n-1)}{2^n}x^{-(2n+1)/2}$. Thus, $\displaystyle\sum_{n=0}^{\infty} (-1)^n \dfrac{1\cdot3\cdot5\cdots\cdots(2n-1)}{n!\,2^n}x^n$. For this series,

$R = 1$ and the interval of convergence is $(-1, 1)$.

33. For a Taylor series, $S_N(x) = P_N(x)$ and $P_N(x) = f(x) - R_N(x)$. Therefore,

$\displaystyle\lim_{N\to\infty} S_N(x) = \lim_{N\to\infty} P_N(x) = \lim_{N\to\infty} \left[f(x) - R_N(x) \right] = f(x) - \lim_{N\to\infty} R_N(x)$. In other words, for fixed x, the

sequence of partial sums of the Taylor series of f converges to f if and only if $\displaystyle\lim_{N\to\infty} R_N(x) = 0$.

35. True. This follows from Theorem 9.

© 2011 Cengage Learning. All Rights Reserved. May not be scanned, copied or duplicated, or posted to a publicly accessible website, in whole or in part.

11.6   Concept Questions   page 745

1. $f(x) = \dfrac{x}{1 + 2x} = x\left[\dfrac{1}{1 - (-2x)}\right]$.

3. $f(x) = \dfrac{2x}{4 + 3x^2} = 2x\left[\dfrac{1}{4 + 3x^2}\right] = 2x \cdot \dfrac{1}{4\left[1 - \left(-\frac{3}{4}x^2\right)\right]}$.

11.6   More on Taylor Series   page 745

1. $f(x) = \dfrac{1}{1 - x} = \dfrac{1}{-1 - (x - 2)} = -\dfrac{1}{1 + (x - 2)}$. Now use the fact that

$\dfrac{1}{1 + u} = 1 - u + u^2 - u^3 + \cdots = \sum_{n=0}^{\infty} (-1)^n u^n$ for $|u| < 1$ with $u = x - 2$ to obtain

$f(x) = \dfrac{1}{1 - x} = \sum_{n=0}^{\infty} (-1)^{n+1} (x - 2)^n$ with an interval of convergence of $(1, 3)$.

3. Let $u = 3x$ in the series $\dfrac{1}{1 + u} = 1 - u + u^2 - u^3 + \cdots = \sum_{n=0}^{\infty} (-1)^n u^n$, where $|u| < 1$. We have

$f(x) = \dfrac{1}{1 + 3x} = 1 - 3x + (3x)^2 - (3x)^3 + \cdots = \sum_{n=0}^{\infty} (-1)^n 3^n x^n$ with $|3x| < 1$, so the interval of convergence

is $\left(-\frac{1}{3}, \frac{1}{3}\right)$.

5. $f(x) = \dfrac{1}{4 - 3x} = \dfrac{1}{4\left(1 - \frac{3}{4}x\right)} = \dfrac{1}{4}\sum_{n=0}^{\infty}\left(\dfrac{3x}{4}\right)^n = \sum_{n=0}^{\infty}\dfrac{3^n}{4^{n+1}}x^n$. We must have $\left|\dfrac{3x}{4}\right| < 1$, so the interval of

convergence is $\left(-\frac{4}{3}, \frac{4}{3}\right)$.

7. $f(x) = \dfrac{1}{1 - x^2} = 1 + \left(x^2\right) + \left(x^2\right)^2 + \left(x^2\right)^3 + \cdots = 1 + x^2 + x^4 + x^6 + \cdots = \sum_{n=0}^{\infty} x^{2n}$. The interval of

convergence is $(-1, 1)$.

9. $f(x) = e^{-x} = 1 + (-x) + \dfrac{(-x)^2}{2!} + \dfrac{(-x)^3}{3!} + \cdots = 1 - x + \dfrac{x^2}{2!} - \dfrac{x^3}{3!} + \cdots = \sum_{n=0}^{\infty} (-1)^n \dfrac{x^n}{n!}$. The interval of

convergence is $(-\infty, \infty)$.

11. $f(x) = x^2 e^{-x^2} = x^2\left[1 - \left(x^2\right) + \dfrac{(-x^2)^2}{2!} + \dfrac{(-x^2)^3}{3!} + \cdots\right] = x^2\left(1 - x^2 + \dfrac{x^4}{2!} - \dfrac{x^6}{3!} + \cdots\right) = \sum_{n=0}^{\infty} (-1)^n \dfrac{x^{2n+2}}{n!}$.

The interval of convergence is $(-\infty, \infty)$.

13. $f(x) = \frac{1}{2}\left(e^x + e^{-x}\right) = \frac{1}{2}\left[\left(1 + x + \dfrac{x^2}{2!} + \dfrac{x^3}{3!} + \cdots\right) + \left(1 - x + \dfrac{x^2}{2!} - \dfrac{x^3}{3!} + \cdots\right)\right]$

$= 1 + \dfrac{x^2}{2!} + \dfrac{x^4}{4!} + \dfrac{x^6}{6!} + \cdots + \dfrac{x^{2n}}{2n!} + \cdots = \sum_{n=0}^{\infty} \dfrac{x^{2n}}{2n!}$.

The interval of convergence is $(-\infty, \infty)$.

© 2011 Cengage Learning. All Rights Reserved. May not be scanned, copied or duplicated, or posted to a publicly accessible website, in whole or in part.

15. $\ln x = (x-1) - \frac{1}{2}(x-1)^2 + \frac{1}{3}(x-1)^3 + \cdots$ Replace x by $1 + 2x$ to obtain

$$f(x) = \ln(1 + 2x) = 2x - \frac{1}{2}(2x)^2 + \frac{1}{3}(2x)^3 - \cdots = \sum_{n=1}^{\infty} (-1)^{n-1} \cdot \frac{2^n x^n}{n}. \text{ We must have}$$

$0 < 1 + 2x \le 2$, so $-1 < 2x \le 1$ and $-\frac{1}{2} < x \le \frac{1}{2}$, and the interval of convergence is $\left(-\frac{1}{2}, \frac{1}{2}\right]$.

17. $f(x) = \ln(1 + x^2) = x^2 - \frac{1}{2}x^4 + \frac{1}{3}x^6 - \cdots$, $f(x) = x^2 - \frac{1}{2}x^4 + \frac{1}{3}x^6 + \cdots + \frac{(-1)^{n+1}}{n}x^{2n} + \cdots = \sum_{n=1}^{\infty} (-1)^{n+1} \frac{x^{2n}}{n}$.

The interval of convergence is $(-1, 1)$.

19. Replace x by $x + 1$ in the formula for $\ln x$ in Table (1), giving $\ln(x+1) = x - \frac{1}{2}x^2 + \frac{1}{3}x^3 - \cdots$

for $-1 < x \le 1$. Next, observe that $x = x - 2 + 2 = 2\left(1 + \frac{x-2}{2}\right)$. Therefore,

$\ln x = \ln 2\left(1 + \frac{x-2}{2}\right) = \ln 2 + \ln\left(1 + \frac{x-2}{2}\right)$. If we replace x in the expression for $\ln(x+1)$ by $\frac{x-2}{2}$,

we obtain $\ln x = \ln 2 + \frac{x-2}{2} - \frac{1}{2}\left(\frac{x-2}{2}\right)^2 + \frac{1}{3}\left(\frac{x-2}{2}\right)^3 - \cdots = \ln 2 + \sum_{n=1}^{\infty} (-1)^{n-1}\left(\frac{1}{n \cdot 2^n}\right)(x-2)^n$.

Therefore, $f(x) = (x-2)\ln x = (x-2)\ln 2 + \sum_{n=1}^{\infty} (-1)^{n-1}\left(\frac{1}{n \cdot 2^n}\right)(x-2)^{n+1}$. To find the interval of

convergence, observe that x must satisfy $-1 < \frac{x-2}{2} < 1$, so $-2 < x - 2 \le 2$ and $0 < x \le 4$. Thus, the interval

of convergence is $(0, 4]$.

21. Replacing x in the formula for $\ln x$ in Table (1) by $1 + x$, we obtain $\ln(1+x) = x - \frac{1}{2}x^2 + \frac{1}{3}x^3 - \frac{1}{4}x^4 + \cdots$.

Differentiating, we find $\frac{1}{1+x} = 1 - x + x^2 - x^3 + \cdots + (-1)^n x^n + \cdots = \sum_{n=0}^{\infty} (-1)^n x^n$.

23. $\frac{1}{1+x} = 1 - x + x^2 - x^3 + \cdots$, so $\int \frac{1}{1+x}\,dx = x - \frac{1}{2}x^2 + \frac{1}{3}x^3 - \frac{1}{4}x^4 + \cdots + (-1)^{n+1}\frac{1}{n}x^n + \cdots + C$.

Therefore, $f(x) = \ln(1+x) = x - \frac{1}{2}x^2 + \frac{1}{3}x^3 - \frac{1}{4}x^4 + \cdots + \frac{(-1)^{n+1}}{n}x^n + \cdots + C$. Now $f(0) = 1$ implies that

$C = 0$, so $f(x) = \ln(1+x) = x - \frac{1}{2}x^2 + \frac{1}{3}x^3 - \frac{1}{4}x^4 + \cdots + \frac{(-1)^{n+1}}{n}x^n + \cdots$.

25. $\int_0^{0.5} \frac{dx}{\sqrt{1+x^2}} \approx \int_0^{.5} \left(1 - \frac{1}{2}x^2 + \frac{3}{8}x^4 - \frac{5}{16}x^6\right)dx = \left(x - \frac{1}{6}x^3 + \frac{3}{40}x^5 - \frac{5}{112}x^7\right)\Big|_0^{0.5} \approx 0.4812.$

27. $\int_0^1 e^{-x^2}\,dx \approx \int_0^1 \left(1 - x^2 + \frac{1}{2}x^4 - \frac{1}{6}x^6 + \frac{1}{24}x^8\right)dx = \left(x - \frac{1}{3}x^3 + \frac{1}{10}x^5 - \frac{1}{42}x^7 + \frac{1}{216}x^9\right)\Big|_0^1 \approx 0.7475.$

29. $\pi = 4\int_0^1 \frac{dx}{1+x^2} \approx 4\int_0^1 \left(1 - x^2 + x^4 - x^6 + x^8\right)dx = 4\left(x - \frac{1}{3}x^3 + \frac{1}{5}x^5 - \frac{1}{7}x^7 + \frac{1}{9}x^9\right)\Big|_0^1$

$= 4\left(1 - 0.333333 + 0.2 - 0.1428571 + 0.1111111\right) \approx 3.34.$

© 2011 Cengage Learning. All Rights Reserved. May not be scanned, copied or duplicated, or posted to a publicly accessible website, in whole or in part.

31. $P(28 \le x \le 32) = \dfrac{1}{10\sqrt{2\pi}} \displaystyle\int_{28}^{32} e^{(-1/2)[(x-30)/10]^2} \, dx$

$$= \frac{1}{10\sqrt{2\pi}} \int_{28}^{32} \left[1 - \frac{1}{2}\left(\frac{x-30}{10}\right)^2 + \frac{1}{8}\left(\frac{x-30}{10}\right)^4 - \frac{1}{48}\left(\frac{x-30}{10}\right)^6 \right] dx$$

$$= \frac{1}{10\sqrt{2\pi}} \left[1 - \frac{1}{6}\left(\frac{x-30}{10}\right)^3 + \frac{1}{40}\left(\frac{x-30}{10}\right)^5 - \frac{1}{336}\left(\frac{x-30}{10}\right)^7 \right]_{28}^{32} \approx 15.85, \text{ or } 15.85\%.$$

11.7   Concept Questions   page 752

1. Pick a number x_0 that is close to the zero being sought. Find the tangent line to f at $(x_0, f(x_0))$, where the tangent line cuts the x-axis is the next estimate of r. Continue the process, to obtain a sequence of approximations $x_1, x_2, x_3, \ldots$ which, under suitable conditions, approaches r.

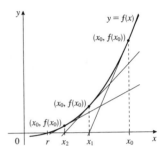

3. No. See Figures 10 and 11.

11.7   Newton's Method   page 753

1. Take $f(x) = x^2 - 3$ so that $f'(x) = 2x$. We have $x_{n+1} = x_n - \dfrac{x_n^2 - 3}{2x_n} = \dfrac{x_n^2 + 3}{2x_n}$. With $x_0 = 1.5$, we find

$x_1 = 1.75$, $x_2 \approx 1.7321429$, and $x_3 \approx 1.7320508$. Therefore, $\sqrt{3} \approx 1.732051$.

3. $x_{n+1} = x_n - \dfrac{x_n^2 - 7}{2x_n} = \dfrac{x_n^2 + 7}{2x_n}$, so with $x_0 = 2.5$, we find $x_1 = 2.65$, $x_2 \approx 2.645755$, and $x_3 \approx 2.645751$.

Therefore, $\sqrt{7} \approx 2.645751$.

5. $x_{n+1} = x_n - \dfrac{x_n^3 - 14}{3x_n^2} = \dfrac{2x_n^3 + 14}{3x_n^2}$, so with $x_0 = 2.5$, we find $x_1 \approx 2.413333$, $x_2 \approx 2.410146$, and

$x_3 \approx 2.410142264$. Therefore, $\sqrt[3]{14} \approx 2.410142$.

7. $f'(x) = \dfrac{d}{dx}(-x^3 - 2x + 2) = -3x^2 - 2$. We use the iteration

$x_{n+1} = x_n - \dfrac{f(x_n)}{f'(x_n)} = x_n - \dfrac{-x_n^3 - 2x_n + 2}{-3x_n^2 - 2} = \dfrac{2x_n^3 + 2}{3x_n^2 + 2}$. With $x_0 = 1$, we find $x_1 = \dfrac{2(1) + 2}{3(1) + 2} = 0.8$,

$x_2 = \dfrac{2(0.8)^3 + 2}{3(0.8)^2 + 2} \approx 0.77143$, and $x_3 \approx 0.77092 \approx x_4$, so the desired root is approximately 0.7709.

© 2011 Cengage Learning. All Rights Reserved. May not be scanned, copied or duplicated, or posted to a publicly accessible website, in whole or in part.

9. $f'(x) = \dfrac{d}{dx}\left(\frac{3}{2}x^4 - 2x^3 - 6x^2 + 8\right) = 6x^3 - 6x^2 - 12x$. We use the iteration

$x_{n+1} = x_n - \dfrac{f(x_n)}{f'(x_n)} = x_n - \dfrac{\frac{3}{2}x_n^4 - 2x_n^3 - 6x_n^2 + 8}{6x_n^3 - 6x_n^2 - 12x_n} = \dfrac{4.5x_n^4 - 4x_n^3 - 6x_n^2 - 8}{6x_n\left(x_n^2 - x_n - 2\right)}$. With $x_0 = 1$, we find $x_1 = 1.125$

and $x_2 \approx 1.12192 \approx x_3$, so one root is approximately 1.1219. With $x_0 = 3$, we find $x_1 \approx 2.70139$, $x_2 \approx 2.59044$, $x_3 \approx 2.57478$, and $x_4 \approx 2.57448 \approx x_5$, so the other root is approximately 2.5745.

11. $f(x) = x^2 - x - 3$, so $f'(x) = 2x - 1$ and $x_{n+1} = x_n - \dfrac{x_n^2 - x_n - 3}{2x_n - 1} = \dfrac{2x_n^2 - x_n - x_n^2 + x_n + 3}{2x_n - 1} = \dfrac{x_n^2 + 3}{2x_n - 1}$.
With $x_0 = 2$, we find $x_1 \approx 2.33333$, $x_2 \approx 2.30303$, $x_3 \approx 2.30278$, and $x_4 \approx 2.30278$, so the zero is approximately 2.30278.

13. $f(x) = x^3 + 2x^2 + x - 5$, so $f'(x) = 3x^2 + 4x + 1$ and $x_{n+1} = x_n - \dfrac{x_n^3 + 2x_n^2 + x_n - 5}{3x_n^2 + 4x_n + 1} = \dfrac{2x_n^3 + 2x_n^2 + 5}{3x_n^2 + 4x_n + 1}$.
With $x_0 = 1$, we find $x_1 \approx 1.53333$, $x_2 \approx 1.19213$, $x_3 \approx 1.11949$, $x_4 \approx 1.11635$, $x_5 \approx 1.11634$, and $x_6 \approx 1.11634$, so the zero is approximately 1.11634.

15. $x_{n+1} = x_n - \dfrac{\sqrt{x_n + 1} - x_n}{\frac{1}{2}(x_n + 1)^{-1/2} - 1} = \dfrac{x_n + 2}{2\sqrt{x_n + 1} - 1}$, so with $x_0 = 1$, we find $x_1 \approx 1.640764$, $x_2 \approx 1.618056$, and
$x_3 \approx 1.618034$. Therefore, the zero is approximately 1.61803.

17. $x_{n+1} = x_n - \dfrac{e^{x_n} - \dfrac{1}{x_n}}{e^{x_n} + \dfrac{1}{x_n^2}} = \dfrac{x_n^2 e^{x_n}(x_n - 1) + 2x_n}{x_n^2 e^{x_n} + 1}$, so with $x_0 = 0.5$, we find $x_1 \approx 0.562187$, $x_2 \approx 0.56712$, and

$x_3 \approx 0.567143$. Therefore, the zero is approximately 0.5671.

19. a. $f(0) = -2$ and $f(1) = 3$. Because $f(x)$ is a polynomial, it is continuous. Furthermore, because f changes sign between $x = 0$ and $x = 1$, we conclude that f has a root in the interval $(0, 1)$.

b. $f(x) = 2x^3 - 9x^2 + 12x - 2$, so $f'(x) = 6x^2 - 18x + 12$ and

$x_{n+1} = x_n - \dfrac{2x_n^3 - 9x_n^2 + 12x_n - 2}{6x_n^2 - 18x_n + 12} = \dfrac{4x_n^3 - 9x_n^2 + 2}{6x_n^2 - 18x_n + 12}$. With $x_0 = 0.5$, we have $x_1 = \dfrac{0.25}{4.5} = 0.055556$,

$x_2 \approx \dfrac{1.972908}{11.0185108} \approx 0.193556$, $x_3 \approx \dfrac{1.734419}{8.969390} \approx 0.193371$, $x_4 \approx \dfrac{1.692391}{0.193371} \approx 0.179054$, and

$x_5 \approx \dfrac{1.691830}{0.193556} \approx 0.193556$. Thus, the root is approximately 0.19356.

21. a. $f(x) = x^3 - 3x - 1$, so $f(1) = -3$ and $f(2) = 1$. Because $f(x)$ has opposite signs at $x = 1$ and $x = 2$, we see that the continuous function f has at least one zero in the interval $(1, 2)$.

b. $f'(x) = 3x^2 - 3$, so the required iteration formula is

$x_{n+1} = x_n - \dfrac{x_n^3 - 3x_n - 1}{3x_n^2 - 3} = \dfrac{3x_n^3 - 3x_n - x_n^3 + 3x_n + 1}{3x_n^2 - 3} = \dfrac{2x_n^3 + 1}{3x_n^2 - 3}$. With $x_0 = 1.5$, we find $x_1 \approx 2.066667$,

$x_2 \approx 1.900876$, $x_3 \approx 1.879720$, $x_4 \approx 1.879385$, and $x_5 \approx 1.879385$, so the required root is approximately 1.87939.

© 2011 Cengage Learning. All Rights Reserved. May not be scanned, copied or duplicated, or posted to a publicly accessible website, in whole or in part.

23.

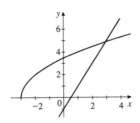

From the graph, the root appears to be approximately 3. Let $F(x) = 2\sqrt{x+3} - 2x + 1$. To solve $F(x) = 0$, we use the iteration $x_{n+1} = x_n - \dfrac{2\sqrt{x_n+3} - 2x_n + 1}{(x_n+3)^{-1/2} - 2}$. With $x_0 = 3$, we find $x_1 \approx 2.93654$, $x_2 \approx 2.93649$, and $x_3 \approx 2.9365$.

25.

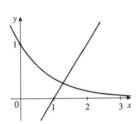

From the graph, the root appears to be approximately 1.2. Let $F(x) = e^{-x} - x + 1$, so $F'(x) = -e^{-x} - 1$ and

$$x_{n+1} = x_n + \frac{e^{-x_n} - x_n + 1}{e^{-x_n} + 1}.$$ With $x_0 = 1.2$, we find

$x_1 \approx 1.2 + \frac{0.1011942}{1.3011942} \approx 1.2777703$,

$x_2 \approx 1.277703 + \frac{0.0000955}{1.2786767} \approx 1.2784499$, and

$x_3 \approx 1.2784499 + \frac{0.0000187}{1.2784686} \approx 1.2784645$. Thus, the root is approximately 1.2785.

27.

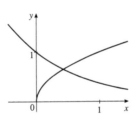

From the graph, the root lies between 0 and 1. To solve $F(x) = e^{-x} - \sqrt{x} = 0$, we use the iteration

$$x_{n+1} = x_n - \frac{e^{-x_n} - x_n^{1/2}}{-e^{-x_n} - \frac{1}{2}x^{-1/2}} = x_n + \frac{2\sqrt{x_n}e^{-x_n} - 2x_n}{2\sqrt{x_n}e^{-x_n} + 1}.$$ With $x_0 = 1$, we have $x_1 \approx 0.271649$, $x_2 \approx 0.411602$, and $x_3 \approx 0.426303 \approx x_4 \approx 0.426303$. Thus, the root is approximately 0.426303.

29. The daily average cost is $\overline{C}(x) = \dfrac{C(x)}{x} = 0.0002x^2 - 0.06x + 120 + \dfrac{5000}{x}$. To find the minimum of

$\overline{C}(x)$, we set $\overline{C}'(x) = 0.0004x - 0.06 - \dfrac{5000}{x^2} = 0$, obtaining $0.0004x^3 - 0.06x^2 - 5000 = 0$, or

$x^3 - 150x^2 - 12{,}500{,}000 = 0$. Write $f(x) = x^3 - 150x^2 - 12{,}500{,}000$ and observe that $f(0) < 0$, whereas $f(500) > 0$. Thus, the root (a critical point of $\overline{C}$) lies between $x = 0$ and $x = 500$. Take $x_0 = 250$ and use the

iteration $x_{n+1} = x_n - \dfrac{x_n^3 - 150x_n^2 - 12{,}500{,}000}{3x_n^2 - 300x_n} = \dfrac{2x_n^3 - 150x_n^2 + 12{,}500{,}000}{3x_n^2 - 300x_n}$. Taking $x_0 = 250$, we find

$x_1 \approx 305.556$, $x_2 \approx 294.818$, $x_3 \approx 294.312$, and $x_4 \approx 294.311$. It can be shown that $\overline{C}(x)$ is concave upward on $(0, \infty)$, so the level of production that minimizes $\overline{C}(x)$ is 294 units per day.

31. We solve the equation $f(t) = 0.05t^3 - 0.4t^2 - 3.8t - 15.6 = 0$. Use the iteration

$t_{n+1} = t_n - \dfrac{0.05t_n^3 - 0.4t_n^2 - 3.8t_n - 15.6}{0.15t_n^2 - 0.8t_n - 3.8} = \dfrac{0.1t_n^3 - 0.4t_n^2 + 15.6}{0.15t_n^2 - 0.8t_n - 3.8}$ with $t_0 = 14$, obtaining $t_1 = 14.6944$,

$t_2 = 14.6447$, and $t_3 = 14.6445$. Thus, the temperature is 0°F at about 8:39 p.m.

© 2011 Cengage Learning. All Rights Reserved. May not be scanned, copied or duplicated, or posted to a publicly accessible website, in whole or in part.

33. Here $f(x) = 120{,}000x^3 - 80{,}000x^2 - 60{,}000x - 40{,}000$, so $f'(x) = 360{,}000x^2 - 160{,}000x - 60{,}000$. Therefore,

$$x_{n+1} = x_n - \frac{120{,}000x_n^3 - 80{,}000x_n^2 - 60{,}000x_n - 40{,}000}{360{,}000x_n^2 - 160{,}000x_n - 60{,}000} = x_n - \frac{6x_n^3 - 4x_n^2 - 3x_n - 2}{18x_n^2 - 8x_n - 3} = \frac{12x_n^3 - 4x_n^2 + 2}{18x_n^2 - 8x_n - 3}.$$

Then with $x_0 = 1.13$, we have $x_1 \approx 1.2981455$, $x_2 \approx 1.2692040$, $x_3 \approx 1.2681893$, $x_4 \approx 1.2681880$, and $x_5 \approx 1.2681880$, so $r \approx 26.819$. Therefore, the rate of return is approximately 26.82% per year.

35. Here $C = 200{,}000$, $R = 2106$, and $N = (12)(25) = 300$. Therefore,

$$r_{n+1} = r_n - \frac{200{,}000r_n + 2106\left[(1+r_n)^{-300} - 1\right]}{200{,}000 - 631{,}800(1+r_n)^{-301}}.$$ With $r_0 = 0.1$, we find $r_1 = 0.0153$, $r_2 = 0.0100044$,

$r_3 = 0.00999747$, and $r_4 = 0.00999747$. Therefore, $r \approx 0.01$ and the interest rate is approximately $12(0.01) = 0.12$, or 12% per year.

37. Here $C = 24{,}000$ (75% of 32,000), $N = 12 \cdot 4 = 48$, and $R = 608.72$. Thus,

$$r_{n+1} = r_n - \frac{24{,}000r_n + 608.72\left[(1+r_n)^{-48} - 1\right]}{24{,}000 - 29{,}218.56(1+r_n)^{-49}}.$$ With $r_0 = 0.01$, we find

$$r_1 = 0.01 - \frac{24{,}000(0.01) + 608.72\left[(1.01)^{-48} - 1\right]}{24{,}000 - 29{,}218.56(1.01)^{-49}} \approx 0.00853956, \quad r_2 \approx 0.0083388, \text{ and } r_3 \approx 0.0083346.$$ Thus, $r \approx 12(0.008335) \approx 0.1000$ or 10% per year.

39. We are required to solve the equation $s(x) = d(x)$. Expanding and simplifying, we have

$$0.1x + 20 = \frac{50}{0.01x^2 + 1}, \quad 0.001x^3 + 0.2x^2 + 0.1x + 20 = 50, \text{ and } 0.001x^3 + 0.2x^2 + 0.1x - 30 = 0,$$

so we set $F(x) = x^3 + 200x^3 + 100x - 30{,}000 = 0$ and use the iteration

$$x_{n+1} = x_n - \frac{x_n^3 + 200x_n^2 + 100x_n - 30{,}000}{3x_n^2 + 400x_n + 100} = \frac{2x_n^3 + 200x_n^2 + 30{,}000}{3x_n^2 + 400x_n + 100}.$$ With $x_0 = 10$, we find $x_1 \approx 11.8182$,

$x_2 \approx 11.6721$, $x_3 \approx 11.6711$, and $x_4 \approx 11.6711$. Therefore, the equilibrium quantity is approximately 11,671 units and the equilibrium price is approximately $p = 0.1(11.671) + 20$, or $21.17/unit.

41. a. Let $f(x) = x^n - a$. We want to solve the equation $f(x) = 0$, $f'(x) = nx^{n-1}$. Therefore, we use the iteration

$$x_{i+1} = x_i - \frac{f(x_i)}{f'(x_i)} = x_i - \frac{x_i^n - a}{nx_i^{n-1}} = x_i - \frac{x_i^n}{nx_i^{n-1}} + \frac{a}{nx_i^{n-1}} = x_i - \frac{1}{n}x_i + \frac{a}{nx_i^{n-1}} = \left(1 - \frac{1}{n}\right)x_i + \frac{a}{nx_i^{n-1}}$$

$$= \left(\frac{n-1}{n}\right)x_i + \frac{a}{nx_i^{n-1}}.$$

b. Using part (a) with $n = 4$, $a = 42$, and initial guess $x_0 = 2$, we find $x_{i+1} = \left(\frac{3}{4}\right)x_i + \frac{42}{4x_i^3}$,

and so $x_1 = \frac{3}{4}(2) + \frac{42}{4(2^3)} = 2.8125$, $x_2 \approx \frac{3}{4}(2.8125) + \frac{42}{4(2.8125)^3} \approx 2.5813$, and

$x_3 \approx \frac{3}{4}(2.5813) + \frac{42}{4(2.5813)^3} \approx 2.5464$.

© 2011 Cengage Learning. All Rights Reserved. May not be scanned, copied or duplicated, or posted to a publicly accessible website, in whole or in part.

CHAPTER 11   **Concept Review**   page 756

1. a. $f(a) + f'(a)(x-a) + \cdots + \dfrac{f^{(n)}(a)}{n!}(x-a)^n$

b. $f(a), f'(a), \dfrac{f^{(n)}(a)}{n!}$

3. a. function, integers, $f(n)$, nth term

b. large, converge

5. a. partial sums

b. geometric, 1, 1

7. a. convergent, divergent

b. $\displaystyle\sum_{n=1}^{\infty} \dfrac{1}{n^p}, p > 1, p \le 1$

9. a. $\displaystyle\sum_{n=0}^{\infty} a_n (x-a)^n$

b. $x - a, x, (a - R, a + R)$, outside

CHAPTER 11   **Review**   page 757

1. $f(x) = \dfrac{1}{x+2} = \dfrac{1}{1+(x+1)}$, so $P_4(x) = 1 - (x+1) + (x+1)^2 - (x+1)^3 + (x+1)^4$.

3. Recall that $\ln(1+x) \approx x - \frac{1}{2}x^2 + \frac{1}{3}x^3 - \frac{1}{4}x^4$, so $P_8(x) = x^2 - \frac{1}{2}\left(x^2\right)^2 + \frac{1}{3}\left(x^2\right)^3 - \frac{1}{4}\left(x^2\right)^4 = x^2 - \frac{1}{2}x^4 + \frac{1}{3}x^6 - \frac{1}{4}x^8$
and $P_4(x) = x^2 - \frac{1}{2}x^4$.

5. $f(x) = x^{1/3}$, so $f'(x) = \frac{1}{3}x^{-2/3}$, $f''(x) = -\frac{2}{9}x^{-5/3}$, $f(8) = 2$, $f'(8) = \frac{1}{12}$, and $f''(8) = -\frac{1}{144}$.

Therefore, $P_2(x) = f(8) + f'(8)(x-8) + \dfrac{f''(8)}{2!}(x-8)^2 = 2 + \frac{1}{12}(x-8) - \frac{1}{288}(x-8)^2$. In particular,

$\sqrt[3]{7.8} = f(7.8) \approx P_2(8) = 2 + \frac{1}{12}(-0.2) - \frac{1}{288}(-0.2)^2 \approx 1.983$.

7. $f(x) = x^{1/3}$, so $f'(x) = \frac{1}{3}x^{-2/3}$, $f''(x) = -\frac{2}{9}x^{-5/3}$, $f'''(x) = \frac{10}{27}x^{-8/3}$,
$f(27) = 3$, $f'(27) = \frac{1}{27}$, and $f''(27) = -\frac{2}{2187}$. Therefore,

$P_2(x) = f(27) + f'(27)(x-27) + \dfrac{f''(27)}{2!}(x-27)^2 = 3 + \frac{1}{27}(x-27) - \frac{1}{2187}(x-27)^2$. We estimate

$\sqrt[3]{26.98} = f(26.98) \approx P_2(26.98) = 3 + \frac{1}{27}(-0.02) - \frac{1}{2187}(-0.02)^2 \approx 2.9992591$. The error is less than

$\dfrac{M}{3!}|x - 27|^3$, where M is a bound for $f'''(x) = \dfrac{10}{27x^{7/3}}$ on $[26.98, 27]$. Observe that $f'''(x)$ is decreasing on the

interval and so $|f'''(x)| \le \dfrac{10}{27(26.98)^{8/3}} < 0.00006$. Thus, the error is less than $\frac{0.0002}{6}(0.02)^3 < 8 \times 10^{-11}$.

9. $e^x = 1 + x + \dfrac{x^2}{2!} + \dfrac{x^3}{3!} + \cdots$. Therefore, $e^{-1} = 1 - 1 + \frac{1}{2} - \frac{1}{6} + \frac{1}{24} - \frac{1}{120} + \cdots \approx 0.367$.

11. $\displaystyle\lim_{n\to\infty} a_n = \lim_{n\to\infty} \dfrac{n}{3n-2} = \lim_{n\to\infty} \dfrac{1}{3 - \frac{2}{n}} = \dfrac{1}{3}$.

13. $\displaystyle\lim_{n\to\infty} \dfrac{2n^2 + 1}{3n^2 - 1} = \lim_{n\to\infty} \dfrac{2 + \frac{1}{n^2}}{3 - \frac{1}{n^2}} = \dfrac{2}{3}$.

© 2011 Cengage Learning. All Rights Reserved. May not be scanned, copied or duplicated, or posted to a publicly accessible website, in whole or in part.

15. $\lim\limits_{n\to\infty} a_n = \lim\limits_{n\to\infty}\left(1-\dfrac{1}{2^n}\right) = \lim\limits_{n\to\infty} 1 - \lim\limits_{n\to\infty}\dfrac{1}{2^n} = 1 - 0 = 1.$

17. $\displaystyle\sum_{n=1}^{\infty}\dfrac{2^n}{3^n} = \sum_{n=1}^{\infty}\left(\tfrac{2}{3}\right)^n = \dfrac{2}{3}\left(\dfrac{1}{1-\frac{2}{3}}\right) = \dfrac{2}{3}\cdot 3 = 2.$

19. $\displaystyle\sum_{n=1}^{\infty}(-1)^{n-1}\left(\tfrac{1}{\sqrt{2}}\right)^n = \dfrac{1}{\sqrt{2}} - \left(\dfrac{1}{\sqrt{2}}\right)^2 + \cdots = \dfrac{\frac{1}{\sqrt{2}}}{1-\left(-\frac{1}{\sqrt{2}}\right)} = \dfrac{1}{\sqrt{2}}\cdot\dfrac{1}{1+\frac{1}{\sqrt{2}}} = \dfrac{1}{1+\sqrt{2}} = \sqrt{2}-1.$

21. $\displaystyle\sum_{n=0}^{\infty}\left(\dfrac{1}{3^n}-\dfrac{1}{4^{n+1}}\right) = \sum_{n=0}^{\infty}\left(\dfrac{1}{3}\right)^n - \dfrac{1}{4}\sum_{n=0}^{\infty}\left(\dfrac{1}{4}\right)^n = \dfrac{1}{1-\frac{1}{3}} - \dfrac{1}{4}\cdot\dfrac{1}{1-\frac{1}{4}} = \dfrac{3}{2} - \dfrac{1}{3} = \dfrac{7}{6}.$

23. $1.424242\ldots = 1 + \dfrac{42}{10^2} + \dfrac{42}{10^4} + \cdots = 1 + \dfrac{42}{100}\left[1 + \dfrac{1}{100} + \left(\dfrac{1}{100}\right)^2 + \cdots\right] = 1 + \dfrac{42}{100}\left(\dfrac{1}{1-\frac{1}{100}}\right)$

$\qquad = 1 + \dfrac{42}{100}\left(\dfrac{100}{99}\right) = 1 + \dfrac{42}{99} = \dfrac{141}{99}.$

25. Let $a_n = \dfrac{n^2+1}{2n^2-1}$. Because $\lim\limits_{n\to\infty}\dfrac{n^2+1}{2n^2-1} = \lim\limits_{n\to\infty}\dfrac{1+\frac{1}{n^2}}{2-\frac{1}{n^2}} = \dfrac{1}{2} \neq 0$, the divergence test implies that $\sum a_n$

diverges.

27. Here $a_n = \dfrac{n}{2n^3+1} \leq \dfrac{n}{2n^3} = \dfrac{1}{2n^2}$. Since $\displaystyle\sum_{n=1}^{\infty}\dfrac{1}{2n^2}$ converges, the comparison test implies that $\displaystyle\sum_{n=1}^{\infty}\dfrac{n}{2n^3+1}$ also

converges.

29. $\displaystyle\sum_{n=1}^{\infty}\left(\dfrac{1}{n}\right)^{1.1} = \sum_{n=1}^{\infty}\dfrac{1}{n^{1.1}}$ is a convergent p-series with $p = 1.1 > 1.$

31. Let $f(x) = \dfrac{1}{x(\ln x)^{3/2}}$ for $x \geq 2$. Then f is nonnegative and

decreasing on $(2,\infty)$. Next, using the substitution $u = \ln x$,

$\displaystyle\int_2^{\infty}\dfrac{dx}{x(\ln x)^{3/2}} = \lim_{b\to\infty}\int_2^b\dfrac{(\ln x)^{-3/2}}{x}\,dx = \lim_{b\to\infty}\left[-2(\ln x)^{-1/2}\right]_2^b = \lim_{b\to\infty}\left(-\dfrac{2}{\sqrt{\ln b}} + \dfrac{2}{\sqrt{\ln 2}}\right) = \dfrac{2}{\sqrt{\ln 2}}.$ Thus,

the integral test implies that the given series converges.

33. $R = \lim\limits_{n\to\infty}\left|\dfrac{a_n}{a_{n+1}}\right| = \lim\limits_{n\to\infty}\dfrac{\frac{1}{n^2+2}}{\frac{1}{(n+1)^2+2}} = \lim\limits_{n\to\infty}\dfrac{(n+1)^2+2}{n^2+2} = \lim\limits_{n\to\infty}\dfrac{n^2+2n+3}{n^2+2} = 1.$ The interval of

convergence is $(-1,1)$.

35. $R = \lim\limits_{n\to\infty}\left|\dfrac{a_n}{a_{n+1}}\right| = \lim\limits_{n\to\infty}\dfrac{(n+1)(n+2)}{n(n+1)} = 1$, so $R = 1$ and the interval of convergence is $(0,2)$.

© 2011 Cengage Learning. All Rights Reserved. May not be scanned, copied or duplicated, or posted to a publicly accessible website, in whole or in part.

37. We have $\dfrac{1}{1+x} = 1 + x + x^2 + x^3 + \cdots$ for $-1 < x < 1$. Replacing x by $2x$ in the expression, we find

$$f(x) = \dfrac{1}{2x-1} = -\dfrac{1}{1-2x} = -\left[1 + 2x + (2x)^2 + (2x)^3 + \cdots\right] = -1 - 2x - 4x^2 - 8x^3 - \cdots - 2^n x^n - \cdots$$

for $-\frac{1}{2} < x < \frac{1}{2}$.

39. We know that $\ln(1+x) = x - \frac{1}{2}x^2 + \frac{1}{3}x^3 - \frac{1}{4}x^4 + \cdots + (-1)^{n+1}\dfrac{x^n}{n}$ for $-1 < x < 1$. Replace x by $2x$ to obtain

$$f(x) = \ln(1+2x) = 2x - 2x^2 + \frac{8}{3}x^3 - \cdots + (-1)^{n+1}\dfrac{2^n x^n}{n} + \cdots. \text{ The interval of convergence is } \left(-\frac{1}{2}, \frac{1}{2}\right].$$

41. $\displaystyle\lim_{n\to\infty}\left|\dfrac{a_{n+1}}{a_n}\right| = \lim_{n\to\infty}\left|\dfrac{(n+1)^{n+1}(x-1)^{n+1}}{(2n+2)!} \cdot \dfrac{(2n)!}{n^n(x-1)^n}\right| = \lim_{n\to\infty}\dfrac{(n+1)\left(\frac{n+1}{n}\right)^n |x-1|}{(2n+2)(2n+1)} = 0$, so the interval of convergence is $(-\infty, \infty)$.

43. $\displaystyle\int \dfrac{e^x-1}{x}\,dx = \int \dfrac{1}{x}\left[1 + x + \dfrac{x^2}{2!} + \dfrac{x^3}{3!} + \cdots + \dfrac{x^n}{n!} + \cdots - 1\right]dx = \int\left(1 + \dfrac{x}{2!} + \dfrac{x^2}{3!} + \cdots + \dfrac{x^{n-1}}{n!} + \cdots\right)dx$

$$= x + \dfrac{x^2}{2!\,2} + \dfrac{x^3}{3!\,3} + \cdots + \dfrac{x^n}{n!\,n} + \cdots + C = \sum_{n=1}^{\infty}\dfrac{x^n}{n!\,n} + C$$

45. $f'(x) = 3x^2 + 2x$. We use the iteration $x_{n+1} = x_n - \dfrac{x_n^3 + x_n^2 - 1}{3x_n^2 + 2x_n} = \dfrac{2x_n^3 + x_n^2 + 1}{3x_n^2 + 2x_n}$. With $x_0 = 0.5$, we find $x_1 \approx 0.857143$, $x_2 \approx 0.764137$, $x_3 \approx 0.754963$, and $x_4 \approx 0.754878$, so the root is approximately 0.7549.

47. We solve the equation $f(t) = 27\left(t + 3e^{-t/3} - 3\right) = 24$ for t. Now $t + 3e^{-t/3} - 3 = \frac{24}{27} = \frac{8}{9}$, so $t + 3e^{-t/3} - \frac{35}{9} = 0$. Let $g(t) = t + 3e^{-t/3} - \frac{35}{9}$ and use Newton's Method to solve $g(t) = 0$. $g'(t) = 1 - e^{-t/3}$, so the iteration is $t_{n+1} = t_n - \dfrac{t_n + e^{-t_n/3} - \frac{35}{9}}{1 - \frac{1}{3}e^{-t_n/3}} = \dfrac{t_n - t_n e^{-t_n/3} - t_n - 3e^{-t_n/3} + \frac{35}{9}}{1 - \frac{1}{3}e^{-t_n/3}} = \dfrac{\frac{35}{9} - (t_n + 3)e^{-t_n/3}}{1 - e^{-t_n/3}}$.

Taking the initial guess $t_0 = 2$, we find $t_1 = 2.71650$, $t_2 = 2.64829$, $t_3 = 2.64775$, and $t_4 = 2.64775$, so it takes approximately 2.65 seconds for the suitcase to reach the bottom.

49. a. The economic impact is

$$(0.92)(10) + (0.92)^2(10) + \cdots = (0.92)(10)\left(1 + 0.92 + 0.92^2 + \cdots\right) = 9.2\left(\dfrac{1}{1-0.92}\right) = 115, \text{ or}$$

$115 billion.

b. The economic impact is $(0.9)(10) + (0.9)^2(10) + \cdots = (0.9)(10)\left(1 + 0.9 + 0.9^2 + \cdots\right) = 9\left(\dfrac{1}{1-0.9}\right) = 90$,

or $90 billion.

© 2011 Cengage Learning. All Rights Reserved. May not be scanned, copied or duplicated, or posted to a publicly accessible website, in whole or in part.

CHAPTER 11   Before Moving On...   page 758

1. $f(x) = xe^{-x}$, so $f'(x) = e^{-x} - xe^{-x} = (1-x)e^{-x}$, $f''(x) = -e^{-x} - (1-x)e^{-x} = -(2-x)e^{-x}$, and
$f'''(x) = e^{-x} + (2-x)e^{-x} = (3-x)e^{-x}$. Thus, $f(0) = 0$, $f'(0) = 1$, $f''(0) = -2$, and $f'''(0) = 3$, so
$$P_0(x) = f(0) = 0, \quad P_1(x) = f(0) + f'(0)x = x, \quad P_2(x) = f(0) + f'(0)x + \frac{f''(0)}{2!}x^2 = x - \frac{2}{2!}x^2 = x - x^2,$$
and $P_3(x) = f(0) + f'(0)x + \dfrac{f''(0)}{2!}x^2 + \dfrac{f'''(0)}{3!}x^3 = x - x^2 + \dfrac{3}{3!}x^3 = x - x^2 + \dfrac{x^3}{2}$.

2. a. $\displaystyle\lim_{n\to\infty} \frac{2n^2}{3n^2 + 2n + 1} = \lim_{n\to\infty} \frac{2}{3 + \dfrac{2}{n} + \dfrac{1}{n^2}} = \frac{2}{3}$.

b. $\displaystyle\lim_{n\to\infty} \frac{2n}{\sqrt{n^2+1}} = \lim_{n\to\infty} \frac{2n\left(\frac{1}{n}\right)}{\frac{1}{n}\sqrt{n^2+1}} = \lim_{n\to\infty} \frac{2}{\sqrt{1 + \frac{1}{n^2}}} = 2$.

3. a. $\displaystyle\sum_{n=0}^{\infty} \frac{3 - 2^n}{5^n} = \sum_{n=0}^{\infty} \left[3\left(\frac{1}{5^n}\right) - \left(\frac{2}{5}\right)^n \right] = 3\sum_{n=0}^{\infty} \left(\frac{1}{5}\right)^n - \sum_{n=0}^{\infty} \left(\frac{2}{5}\right)^n = 3 \cdot \frac{1}{1 - \frac{1}{5}} - \frac{1}{1 - \frac{2}{5}} = \frac{15}{4} - \frac{5}{3} = \frac{25}{12}$.

$$S_N = \sum_{n=1}^{\infty} \left(\frac{1}{n+2} - \frac{1}{n+3} \right) = \left(\frac{1}{3} - \frac{1}{4} \right) + \left(\frac{1}{4} - \frac{1}{5} \right) + \left(\frac{1}{5} - \frac{1}{6} \right) + \cdots + \left(\frac{1}{N+2} - \frac{1}{N+3} \right) = \frac{1}{3} - \frac{1}{N+3}.$$

Therefore, $\displaystyle\sum_{n=1}^{\infty} \left(\frac{1}{n+2} - \frac{1}{n+3} \right) = \lim_{N\to\infty} S_N = \lim_{N\to\infty} \left(\frac{1}{3} - \frac{1}{N+3} \right) = \frac{1}{3}$.

4. a. $a_n = \dfrac{1}{2n^2 + n + 3} < \dfrac{1}{2n^2} = \dfrac{1}{2} \cdot \dfrac{1}{n^2}$. Because $\displaystyle\sum_{n=1}^{\infty} \frac{1}{n^2}$ converges, so does $\displaystyle\sum_{n=1}^{\infty} \frac{1}{2n^2 + n + 3}$, by the comparison
test.

b. Let $f(x) = \dfrac{(\ln x)^2}{x}$ for $x \geq 2$. Then f is nonnegative. Next,

$$f'(x) = \frac{x \cdot 2\ln x \left(\frac{1}{x}\right) - (\ln x)^2}{x^2} = \frac{(2 - \ln x)\ln x}{x^2} < 0 \text{ if } x > e^2. \text{ Therefore, } f(x) = \frac{(\ln x)^2}{x} \text{ is}$$

nonnegative and decreasing on $[9, \infty)$. The integral test applies, and we calculate

$\displaystyle\int_9^{\infty} \frac{(\ln x)^2}{x}\, dx = \lim_{b\to\infty} \int_9^b \frac{(\ln x)^2}{x}\, dx = \lim_{b\to\infty} \frac{1}{3}(\ln x)^3 \Big|_9^b = \infty$. Therefore, $\displaystyle\int_1^{\infty} \frac{(\ln x)^2}{x}\, dx$ diverges, as does the
given series.

5. $f(x) = \dfrac{1}{1+x} = \dfrac{1}{1 + (x+2) - 2} = \dfrac{1}{-1 + (x+2)} = \dfrac{-1}{1 - (x+2)} = -\sum_{n=0}^{\infty} (x+2)^n$,

$R = \displaystyle\lim_{n\to\infty} \left| \frac{a_n}{a_{n+1}} \right| = \lim_{n\to\infty} \left| \frac{-1}{-1} \right| = 1$.

© 2011 Cengage Learning. All Rights Reserved. May not be scanned, copied or duplicated, or posted to a publicly accessible website, in whole or in part.

12 TRIGONOMETRIC FUNCTIONS

12.1 Problem-Solving Tips

1. Familiarize yourself with the radian and degree measures of common angles:

Degrees	0	30°	45°	60°	90°	120°	135°	150°	180°	270°	360°
Radians	0	$\frac{\pi}{6}$	$\frac{\pi}{4}$	$\frac{\pi}{3}$	$\frac{\pi}{2}$	$\frac{2\pi}{3}$	$\frac{3\pi}{4}$	$\frac{5\pi}{6}$	π	$\frac{3\pi}{2}$	2π

2. **To convert degrees to radians,** use the formula $f(x) = \frac{\pi}{180}x$, where x is the number of degrees and $f(x)$ is the number of radians.

3. **To convert radians to degrees**, use the formula $g(x) = \frac{180}{\pi}x$, where x is the number of radians and $g(x)$ is the number of degrees.

12.1 Concept Questions page 763

1. **a.** A degree is the measure of the angle formed by $\frac{1}{360}$ of one complete revolution.

   b. A radian is the measure of the central angle subtended by an arc equal in length to the radius of the circle.

3. **a.** To convert degrees to radians, we use the formula $f(x) = \frac{\pi}{180}x$, where x is the number of degrees and $f(x)$ is the number of radians.

   b. To convert radians to degrees, we use the formula $g(x) = \frac{180}{\pi}x$, where x is the number of radians and $g(x)$ is the number of degrees.

12.1 Measurement of Angles page 763

1. $450° = \frac{450}{180}\pi = \frac{5\pi}{2}$ rad.

3. $-270° = -\frac{270}{180}\pi = -\frac{3\pi}{2}$ rad.

5. **a.** III **b.** III **c.** II **d.** I

7. $f(x) = \frac{\pi}{180}x$ rad, $f(75) = \frac{\pi}{180}(75)$ rad $= \frac{5\pi}{12}$ rad.

9. $f(x) = \frac{\pi}{180}x$ rad, $f(160) = \frac{\pi}{180}(160)$ rad $= \frac{8\pi}{9}$ rad.

11. $f(630) = \frac{\pi}{180}(630)$ rad $= \frac{7\pi}{2}$ rad.

13. $g\left(\frac{2\pi}{3}\right) = \left(\frac{180}{\pi}\right)\left(\frac{2\pi}{3}\right) = 120°.$

15. $g\left(-\frac{3\pi}{2}\right) = \left(\frac{180}{\pi}\right)\left(-\frac{3\pi}{2}\right) = -270°.$

© 2011 Cengage Learning. All Rights Reserved. May not be scanned, copied or duplicated, or posted to a publicly accessible website, in whole or in part.

17. $g\left(\frac{22\pi}{18}\right) = \left(\frac{180}{\pi}\right)\left(\frac{22\pi}{18}\right) = 220°$.

19.

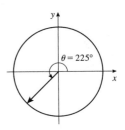

21.

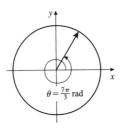

23. $\frac{5\pi}{6}$ rad $= 150°$. The coterminal angle is $-210°$.

25. $-\frac{\pi}{4}$ rad $= -45°$. The coterminal angle is $360° - 45° = 315°$.

27. True. $3630 = (360)(10) + 30$, and the result is evident.

29. True. Adding $n(360)$ degrees to θ changes the angle by $\theta|n|$ revolutions, clockwise if n is positive and counterclockwise if n is negative.

12.2   Concept Questions   page 771

1. If P is a point on the unit circle and the coordinates of P are (x, y), then $\cos\theta = x$, $\sin\theta = y$, $\tan\theta = \frac{y}{x}$ (for $x \neq 0$), $\csc\theta = \frac{1}{y}$ (for $y \neq 0$), $\sec\theta = \frac{1}{x}$ (for $x \neq 0$), and $\cot\theta = \frac{x}{y}$ (for $y \neq 0$).

3. a. 1   **b.** $\tan^2\theta + 1$ **c.** $\csc^2\theta$

5. a. $\cos A \cos B \mp \sin A \sin B$ **b.** $2\sin A \cos A$ **c.** $\cos^2 A - \sin^2 A$

12.2   Trigonometric Functions   page 771

1. $\sin 3\pi = 0$.

3. $\sin\frac{9\pi}{2} = 1$.

5. $\sin\left(-\frac{4\pi}{3}\right) = \sin\frac{\pi}{3} = \frac{\sqrt{3}}{2}$.

7. $\tan\frac{\pi}{6} = \frac{\sqrt{3}}{3}$.

9. $\sec\left(-\frac{5\pi}{8}\right) = \sec\frac{5\pi}{8} \approx -2.6131$.

11. $\sin\frac{\pi}{2} = 1$, $\cos\frac{\pi}{2} = 0$, $\tan\frac{\pi}{2}$ is undefined, $\sec\frac{\pi}{2}$ is undefined, $\csc\frac{\pi}{2} = 1$, $\cot\frac{\pi}{2} = 0$.

© 2011 Cengage Learning. All Rights Reserved. May not be scanned, copied or duplicated, or posted to a publicly accessible website, in whole or in part.

13. $\sin\frac{5\pi}{3} = -\frac{\sqrt{3}}{2}$, $\cos\frac{5\pi}{3} = \frac{1}{2}$, $\tan\frac{5\pi}{3} = -\sqrt{3}$, $\csc\frac{5\pi}{3} = -\frac{2\sqrt{3}}{3}$, $\sec\frac{5\pi}{3} = 2$, and $\cot\frac{5\pi}{3} = -\frac{\sqrt{3}}{3}$.

15. $\theta = \frac{7\pi}{6}$ or $\frac{11\pi}{6}$.   **17.** $\theta = \frac{5\pi}{6}$ or $\frac{11\pi}{6}$.   **19.** $\theta = \pi$.

21. $\sin\theta = \sin\left(-\frac{4\pi}{3}\right) = \sin\frac{2\pi}{3} = \sin\frac{\pi}{3}$, so $\theta = \frac{2\pi}{3}$ or $\frac{\pi}{3}$.

23. $y = \csc x$ **25.** $y = \cot x$

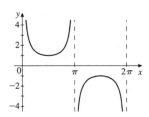   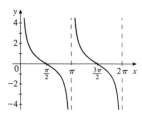

27. $y = \sin 2x$ **29.** $y = -\sin x$

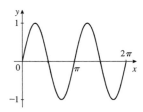   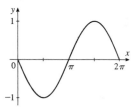

31. $\cos^2\theta - \sin^2\theta = \cos^2\theta - \left(1 - \cos^2\theta\right) = 2\cos^2\theta - 1$.

33. $(\sec\theta + \tan\theta)(1 - \sin\theta) = \sec\theta + \tan\theta - \sin\theta\sec\theta - \tan\theta\sin\theta = \sec\theta + \tan\theta - \tan\theta - \tan\theta\sin\theta$

$$= \frac{1}{\cos\theta} - \frac{\sin^2\theta}{\cos\theta} = \frac{1}{\cos\theta}\left(1 - \sin^2\theta\right) = \frac{\cos^2\theta}{\cos\theta} = \cos\theta.$$

35. $\left(1 + \cot^2\theta\right)\tan^2\theta = \csc^2\theta\tan^2\theta = \frac{1}{\sin^2\theta}\cdot\frac{\sin^2\theta}{\cos^2\theta} = \frac{1}{\cos^2\theta} = \sec^2\theta.$

37. $\dfrac{\csc\theta}{\tan\theta + \cot\theta} = \dfrac{\frac{1}{\sin\theta}}{\frac{\sin\theta}{\cos\theta} + \frac{\cos\theta}{\sin\theta}} = \dfrac{\frac{1}{\sin\theta}}{\frac{\sin^2\theta + \cos^2\theta}{\cos\theta\sin\theta}} = \frac{1}{\sin\theta}\cdot\frac{\cos\theta}{1}\cdot\frac{\sin\theta}{1} = \cos\theta.$

39. The results follow by similar triangles.

41. $|AB| = \sqrt{169 - 25} = 12$, so $\sin\theta = \frac{5}{13}$, $\cos\theta = \frac{12}{13}$, $\tan\theta = \frac{5}{12}$, $\csc\theta = \frac{13}{5}$, $\sec\theta = \frac{13}{12}$, $\cot\theta = \frac{12}{5}$.

43. a. $P(t) = 100 + 20\sin 6t$. The maximum value of P occurs when $\sin t = 1$, and $P(t) = 100 + 20 = 120$. The minimum value of P occurs when $\sin 6t = -1$, and $P(t) = 100 - 20 = 80$.

    b. $\sin 6t = 1$ implies that $6t = \frac{\pi}{2}, \frac{3\pi}{2}, \ldots$; that is, $t = \dfrac{\pi(4n + 1)}{12}$ for $n = 0, 1, 2, \ldots$. $\sin 6t = -1$ implies that $6t = \frac{3\pi}{2}, \frac{7\pi}{2}, \ldots$; that is, $t = \dfrac{\pi(4n + 3)}{12}$ for $n = 0, 1, 2, \ldots$.

45. False. In fact, $\sin\theta = -\frac{\sqrt{3}}{2}$ if $\theta = \frac{4\pi}{3}$ or $\frac{5\pi}{3}$.

© 2011 Cengage Learning. All Rights Reserved. May not be scanned, copied or duplicated, or posted to a publicly accessible website, in whole or in part.

47. True. $\cos 2\theta = \cos^2 \theta - \sin^2 \theta = \cos^2 \theta - \left(1 - \cos^2 \theta\right) = 2\cos^2 \theta - 1$.

12.3 Problem-Solving Tips

Here are the derivative rules you should memorize as you work through these exercises.

1. $\frac{d}{dx}\left(\sin u\right) = \cos u \frac{du}{dx}$

2. $\frac{d}{dx}\left(\cos u\right) = -\sin u \frac{du}{dx}$

3. $\frac{d}{dx}\left(\tan u\right) = \sec^2 u \frac{du}{dx}$

4. $\frac{d}{dx}\left(\csc u\right) = -\csc u \cot u \frac{du}{dx}$

5. $\frac{d}{dx}\left(\sec u\right) = \sec u \tan u \frac{du}{dx}$

6. $\frac{d}{dx}\left(\cot u\right) = -\csc^2 u \frac{du}{dx}$

12.3 Concept Questions   page 780

1. $\frac{d}{dx}\left(\sin x\right) = \cos x$ and $\frac{d}{dx}\left(\cos x\right) = -\sin x$.

3. $\frac{d}{dx}\left(\tan x\right) = \sec^2 x$ and $\frac{d}{dx}\left[\tan f\left(x\right)\right] = \left[\sec^2 f\left(x\right)\right] f'\left(x\right)$.

12.3 Differentiation of Trigonometric Functions   page 781

1. $f\left(x\right) = \cos 3x$, so $f'\left(x\right) = -3\sin 3x$.

3. $f\left(x\right) = 2\cos \pi x$, so $f'\left(x\right) = \left(-2\sin \pi x\right)\frac{d}{dx}\left(\pi x\right) = -2\pi \sin \pi x$.

5. $f\left(x\right) = \sin\left(x^2 + 1\right)$, so $f'\left(x\right) = \cos\left(x^2 + 1\right)\frac{d}{dx}\left(x^2 + 1\right) = 2x\cos\left(x^2 + 1\right)$.

7. $f\left(x\right) = \tan 2x^2$, so $f'\left(x\right) = \left(\sec^2 2x^2\right)\frac{d}{dx}\left(2x^2\right) = 4x\sec^2 2x^2$.

9. $f\left(x\right) = x\sin x$, so $f'\left(x\right) = \sin x\frac{d}{dx}\left(x\right) + x\frac{d}{dx}\left(\sin x\right) = \sin x + x\cos x$.

11. $f\left(x\right) = 2\sin 3x + 3\cos 2x$, so $f'\left(x\right) = \left(2\cos 3x\right)\left(3\right) - \left(3\sin 2x\right)\left(2\right) = 6\left(\cos 3x - \sin 2x\right)$.

13. $f\left(x\right) = x^2 \cos 2x$, so $f'\left(x\right) = 2x\cos 2x + x^2\left(-\sin 2x\right)\left(2\right) = 2x\left(\cos 2x - x\sin 2x\right)$.

15. $f\left(x\right) = \sin\sqrt{x^2 - 1} = \sin\left(x^2 - 1\right)^{1/2}$, so

$$f'\left(x\right) = \cos\sqrt{x^2 - 1}\frac{d}{dx}\left(x^2 - 1\right)^{1/2} = \left(\cos\sqrt{x^2 - 1}\right)\left(\tfrac{1}{2}\right)\left(x^2 - 1\right)^{-1/2}\left(2x\right) = \frac{x\cos\sqrt{x^2 - 1}}{\sqrt{x^2 - 1}}.$$

17. $f\left(x\right) = e^x \sec x$, so $f'\left(x\right) = e^x \sec x + e^x\left(\sec x\tan x\right) = e^x \sec x\left(1 + \tan x\right)$.

19. $f\left(x\right) = x\cos\dfrac{1}{x}$, so $f'\left(x\right) = \cos\dfrac{1}{x} + x\left(-\sin\dfrac{1}{x}\right)\dfrac{d}{dx}\left(\dfrac{1}{x}\right) = \cos\dfrac{1}{x} - x\left(-\dfrac{1}{x^2}\right)\sin\dfrac{1}{x} = \cos\dfrac{1}{x} + \dfrac{1}{x}\sin\dfrac{1}{x}$.

© 2011 Cengage Learning. All Rights Reserved. May not be scanned, copied or duplicated, or posted to a publicly accessible website, in whole or in part.

21. $f(x) = \dfrac{x - \sin x}{1 + \cos x}$, so

$$f'(x) = \frac{(1 + \cos x)(1 - \cos x) - (x - \sin x)(-\sin x)}{(1 + \cos x)^2} = \frac{1 - \cos^2 x + x \sin x - \sin^2 x}{(1 + \cos x)^2} = \frac{x \sin x}{(1 + \cos x)^2}.$$

23. $f(x) = \sqrt{\tan x}$, so $f'(x) = \frac{1}{2}(\tan x)^{-1/2} \sec^2 x = \dfrac{\sec^2 x}{2\sqrt{\tan x}}$.

25. $f(x) = \dfrac{\sin x}{x}$, so $f'(x) = \dfrac{x \cos x - \sin x}{x^2}$.

27. $f(x) = \tan^2 x$, so $f'(x) = 2 \tan x \sec^2 x$.

29. $f(x) = e^{\cot x}$, so $f'(x) = e^{\cot x}(-\csc^2 x) = -\csc^2 x \cdot e^{\cot x}$.

31. $f(x) = \cot 2x$, so $f'(x) = -2 \csc^2 2x$. Therefore, $f'\left(\frac{\pi}{4}\right) = -\dfrac{2}{\sin^2\left(\frac{\pi}{2}\right)} = -2$. Then $y - 0 = -2\left(x - \frac{\pi}{4}\right)$, or

$y = -2x + \frac{\pi}{2}$.

33. $f(x) = e^x \cos x$, so $f'(x) = e^x \cos x + e^x(-\sin x) = e^x(\cos x - \sin x)$. Setting $f'(x) = 0$ gives

$\cos x - \sin x = 0$, or $\tan x = 1$. Thus, $x = \frac{\pi}{4}$ or $\frac{5\pi}{4}$. From the sign diagram for f', we see that f is increasing on $\left(0, \frac{\pi}{4}\right) \cup \left(\frac{5\pi}{4}, 2\pi\right)$ and decreasing on $\left(\frac{\pi}{4}, \frac{5\pi}{4}\right)$.

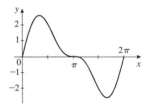

35. $f(x) = \sin x + \cos x$

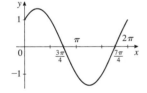

37. $f(x) = 2 \sin x + \sin 2x$

39. a. $f(x) = \sin x$, so $f'(x) = \cos x$, $f''(x) = -\sin x$, $f'''(x) = -\cos x$, $f^{(4)}(x) = \sin x, \ldots$.

Thus, $f(0) = 0$, $f'(0) = 1$, $f''(0) = 0$, $f'''(0) = -1$, $f^{(4)}(0) = 0, \ldots$, and so

$$f(x) = \sin x = x - \frac{x^3}{3!} + \frac{x^5}{5!} - \frac{x^7}{7!} + \cdots + (-1)^n \frac{x^{2n+1}}{(2n+1)!} + \cdots.$$

b. $\displaystyle\lim_{x \to 0} \frac{\sin x}{x} = \lim_{x \to 0}\left(1 - \frac{x^2}{3!} + \frac{x^4}{5!} - \cdots\right) = 1.$

41. $P_1(t) = 8000 + 1000 \sin \frac{\pi t}{24}$, so $P_1'(t) = 1000\left(\frac{\pi}{24}\right) \cos \frac{\pi t}{24}$ and $P_1'(12) = \frac{1000\pi}{24} \cos \frac{\pi}{2} = 0$; that is, the wolf population is not changing during the twelfth month.

$P_2(t) = 40{,}000 + 12{,}000 \cos \frac{\pi t}{24}$, so $P_2'(t) = -12{,}000\left(\frac{\pi}{24}\right) \sin \frac{\pi t}{24}$ and $P_2'(12) = -500 \sin \frac{\pi}{2} = -500\pi$; that is, the caribou population is decreasing at the rate of 1571/month.

43. $f(t) = 3 \sin \frac{2\pi}{365}(t - 79) + 12$, so $f'(t) = 3 \cos \frac{2\pi}{365}(t - 79) \cdot \frac{2\pi}{365}$. Therefore, $f'(79) = \frac{6\pi}{365} \approx 0.05164$. The number of hours of daylight is increasing at the rate of 0.05 hours per day on March 21.

© 2011 Cengage Learning. All Rights Reserved. May not be scanned, copied or duplicated, or posted to a publicly accessible website, in whole or in part.

45. $T = 62 - 18\cos\dfrac{2\pi(t-23)}{365}$, so $T' = \dfrac{36\pi}{365}\sin\left(\dfrac{-46\pi + 2\pi t}{365}\right)$. Setting $T' = 0$, we obtain either

$\dfrac{-46\pi + 2\pi t}{365} = 0$, in which $-46\pi = -2\pi t$ and $t = 23$, or $\dfrac{-46\pi + 2\pi t}{365} = \pi$, in which case $-46\pi + 2\pi t = 365\pi$,

$2\pi t = 411\pi$, and $t = 205.5$. From the sign diagram, we see
that a minimum occurs at $t = 23$ and a maximum occurs at
$t = 205.5$. We conclude that the warmest day is July 25th and
the coldest day is January 23rd.

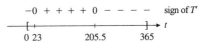

47. $R(t) = 2\left(5 - 4\cos\dfrac{\pi t}{6}\right)$, so $R'(t) = \dfrac{4\pi}{3}\sin\left(\dfrac{\pi t}{6}\right)$. Setting $R' = 0$, we obtain $\dfrac{\pi t}{6} = 0$ and conclude that $t = 0, 6$,
and 12 are critical points of R. From the sign diagram for R',
we conclude that a maximum occurs at $t = 6$.

49. From the solution to Exercise 46, we have $V'(t) = \dfrac{6}{5\pi}\left(\sin\dfrac{\pi t}{2}\right)\left(\dfrac{\pi}{2}\right) = \dfrac{3}{5}\sin\dfrac{\pi t}{2}$ and $V''(t) = \dfrac{3\pi}{10}\cos\dfrac{\pi t}{2}$. Setting
$V''(t) = 0$ gives $t = 1, 3, 5, 7, \ldots$ as critical points. Evaluating $V''(t)$ at each of these points, we see that the rate
of flow of air is maximized when $t = 1, 5, 9, 13, \ldots$ and minimized when $t = 3, 7, 11, 15, \ldots$.

51. $\tan\theta = \dfrac{y}{20}$, so $\sec^2\theta \cdot \dfrac{d\theta}{dt} = \dfrac{1}{20}\cdot\dfrac{dy}{dt}$. We want to find $\dfrac{dy}{dt}$ when $z = 30$. But when $z = 30$,

$y = \sqrt{900 - 400} = \sqrt{500} = 10\sqrt{5}$ and $\sec\theta = \dfrac{30}{20} = \dfrac{3}{2}$. Therefore, with $\dfrac{d\theta}{dt} = \dfrac{\pi}{2}$ radians per second, we find

$\dfrac{dy}{dt} = 20\sec^2\theta \cdot \dfrac{d\theta}{dt} = 20\left(\dfrac{3}{2}\right)^2 \cdot \dfrac{\pi}{2} = \dfrac{45\pi}{2}$, or approximately 70.7 ft/sec.

53. The area of the cross-section is $A = (2)\left(\dfrac{1}{2}\right)(5\cos\theta)(5\sin\theta) + 5(5\sin\theta) = 25(\cos\theta\sin\theta + \sin\theta)$. Thus,

$\dfrac{dA}{d\theta} = 25\left(-\sin^2\theta + \cos^2\theta + \cos\theta\right) = 25\left(\cos^2\theta - 1 + \cos^2\theta + \cos\theta\right) = 25\left(2\cos^2\theta + \cos\theta - 1\right)$

$= 25(2\cos\theta - 1)(\cos\theta + 1)$.

Setting $\dfrac{dA}{d\theta} = 0$ gives $\cos\theta = \dfrac{1}{2}$ or $\cos\theta = -1$; that is, $\theta = \dfrac{\pi}{3}$ or π. Now $A(0) = 0$, $A\left(\dfrac{\pi}{3}\right) = \dfrac{75\sqrt{3}}{4}$, and
$A(\pi) = 0$, so the absolute maximum of A occurs at $\theta = \dfrac{\pi}{3}$. Therefore, the angle should be 60°.

55. Let $f(\theta) = \theta - 0.5\sin\theta - 1$. Then $f'(\theta) = 1 - 0.5\cos\theta$, and so Newton's Method leads to the iteration

$\theta_{i+1} = \theta_i - \dfrac{f(\theta_i)}{f'(\theta_i)} = \theta_i - \dfrac{\theta_i - 0.5\sin\theta_i - 1}{1 - 0.5\cos\theta_i} = \dfrac{\theta_i - 0.5\theta_i\cos\theta_i - \theta_i + 0.5\sin\theta_i + 1}{1 - 0.5\cos\theta_i}$

$= \dfrac{1 + 0.5(\sin\theta_i - \theta_i\cos\theta_i)}{1 - 0.5\cos\theta_i}$.

With $\theta_i = 1.5$, we find $\theta_1 = \dfrac{1 + 0.5(\sin 1.5 - 1.5\cos 1.5)}{1 - 0.5\cos 1.5} = \dfrac{1.4456946}{0.9646314} \approx 1.4987$ and

$\theta_2 = \dfrac{1 + 0.5(\sin 1.4987 - 1.4987\cos 1.4987)}{1 - 0.5\cos 1.4987} = \dfrac{1.4447225}{0.9639831} \approx 1.4987$.

57. True. $\displaystyle\lim_{x\to 0}\dfrac{\tan x}{x} = \lim_{x\to 0}\dfrac{\frac{\sin x}{\cos x}}{x} = \lim_{x\to 0}\left(\dfrac{\sin x}{x}\cos x\right) = 1\cdot 1 = 1$.

59. False. Take $x = \pi$. Then f has a relative minimum at $x = \pi$, but g does not have a relative maximum at $x = \pi$.

61. $h(x) = \csc f(x) = \dfrac{1}{\sin f(x)}$, so $h'(x) = \dfrac{-\cos f(x)\cdot f'(x)}{\sin^2 f(x)} = -[\csc f(x)][\cot f(x)]f'(x)$.

© 2011 Cengage Learning. All Rights Reserved. May not be scanned, copied or duplicated, or posted to a publicly accessible website, in whole or in part.

63. $h(x) = \cot f(x) = \dfrac{\cos f(x)}{\sin f(x)}$, so

$$h'(x) = \frac{\sin f(x)\left[-\sin f(x)\right]f'(x) - \left[\cos f(x)\right]\left[\cos f(x)\right]f'(x)}{\sin^2 f(x)} = \frac{\left[-\sin^2 f(x) - \cos^2 f(x)\right]f'(x)}{\sin^2 f(x)}$$

$$= \frac{-1 \cdot f'(x)}{\sin^2 f(x)} = -\left[\csc^2 f(x)\right]f'(x).$$

12.3   Using Technology   page 785

1. 1.2038   **3.** 0.7762   **5.** −0.2368   **7.** 0.8415, −0.2172   **9.** 1.1271, 0.2013

11. a.

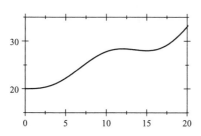

13.

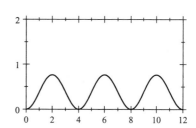

b. Approximately $0.90.

c. Approximately $37.86

15. Approximately 0.006 ft.

12.4   Problem-Solving Tips

Here are the integrals of the trigonometric functions that you should be familiar with as you work through these exercises.

1. $\int \sin u \, du = -\cos u + C$

2. $\int \cos u \, du = \sin u + C$

3. $\int \sec^2 u \, du = \tan u + C$

4. $\int \csc^2 u \, du = -\cot u + C$

5. $\int \sec u \tan u \, du = \sec u + C$

6. $\int \csc u \cot u \, du = -\csc u + C$

7. $\int \tan u \, du = -\ln|\cos u| + C$

8. $\int \sec u \, du = \ln|\sec u + \tan u| + C$

9. $\int \csc u \, du = \ln|\csc u - \cot u| + C$

10. $\int \cot u \, du = \ln|\sin u| + C$

© 2011 Cengage Learning. All Rights Reserved. May not be scanned, copied or duplicated, or posted to a publicly accessible website, in whole or in part.

12.4   Concept Questions   page 789

1. $\int \sin x \, dx = -\cos x + C$, $\int \cos x \, dx = \sin x + C$, and $\int \sec^2 x \, dx = \tan x + C$.

12.4   Integration of Trigonometric Functions   page 789

1. $\int \sin 3x \, dx = -\frac{1}{3} \cos 3x + C$.

3. $\int (3 \sin x + 4 \cos x) \, dx = -3 \cos x + 4 \sin x + C$.

5. Let $u = 2x$, so $du = 2 \, dx$. Then $\int \sec^2 2x \, dx = \frac{1}{2} \int \sec^2 u \, du = \frac{1}{2} \tan u + C = \frac{1}{2} \tan 2x + C$.

7. Let $u = x^2$, so $du = 2x \, dx$ and $x \, dx = \frac{1}{2} du$. Then $\int x \cos x^2 dx = \frac{1}{2} \int \cos u \, du = \frac{1}{2} \sin u + C = \frac{1}{2} \sin x^2 + C$.

9. Let $u = \pi x$, so $du = \pi \, dx$ and $dx = \frac{1}{\pi} du$. Then
$\int \csc \pi x \cot \pi x \, dx = \frac{1}{\pi} \int \csc u \cot u \, du = -\frac{1}{\pi} \csc u + C = -\frac{1}{\pi} \csc \pi x + C$.

11. $\int_{-\pi/2}^{\pi/2} (\sin x + \cos x) \, dx = (-\cos x + \sin x)|_{-\pi/2}^{\pi/2} = 1 - (-1) = 2$.

13. $\int_0^{\pi/6} \tan 2x \, dx = -\frac{1}{2} \ln |\cos 2x| \Big|_0^{\pi/6} = -\frac{1}{2} \ln \frac{1}{2}$.

15. Let $u = \sin x$, so $du = \cos x \, dx$. Then $\int \sin^3 x \cos x \, dx = \int u^3 du = \frac{1}{4} u^4 + C = \frac{1}{4} \sin^4 x + C$.

17. Let $u = \pi x$, so $du = \pi \, dx$ and $dx = \frac{1}{\pi} du$. Then $\int \sec \pi x \, dx = \frac{1}{\pi} \int \sec u \, du = \frac{1}{\pi} \ln |\sec u + \tan u| + C$ [by
Formula (8)] $= \frac{1}{\pi} \ln |\sec \pi x + \tan \pi x| + C$.

19. Let $u = 3x$, so $du = 3 \, dx$ and $dx = \frac{1}{3} du$. If $x = 0$, then $u = 0$, and if $x = \frac{\pi}{12}$, then $u = \frac{\pi}{4}$. Thus,
$\int_0^{\pi/12} \sec 3x \, dx = \frac{1}{3} \int_0^{\pi/4} \sec u \, du = \frac{1}{3} \ln |\sec u + \tan u| \Big|_0^{\pi/4} = \frac{1}{3} [\ln (\sqrt{2} + 1) - \ln 1] = \frac{1}{3} \ln (\sqrt{2} + 1)$.

21. Let $u = \cos x$, so $du = -\sin x \, dx$. Then $\int \sqrt{\cos x} \sin x \, dx = -\int \sqrt{u} \, du = -\frac{2}{3} u^{3/2} + C = -\frac{2}{3} (\cos x)^{3/2} + C$.

23. $I = \int \cos 3x \, (1 - 2 \sin 3x)^{1/2} \, dx$. Put $u = \sin 3x$, so $du = 3 \cos 3x \, dx$. Then
$I = \frac{1}{3} \int (1 - 2u)^{1/2} \, du = -\frac{1}{6} \cdot \frac{2}{3} (1 - 2u)^{3/2} + C = -\frac{1}{9} (1 - 2 \sin 3x)^{3/2} + C$.

25. Let $u = \tan x$, so $du = \sec^2 x \, dx$. Then $\int \tan^3 x \sec^2 x \, dx = \int u^3 \, du = \frac{1}{4} u^4 + C = \frac{1}{4} \tan^4 x + C$.

27. Let $u = \cot x - 1$, so $du = -\csc^2 x \, dx$. Then
$\int \csc^2 x \, (\cot x - 1)^3 \, dx = -\int u^3 \, du = -\frac{1}{4} u^4 + C = -\frac{1}{4} (\cot x - 1)^4 + C$.

29. Put $u = \ln x$, so $du = \dfrac{dx}{x}$. Then $\displaystyle\int \dfrac{\sin (\ln x)}{x} \, dx = \int \sin u \, du = -\cos u + C = -\cos (\ln x) + C$. Thus,
$\displaystyle\int_1^{e^\pi} \dfrac{\sin (\ln x)}{x} \, dx = -\cos (\ln x)|_1^{e^\pi} = -\cos \pi + \cos 0 = 2$.

© 2011 Cengage Learning. All Rights Reserved. May not be scanned, copied or duplicated, or posted to a publicly accessible website, in whole or in part.

31. Let $I = \int \sin(\ln x)\, dx$ and put $u = \sin(\ln x)$ and $dv = dx$, so $du = \dfrac{\cos(\ln x)}{x}\, dx$ and $v = x$.

Integrating by parts, we have $I = x\sin(\ln x) - \int \cos(\ln x)\, dx$. Now let $J = \int \cos(\ln x)\, dx$ and put

$u = \cos(\ln x)$ and $dv = dx$, so $v = x$ and $du = -\dfrac{\sin(\ln x)}{x}\, dx$. Integrating by parts again, we have

$J = x\cos(\ln x) + \int \sin(\ln x)\, dx = x\cos(\ln x) + I$. Therefore, $I = x[\sin(\ln x) - \cos(\ln x)] - I$, so

$2I = x[\sin(\ln x) - \cos(\ln x)]$ and thus $I = \int \sin(\ln x)\, dx = \frac{1}{2}x[\sin(\ln x) - \cos(\ln x)] + C$.

33. $A = \int_0^\pi \cos\frac{x}{4}\, dx = 4\sin\frac{x}{4}\Big|_0^\pi = 4\left(\frac{\sqrt{2}}{2}\right) = 2\sqrt{2}$.

35. $A = \int_0^{\pi/4} \tan x\, dx = -\ln|\cos x|\Big|_0^{\pi/4} = -\ln\frac{\sqrt{2}}{2} = -\ln\frac{1}{\sqrt{2}} = \ln\sqrt{2} = \frac{1}{2}\ln 2$.

37. $A = \int_0^\pi (x - \sin x)\, dx = \left(\frac{1}{2}x^2 + \cos x\right)\Big|_0^\pi = \frac{1}{2}\pi^2 - 1 - 1 = \frac{1}{2}\pi^2 - 2 = \frac{1}{2}\left(\pi^2 - 4\right)$.

39. The average is $A = \frac{1}{15}\int_0^{15}\left(80 + 3t\cos\frac{\pi t}{6}\right)dt = 80 + \frac{1}{5}\int_0^{15} t\cos\frac{\pi t}{6}\, dt$.

Integrating by parts, we find $\int t\cos\frac{\pi t}{6}\, dt = \left(\frac{6}{\pi}\right)^2\left[\cos\frac{\pi t}{6} + \frac{\pi t}{6}\sin\frac{\pi t}{6}\right]$, so

$A = 80 + \left[\frac{1}{5}\left(\frac{6}{\pi}\right)^2\left(\cos\frac{\pi t}{6} + \frac{\pi t}{6}\sin\frac{\pi t}{6}\right)\right]_0^{15} = 80 + \frac{1}{5}\left(\frac{6}{\pi}\right)^2\left(\frac{15\pi}{6} - 1\right) \approx 85$, or approximately \$85 per share.

41. $R = \int_0^{12} 2\left(5 - 4\cos\frac{\pi t}{6}\right)dt = \left[10t - 8\left(\frac{6}{\pi}\right)\sin\frac{\pi t}{6}\right]_0^{12} = 120$, or \$120,000.

43. The required volume is $V = \int R(t)\, dt = 0.6\int \sin\frac{\pi t}{2}\, dt = (0.6)\left(\frac{2}{\pi}\right)\cos\frac{\pi t}{2} + C = -\frac{1.2}{\pi}\cos\frac{\pi t}{2} + C$. When

$t = 0$, $V = 0$, we have $0 = -\frac{1.2}{\pi} + C$, and so $C = \frac{1.2}{\pi}$. Therefore, $V = \frac{1.2}{\pi}\left(1 - \cos\frac{\pi t}{2}\right)$.

45. $\frac{dQ}{dt} = 0.0001(4 + 5\cos 2t)Q(400 - Q)$, so $\dfrac{dQ}{Q(400 - Q)} = 0.0001(4 + 5\cos 2t)\, dt$. Thus,

$\int\left(\frac{1}{Q} + \frac{1}{400 - Q}\right)dQ = \int 0.04(4 + 5\cos 2t)\, dt$, $\ln Q - \ln|400 - Q| = 0.04(4t + 2.5\sin 2t) + C_1$,

$\ln\left|\frac{Q}{400 - Q}\right| = 0.04(4t + 2.5\sin 2t) + C_1$, and $\frac{Q}{400 - Q} = Ce^{0.04(4t + 2.5\sin 2t)}$ (where $C = e^{C_1}$ and $0 < Q < 400$).

Using the condition $Q(0) = 10$, we have $\frac{10}{390} = C$, so $\frac{Q}{400 - Q} = \frac{10}{390}e^{0.04(4t + 2.5\sin 2t)}$. When $t = 20$, we have

$\frac{Q}{400 - Q} = \frac{10}{390}e^{0.04(80 + 2.5\sin 40)} \approx 0.6777$, so $Q = 0.6777[400 - Q] = 271.08 - 0.6777Q$, $1.6777Q \approx 271.08$, so

$Q(20) \approx \frac{271.08}{1.6777} \approx 161.578$, or approximately 162 flies.

47. $\displaystyle\int_0^1 \cos t^2\, dt = \int_0^1\left(1 - \frac{t^4}{2} + \frac{t^8}{24} - \frac{t^{12}}{720}\right)dt = \left(t - \frac{t^5}{10} + \frac{t^9}{216} - \frac{t^{13}}{9360}\right)\Big|_0^1 = 1 - \frac{1}{10} + \frac{1}{216} - \frac{1}{9360} \approx 0.904523$.

49. True.

$\int_a^{b+2\pi}\cos x\, dx = \sin x\big|_a^{b+2\pi} = \sin(b + 2\pi) - \sin a = \sin b\cos 2\pi + \cos b\sin 2\pi - \sin a = \sin b - \sin a$

$= \int_a^b \cos x\, dx$.

51. True. $\int_{-\pi/2}^{\pi/2}|\sin x|\, dx = 2\int_0^{\pi/2}\sin x\, dx = -2\cos x\big|_0^{\pi/2} = 2$ and

$\int_{-\pi/2}^{\pi/2}|\cos x|\, dx = 2\int_0^{\pi/2}\cos x\, dx = -2\sin x\big|_0^{\pi/2} = 2$.

© 2011 Cengage Learning. All Rights Reserved. May not be scanned, copied or duplicated, or posted to a publicly accessible website, in whole or in part.

12.4   Using Technology   page 792

1. 0.5419   **3.** 0.7544   **5.** 0.2231

7. 0.6587   **9.** −0.2032   **11.** 0.9045

13. a.

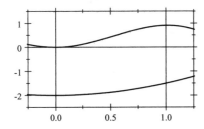

15. a.

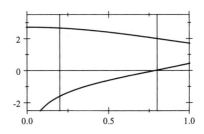

b. 2.2687   **b.** 1.8239

17. a.

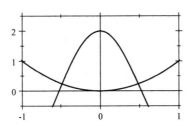

19. a.

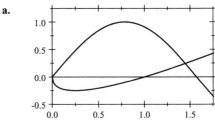

b. 1.2484   **b.** 1.0983

21. The average water level is approximately 7.6 ft.

CHAPTER 12   Concept Review   page 793

1. a. radian   **b.** 2π, 360

3. $x, y, \frac{y}{x}, \frac{1}{y}, \frac{1}{x}, \frac{x}{y}$

5. a. $\cos 2\theta$ **b.** +   **c.** $\sin A \cos A$ **d.** $\sin^2 A$

7. a. $-\cos x + C$ **b.** $\sin x + C$ **c.** $\tan x + C$

   d. $-\cot x + C$ **e.** $\sec x + C$ **f.** $-\csc x + C$

© 2011 Cengage Learning. All Rights Reserved. May not be scanned, copied or duplicated, or posted to a publicly accessible website, in whole or in part.

CHAPTER 12   Review   page 794

1. $120° = \frac{2\pi}{3}$ rad.

3. $-225° = -\frac{225}{180}\pi = -\frac{5\pi}{4}$ rad.

5. $-\frac{5\pi}{2}$ rad $= -\frac{5}{2}(180) = -450°$.

7. $\cos\theta = \frac{1}{2}$ implies that $\theta = \frac{\pi}{3}$ or $\frac{5\pi}{3}$.

9. $f(x) = \sin 3x$, so $f'(x) = 3\cos 3x$.

11. $f(x) = 2\sin x - 3\cos 2x$, so $f'(x) = 2\cos x + 3(\sin 2x)(2) = 2(\cos x + 3\sin 2x)$.

13. $f(x) = e^{-x}\tan 3x$, so $f'(x) = e^{-x}\tan 3x + e^{-x}(\sec^2 3x)(3) = e^{-x}(3\sec^2 3x - \tan 3x)$.

15. $f(x) = 4\sin x \cos x$, so $f'(x) = 4[\sin x(-\sin x) + \cos x(\cos x)] = 4(\cos^2 x - \sin^2 x) = 4\cos 2x$.
Another method: Notice that $f(x) = 2\sin 2x$, so $f'(x) = 4\cos 2x$.

17. $f(x) = \dfrac{1 - \tan x}{1 - \cot x}$, so

$$f'(x) = \frac{(1 - \cot x)(-\sec^2 x) - (1 - \tan x)(\csc^2 x)}{(1 - \cot x)^2} = \frac{-\sec^2 x + \sec^2 x \cot x - \csc^2 x + \csc^2 x \tan x}{(1 - \cot x)^2}$$

$$= \frac{(\cot x - 1)\sec^2 x - (1 - \tan x)\csc^2 x}{(1 - \cot x)^2} = -\sec^2 x.$$

19. $f(x) = \sin(\sin x)$, so $f'(x) = \cos(\sin x) \cdot \cos x$.

21. $f(x) = \tan^2 x$, so $f'(x) = 2\tan x \sec^2 x$. Thus, the slope is $f'\left(\frac{\pi}{4}\right) = 2(1)\left(\sqrt{2}\right)^2 = 4$, and an equation of the tangent line is $y - 1 = 4\left(x - \frac{\pi}{4}\right)$, or $y = 4x + 1 - \pi$.

23. Let $u = \frac{2}{3}x$, so $du = \frac{2}{3}dx$ and $dx = \frac{3}{2}du$. Then $\int \cos\frac{2}{3}x\,dx = \frac{3}{2}\int \cos u\,du = \frac{3}{2}\sin u + C = \frac{3}{2}\sin\frac{2}{3}x + C$.

25. $I = \int x\csc x^2 \cot x^2\,dx$. Put $u = x^2$, so $du = 2x\,dx$. Then
$$\int x\csc x^2 \cot x^2\,dx = \frac{1}{2}\int \csc u \cot u\,du = -\frac{1}{2}\csc u + C = -\frac{1}{2}\csc x^2 + C.$$

27. Let $u = \sin x$, so $du = \cos x\,dx$. Then $\int \sin^2 x \cos x\,dx = \int u^2 du = \frac{1}{3}\sin^3 x + C$.

29. Let $u = \sin x$, so $du = \cos x\,dx$.   Then
$$\int \frac{\cos x}{\sin^2 x}\,dx = \int \frac{du}{u^2} = \int u^{-2}\,du = -\frac{1}{u} + C = -\frac{1}{\sin x} + C = -\csc x + C.$$

31. $\displaystyle\int_{\pi/6}^{\pi/2} \frac{\cos x}{1 - \cos^2 x}\,dx = \int_{\pi/6}^{\pi/2} \frac{\cos x}{\sin^2 x}\,dx = \int_{\pi/6}^{\pi/2} \sin^{-2} x \cos x\,dx = -\frac{1}{\sin x}\Big|_{\pi/6}^{\pi/2} = -1 + 2 = 1.$

33. $A = \int_{\pi/4}^{5\pi/4}(\sin x - \cos x)\,dx = (-\cos x - \sin x)\big|_{\pi/4}^{5\pi/4} = \left(\frac{\sqrt{2}}{2} + \frac{\sqrt{2}}{2}\right) - \left(-\frac{\sqrt{2}}{2} - \frac{\sqrt{2}}{2}\right) = 2\sqrt{2}.$

© 2011 Cengage Learning. All Rights Reserved. May not be scanned, copied or duplicated, or posted to a publicly accessible website, in whole or in part.

35. $R(t) = 60 + 37 \sin^2\left(\frac{\pi t}{12}\right)$, so $R'(t) = 37(2)\sin\left(\frac{\pi t}{12}\right) \cdot \cos\left(\frac{\pi t}{12}\right)\left(\frac{\pi}{12}\right) = \frac{37\pi}{6}\sin\left(\frac{\pi t}{6}\right)$ (since $\sin 2\theta = 2\sin\theta\cos\theta$).
Setting $R'(t) = 0$ gives $\frac{\pi t}{6} = 0, \pi, 2\pi, \ldots$, so $t = 0, 6, 12, \ldots$ are the critical points of R. In particular, $t = 6$ is the
critical point of R in $(0, 12)$. Now $R''(t) = \frac{37\pi}{6}\cos\left(\frac{\pi t}{6}\right) \cdot \frac{\pi}{6} = \frac{37\pi^2}{36}\cos\left(\frac{\pi t}{6}\right)$, so because $R''(6) = -\frac{37\pi^2}{36} < 0$,
the second derivative test implies that the occupancy rate is highest when $t = 6$ (at the beginning of December).
Next, we set $R''(t) = 0$, giving $\frac{\pi t}{6} = \frac{\pi}{2}$, so $t = 3$. Because $R''(3) < 0$, we see that $R'(t)$ is maximized at $t = 3$, so
the occupancy rate is increasing most rapidly at the beginning of September.

CHAPTER 12	Before Moving On...	page 794

1. The period of f is $\frac{2\pi}{2} = \pi$.

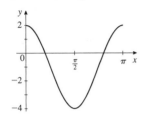

2. The right-hand side is

$$\frac{1 + \cos 2x}{2} = \frac{1 + \left(\cos^2 x - \sin^2 x\right)}{2}$$

$$= \frac{1 + \cos^2 x - \left(1 - \cos^2 x\right)}{2} = \cos^2 x.$$

3. $f(x) = e^{-2x}\sin 3x$, so

$$f'(x) = e^{-2x}\frac{d}{dx}\sin 3x + \sin 3x\frac{d}{dx}e^{-3x}$$

$$= 3e^{-2x}\cos 3x - 2\sin 3xe^{-2x} = (3\cos 3x - 2\sin 3x)\,e^{-2x}.$$

4. $\displaystyle\int_0^{\pi/4}\cos^2 x\,dx = \int_0^{\pi/4}\frac{1 + \cos 2x}{2}\,dx = \left(\frac{1}{2}x + \frac{1}{4}\sin 2x\right)\Big|_0^{\pi/4} = \frac{\pi}{8} + \frac{1}{4} = \frac{\pi + 2}{8}.$

5. $A = \int_0^{\pi/4}\sin^2 x\cos x\,dx = \frac{1}{3}\sin^3 x\Big|_0^{\pi/4} = \frac{1}{3}\left(\frac{\sqrt{2}}{2}\right)^3 = \frac{2\sqrt{2}}{3 \cdot 8} = \frac{\sqrt{2}}{12}.$

© 2011 Cengage Learning. All Rights Reserved. May not be scanned, copied or duplicated, or posted to a publicly accessible website, in whole or in part.